Plant Responses to Drought Stress

Ricardo Aroca
Editor

Plant Responses to Drought Stress

From Morphological to Molecular Features

Springer

Editor
Dr. Ricardo Aroca
Departamento de Microbiología del Suelo y
Sistemas Simbióticos
Estación Experimental del Zaidín (CSIC)
Granada
Spain

ISBN 978-3-642-32652-3 ISBN 978-3-642-32653-0 (eBook)
DOI 10.1007/978-3-642-32653-0
Springer Heidelberg New York Dordrecht London

Library of Congress Control Number: 2012949708

Printed on acid-free paper

Springer is part of Springer Science+Business Media (www.springer.com)

Preface

Since plants are sessile organisms, they to have develop multiple strategies to cope with environmental constrains. One of the most common and damaging environmental stress is soil drought. The soil drought characteristics may vary from intervals of water scarcity and water depletion to prolonged periods of water deprivation or to long periods of soil water contents below the full capacity. Hence, the drought experienced by dessert plants is not the same that for agricultural ones, or plants grown in Mediterranean climatic areas. So, each specific plant is adapted to their specific water soil conditions. At the same time, the responses of plants to drought varied from morphological ones to molecular, including physiological and biochemical ones too.

This book is intended to complete a comprehensive review about all aspects of the response of plants to drought. In each chapter a basic concepts will be first exposed, followed by the last findings of each topic. The first chapter is an overview of the effects and responses of plants to drought stress. The following chapters are subdivided in five parts: Morphological and anatomical responses, Physiological responses, Biochemical and Molecular responses, Ecophysiological responses, and Field responses. Chapters 2 and 3 will deal about the morphological and anatomical adaptations of plants in response to drought. Chapter 4 will deal with how water is up taken from the soil. Chapters 5 and 6 will focus on how photosynthesis and water use efficiency is regulated under drought conditions. Chapter 7 will deal about how drought stress affects nutrients uptake and assimilation. Chapters 8–10 will tackle different biochemical drought responses such as osmotic adjustment, antioxidant systems or hormones. Chapter 11 will deal with molecular aspects of the drought response. Chapter 12 will tackle the particularities of tree response to drought. Chapters 13–15 will review how different soil beneficial microorganisms change the response of plant to drought. Finally, Chap. 16 and 17 will deal with the plant responses to drought under field conditions.

The potential readers of this book will be any graduate student or established researcher who wants to know basic concepts of plant responses to drought, as well as such researchers specialized in studies of plant response to drought stress.

Dr. Ricardo Aroca

Contents

Part V Field Responses

Chapter 1
Drought Stress in Plants: An Overview

M. Farooq, M. Hussain, Abdul Wahid and K. H. M. Siddique

Abstract Drought is one of the major constraints limiting crop production worldwide. Crop growth models predict that this issue will be more severe in future. Drought impairs normal growth, disturbs water relations, and reduces water use efficiency in plants. Plants, however, have a variety of physiological and biochemical responses at cellular and whole organism levels, making it a more complex phenomenon. The rate of photosynthesis is reduced mainly by stomatal closure, membrane damage, and disturbed activity of various enzymes, especially those involved in ATP synthesis. Plants display a range of mechanisms to withstand drought, such as reduced water loss by increased diffusive resistance,

M. Farooq
Institute of Plant Nutrition, Justus-Liebig-University, Heinrich-Buff-Ring 26-32, D-35392 Giessen, Germany

M. Farooq
Department of Agronomy, University of Agriculture, Faisalabad, Pakistan

M. Farooq (✉) · K. H. M. Siddique
The UWA Institute of Agriculture, The University of Western Australia, Crawley, WA 6009, Australia
e-mail: farooqcp@gmail.com

M. Hussain
Department of Agronomy, Bahauddin Zakariya University, Multan, Pakistan

M. Hussain
Department of Crop Science and Biotechnology, Dankook University, Chungnam 330-714, South Korea

A. Wahid
Department of Botany, University of Agriculture, Faisalabad, Pakistan

K. H. M. Siddique
College of Food and Agricultural Sciences, King Saud University, 2460 Riyadh 11451, Saudi Arabia

R. Aroca (ed.), *Plant Responses to Drought Stress*,
DOI: 10.1007/978-3-642-32653-0_1,

increased water uptake with prolific and deep root systems, and smaller and succulent leaves to reduce transpirational loss. Low-molecular-weight osmolytes, including glycinebetaine, proline and other amino acids, organic acids, and polyols also play vital roles in sustaining cellular functions under drought. Plant growth substances such as salicylic acid, auxins, gibberellins, cytokinins, and abscisic acid modulate plant responses toward drought. Polyamines, citrulline, and several enzymes act as antioxidants and reduce adverse effects of water deficit. Plant drought stress can be managed by adopting strategies such as mass screening and breeding, marker-assisted selection, and exogenous application of hormones and osmoprotectants to seeds or growing plants, as well as engineering for drought resistance. Here, we provide an overview of plant drought stress, its effects on plants' resistance mechanisms and management strategies to cope with drought stress.

Abbreviations

ABA	Abscisic acid
ADC2	Arginine decarboxlase 2 gene
A_{max}	Maximum photosynthetic efficiency
APX	Ascorbate peroxidase
BRs	Brassinolides
CAT	Catalase
chl	Chlorophyll
Cks	Cytokinins
DRE/CRT	Dehydration-responsive element/C-repeat
DREB	Dehydration-responsive element binding proteins
EBR	Epibrassinolide
ETC	Electron transport chain
GA_3	Gibberellins
GB	Glycinebetaine
GR	Glutathione reductase
H^+-ATPase	Hydrogen pump ATPase protein
H_2O_2	Hydrogen peroxide
IAA	Indole acetic acid
K	Potassium
LAI	Leaf area index
LEA	Late embryogenesis abundant
N	Nitrogen
O_2^-	Superoxide radicals
O_2^1	Single oxygen
OH^-	Hydroxyl radicals
OsRDCPs	*Oryza sativa* RING domain-containing proteins
P	Phosphorous
PA	Polyamine
PAL	Phenylalanine ammonia-lyase

POX	Peroxidase
PPO	Polyphenol oxidase
PSI	Photosystem I
PSII	Photosystem II
QTL	Quantitative trait loci
RO	Alkoxy radicals
ROS	Reactive oxygen species
Rubisco	Ribulose-1,5-bisphosphate carboxylase/oxygenase
RuBP	Ribulose-1,5-bisphosphate
RWC	Relative water contents
SA	Salicylic acid
Si	Silicon
SOD	Superoxide dismutase
TcADC	Arginine decarboxylase
TcODC	Ornithine decarboxylase
TcSAMDC	S-adenosylmethionine decarboxylase
TcSPDS	Spermidine synthase
TcSPMS	Spermine synthase
$V_{c,max}$	Carboxylation velocity of Rubisco
WUE	Water use efficiency

1.1 Introduction

Crop plants are exposed to several environmental stresses, all affecting plant growth and development, which consequently hampers the productivity of crop plants (Seki et al. 2003; Farooq et al. 2009a, b, 2011). Drought is considered the single most devastating environmental stress, which decreases crop productivity more than any other environmental stress (Lambers et al. 2008).

A continuous shortfall in precipitation (meteorological drought) coupled with higher evapotranspiration demand leads to agricultural drought (Mishra and Cherkauer 2010). Agricultural drought is the lack of ample moisture required for normal plant growth and development to complete the life cycle (Manivannan et al. 2008). Drought severely affects plant growth and development with substantial reductions in crop growth rate and biomass accumulation. The main consequences of drought in crop plants are reduced rate of cell division and expansion, leaf size, stem elongation and root proliferation, and disturbed stomatal oscillations, plant water and nutrient relations with diminished crop productivity, and water use efficiency (WUE) (Li et al. 2009; Farooq et al. 2009a). Climate models have predicted increased severity and frequency of drought under the ongoing global climate change scenarios (IPCC 2007; Walter et al. 2011).

Water deficit accelerates abscisic acid (ABA) biosynthesis, which decreases stomatal conductance to minimize transpirational losses (Yamaguchi-Shinozaki and Shinozaki 2006). To cope with such challenges, understanding the effects of drought on plants and morphological and physiological adaptations is crucial (Yamaguchi-Shinozaki and Shinozaki 2006). This chapter presents an overview of the effects of drought on morphology, water relations, nutrient uptake, and assimilation in crop plants; morphological and physiological mechanisms of drought resistance; and suggests some pragmatic options and strategies to cope with this global challenge.

1.2 Effects of Drought Stress

Deficit water supply at any growth stage poses detrimental effects on crop growth and development in general but varies depending on the severity of stress and the crop growth stage. Effects of drought on morphological, physiological, and biochemical processes in plants are discussed below.

1.2.1 Plant Growth and Productivity

Establishment of an early and optimum crop stand is important for harvesting maximum productivity. However, if the crop experiences an early drought, thereby affecting germination, then the suboptimal plant population is the major cause of low grain yield. Early season drought severely reduces germination and stand establishment principally due to reduced water uptake during the imbibition phase of germination, reduced energy supply, and impaired enzyme activities (Okcu et al. 2005; Taiz and Zeiger 2010).

Growth is an irreversible increase in volume, size, or weight, which includes the phases of cell division, cell elongation, and differentiation. Both cell division and cell enlargement are affected under drought owing to impaired enzyme activities, loss of turgor, and decreased energy supply (Kiani et al. 2007; Farooq et al. 2009a; Taiz and Zeiger 2010). For example, drought decreases growth and productivity of sunflower (*Heliantus annuus* L.) owing to reductions in leaf water potential, rate of cell division, and enlargement primarily due to loss of turgor (Kiani et al. 2007; Hussain et al. 2009). Under drought, reduced dry matter accumulation occurs in all plant organs, although different organs manifest varying degrees of reduction. For instance, drought decreased shoot and flower fresh and dry weights of marigold (*Tagetes erecta* L.) plants (Asrar and Elhindi 2011). Likewise, drought considerably reduced shoot and root dry weights in Asian red sage (*Salvia miltiorrhiza* L.), although roots were less affected than shoots (Liu et al. 2011). Drought also decreased leaf area owing to loss of turgor and reduced leaf numbers (Farooq et al. 2010a).

Leaf area index (LAI) is the ratio of leaf area to ground area, which denotes the extent of assimilatory power of crops under field conditions. Drought decreases LAI in crop plants in general. For instance, Hussain et al. (2009) reported decline in LAI of sunflower exposed to drought at budding and flowering stages. Drought also suppresses leaf expansion and tillering (Kramer and Boyer 1995), and reduces leaf area due to early senescence (Nooden 1988). All these factors contribute to reduced dry matter accumulation and grain yield under drought.

The study of different growth and developmental events in crop plants with respect to time is called crop phenology. Drought strongly affects crop phenology by shortening the crop growth cycle with a few exceptions. Limited water supply triggers a signal to cause an early switching of plant development from the vegetative to reproductive phase (Desclaux and Roumet 1996). For instance, total growth duration of both bread wheat (*Triticum aestivum* L.) and barley (*Hordeum vulgare* L.) decreased under drought (McMaster and Wilhelm 2003), which generally results in substantial yield reductions. The effect of drought is phase specific in most cases. For example, drought at pre-anthesis delayed flowering in quinoa (*Chenopodium quinoa* Wild.) and bread wheat plants (Majid et al. 2007; Geerts et al. 2008). Likewise, drought at anthesis commonly delays flowering in rice (*Oryza sativa* L.); interestingly, the longer the delay, the higher the yield penalty (Fukai 1999). In soybean (*Glycine max* L.), drought during grain filling hastened maturity but yield was down due to smaller grains (Desclaux and Roumet 1996).

Different crops respond to drought differently. For instance, upon exposure to drought flowering is delayed in maize (*Zea mays* L.) (Abrecht and Carberry 1993), quinoa (Geerts et al. 2008), and rice (Fukai 1999), whereas in soybean (Desclaux and Roumet 1996), wheat, and barley (McMaster and Wilhelm 2003) drought hastened flowering and physiological maturity.

While drought occurs during the vegetative period of crop growth, it may substantially decrease economic yield. Drought stress during reproductive and grain filling phases is more devastating (Table 1.1; Reddy et al. 2003; Vijay 2004; Yadav et al. 2004; Lafitte et al. 2007). Drought at flowering is critical as it can increase pollen sterility resulting in hampered grain set. In sunflower, for example, under drought at flowering, achene yield declined primarily due to less achenes (Hussain et al. 2008). In pearl millet (*Pennisetum glaucum* L. Leeke), drought at flowering increased the rate of ear abortion due to a decline in assimilate supply to developing ears (Yadav et al. 2004). In drought-stressed maize, kernel set was lost leading to low grain yield (Schussler and Westgate 1995). Likewise, water deficit at anthesis increased pod abortion which reduced yield in soybean (Liu et al. 2003).

1.2.2 Plant Water Relations

Relative water contents (RWC), leaf water potential, osmotic potential, pressure potential, and transpiration rate are the major attributes of plant water relations (Kirkham 2005), which are significantly affected under water deficit owing to

Table 1.1 Decrease in grain yield in different crops by drought stress

Growth stage	Crop	Stress type	Yield reduction (%)	Reference
Reproductive	Rice	Mild stress	54	Lafitte et al. (2007)
Reproductive	Rice	Severe stress	94	Lafitte et al. (2007)
Reproductive	Rice	–	24–84	Venuprasad et al. (2007)
Flowering	Rice	Short severe stress	54	Lanceras et al. (2004)
Flowering and grain filling	Rice	Prolonged severe stress	84	Lanceras et al. (2004)
Flowering and grain filling	Rice	Prolonged mild stress	52	Lanceras et al. (2004)
Reproductive	Wheat	Prolonged mild stress	50–66	Kiliç and Yağbasanlar (2010)
Pre-anthesis	Wheat	Prolonged mild stress	18–53	Majid et al. (2007)
Post-anthesis	Wheat	Prolonged mild stress	13–38	Majid et al. (2007)
Terminal	Wheat	Prolonged mild stress	32–63	Majid et al. (2007)
Flowering and grain filling	Wheat	Prolonged mild stress	58–92	Dhanda and Sethi (2002)
Stem elongation	Wheat	Mild stress	18	Akram (2011)
Anthesis	Wheat	Mild stress	8	Akram (2011)
Stem elongation + anthesis	Wheat	Mild stress	22	Akram (2011)
Vegetative growth stage	Wheat	–	40	Allahmoradi et al. (2011)
Reproductive growth stage	Wheat	–	4	Allahmoradi et al. (2011)
Immediately prior anthesis	Sunflower	Mild stress	5–56	Chimenti et al. (2002)

decrease in water supply. Drought lowered RWC in tomato (*Solanum lycopersicum* L.) and caper bush (*Capparis spinosa* L.) (Subramanian et al. 2006; Ozkur et al. 2009). Water potential significantly declined in soybean roots, leaves, and pods under drought in general; however, root water potential dropped much earlier than leaves and pods (Liu et al. 2004). In sunflower, RWC, leaf water potential and osmotic potential were affected by drought (Tezara et al. 2002). However, different genotypes behaved differently; drought-tolerant genotypes maintained higher leaf water potential for longer and wilted later than sensitive genotypes upon exposure to drought (Ouvrard et al. 1996). Effects of drought also depend on the intensity and duration of drought. Tissue water contents decreased linearly with increased severity of drought (Reddy et al. 2004).

Transpiration not only helps to maintain leaf temperature but also drives water and nutrient uptake and CO_2 influx. Rise in leaf temperature of bread wheat and rice plants is reported under drought owing to reduced transpiration rates (Siddique et al. 2001). de Campos et al. (2011) reported reduced turgor pressure and transpiration rate in citrumelo (*Citrus trifoliata* L.) rootstocks grown under drought.

Dry matter produced per unit of water consumed is termed WUE. WUE of genotypes and crops varies under drought. Abbate et al. (2004) and Subramanian et al. (2006) reported higher WUE in wheat and tomato under drought than well-watered controls mainly due to reduced transpiration rates under drought. However, in potato (*Solanum tuberosum* L.), early season drought substantially reduced WUE (Costa et al. 1997). Crop stage is also important in defining the effect of drought on WUE. For instance, drought stress decreased WUE in sunflower; however, the extent of the reduction was significantly higher when stress was imposed at flowering than at budding (Hussain et al. 2009).

1.2.3 Mineral Uptake and Assimilation

Nutrients used for plant growth and biomass production generally come from the internal cycling of reserve materials, which require water for their solubilization and translocation (Singh and Singh 2004). Limited nutrient uptake is a general phenomenon in crop plants grown under water deficit. Subramanian et al (2006) reported reduced nitrogen (N) and phosphorous (P) contents in roots and shoots of tomato seedlings grown under drought. Similarly, McWilliams (2003) reported reduced N and potassium (K) uptake in cotton (*Gossypium hirsutum* L.) under drought. In marigold seedlings, P content under drought was severely reduced (Asrar and Elhindi 2011).

Nutrient absorption is governed by interactions at the soil–root interface, including (1) root morphology and growth rate, (2) nutrient absorption kinetics of the roots; and (3) soil nutrient supply (Gutierrez-Boemand and Thomas 1999). Decreased soil water availability affects the rate of diffusion in many plant nutrients and finally the composition and concentration of soil solution (Singh and Singh 2004). With limited water supply, nutrient uptake by roots decreases because a decline in soil-water potential slows the diffusion rate of nutrients between the soil matrix and root surface (Farooq et al. 2009a). Lower transpiration rate and impaired active transport, due to a lack of energy input and altered membrane permeability, decreases root nutrient adsorbing power of crop plants under drought (Kramer and Boyer 1995; Baligar et al. 2001).

Impaired enzyme activity involved in nutrient assimilation under drought stress also disturbs nutrient acquisition. The activity of nitrate reductase in leaves and nodules of common bean (*Phaseolus vulgaris* L.) and dhainicha (*Sesbania aculeata* L.) is substantially decreased under drought (Ashraf and Iram 2005).

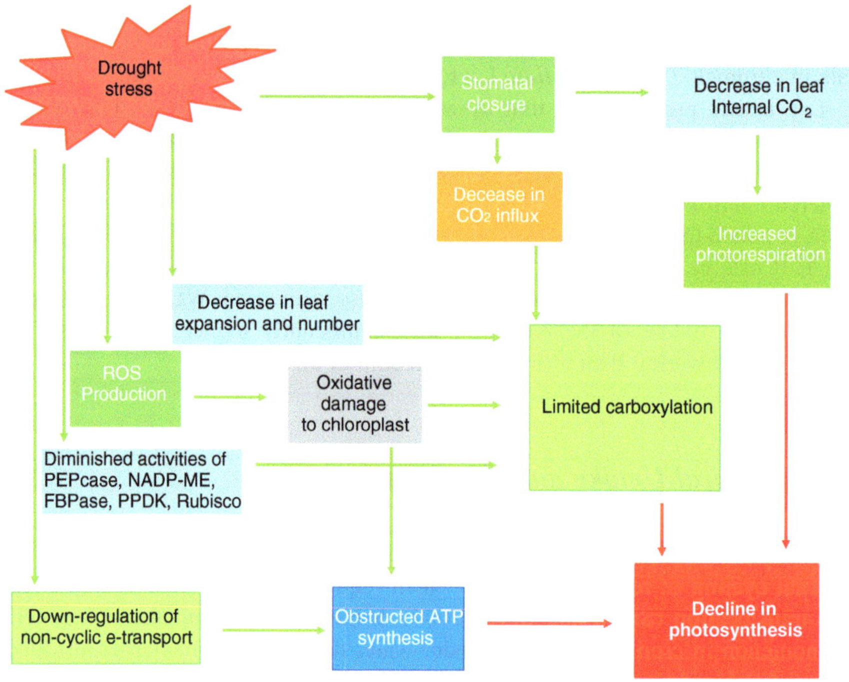

Fig. 1.1 Possible mechanism in which photosynthesis declines under drought. Drought disturbs the balance between production of ROS and antioxidant defense causing accumulation of ROS, which induces oxidative stress. With stomatal closure, CO_2 influx, and leaf internal CO_2 decrease which not only reduces carboxylation directly but also directs more electrons to form ROS and promotes photorespiration. Severe drought limits photosynthesis due to a decline in activities of ribulose-1, 5-bisphosphate carboxylase/oxygenase (*Rubisco*), phosphoenolpyruvate carboxylase (*PEPCase*), NADP-malic enzyme (*NADP-ME*), fructose-1, 6-bisphosphatase (*FBPase*), and pyruvate orthophosphate dikinase (*PPDK*). Drought limits photosynthetic area owing to reduced leaf expansion and leaf number. Moreover, non-cyclic electron transport is downregulated to match the reduced requirements of NADPH production and thus reduces ATP synthesis

1.2.4 Light Harvesting and Carbon Fixation

Efficient fixation of CO_2 by crop plants is vital for growth and productivity. Plants fix atmospheric CO_2 to organic compounds in photosynthesis. Photosynthesis is very sensitive to drought, as drought directly influences the photosynthetic capacity of crops. Reductions in leaf area (size and number) and stomatal closure, impaired activities of carboxylation enzymes and ATP synthesis, and destruction of photosynthetic apparatus are among the key factors lowering carbon fixation under drought (Fig. 1.1; Barlow 1988; Yamance et al. 2003). In general, effects of drought on carbon fixation can be divided into stomatal and non-stomatal limitations. Stomatal closure is one of the primary responses of crop plants upon exposure to drought. Stomatal closure helps reduce transpirational water loss with a simultaneous decrease in stomatal CO_2 influx (Medrano et al. 2002; Flexas et al. 2004). Miyashita et al. (2005) reported decreased photosynthesis and transpiration rates due to reduced stomatal conductance in kidney beans (*Phaseolus vulgaris* L). Likewise, drought stress progressively decreased CO_2 assimilation rates owing to reduced stomatal conductance in coconut (*Cocos nucifera* L.) plants (Gomes et al. 2010).

Reduced stomatal size is the dominant limitation to photosynthesis under mild to moderate drought stress. Reduced water potential and loss of turgor are responsible for reduced stomatal aperture; a decline in stomatal conductance due to diminished water and turgor potential is noted in sunflower under drought (Correia et al. 2006); nonetheless, stomata may also be closed without leaf dehydration. Carbon assimilation and stomatal conductance of intact sunflower leaves decreased with increasing degree of drought (Iqbal et al. 2009). Decreased stomatal aperture limits CO_2 influx resulting in a decline in the rate of photosynthesis. Although stomatal limitations may be the key factor responsible for the decline in rate of photosynthesis under some circumstances, non-stomatal limitations may be more important in other situations (Farooq et al. 2009a; Signarbieux and Feller 2011). Even avoiding stomatal limitations is not enough to overcome the detrimental effects of reduced stomatal conductance on photosynthesis under drought indicating the involvement of non-stomatal limitations (Tezara et al. 1999). Decreased synthesis and altered activities of essential enzymes and photosynthetic pigments, impaired ATP synthesis, photorespiration and heavy oxidative load are among the major non-stomatal limitations of carbon fixation (Farooq et al. 2009a).

Impaired activities of essential photosynthetic enzymes such as ribulose-1,5-bisphosphate carboxylase/oxygenase (Rubisco), phosphoenolpyruvate carboxylase, pyruvate phosphate dikinase, NADP-malate dehydrogenase, and NADP-malic enzyme are responsible for reduced rates of photosynthesis under drought (Reddy et al. 2004; Farooq et al. 2009a, b). Progressive downregulation and inhibition of metabolic processes leads to decreased ribulose-1,5-bisphosphate (RuBP) contents, which becomes the dominant limitation under severe drought, and thereby inhibits photosynthetic CO_2 assimilation (Flexas and Medrano 2002). The conversion of RuBP to 3-phosphoglyceric acid (3-PGA) decreases with decreasing leaf RWC suggesting that regeneration of RuBP is substantially inhibited under drought (Flexas and Medrano 2002). Therefore, declining RuBP contents and its regeneration are responsible for the rapid decline in photosynthesis under drought (Reddy et al. 2004). Based on maximum photosynthetic efficiency (A_{max}) and carboxylation velocity of Rubisco ($V_{c,max}$), Signarbieux and Feller (2011) reported strong non-stomatal limitations to photosynthesis in different grasslands species under drought at three different altitudes. Further, they highlighted that $V_{c,max}$ is highly sensitive to water deficit and might represent a tool to evaluate the drought response of various species for grassland performance.

Photorespiration is also responsible for the decline in carbon fixation especially in C_3 plants under drought. As Rubisco is the key enzyme in CO_2 assimilation in C_3 plants and, due to its dual nature, it may act as carboxylase or oxygenase depending upon leaf internal concentration of CO_2 or O_2. Therefore, at moderate water stress, Rubisco acts as oxygenase due to higher internal O_2 than CO_2 contents owing to stomatal closure, and increase photorespiration at the expense of carbon-fixation (Ghannoum 2009). Under acute water deficit, photorespiration may be beneficial in three ways: (1) it is involved in energy dissipation and consequently reduces photoinhibition (Flexas and Medrano 2002), (2) it produces glycine (amino acid) that is used for glutathione synthesis, a component of antioxidant defense (Noctor et al. 1999), and (3) it enhances the RuBP supply to Calvin cycle (Wingler et al. 2000). In parallel to these benefits, photorespiration amplifies oxidative stress in photosynthetic tissues with elevated production of H_2O_2 in the peroxisome through glycolate oxidase; and more than 70 % H_2O_2 generation in C_3 plants under drought is related to photorespiration (Noctor et al. 2002).

The role of photosynthetic pigments such as chlorophyll (chl) contents, carotenoids, and xanthophylls are also vital in carbon fixation, as they are involved in capturing solar radiation to drive the photosynthetic mechanism. Drought stress severely decreased chl a and chl b contents in marigold (Asrar and Elhindi 2011). Similarly, water stress diminished the photosynthetic rate with declining photochemical efficiency (Fv/Fm) of chl fluorescence in primary leaves of kidney beans (Miyashita et al. 2005).

As mentioned earlier, low CO_2 influx under moderate and severe drought impaired RuBP contents, Rubisco activities or ATP synthesis which downregulates carbon fixation; consequently, oxidation of reduced nicotinamide adenine dinucleotide (NADPH) in the Calvin cycle is lowered. As a result, $NADP^+$ (primary electron acceptor) is not sufficiently available. Therefore, when drought-stressed plants are exposed to high irradiance, overproduction of ROS results in photoinhibition (Flexas and Medrano 2002). In canary pine (*Pinus canariensis* L.) exposed to moderate drought, for instance, rate of net photosynthesis was significantly reduced primarily owing to overproduction of ROS and reduced stomatal conductance (Tausz et al. 2001). Similarly, Pompelli et al. (2010) reported higher lipid peroxidation in jatropha (*Jatropha curcas* L.) plants due to decreased stomatal conductance under drought stress.

Plants exposed to almost all kinds of abiotic stresses including drought leads to elevated oxidative stress with overproduction of ROS, which are highly toxic and cause damage to proteins, lipids, carbohydrates, and DNA (Farooq et al. 2009a, b, 2011). Superoxide radicals (O_2^-), single oxygen (O_2^1), hydrogen peroxide (H_2O_2), alkoxy radicals (RO), and hydroxyl radicals (OH^-) are among the major ROS generated in plants under abiotic stresses (Apel and Hirt 2004). These ROS are highly reactive and deteriorate normal plant metabolism through oxidative damage to lipids, protein, and other macromolecules in the absence of any protective mechanism (Rout and Shaw 2001). In chloroplasts, PSI and PSII and in mitochondria, complex I, ubiquinone and complex III of electron transport chain (ETC.) are the major sites of generation of ROS (Gill and Tuteja 2010). Therefore, impaired activities of essential enzymes and ATP synthesis due to oxidative damage hamper photosynthetic and respiratory activities. Increased lipid peroxidation due to overproduction of ROS is reported in caper bush seedlings under drought (Ozkur et al. 2009). Likewise, Qiu et al. (2008) reported higher lipid peroxidation (as indicated by MDA) in young wheat seedlings exposed to drought due to build up of O_2^- and H_2O_2.

1.3 Drought Resistance Mechanisms

Plants undergo several morphological and biochemical adaptations at subcellular, cellular, and organ level to survive under drought. As far as crop production in a farmer's field is concerned, survival without a rational yield for subsistence requirement is not sufficient. Only a genotype producing higher grain yield compared with others under drought is truly drought resistant (Fukai and Cooper 1995; Kiliç and Yağbasanlar 2010). However, drought tolerance is a complex phenomenon associated with cuticle thickness, stomatal regulation, root system, hormonal balances, antioxidant defense system, osmotic adjustment, and maintenance of tissue water contents, etc. An overview of drought resistance mechanisms at various levels in crop plants is briefly described.

1.3.1 Morphological Adaptations

Drought escape, dehydration avoidance, and dehydration tolerance are important adaptive mechanisms of plants exposed to drought; a brief account of these adaptations is given below.

1.3.1.1 Drought Escape

Drought escape is the ability of a plant to complete its life cycle before the onset of drought and to undergo dormancy before the onset of the dry season. It is a general phenomenon in some desert plants, which exhibit extremely short life cycles and produce seeds during short rainy seasons in order to save them from extinction (Levitt 1980). Some extremely short duration crops and even genotypes within a species exhibit drought escape, but this always taxes yield (Turner et al. 2001). Drought escape is more advantageous where chances of terminal drought are more recurrent; as short-duration cultivars frequently escape terminal drought compared with late maturing cultivars (Meyre et al. 2001).

Matching crop phenology to prevailing growing conditions seems advantageous especially in terminal drought-prone areas. However, flowering may be critical as flowering time is a major trait of crop adaptation in areas where the growing season is restricted by terminal drought (Meyre et al. 2001).

1.3.1.2 Drought Avoidance and Phenotypic Flexibility

Drought avoidance is the ability of a plant to sustain high plant water status or cellular hydration under drought (Blum 2005). Plants accomplish this mechanism either by catching more water from the soil or by minimizing water loss through transpiration (Blum 2005).

Plants undergo certain morphological and physiological changes to minimize stress-induced losses; and considerable flexibility among crop species and even within species exists. Reduced biomass is a common phenomenon under drought; however, the extent of the reduction varies as different plant organs are affected differently. Drought reduced both shoot and root dry weight in Asian red sage (*Salvia miltiorrhiza* Bunge) but the effect was more severe on shoots, which enhanced the root-shoot dry weight ratio (Liu et al. 2011).

Crops plants more able to extract water from the entire soil profile can better resist drought. Root plasticity is the ability of a genotype to regulate its root growth pattern in accordance with prevailing constraints (Yamauchi et al. 1996). More rooting depth, root proliferation, and root length density commensurate with fetching more water are considered drought avoidance traits (Matsui and Singh 2003; Wang and Yamauchi 2006). A more prolific and deeper root system is imperative under drought as roots are the only organ capable of extracting water from the soil profile (Kavar et al. 2007; Gowda et al. 2011). Genotypes with more root growth in stress-prone environments are preferred. Drought stress inhibited root growth in general, even in tolerant genotypes, but the effect was more prominent on sensitive wheat genotypes due to an overall decrease in newly synthesized cell wall polysaccharides such as pectins, hemicelluloses, and cellulose (Piro et al. 2003). Likewise, peanut (*Arachis hypogaea* L.) genotypes with more root dry weight and root length density in deeper soil layers produced more pod yield than genotypes with less root dry weight and length exposed to water stress at pre-flowering (Jongrungklang et al. 2011).

Reduced leaf size is well correlated with drought stress; many xerophytes have small leaves as their adaptation to survive in harsh environments. As earlier mentioned, a small leaf area is advantageous to restricted water use and also responsible for low productivity of crops (Sinclair and Muchow 2001) but different crops or genotypes behave differently.

Reduced transpirational water loss is another important adaptation of drought avoidance. To attain that, plants have developed several morphological adaptations such as higher root–shoot ratios with fewer and smaller leaves to withstand water-deficit conditions by maintaining water budget with higher uptake and minimal loss (Lei et al. 2006). Increased stomatal and cuticular resistance, less small stomata, smaller leaf area, and vertical leaf orientation are among the major drought avoidance traits to minimize transpiration in order to save water under stress conditions (Sinclair and Muchow 2001; Wang and Yamauchi 2006). However, reduced leaf area and plant stature are advantageous for restricted water use but may also result in low productivity of crop plants (Sinclair and Muchow 2001).

1.3.2 Physiological Adaptations

Dehydration-tolerant plants maintain metabolic activities at low tissue water potential. Osmotic adjustment, antioxidant defense system, and changed dynamics of phytohormones are among the major physiological adaptations of plants under drought.

1.3.2.1 Osmotic Adjustment

Inadequate water availability in drought-prone environments affects the growth and productivity of crops by lowering tissue water status and turgor (Kiani et al. 2007; Hussain et al. 2009). Accumulation of organic and inorganic solutes under drought and/or salinity, which help lower water potential without decreasing actual water contents, is referred to as osmotic adjustment or osmoregulation (Serraj and Sinclair 2002). Nonetheless, these solutes do not pose any detrimental effects on membranes, enzymes, and other macromolecules, even at higher concentration, and are also called compatible solutes (Cechin et al. 2006; Kiani et al. 2007). These compatible solutes include soluble sugars, sugar alcohols, proline, glycinebetaine (GB), organic acids, trehalose, etc., (Cechin et al. 2006; Kiani et al. 2007; Farooq et al. 2008, 2009a, b). These compatible solutes not only help to maintain turgor pressure but also to protect the enzymes and macro molecules of cells from the damaging effects of ROS (Farooq et al. 2009a, b).

Osmotic adjustment is the key adaptation of plants at the cellular level to minimize the effects of drought-induced damage in crop plants (Blum 2005) and helps plants under drought in two ways: (1) it helps maintain leaf turgor to improve stomatal conductance for efficient intake of CO_2 (Kiani et al. 2007), and (2) it promotes the root's ability to uptake more water (Chimenti et al. 2006).

Under drought, greater accumulation of compatible solutes such as free proline, free amino acids, and GB may take place (Manivannan et al. 2007; Farooq et al. 2008). Proline is an important compatible solute which accumulates in plants exposed to dehydration stress, particularly in younger leaves (Perez-Perez et al. 2009). High levels of free proline allow the plant to maintain low water potentials (Jalil et al. 2007; Sankar et al. 2007) and derive water from the environment. It is not only involved in osmoregulation but also protects plants from osmotic stress (Sankar et al. 2007). Recently, de Campos et al. (2011) explored the role of elevated endogenous proline levels on gas exchange, antioxidant enzymatic activities, and water relations in leaves of transgenic citrumelo rootstocks equipped with *P5CSF129A* gene coding for the synthesis of key enzyme for proline synthesis under drought stress. Better turgor pressure highlighted the importance of osmotic adjustment with elevated endogenous proline accumulation. Higher turgor pressure in leaves of transgenic plants promoted stomatal conductance, photosynthetic rate, and transpiration rate compared with non-transgenic plants. They suggested that higher endogenous proline contents in transgenic plants are not only involved in osmoregulation but also participate in leaf gas exchange and nullify the lethal effects of oxidative stress under drought conditions.

In addition to its function in osmoregulation (Wyn Jones et al. 1977), GB has been shown to protect functional proteins, enzymes (e.g. Rubisco), and lipids of the photosynthetic apparatus, and to maintain electron flow through thylakoid membranes (Xing and Rajashekar 1999). Therefore, osmotic adjustment by accumulating a variety of organic and inorganic solutes, along with higher activity of antioxidant enzymes in leaves, is among the imperative physiological adaptations of plants grown in drought-prone conditions (Lei et al. 2006).

1.3.2.2 Plant Growth Substances

Plant growth and development is regulated by certain growth substances produced internally called phytohormones or, if applied externally, plant growth regulators (PGRs). Phytohormones are needed in minute concentrations to modulate plant growth (Farooq et al. 2009a). Auxins, gibberellins (GA_3), cytokinins (Cks), ethylene, and ABA are the most studied phytohormones in higher plants. Of these, GA_3 and Cks promote plant growth (growth promoters), while ethylene and ABA have inhibitory effects (growth retardants) (Taiz and Zeiger 2010). Drought stress alters the endogenous synthesis of these growth substances. Generally, under stress conditions, the concentration of growth retardants increases at the expense of growth promoters to regulate plant water budget (Farooq et al. 2009a). For instance, drought altered the hormonal balance in common bean leaves, which included increased ABA content, a small decline in indole acetic acid (IAA) and GA_3, and a sharp drop in zeatin content. Furthermore, a sharp decline in endogenous Cks contents amplified the response of shoots to increasing ABA content under drought (Figueiredo et al. 2008). The ABA and polyamine (PA) contents increased, while that of Cks decreased during a drought stress period in sunflower (Bano et al. 1994).

Under drought, Cks regulate plant responses via root-to-shoot signals (Rivero et al. 2007). Elevated ABA contents in plant organs under drought stress lead to many physiological changes to modulate plant growth. Higher ABA accumulation in roots under limited water supply signals the leaves to induce stomatal closure and avoid water loss via transpiration (Davies and Zhang 1991). ABA treatment on intact sunflower plants showed that K^+-starved plants displayed faster water uptake, both in light and dark but the ABA treatment significantly inhibited water uptake during darkness over a 24 h cycle (Fournier et al. 2005).

Tomato seedlings treated with 1 μM 24-epibrassinolide (EBR) nullified the negative effects of drought due to enhanced synthesis of endogenous ABA (Yuan et al. 2010). Salicylic acid (SA) is also a secondary metabolite accumulated in plants under drought stress, which is involved in inducing drought tolerance in plants by regulating several physiological processes through signaling. For instance, SA regulates drought-induced leaf senescence in perennials (Abreu and Munne-Bosch 2008). Leaf senescence in drought-stressed plants contributes to nutrient remobilization thus allowing the rest of the plant to benefit from the nutrients accumulated during the lifespan of the leaf. Higher dry matter partitioning from stems and leaves to developing grains was noted in cultivars with more leaf senescence during grain filling under severe drought stress that ultimately helps to maintain grain yields (Kumar et al. 2006).

1.3.2.3 Antioxidant Defense System

Limited water supply under drought promotes oxidative stress with overproduction of ROS. Declined CO_2 influx with stomatal closure or impaired activities of enzymes and damaged photosynthetic apparatus under water stress downregulate photosynthesis leading to the generation of a variety of ROS such as O_2^-, O_2^1, H_2O_2, RO and OH^- (Flexas and Medrano 2002; Lawlor and Cornic 2002). These ROS are highly reactive and deteriorate normal plant metabolism through oxidative damage to lipids, protein, and other macromolecules in the absence of any protective mechanism (Rout and Shaw 2001). Plants douse the ROS through their antioxidant defense system with enzymatic and non-enzymatic components (Li 2008; Simova-Stoilova 2008; Hussain et al. 2008).

Of the enzymatic antioxidants, superoxide dismutase (SOD), catalase (CAT), peroxidase (POX), glutathione reductase (GR), and ascorbate peroxidase (APX), and non-enzymatic antioxidants, ascorbic acid, α-tocopherol, reduced glutathione, β-carotene, PAs, salicylates, compatible solutes such as proline and GB and zeaxanthin accumulate in higher plants under drought stress to avoid oxidative damages (Scandalios 2005; Li 2008; Ozkur et al. 2009; 2010b).

Plants enhance the production of antioxidants in order to minimize detrimental effects of oxidative stress to normalize their metabolic activities under drought-induced oxidative stress (Fig. 1.2). It is generally accepted that O_2^- might be converted to H_2O_2 and then metabolized to water by APX and GR in plants to maintain membrane structures (Foyer and Fletcher 2001). Likewise, several other antioxidant enzyme molecules are responsible to counteract the deleterious effects of ROS. Initially, SOD catalyzes the conversion of O_2^- to H_2O_2 that is further reduced to water by APX by using ascorbate as an electron donor (Scandalios 2005). Elevated accumulation of antioxidant enzymes such as SOD, CAT, GR, APX, and POX are involved in lowering oxidative injury in caper bush seedlings under drought stress (Ozkur et al. 2009). Improved activities of CAT, APX, and SOD by brassinolides (BRs) application decreased H_2O_2 and MDA contents in drought-prone tomato seedlings (Yuan et al. 2010).

The ascorbate–glutathione cycle is a more efficient enzymatic way to breakdown H_2O_2 than CAT and POX as indicated in sorghum and sunflower (Cakmak et al. 1993). Drought decreased cytosolic activities of monodehydroascorbate reductase, and POX, and increased cytosolic dehydroascorbate reductase activity; however, chloroplastic activities of monodehydroascorbate reductase and dehydroascorbate reductase as well as cytosolic activities of CAT and SOD were not affected by drought in sorghum. Activities of the enzymes involved in the ascorbate–glutathione cycle were higher in the cytosolic fraction than the

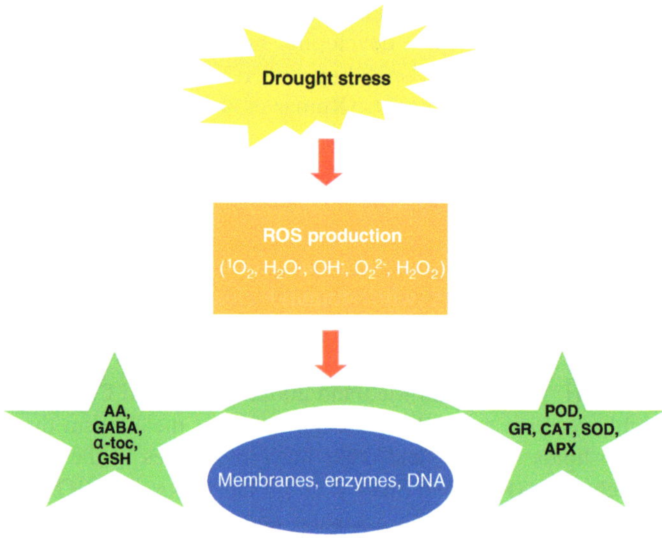

Fig. 1.2 Role of antioxidant enzymes in the ROS scavenging mechanism. Exposure to drought leads to generation of ROS, including singlet oxygen ($^1O_2$), perhydroxyl radical ($H_2O^\cdot$), hydroxyl radicals (O_2^{2-}), hydrogen peroxide (H_2O_2), and alkoxy radical (RO). The ROS attack membranes, enzymes and DNA, causing oxidative damage and impairing normal functions of cells. The antioxidant defense system in the plant cell includes both enzymatic and non-enzymatic constituents. Among the enzymatic components are peroxidase (POD), glutathione reductase (*GR*), catalase (*CAT*), superoxide dismutase (*SOD*), and ascorbate peroxidase (*APX*), whereas major constituents of the non-enzymatic antioxidant system are ascorbic acid (*AA*), γ-amino butyric acid (*GABA*), α-tocopherol (α-toc), and reduced glutathione (GSH). Upon exposure to drought, tolerant cells activate their antioxidant defense system, which then starts scavenging ROS and protecting the cell

chloroplastic fraction in sorghum. But in sunflower, chloroplasts had higher enzymatic activities than cytosol. Because of differential localization of enzymes in cell fractions, it can be inferred that to detoxify H_2O_2 sorghum (C_4 plant) mainly uses the cytosolic ascorbate–glutathione cycle, whereas sunflower (C_3 plant) primarily uses the chloroplastic ascorbate–glutathione cycle (Zhang and Kirkham 1996).

Higher free proline accumulation (1.5–2.1 times higher than control) in leaflets of coconut tree under drought stress decreased membrane damage (Gomes et al. 2008, 2010). Elevated accumulation of amino acids, proline, and GB accompanied by higher productivity in sunflower under drought stress highlights the protective role of these non-enzymatic antioxidants against oxidative damages under drought conditions (Manivannan et al. 2008; Hussain et al. 2008). Similarly, a decline in soluble sugar contents in rice anthers under drought stress impaired the ability to scavenge ROS and further intensify oxidative stress (Guan-fu et al. 2011).

1.3.3 Molecular Mechanism

Plants challenged with drought undergo many adaptive mechanisms at molecular levels to modulate water balance. Among them, up- and downregulation of many gene transcripts and accumulation of stress proteins is important (Kavar et al. 2007). A considerable rise in CDSP 32 (chloroplastic drought-induced stress protein) mRNA and protein was noted in potato under drought, which protected the chloroplast from drought-induced oxidative damage (Broin et al. 2000). Aquaporins are an important group of intrinsic membrane proteins able to assist passive exchange of water across membranes; these potentiate a 10- to 20-fold increase in water permeability by regulating hydraulic conductivity of membranes (Tyerman et al. 2002).

Many dehydration-responsive element-binding genes are also involved in signaling pathways in response to abiotic stresses including drought (Agarwal et al. 2006). The dehydration-responsive element/C-repeat (DRE/CRT) *cis*-acting element and its DNA-binding protein are a major transcription system modulating ABA-independent gene expression in response to drought and includes dehydration-responsive element binding proteins (DREB)/C-repeat binding factors (CBF) family of proteins. DREB2 subclass of DREB/CBF family proteins are expressed under drought to articulate genes involved in stress tolerance (Seki et al. 2003).

Signal transduction pathways are also induced in plants under drought to regulate growth. An early-warning response mechanism exists in plant roots to activate the hydrogen pump ATPase protein (H^+-ATPase) on plasma membrane of root hairs before a substantial decline in plant RWC. The activation of root hair cell plasma membrane H^+-ATPase triggers amplified biosynthesis of key osmolytes such as leaf proline and GB to maintain the water budget of plants. Moreover, interspecific and intraspecific differences in the timing of triggering early responses may exist and drought-tolerant cultivars to initiate warming responses much earlier than sensitive cultivars (Gong et al. 2010). PAs have been associated with the response of plants to drought via signaling, in addition to their roles in responses to many other stresses (Bae et al. 2008). Ornithine decarboxylase (TcODC), arginine decarboxylase (TcADC), S-adenosylmethionine decarboxylase (TcSAMDC), spermidine synthase (TcSPDS), and spermine synthase (TcSPMS) are the expression patterns of genes encoding enzymes involved in PA in cacao (*Theobroma cacao* L.) leaves. Expression of TcODC, TcADC, and TcSAMDC is induced at the start of drought which modulates stomatal conductance, photosynthesis, photosystem II efficiency, and leaf water potential. Induction of TcSAMDC in leaves is most closely correlated with changes in water potential. The earliest measured responses to drought, detected in cacao leaves 13 days after the onset of drought, were enhanced expression of TcADC and TcSAMDC in roots along with decreases in stomatal conductance, photosynthesis, and PS II efficiency due to elevated levels of putrescine, spermidine, and spermine (Bae et al. 2008).

1.4 Management of Drought Stress

Water-wise cultivation augmented with drought-tolerant crops and genotypes is the only option to feed the escalating world population. Development of crop genotypes requiring less water to produce unit biomass along with an accurate site-specific package of production technology is urgently needed to sustain crop productivity in drought-prone areas. Advancements in the field of genomics and biotechnology along with traditional breeding approaches seem viable to introduce transgenic plants to perform better under stress conditions. Nonetheless, exogenous application of certain osmoprotectants, growth regulators, and minerals are helpful to induce drought tolerance in otherwise susceptible plants.

1.4.1 Selection and Breeding Approaches

Mass screening and breeding approach is often used to explore genetic variability for drought tolerance among crop genotypes for desired agronomic traits to then breed genotypes better able to perform in drought-prone areas (Ashraf 2010). Several strategies such as induction of earliness for drought escape, modification in morphology that leads toward drought resistance, and introduction of drought-tolerant traits associated with high yield can be targeted in breeding programs to induce drought tolerance (Rauf 2008).

In most regions of arable cropping, droughts are unpredictable but do not generally occur every year. Selection and breeding efforts should therefore be targeted to develop genotypes producing good yield under both drought and well-watered conditions. Conventional breeding efforts focus on direct selection for yield under target environments (Atlin and Lafitte 2002). Since yield is a quantitative trait with low heritability and high genotype × environment interactions (Babu et al. 2003), this approach is no more an attractive option (Farooq et al. 2009a). Mass screening of genotypes may therefore be done under managed or natural environments for certain stable secondary traits with highly heritable, easy to measure genetic associations with grain yield under drought but no association with yield reduction under optimal environments (Edmeades et al. 2001). However, some studies have reported the effectiveness of direct selection for grain yield under drought (Venuprasad et al. 2007; Kumar et al. 2008).

Developing short-duration genotypes often seems useful to avoid yield losses from late-season drought, as short life cycles help to avoid periods of stress (Kumar and Abbo 2001). However, these short-duration cultivars do not do well in a prolonged growing season. Therefore, genotypes with developmental plasticity are better under drought stress conditions, as they can to some extent adjust their life cycle according to the prevailing conditions (Levitt 1980).

Root characteristics, drought susceptibility index, and relative yield are among the important secondary traits, which may be used to select drought-tolerant genotypes (Chimenti et al. 2002; Blum 2005; Kiani et al. 2007). According to Biao-lin et al. (2011), root length and number, root fresh weight, and relative water contents significantly influence drought resistance, which could thus be used as a comprehensive index for drought resistance at the seedling stage. Gowda et al. (2011) also opined to select genotypes with deep roots for improving drought resistance. Signarbieux and Feller (2011) reported that carboxylation velocity of Rubisco provides a rapid tool for screening genotypes for drought tolerance. Manivannan et al. (2008) reported osmotic adjustment as an important adaptive trait nullifying the bad effects of drought stress in sunflower.

1.4.2 Marker-Assisted Selection

Advancements in the field of biology and proteomics make it possible to identify various quantitative trait loci (QTL) and proteins associated with drought tolerance in crop plants; and these drought-related QTL and proteins can be used as markers in breeding programs to tailor drought-tolerant genotypes (Salekdeh et al. 2002; Lanceras et al. 2004; Farooq et al. 2009a, b; Ashraf 2010). After mapping QTLs for grain yield and related traits with simple sequence repeat/expressed sequence tag marker map, Kirigwi et al. (2007) explained substantial variation in chromosome 4A of wheat. A QTL located on rice chromosome 1 near sd1 contributes 4–32 % of total genetic variation for grain yield under drought and optimal conditions, respectively, suggesting that this QTL is more responsible for varied grain yield of rice under stressful environments than under normal situations (Kumar et al. 2007).

Ke et al. (2009) identified 10 drought-responsive phosphoproteins in rice using proteomics approach. Similarly, Salekdeh et al. (2002) identified 16 drought-responsive proteins in rice; and among them, actin depolymerizing factor, S-like RNase homolog, and RuBP activase were upregulated and isoflavone reductase-like protein was downregulated under drought stress. LEA proteins accumulate in various crop plants exposed to drought-stressed environments (Babu et al. 2004: Gosal et al. 2009), acting as molecular chaperones and protecting the biological membrane from the damaging effects of dehydration stress (Browne et al. 2002; Babu et al. 2004). Substantial increases in chloroplastic drought-induced stress protein (CDSP 32) mRNA and protein was reported in potato under drought, which may preserve chloroplastic structures against oxidative injury upon drought (Broin et al. 2000).

Recently, Bae et al. (2011) identified five putative proteins in the rice genome database (http://signal.salk.edu/cgi-bin/RiceGE) named as *Oryza sativa* RING domain-containing proteins (OsRDCPs) possessing a single RING motif in their N-terminal regions. Of five identified proteins, OsRDCP1 is involved in physiological responses to neutralize the effects of drought stress. Nonetheless, OsRDCP1-overexpressing transgenic rice lines exhibited superior growth and dehydration tolerance than wild and OsRDCP1-mutant lines under water deficit.

Therefore, OsRDCP-1 can be used as a target protein for developing drought-tolerant rice genotypes in future. Proton-ATPase (H^+-ATPase) is also an important protein located on the plasma membrane of root hair cells mediating root-to-shoot signaling under limited water supply and is actively involved in osmoregulation, maintaining water contents and other water stress-related mechanisms (Liu et al. 2005, 2008). Likewise, PebC1 a novel protein elicitor isolated and purified from *Botrytis cinerea* strain BC-4-2-2-1 (mold fungus) notably improved wheat seedling growth and the drought resistance integrated index from 36.53 to 57.08 under two cycles of drought stress. Nevertheless, phenylalanine ammonia-lyase (PAL), POD and polyphenol oxidase (PPO)-related resistance mechanisms also perked up after PebC1 treatment (Zhang et al. 2010).

Molecular and genetic approaches are also used to explore signal transduction pathways induced in plants in response to drought. Hydraulic properties of roots are regulated by root-sourced signals such as ABA (Mahdieh and Mostajeran 2009). DRE/CRT cis-acting element and its DNA-binding protein is a major transcription system including a DREB/CBF family protein that modulates ABA-independent gene expression under drought. DREB1/CBF and DREB2 are two subclasses of DREB/CBF family proteins generated under cold and drought stress, respectively, to articulate various genes involved in stress tolerance (Seki et al. 2003). Therefore, combined knowledge of traditional breeding along with marker-assisted selection makes it easier and more efficient to induce drought tolerance in crop plants.

1.4.3 Biotechnology and Functional Genomics

Advancements in biotechnology made it possible to identify drought-responsive genes to tailor plants with superior drought tolerance using the transgenic approach (Hadiarto and Tran 2011). Transfer of one or more genes from one species to another to induce some desired qualitative and quantitative traits is called the transgenic approach and plants are called transgenic plants. Compared with conventional breeding, the transgenic approach ensures induction of only desired genes and confines the entry of unwanted genes into an organism from the donor organism (Gosal et al. 2009). Cloning of stress tolerance-related genes and identifying their functions are crucial to augment crop tolerance to abiotic stresses including drought (Ashraf 2010). Plants with C_4 pathways exhibit higher photosynthesis rates, water use efficiency and higher biomass production, especially at low internal leaf CO_2 concentration, high temperature and drought stress. Efforts are underway to introduce this feature to C_3 plants by overproducing PEP carboxylase to improve photosynthetic efficiency and yield potential under drought (Bao-Yuan et al. 2011). Transgenic rice plants producing PEP carboxylase resist drought stress with a slow decline in photosynthetic rate. It is supposed that overproduction of PEP carboxylase might be involved in drought resistance to decrease the effect of drought stress on photosynthesis in rice (Bao-Yuan et al. 2011).

Elevated accumulation of compatible solutes such as proline and GB play a pivotal role in enhancing drought tolerance among a variety of crops with superior osmotic adjustment (Ashraf and Foolad 2007) and transferring genes to increase the endogenous accumulation of such solutes can improve drought tolerance of transgenic plants (Table 1.2). Recently, de Campos et al. (2011) studied the performance of transgenic 'Swingle' citrumelo rootstocks equipped with *P5CSF129A* gene to code the key enzyme for proline synthesis. Maintenance of turgor pressure in leaves of transgenic plants led to higher stomatal conductance, photosynthetic, and transpiration rates when compared to non-transgenic plants. Results suggest that transgenic plants coped better with water deficit than non-transgenic controls, as high endogenous proline levels acted not only by mediating osmotic adjustment, but also by contributing to gas exchange parameters and ameliorating deleterious effects of drought-induced oxidative stress by overproduction of APX and SOD.

Similarly, encoding genes to perk up endogenous accumulation of polyamines might help to alleviate damaging effects of water stress. *Arabidopsis thaliana* transgenic lines overexpressing arginine decarboxlase 2 gene (ADC2) contain high levels of Put without disturbing spermidine and spermine levels, even under drought stress, suggesting the role of Put in dehydration resistance. A notable reduction in transpiration rate with elevated stomatal closure is observed in the ADC2 over-expresser lines compared with wild type, while the number of stomata remained unchanged. These findings confirm that the role of Put in drought induction in transgenic plants is related to avoiding the water loss via transpiration by regulating stomatal opening (Alcázar et al. 2010). However, this may tax crop yield since most probably photosynthesis was decreased.

1.4.4 Exogenous Application of Hormones and Osmoprotectants

Exogenous application of various PGRs and osmoprotectants can also enhance drought tolerance of crop plants. Foliar application of plant hormones and osmoprotectants such as GA_3, Cks, ABA, proline, GB, brassinolids, PAs, and SA has proven worthwhile to ameliorate stress effects with elevated osmotic adjustment to maintain turgor and antioxidants accumulation to detoxify ROS to maintain stability of membrane structures, enzymes, and other macromolecules under drought stress (Manivannan et al. 2008; Farooq et al. 2009c, d; Yuan et al. 2010; Alcázar et al. 2010; Anjum et al. 2011).

SA is a secondary metabolite that induces drought tolerance in plants by regulating several physiological processes through signaling. It plays an important role in producing tolerance against drought (Senaratna et al. 2000; Singh and Usha 2003). Exogenous application of methyl salicylic acid on leaves of water-stressed perennial plants promotes leaf senescence; as programmed leaf senescence in drought-stressed plants contributes to nutrient remobilization thus allowing the rest of the plant to benefit from the nutrients accumulated during the life span of the leaf (Abreu and Munne-Bosch 2008). Likewise, exogenous application of SA

Table 1.2 Transgenic plants for drought resistance

Genes	Source	Host	Mechanism of action	Reference
Farnesyl-transferase *(ERA1)*	Arabidopsis	Canola	Down-regulation of farnesyltransferase enhances plant's response to ABA and drought tolerance reducing stomatal conductance	Wang et al. (2005a)
Mannitol-1-phosphate dehydrogenase (*mtlD*)	*Escherichia coli*	Wheat	Improved fresh and dry weights, plant height, and flag leaf length	Abebe et al. (2003)
Late embryogenesis abundant (LEA) abscisic acid-regulated gene (*HVA1*)	Barley	Wheat	Overaccumulation of LEA proteins increases drought tolerance	Sivamani et al. (2000) Bahieldin et al. (2005)
Pyrroline-5-carboxylate synthetas (*P5CS*)	*Vigna aconififoli*	Tobacco	Enhanced accumulation of proline increases drought tolerance	Kavi Kishor et al. (1995)
OsLEA3	–	Rice	Over-accumulation of LEA increases drought tolerance	Hu et al. (2008)
Beta	*Escherichia coli*	Wheat	Encoding choline dehydrogenase results in accumulation of glycinebetaine, which increases drought tolerance	He et al. (2011)
Bifunctional trehalose-6-phosphate synthase or phosphatase (TPSP) as a fusion gene of *otsA* and *otsB*	*Escherichia coli*	Rice	Higher amounts of trehalose and less photo-oxidative damage	Wu and Garg (2003)
MnSOD	Pea	Rice	Improved expression of superoxide dismutase (SOD) results in better (reactive oxygen species) ROS scavenging system in plant chloroplasts	Wang et al. (2005b)
NADP-ME	Maize	Tobacco	Increased activity of NADP-malic enzyme (ME), decreased stomatal conductance and improved water use efficiency	Laporte et al. (2002)
Nicotiana protein kinase (*NPK1*)	Tobacco	Maize	Protects photosynthesis machinery from dehydration damage	Shou et al. (2004)

(continued)

Table 1.2 (continued)

Genes	Source	Host	Mechanism of action	Reference
Cox	Arthrobactr pascens	Indian mustard	Over accumulation of glycinebetaine increases drought tolerance	Huang et al. (2000)
Cox	Arthrobactr pascens	Arabidopsis	Over accumulation of glycinebetaine increases drought tolerance	Huang et al. (2000)
Cox	Arthrobactr pascens	Tobacco	Over accumulation of glycinebetaine increases drought tolerance	Huang et al. (2000)
beta	–	Cotton	Over accumulation of glycinebetaine increases drought tolerance	Lv et al. (2007)
TPSP	*Escherichia coli*	Rice	Protection of photosynthesis machinery	Waditee et al. (2005)
Trehalose-6 phosphate synthase gene (TPSJ)	Yeast	tobacco	Over accumulation of trehalose increases drought tolerance	Romero et al. (1997)
otsA, *otsB*	*Escherichia coli*	Rice	Encoding choline dehydrogenase results in accumulation of glycinebetaine and trehalose, which increases drought tolerance	Holmström et al. (1996)

enhanced growth and productivity of sunflower under drought stress with notable improvement in turgor pressure due to elevated accumulation of endogenous proline and GB than control plants (Hussain et al. 2008, 2009).

Polyamines are associated with the response of plants to drought via signaling (Bae et al. 2008). Spermine-deficient mutants of arabidopsis were more sensitive to drought compared with wild types with 40 % extra reduction in chl contents and more water loss due to opened stomata under deficit water conditions. However, exogenous applications of PAs other than spermine did not improve growth; while spermine application perks up growth and maintained higher chl contents such as wild types under water stress. Nevertheless, spermine application also regulates stomatal conductance by modulating Ca^{2+}-permeable channels to lift cytoplasmic Ca^{2+} contents to inactivate the K^{+} inward rectifier plasma membrane to motivate stomatal closure (Yamaguchi et al. 2007).

Elevated levels of putrescine, spermidine and spermine were detected in cacao leaves 13 days after the onset of drought along with declined stomatal conductance, suggesting the possible role of PA in enhanced tolerance to drought (Bae et al. 2008). Exogenous spermidine application before the onset of drought notably improved the performance of barley under stress conditions (kubis 2003). Exogenous Ca^{2+} may have a role in modifying endogenous PAs levels under drought stress leading to increased drought tolerance in bread wheat (Ma et al. 2005).

Brassinosteroids are naturally occurring substances with the potential to regulate plant growth and perk up the plant's potential to perform well in drought-prone conditions (Bajguz and Hayat 2009). Tomato seedlings treated with 1 μM 24-epibrassinolide successfully ameliorated the negative effects of drought due to elevated synthesis of endogenous ABA along with higher activities of antioxidant enzymes such as CAT, APX, and superoxide; as a consequence, H_2O_2 and MDA contents decreased along with elevated RWC and net photosynthesis in tomato seedlings under drought stress (Yuan et al. 2010). Likewise, exogenous application of BR amazingly improved gas exchange attributes, plant height, leaf area, seedling dry weight, and cobs per plant under both drought and well-watered conditions in maize due to greater protein synthesis, maintained tissue water potential and higher activities of antioxidant enzymes, which lowered lipid peroxidation (Anjum et al. 2011).

Externally applied amines have also proven beneficial to ameliorate the damaging effects of water deficit in field-grown crops. Barley plants treated with 2-aminoethanol maintained higher growth and accumulated more biomass due to superior chlorophyll contents and enhanced activity of superoxide dismutase, which provides protection against oxidative stress to barley plants under drought. Nonetheless, under drought stress, 2-aminoethanol pretreatment increased grain yield of barley by about 5–30 % (Mascher et al. 2005). Likewise, applications of trizoles regulate plant growth under water deficit by detoxifying ROS due to higher accumulation of quaternary ammonium compounds and antioxidants.

Externally applied GB is involved in osmoregulation and also acts as an osmoprotectant to save membranes and enzymes from oxidative stress under drought stress (Ma et al. 2006). Exogenous GB application protects photosynthetic machinery in maize, wheat, and sorghum thereby increasing final yield (Agboma et al. 1997). Foliar application of GB helped to maintain a higher net photosynthetic rate during drought stress than non-GB treated plants. Moreover, GB also triggered antioxidant defense (Ma et al. 2006). Foliage-applied GB enhanced the yield of tomato plants not only in stress-prone environments but also in optimal conditions in greenhouse (Makela et al. 1998). Foliar applications of GB at vegetative and flowering stages of sunflower alleviated the detrimental effects of water stress and improved turgor pressure, productivity, and WUE compared with non-applied plants (Iqbal et al. 2005; Hussain et al. 2009).

Silicon (Si) is the second most plentiful mineral element and has never been reported deficient for plant growth so is considered a non-essential element for plant growth. Several reports highlight that external silicon application promotes plant growth under drought stress with notable improvement in nutrient uptake and antioxidants and other osmoprotectants (Hattori et al. 2005; Gunes et al. 2008; Shen et al. 2010).

1.5 Conclusion

Under the ongoing global climate change scenario, with escalating emissions of greenhouse gases, increase in severity and frequency of drought has been predicted to further increase in the near future. Some plants escape drought by reducing growth accompanied by a yield penalty. In dehydration tolerance, plants potentiate to maintain metabolic activities at low tissue water potential. Osmotic adjustment, antioxidant activities, and altered growth regulators are among the major physiological adaptations of plants under drought stress. Increased accumulation of osmoprotectants such as proline, GB, amino acid, and sugars are involved in osmoregulation. Scavenging of ROS by enzymatic and non-enzymatic systems, cell membrane stability, expression of aquaporin, and stress proteins such as LEA are also vital mechanisms of dehydration tolerance. Combined knowledge of traditional breeding along with marker-assisted selection makes it easier and more efficient to induce drought tolerance in crop plants to enhance and sustain productivity in drought-prone environments.

References

Abbate PE, Dardanellib JL, Cantareroc MG, Maturanoc M, Melchiorid RJM, Sueroa EE (2004) Climatic and water availability effects on water-use efficiency in wheat. Crop Sci 44:474–483

Abebe T, Guenzi AC, Martin B, Cushman JC (2003) Tolerance of mannitol-accumulating transgenic wheat to water stress and salinity. Plant Physiol 131:1748–1755

Abrecht DG, Carberry PS (1993) The influence of water deficit prior to tassel initiation on maize growth, development and yield. Field Crops Res 31:55–69

Abreu ME, Munne-Bosch S (2008) Salicylic acid may be involved in the regulation of drought-induced leaf senescence in perennials: a case study in field-grown *Salvia officinalis* L. plants. Environ Exp Bot 64:105–112

Agarwal PK, Agarwal P, Reddy MK, Sopory SK (2006) Role of DREB transcription factors in abiotic and biotic stress tolerance in plants. Plant Cell Rep 25:1263–1274

Agboma P, Jones MGK, Peltonen-Sainio P, Rita H, Pehu E (1997) Exogenous glycinebetaine enhances grain yield of maize, sorghum and wheat grown under two watering regimes. J Agron Crop Sci 178:29–37

Akram M (2011) Growth and yield components of wheat under water stress of different growth stages. Bangladesh J Agric Res 36:455–468

Alcázar R, Planas J, Saxena T, Zarza X, Bortolotti C, Cuevas J, Bitrián M, Tiburcio AF, Altabella T (2010) Putrescine accumulation confers drought tolerance in transgenic Arabidopsis plants over expressing the homologous Arginine decarboxylase2 gene. Plant Physiol Biochem 48:547–552

Allahmoradi P, Ghobadi M, Taherabadi S, Taherabadi S (2011) Physiological aspects of mungbean (*Vigna radiata* L.) in response to drought stress. In: International conference on food engineering and biotechnology, pp 272–275

Anjum SA, Wang LC, Farooq M, Hussain M, Xue LL, Zou CM (2011) Brassinolide application improves the drought tolerance in maize through modulation of enzymatic antioxidants and leaf gas exchange. J Agron Crop Sci 197:177–185

Apel K, Hirt H (2004) Reactive oxygen species: metabolism, oxidative stress, and signal transduction. Annu Rev Plant Biol 55:373–399

Ashraf M (2010) Inducing drought tolerance in plants: recent advances. Biotechnol Adv 28: 169–183

Ashraf M, Foolad MR (2007) Roles of glycinebetaine and proline in improving plant abiotic stress resistance. Environ Exp Bot 59:206–216

Ashraf M, Iram A (2005) Drought stress induced changes in some organic substances in nodules and other plant parts of two potential legumes differing in salt tolerance. Flora 200:535–546

Asrar A-WA, Elhindi KM (2011) Alleviation of drought stress of marigold (*Tagetes erecta*) plants by using arbuscular mycorrhizal fungi. S J Biol Sci 18:93–98

Atlin GN, Lafitte HR (2002) Marker-assisted breeding versus direct selection for drought tolerance in rice. In: Saxena NP, O'Toole JC (eds.) Field screening for drought tolerance in crop plants with emphasis on rice. In: Internationak workshop on field screening for drought tolerance in Rice, Patancheru, India, 11–14 Dec 2000, ICRISAT, Patancheru, India, and The Rockefeller Foundation, New York, p 208

Babu RC, Nguyen BD, Chamarerk VP, Shanmugasundaram P, Chezhian P, Jeyaprakash SK, Ganesh A, Palchamy S, Sadasivam S, Sarkarung S, Wade LJ, Nguyen HT (2003) Genetic analysis of drought resistance in rice by molecular markers. Crop Sci 43:1457–1469

Babu RC, Zhang J, Blum A, Ho THD, Wu R, Nguyen HT (2004) HVA1, a LEA gene from barley confers dehydration tolerance in transgenic rice (*Oryza sativa* L.) via cell membrane protection. Plant Sci 166:855–862

Bae H, Kim S-H, Kim MS, Sicher RC, Lary D, Strem MD, Natarajan S, Bailey BA (2008) The drought response of *Theobroma cacao* (cacao) and the regulation of genes involved in polyamine biosynthesis by drought and other stresses. Plant Physiol Biochem 46:174–188

Bae H, Kim SK, Cho SK, Kang BG, Kima WT (2011) Overexpression of OsRDCP1, a rice RING domain-containing E3 ubiquitin ligase, increased tolerance to drought stress in rice (*Oryza sativa* L.). Plant Sci 180:775–782

Bahieldin A, Hesham HT, Eissa HF, Saleh OM, Ramadan AM, Ahmed IA, Dyer WE, El-Itriby HA, Madkour MA (2005) Field evaluation of transgenic wheat plants stably expressing the HVA1 gene for drought tolerance. Physiol Plant 123:421–427

Bajguz A, Hayat S (2009) Effects of brassinosteroids on the plant responses to environmental stresses. Plant Physiol Biochem 47:1–8

Baligar VC, Fageria NK, He ZL (2001) Nutrient use efficiency in plants. Commun Soil Sci Plant Anal 32:921–950

Bano A, Hansen H, Dorffling K, Hahn H (1994) Changes in the contents of free and conjugated abscisic acid, phaseic acid and cytokinins in xylem sap of drought stressed sunflower plants. Phytochem 37:345–347

Bao-Yuan Z, Zai-Song D, Ming Z (2011) Alleviation of drought stress inhibition on photosynthesis by overexpression of PEPC in rice. Acta Agron Sin 37:112–118

Barlow EWK (1988) The growth and functioning of leaves. Cambridge University Press, London, pp 314–345

Biao-lin H, Xue-qin F, Tao Z, Yong W, Xia L, Yun-hong H, Liang-fang D, Xiang-dong L, Jian-kun X (2011) Genetic analysis on characteristics to measure drought resistance using dongxiang wild rice (*Oryza rufupogon* Griff.) and its derived backcross inbred lines population at seedling stage. Agric Sci China 10:1653–1664

Blum A (2005) Drought resistance, water-use efficiency, and yield potential—are they compatible, dissonant, or mutually exclusive? Aust J Agric Res 56:1159–1168

Broin M, Cuine S, Peltier G, Rey P (2000) Involvement of CDSP 32, a drought-induced thioredoxin, in the response to oxidative stress in potato plants. FEBS Lett 467:245–248

Browne J, Tunnacliffe A, Burnell A (2002) Anhydrobiosis-plant desiccation gene found in a nematode. Nat 416:38

Cakmak I, Strbac D, Marschner H (1993) Activities of hydrogen peroxide-scavenging enzymes in germinating wheat seeds. J Exp Bot 44:127–132

Cechin I, Rossi SC, Oliveira VC, Fumis TF (2006) Photosynthetic responses and proline content of mature and young leaves of sunflower plants under water deficit. Photosynthetica 44: 143–146

Chimenti CA, Marcantonio M, Hall AJ (2006) Divergent selection for osmotic adjustment results in improved drought tolerance in maize (*Zea mays* L.) in both early growth and flowering phases. Field Crops Res 95:305–315
Chimenti CA, Pearson J, Hall AJ (2002) Osmotic adjustment and yield maintenance under drought in sunflower. Field Crops Res 75:235–246
Correia MJ, Osorio ML, Osorio J, Barrote I, Martins M, David MM (2006) Influence of transient shade periods on the effects of drought on photosynthesis, carbohydrate accumulation and lipid peroxidation in sunflower leaves. J Environ Exp Bot 58:75–84
Costa LD, Vedove GD, Gianquinto G, Giovanardi R, Peressotti A (1997) Yield, water use efficiency and nitrogen uptake in potato: influence of drought stress. Potato Res 40:19–34
Davies W, Zhang J (1991) Root signals and the regulation of growth and the development of plants in drying soil. Annu Rev Plant Physiol Plant Mol Biol 42:55–76
de Campos MKF, de Carvalho K, de Souza FS, Marur CJ, Pereira LFP, Filho JCB, Vieira LGE (2011) Drought tolerance and antioxidant enzymatic activity in transgenic 'Swingle'citrumelo plants over-accumulating proline. Environ Exp Bot 72:242–250
Desclaux D, Roumet P (1996) Impact of drought stress on the phenology of two soybean (*Glycine max* L. Merr) cultivars. Field Crops Res 46:61–70
Dhanda SS, Sethi GS (2002) Tolerance to drought stress among selected Indian wheat cultivars. J Agric Sci 139:319–326
Edmeades GO, Cooper M, Lafitte R, Zinselmeier C, Ribaut JM, Habben JE, Löffler C, Bänziger M (2001) Abiotic stresses and staple crops. In: Proceedings of the third international crop science congress, 18–23 Aug
Farooq M, Basra SMA, Wahid A, Ahmad N, Saleem BA (2009a) Improving the drought tolerance in rice (*Oryza sativa* L.) by exogenous application of salicylic acid. J Agron Crop Sci 195:237–246
Farooq M, Basra SMA, Wahid A, Cheema ZA, Cheema MA, Khaliq A (2008) Physiological role of exogenously applied glycinebetaine in improving drought tolerance of fine grain aromatic rice (*Oryza sativa* L.). J Agron Crop Sci 194:325–333
Farooq M, Bramley H, Palta JA, Siddique KHM (2011) Heat stress in wheat during reproductive and grain filling phases. Crit Rev Plant Sci 30:491–507
Farooq M, Kobayashi N, Ito O, Wahid A, Serraj R (2010a) Broader leaves result in better performance of indica rice under drought stress. J Plant Physiol 167:1066–1075
Farooq M, Wahid A, Cheema SA, Lee DJ, Aziz T (2010b) Comparative time course action of the foliar applied glycinebetaine, salicylic acid, nitrous oxide, brassinosteroids and spermine in improving drought resistance of rice. J Agron Crop Sci 196:336–345
Farooq M, Wahid A, Ito O, Lee DJ, Siddique KHM (2009b) Advances in drought resistance of rice. Crit Rev Plant Sci 28:199–217
Farooq M, Wahid A, Kobayashi N, Fujita D, Basra SMA (2009c) Plant drought stress: effects, mechanisms and management. Agron Sustain Devel 29:185–212
Farooq M, Wahid A, Lee D-J (2009d) Exogenously applied polyamines increase drought tolerance of rice by improving leaf water status, photosynthesis and membrane properties. Acta Physiol Plant 31:937–945
Figueiredo MVB, Buritya AH, Martınez CR, Chanway CP (2008) Alleviation of drought stress in the common bean (*Phaseolus vulgaris* L.) by co-inoculation with *Paenibacillus polymyxa* and *Rhizobium tropici*. App Soil Ecol 40:182–188
Flexas J, Bota J, Loreto F, Comic G, Sharkey TD (2004) Diffusive and metabolic limitations to photosynthesis under drought and salinity in C_3 plants. Plant Biol 6:269–279
Flexas J, Medrano H (2002) Energy dissipation in C_3 plants under drought. Funct Plant Biol 29:1209–1215
Fournier JM, Roldan AM, Sanchez C, Alexandre G, Benlloch M (2005) K^+ starvation increases water uptake in whole sunflower plants. Plant Sci 168:823–829
Foyer CH, Fletcher JM (2001) Plant antioxidants: colour me healthy. Biologist 48:115–120
Fukai S (1999) Phenology in rainfed lowland rice. Field Crops Res 64:51–60

Fukai S, Cooper M (1995) Development of drought-resistant cultivars using physio-morphological traits in rice. Field Crops Res 40:67–86
Geerts S, Raes D, Garcia M, Mendoza J, Huanca R (2008) Crop water use indicators to quantify the flexible phenology of quinoa (*Chenopodium quinoa* Willd.) in response to drought stress. Field Crops Res 108:150–156
Ghannoum O (2009) C_4 photosynthesis and water stress. Ann Bot 103:635–644
Gill SS, Tuteja N (2010) Reactive oxygen species and antioxidant machinery in abiotic stress tolerance in crop plants. Plant Physiol Biochem 48:909–930
Gomes FP, Oliva MA, Mielke MS, Almeida A-AF, Aquino LA (2010) Osmotic adjustment, proline accumulation and cell membrane stability in leaves of *Cocos nucifera* submitted to drought stress. Sci Hort 126:379–384
Gomes FP, Oliva MA, Mielke MS, Almeida A-AF, Leite HG, Aquino LA (2008) Photosynthetic limitations in leaves of young Brazilian green dwarf coconut (*Cocos nucifera* L. 'nana') palm under well-watered conditions or recovering from drought stress. Environ Exp Bot 62:195–204
Gong D-S, Xiong Y-C, Ma B-L, Wang T-M, Ge J-P, Qin X-L, Li P-F, Kong H-Y, Li Z-Z, Li F-M (2010) Early activation of plasma membrane H^+-ATPase and its relation to drought adaptation in two contrasting oat (*Avena sativa* L.) genotypes. Environ Exp Bot 69:1–8
Gosal SS, Wani SH, Kang MS (2009) Biotechnology and drought tolerance. J Crop Imp 23:19–54
Gowda VRP, Henry A, Yamauchi A, Shashidhar HE, Serraj R (2011) Root biology and genetic improvement for drought avoidance in rice. Field Crops Res 122:1–13
Guan-fu F, Jian S, Jie X, Yu-rong L, Hui-zhe C, Ming-kai L, Long-xing T (2011) Changes of oxidative stress and soluble sugar in anthers involve in rice pollen abortion under drought stress. Agric Sci China 10:1016–1025
Gunes A, Kadioglu YK, Pilbeam DJ, Inal A, Coban S, Aksu A (2008) Influence of silicon on sunflower cultivars under drought stress, II: essential and non essential element uptake determined by polarized energy dispersive X-ray fluorescence. Comm Soil Sci Plant Anal 39:1904–1927
Gutierrez-Boem FH, Thomas GW (1999) Phosphorus nutrition and water deficits in field grown soybeans. Plant Soil 207:87–96
Hadiarto T, Tran L-SP (2011) Progress studies of drought-responsive genes in rice. Plant Cell Rep 30:297–310
Hattori T, Inanaha S, Araki H, An P, Morita S, Luxova M, Lux A (2005) Application of silicon enhanced tolerance in *Sorghum bicolor*. Physiol Plant 123:459–466
He C, Zhang CW, Gao Q, Yang A, Hu X, Zhang J (2011) Enhancement of drought resistance and biomass by increasing the amount of glycine betaine in wheat seedlings. Euphytica 177: 16–151
Holmström K-O, Mäntylä E, Welin B, Mandal A, Palva ET, Tunnela O, Londesborough J (1996) Drought tolerance in tobacco. Nature 379:683–684
Hu TZ (2008) *OsLEA3*, a late embryogenesis abundant protein gene from rice, confers tolerance to water deficit and salt stress to transgenic rice. Russ J Plant Physiol 55:530–537
Huang J, Hirji R, Adam L, Rozwadowski KL, Hammerlindl JK, Keller WA, Selvaraj G (2000) Genetic engineering of glycinebetaine production toward enhancing stress tolerance in plants: metabolic limitations. Plant Physiol 122:747–756
Hussain M, Malik MA, Farooq M, Ashraf MY, Cheema MA (2008) Improving drought tolerance by exogenous application of glycinebetaine and salicylic acid in sunflower. J Agron Crop Sci 194:193–199
Hussain M, Malik MA, Farooq M, Khan MB, Akram M, Saleem MF (2009) Exogenous glycinebetaine and salicylic acid application improves water relations, allometry and quality of hybrid sunflower under water deficit conditions. J Agron Crop Sci 195:98–109
IPCC (2007) Climate change 2007: the physical science basis, vol 1009. Cambridge University Press, Cambridge
Iqbal N, Ashraf M, Ashraf MY (2009) Influence of exogenous glycine betaine on gas exchange and biomass production in sunflower (*Helianthus annuus* L.) under water limited conditions. J Agron Crop Sci 195:420–426

Iqbal N, Ashraf M, Ashraf MY, Azam F (2005) Effect of exogenous application of glycinebetaine on capitulum size and achene number of sunflower under water stress. Int J Biol Biotech 2:765–771

Jalil CA, Gopi R, Sankar B, Manivannan P, Kishorkumar A, Sridharan R (2007) Studies on germination, seedling vigor, lipid peroxidation and proline metabolism in *Catharanthus roseus* seedlings under salt stress. South Afr J Bot 73:190–195

Jongrungklang N, Toomsan B, Vorasoot N, Jogloy S, Boote KJ, Hoogenboom G, Patanothai A (2011) Rooting traits of peanut genotypes with different yield responses to pre-flowering drought stress. Field Crops Res 120:262–270

Kavar T, Maras M, Kidric M, Sustar-Vozlic J, Meglic V (2007) Identification of genes involved in the response of leaves of *Phaseolus vulgaris* to drought stress. Mol Breed 21:159–172

Kavi Kishor PB, Hong Z, Miao GH, Hu CAA, Verma DPS (1995) Over-expression of d-pyrroline-5-carboxylate synthetase increases proline production and confers osmotolerance in transgenic plants. Plant Physiol 25:1387–1394

Ke Y, Han G, He H, Li J (2009) Differential regulation of proteins and phosphoproteins in rice under drought stress. Biochem Biophy Res Commun 379:133–138

Kiani SP, Talia P, Maury P, Grieu P, Heinz R, Perrault A, Nishinakamasu V, Hopp E, Gentzbittel L, Paniego N, Sarrafi A (2007) Genetic analysis of plant water status and osmotic adjustment in recombinant inbred lines of sunflower under two water treatments. Plant Sci 172:773–787

Kiliç H, Yağbasanlar T (2010) The effect of drought stress on grain yield, yield components and some quality traits of durum wheat (*Triticum turgidum* ssp. durum) cultivars. Not Bot Hort Agrobot Cluj 38:164–170

Kirigwi FM, Ginkel MV, Brown-Guedira G, Gill BS, Paulsen GM, Fritz AK (2007) Markers associated with a QTL for grain yield in wheat under drought. Mol Breed 20:401–413

Kirkham MB (2005) Principles of soil and plant water relations. Elsevier, The Netherlands

Kramer PJ, Boyer JS (1995) Water relations of plants and soils. Academic, San Diego

Kubis J (2003) Polyamines and "scavenging system": influence of exogenous spermidine on catalase and guaiacol peroxidase activities, and free polyamine level in barley leaves under water deficit. Acta Physiol Plant 25:337–343

Kumar J, Abbo S (2001) Genetics of flowering time in chickpea and its bearing on productivity in semiarid environments. Adv Agron 72:107–138

Kumar R, Sarawgi AK, Ramos C, Amarante ST, Ismail AM, Wade LJ (2006) Partitioning of dry matter during drought stress in rainfed lowland rice. Field Crops Res 98:1–11

Kumar R, Venuprasad R, Atlin GN (2007) Genetic analysis of rainfed lowland rice drought tolerance under naturally-occurring stress in eastern India: heritability and QTL effects. Field Crops Res 103:42–52

Kumar A, Bernier J, Verulkar S, Lafitte HR, Atlin GN (2008) Breeding for drought tolerance: direct selection for yield, response to selection and use of drought-tolerant donors in upland and lowland-adapted populations. Field Crops Res 107:221–231

Lafitte HR, Yongsheng G, Yan S, Li ZK (2007) Whole plant responses, key processes, and adaptation to drought stress: the case of rice. J Exp Bot 58:169–175

Lambers H, Chapin FS, Pons TL (2008) Plant physiological ecology, 2nd edn. Springer, New York

Lanceras J, Pantuwan G, Jongdee B, Toojinda T (2004) Quantitative trait loci associated with drought tolerance at reproductive stage in rice. Plant Physiol 135:384–399

Laporte MM, Shen B, Tarczynski MC (2002) Engineering for drought avoidance: expression of maize NADP-malic enzyme in tobacco results in altered stomatal function. J Exp Bot 369:699–705

Lawlor DW, Cornic G (2002) Photosynthetic carbon assimilation and associated metabolism in relation to water deficits in higher plants. Plant, Cell Environ 25:275–294

Lei YB, Yin CY, Li CY (2006) Differences in some morphological, physiological and biochemical responses to drought stress in two contrasting populations of *Populus przewalskii*. Physiol Plant 127:182–191

Levitt J (1980) Responses of plants to environmental stresses. In: Kozlowski TT (ed) Water, radiation, salt and other stresses, vol 2, 2nd edn. Academic, New York, pp 93–186

Li Y (2008) Kinetics of the antioxidant response to salinity in the halophyte *Limonium bicolor*. Plant Soil Environ 54:493–497

Li YP, Ye W, Wang M, Yan XD (2009) Climate change and drought: a risk assessment of crop-yield impacts. Climate Res 39:31–46

Liu F, Andersen MN, Jensen CR (2003) Loss of pod set caused by drought stress is associated with water status and ABA content of reproductive structures in soybean. Funct Plant Biol 30:271–280

Liu F, Jensen CR, Andersen MN (2004) Drought stress effect on carbohydrate concentration in soybean leaves and pods during early reproductive development: its implication in altering pod set. Field Crops Res 86:1–13

Liu H, Wang X, Wang D, Zou Z, Liang Z (2011) Effect of drought stress on growth and accumulation of active constituents in *Salvia miltiorrhiza* Bunge. Ind Crops Prod 33:84–88

Liu HP, Yu BJ, Zhang WH, Liu YL (2005) Effect of osmotic stress on the activity of H^+-ATPase and the levels of covalently and noncovalently conjugated polyamines in plasma membrane preparation from wheat seedling roots. Plant Sci 168:1599–1607

Liu YP, Zhang JH, Liu HT, Huang WD (2008) Salicylic acid or heat acclimation pre-treatment enhances the plasma membrane-associated ATPase activities in young grape plants. Sci Hort 119:21–27

Lv S, Yang A, Zhang K, Wang L, Zhang J (2007) Increase of glycinebetaine synthesis improves drought tolerance in cotton. Mol Breed 20:233–248

Ma QQ, Wei W, Yong-hua L, De-Quan L, Zoa Q (2006) Alleviation of photoinhibition in drought-stressed wheat (*Triticum aestivum*) by foliar- applied glycinebetaine. J Plant Physiol 163:65–175

Ma R, Zhang M, Li B, Du G, Wang J, Chen J (2005) The effects of exogenous Ca^{2+} on endogenous polyamine levels and drought-resistant traits of spring wheat grown under arid conditions. J Arid Environ 63:177–190

Mahdieh M, Mostajeran A (2009) Abscisic acid regulates root hydraulic conductance via aquaporin expression modulation in *Nicotiana tabacum*. J Plant Physiol 166:1993–2003

Majid SA, Asghar R, Murtaza G (2007) Yield stability analysis conferring adaptation of wheat to pre- and post-anthesis drought conditions. Pak J Bot 39:1623–1637

Makela P, Jokinen K, Kontturi M, Peltonen-Sainio P, Pehu E, Somersalo S (1998) Foliar application of glycinebetaine—a novel product from sugar beet—as an approach to increase tomato yield. Indust Crops Prod 7:139–148

Manivannan P, Jaleel CA, Kishorekumar A, Sankar B, Somasundaram R, Sridharan R, Panneerselvam R (2007) Drought stress induced changes in the biochemical parameters and photosynthetic pigments of cotton (*Gossypium hirsutum* L.). Indian J Appl Pure Biol 22:369–372

Manivannan P, Jaleel CA, Somasundaram R, Panneerselvam R (2008) Osmoregulation and antioxidant metabolism in drought-stressed *Helianthus annuus* under triadimefon drenching. C R Biologies 331:418–425

Mascher R, Nagy E, Lippmann B, Hornlein S, Fischer S, Scheiding W, Neagoe A, Bergmann H (2005) Improvement of tolerance to paraquat and drought in barley (*Hordeum vulgare* L.) by exogenous 2-aminoethanol: effects on superoxide dismutase activity and chloroplast ultrastructure. Plant Sci 168:691–698

Matsui T, Singh BB (2003) Root characteristics in cowpea related to drought tolerance at the seedling stage. Exp Agric 39:29–38

McMaster GS, Wilhelm WW (2003) Phenological responses of wheat and barley to water and temperature: improving simulation models. J Agric Sci 141:129–147

McWilliams D (2003) Drought strategies for cotton, cooperative extension service circular 582. College of Agriculture and Home Economics, New Mexico State University, USA

Medrano H, Escalona JM, Bota J, Gulías J, Flexas J (2002) Regulation of photosynthesis of C_3 plants in response to progressive drought: stomatal conductance as a reference parameter. Ann Bot 89:895–905

Meyre D, Leonardi A, Brisson G, Vartanian N (2001) Drought-adaptive mechanisms involved in the escape/tolerance strategies of Arabidopsis *Landsberg erecta* and Columbia ecotypes and their F1 reciprocal progeny. J Plant Physiol 158:1145–1152

Mishra V, Cherkauer KA (2010) Retrospective droughts in the crop growing season: implications to corn and soybean yield in the Midwestern United States. Agric Forest Met 150:1030–1045

Miyashita K, Tanakamaru S, Maitani T, Kimura K (2005) Recovery responses of photosynthesis, transpiration, and stomatal conductance in kidney bean following drought stress. Environ Exp Bot 53:205–214

Noctor G, Arisi A-CM, Jouanin L, Foyer CH (1999) Photorespiratory glycine enhances glutathione accumulation in both the chloroplastic and cytosolic compartments. J Exp Bot 50:157–1167

Noctor G, Veljovic-Javanovic S, Driscoll S, Novitskaya L, Foyer CH (2002) Drought and oxidative load in the leaves of C_3 plants: a predominant role for photorespiration. Ann Bot 89:841–850

Nooden LD (1988) The phenomena of senescence and aging. In: Nooden LD, Leopald AC (eds) Senescence and aging in plants. Academic, USA, pp 1–50

Okcu G, Kaya MD, Atak M (2005) Effects of salt and drought stresses on germination and seedling growth of pea (*Pisum sativum* L.). Turk J Agric For 29:237–242

Ouvrard O, Cellier F, Ferrare K, Tousch D, Lamaze T, Dupuis JM, Casse-Delbart F (1996) Identification and expression of water stress and abscisic acid -regulated genes in a drought-tolerant sunflower genotype. Plant Mol Bio 31:819–829

Ozkur O, Ozdemir F, Bor M, Turkan I (2009) Physiochemical and antioxidant responses of the perennial xerophyte *Capparis ovata* Desf. to drought. Environ Exp Bot 66:487–492

Perez-Perez JG, Robles JM, Tovar JC, Botia P (2009) Response to drought and salt stress of lemon 'Fino 49' under field conditions: Water relations, osmotic adjustment and gas exchange. Sci Hort 122:83–90

Piro G, Leucci MR, Waldron K, Dalessandro G (2003) Exposure to water stress causes changes in the biosynthesis of cell wall polysaccharides in roots of wheat cultivars varying in drought tolerance. Plant Sci 165:559–569

Pompelli MF, Barata-Luıs R, Vitorino HS, Goncalves ER, Rolim EV, Santos MG, Almeida-Cortez JS, Lemos EE, Ferreira VM, Endres L (2010) Photosynthesis, photoprotection and antioxidant activity of purging nut under drought deficit and recovery. Biomass Bioenergy 34:1207–1215

Qiu Z-B, Liu X, Tian X-J, Yue M (2008) Effects of CO_2 laser pretreatment on drought stress resistance in wheat. J Photochem Photobiol B: Bio 90:17–25

Rauf S (2008) Breeding sunflower (*Helianthus annuus* L.) for drought tolerance. Commun Biomet Crop Sci 3:29–44

Reddy AR, Chaitanya KV, Vivekanandan M (2004) Drought-induced responses of photosynthesis and antioxidant metabolism in higher plants. J Plant Physiol 161:1189–1202

Reddy GKM, Dangi KS, Kumar SS, Reddy AV (2003) Effect of moisture stress on seed yield and quality in sunflower (*Helianthus annuus* L.). J Oilseeds Res 20:282–283

Rivero RM, Kojima M, Gepstein A, Sakakibara H, Mittler R, Gepstein S, Blumwald E (2007) Delayed leaf senescence induces extreme drought tolerance in a flowering plant. Proc Natl Acad Sci USA 104:19631–19636

Romero C, Belles JM, Vaya JL, Serrano R, Culianez-Macia FA (1997) Expression of the yeast trehalose-6-phosphate synthase gene transgenic tobacco plants: pleiotropic phenotypes include droughttolerance. Planta 201:293–297

Rout NP, Shaw BP (2001) Salt tolerance in aquatic macrophytes: possible involvement of the antioxidative enzymes. Plant Sci 160:415–423

Salekdeh GH, Siopongco J, Wade LJ, Ghareyazie B, Bennett J (2002) Proteomic analysis of rice leaves during drought stress and recovery. Proteomics 2:1131–1145

Sankar B, Jaleel CA, Manivannan P, Kishorekumar A, Somasundram R, Panneerselvam R (2007) Drought-induced biochemical modifications and proline metabolism in *Abelmoschus esculentus* (L.) Moench. Acta Bot Croat 66:43–56

Scandalios JG (2005) Oxidative stress: molecular perception and transduction of signals triggering antioxidant gene defenses. Braz J Med Biol Res 38:995–1014
Schussler JR, Westgate ME (1995) Assimilate flux determines kernel set at low water potential in maize. Crop Sci 35:1196–1203
Seki M, Kameiy A, Yamaguchi-Shinozaki K, Shinozaki K (2003) Molecular responses to drought, salinity and frost: common and different paths for plant protection. Curr Opin Biotechn 14:194–199
Senaratna T, Touchell D, Bunn E, Dixon K (2000) Acetyl salicylic acid (aspirin) and salicylic acid induce multiple stress tolerance in bean and tomato plants. Plant Growth Reg 30:157–161
Serraj R, Sinclair TR (2002) Osmolyte accumulation: can it really help increase crop yield under drought conditions? Plant, Cell Environ 25:333–341
Shen X, Zhou Y, Duan L, Li Z, Eneji AE, Li J (2010) Silicon effects on photosynthesis and antioxidant parameters of soybean seedlings under drought and ultraviolet-B radiation. J Plant Physiol 167:1248–1252
Shou H, Bordallo P, Wang K (2004) Expression of the Nicotiana protein kinase (NPK1) enhanced drought tolerance in transgenic maize. J Exper Bot 55:1013–1019
Siddique KHM, Regan KL, Tennant D, Thomson BD (2001) Water use and water use efficiency of cool season grain legumes in low rainfall Mediterranean-type environments. Eur J Agron 15:267–280
Signarbieux C, Feller U (2011) Non-stomatal limitations of photosynthesis in grassland species under artificial drought in the field. Environ Exp Bot 71:192–197
Simova-Stoilova L, Demirevska K, Petrova T, Tsenov N, Feller U (2008) Antioxidative protection in wheat varieties under severe recoverable drought at seedling stage. Plant Soil Environ 54:529–536
Sinclair TR, Muchow RC (2001) System analysis of plant traits to increase grain yield on limited water supplies. Agron J 93:263–270
Singh B, Singh G (2004) Influence of soil water regime on nutrient mobility and uptake by *Dalbergia sissoo* seedlings. Trop Ecol 45:337–340
Singh B, Usha K (2003) Salicylic acid induced physiological and biochemical changes in wheat seedlings under water stress. Plant Growth Reg 39:137–141
Sivamani E, Bahieldin A, Wraith JM, Al-Niemi T, Dyer WE, Ho THD, Qu R (2000) Improved biomass productivity and water use efficiency under water deficit conditions in transgenic wheat constitutively expressing the barley HVA1 gene. Plant Sci 155:1–9
Subramanian KS, Santhanakrishnan P, Balasubramanian P (2006) Responses of field grown tomato plants to arbuscular mycorrhizal fungal colonization under varying intensities of drought stress. Sci Hort 107:245–253
Taiz L, Zeiger E (2010) Plant Physiology, 5th edn. Sinauer Associates Inc. Publishers, Massachusetts
Tausz M, Wonisch A, Peters J, Jimenez MS, Morales D, Grill D (2001) Short-term changes in free-radical scavengers and chloroplast pigments in *Pinus canariensis* needles as affected by mild drought stress. J Plant Physiol 158:213–219
Tezara W, Mitchell VJ, Driscoll SD, Lawlor DW (1999) Water stress inhibits plant photosynthesis by decreasing coupling factor and ATP. Nature 401:914–917
Tezara W, Mitchell VJ, Driscoll SD, Lawlor DW (2002) Effects of water deficit and its interaction with CO_2 supply on the biochemistry and physiology of photosynthesis in sunflower. J Exp Bot 53:1781–1791
Turner NC, Wright GC, Siddique KHM (2001) Adaptation of grain legumes (pulses) to water-limited environments. Adv Agron 71:193–271
Tyerman SD, Niemietz CM, Brameley H (2002) Plant aquaporins: multifunctional water and solute channels with expanding roles. Plant, Cell Environ 25:173–194
Venuprasad R, Lafitte HR, Atlin GN (2007) Response to direct selection for grain yield under drought stress in rice. Crop Sci 47:285–293
Vijay KL (2004) Irrigation strategies for crop production under water scarcity. International Commission on Irrigation and Drainage New Delhi 110–021:89–109

Waditee R, Bhuiyan MN, Rai V, Aoki K, Tanaka Y, Hibino T, Suzuki S, Takano J, Jagendorf AT, Takabe T et al (2005) Genes for direct methylation of glycine provide high levels of glycinebetaine and abiotic-stress tolerance in Synechococcus and Arabidopsis. Proc Natl Acad Sci USA 102:1318–1323

Walter J, Nagy L, Hein R, Rascher U, Beierkuhnlein C, Willner E, Jentsch A (2011) Do plants remember drought? Hints towards a drought-memory in grasses. Environ Exp Bot 71:34–40

Wang FZ, Wang Q-B, Kwon S-Y, Kwak S-S, Su W-A (2005a) Enhanced drought tolerance of transgenic rice plants expressing a pea manganese superoxide dismutase. J Plant Physiol 162:465–472

Wang H, Yamauchi A (2006) Growth and functions of roots under abiotic stress in soil. In: Huang B (ed) Plant–environment interactions, 3rd edn. CRC Press, New York, pp 271–320

Wang Y, Ying J, Kuzma M, Chalifoux M, Sample A, McArthur C, Uchacz T, Sarvas C, Wan J, Tennis DT, McCourt P, Huang Y (2005b) Molecular tailoring of farnesylation for plant drought tolerance and yield protection. Plant J 43:413–424

Wingler A, Lea PJ, Quick WP, Leegood RC (2000) Photorespiration: metabolic pathways and their role in stress protection. Philos Trans R Soc Lond B Biol Sci 355:1517–1529

Wu R, Garg A (2003) Engineering rice plants with trehalose-producing genes improves tolerance to drought, salt, and low temperature. ISB News Report

Wyn Jones RG, Storey R, Leigh PA (1977) A hypothesis on cytoplasmic osmoregulation. In: Marre E, Ciferri O (eds) Regulation of cell membrane activities in higher plants. Elsevier, Amsterdam, pp 121–136

Xing W, Rajashekar CB (1999) Alleviation of water stress in beans by exogenous glycinebetaine. Plant Sci 148:185–195

Yadav RS, Hash CT, Bidinger FR, Devos KM, Howarth CJ (2004) Genomic regions associated with grain yield and aspects of post flowering drought tolerance in pearl millet across environments and tester background. Euphytica 136:265–277

Yamaguchi K, Takahashi Y, Berberich T, Imai A, Takahashi T, Michael AJ, Kusano T (2007) A protective role for the polyamine spermine against drought stress in Arabidopsis. Biochem Biophys Res Commun 352:486–490

Yamaguchi-Shinozaki K, Shinozaki K (2006) Transcriptional regulatory networks in cellular responses and tolerance to dehydration and cold stresses. Annu Rev Plant Biol 57:781–803

Yamance K, Hayakawa K, Kawasaki M (2003) Bundle sheath chloroplasts of rice are more sensitive to drought stress than mesophyll chloroplasts. J Plant Physiol 160:1319–1327

Yamauchi Y, Pardales JR, Kono Y (1996) Root system structure and its relation to stress tolerance. In: Ito O et al (eds) Roots and nitrogen in cropping systems of the semi-arid tropics. JIRCAS Publication, Tsukuba

Yuan G-F, Jia C-G, Li Z, Sun B, Zhang L-P, Liu N, Wang Q-M (2010) Effect of brassinosteroids on drought resistance and abscisic acid concentration in tomato under water stress. Sci Hort 126:103–108

Zhang J, Kirkham MB (1996) Enzymatic responses of the ascorbate-glutathione cycle to drought in sorghum and sunflower plants. Plant Sci 113:139–147

Zhang Y, Yang X, Liu Q, Qiu D, Zhang Y, Zeng H, Yuan J, Mao J (2010) Purification of novel protein elicitor from *Botrytis cinerea* that induces disease resistance and drought tolerance in plants. Microbiol Res 165:142–151

Part I
Morphological and Anatomical Responses

Chapter 2
Morpho-Anatomical Traits for Plant Adaptation to Drought

Veronica De Micco and Giovanna Aronne

Abstract Plant resistance to drought relies on adaptive strategies based on the timing of phenophases and on the presence of structural traits mainly related to: (1) increase of water uptake and storage; (2) reduction of water loss during dry periods; and (3) mechanical reinforcement of tissues to prevent wilting that may lead to irreversible collapse and damage of cells. In this chapter, after a few evolutionary considerations, we focus on the adaptive value of the main phenological, morphological and anatomical properties. We report the common existence of such traits in both desert and semiarid environments, especially in Mediterranean-type ecosystems. All morpho-anatomical characteristics are interpreted considering that plant resistance to drought also depends on the ability to respond to multiple stressors. We conclude that various combinations of anatomical features can contribute in different degrees to the adaptive capacity of plants to drought.

2.1 Introduction

Shortage of water is a constraint to life that recurs more and more in many regions of the world due to global climate change (IPCC 2007). Increasing drought has an impact on the survival of plants in natural ecosystems, while it also results in reduced yields in crops.

Climate change in semiarid regions is expected to trigger desertification processes; these can have serious consequences considering that semiarid woody

V. De Micco (✉) · G. Aronne
Faculty of Agriculture, Department of Arboriculture, Botany and Plant Pathology, University of Naples Federico II, via Università 100, I-80055, Portici (Naples), Italy
e-mail: demicco@unina.it

R. Aroca (ed.), *Plant Responses to Drought Stress*,
DOI: 10.1007/978-3-642-32653-0_2, © Springer-Verlag Berlin Heidelberg 2012

ecosystems sustain a considerable part of the world's terrestrial biomass, net primary productivity and biodiversity (Atjay et al. 1979). Mediterranean ecosystems are particularly fragile and characterised by two stress periods during the year, namely summer drought and winter cold (Mitrakos 1980). In these ecosystems, the impacts of human-induced climate change are estimated to be highest (Hulme et al. 1999) and future changes are likely to aggravate significantly the existing environmental, structural and socio-economical problems. Here, global climate change will cause significant alteration in temperature regimes and precipitation patterns as well as an increase in both the frequency and intensity of extreme events (e.g. heat waves, floods, droughts) (IPCC 2007). Understanding the response of plants to increasing drought would be desirable in the light of global and regional changes not only to forecast population dynamics in natural ecosystems, but also to adjust management practices in agriculture. All biogeochemical processes are climate driven, and thus, increasing drought could not only impact the C-gains and C-losses of ecosystems, but may influence plant growth by affecting phenology and determining modifications at biochemical, physiological and anatomical levels (Schwartz 1999; Morison and Morecroft 2006).

Every plant organ is ideally designed to fulfil metabolic and physiological processes in specific environmental conditions. In environments characterised by arid conditions, plant survival depends upon the ability to harmonise structure and function to withstand desiccation without permanent damage (Maximov 1931). These plants show structural alterations that are mainly related to water saving (e.g. water storage and reduction of water losses) and mechanical reinforcement of tissues (e.g. thickening and straightening of cell walls) (Shields 1950; Fahn 1964; De Micco and Aronne 2007). Mechanical reinforcement may prevent phenomena such as collapse and damage of cells, associated with irreversible wilting, and hence the loss of functionality during water deficit.

In this chapter, we discuss the adaptive value of some morphological and anatomical traits in the various organs of plants growing under dry conditions. A specific focus is on woody plants of semiarid Mediterranean-type ecosystems. More specifically, we start from general considerations about phenological features, also in relation to growth forms; subsequently, we discuss plant structural attributes linked with water storage, reduced water loss and efficiency and safety of water transport in the xylem. The role of some attributes is considered at the ultrastructural level. Moreover, structures that reinforce tissues and defend them against predators and excess of light are pondered considering that plant resistance to drought is also associated with the ability to respond to the action of multiple stressors.

All morpho-anatomical attributes are interpreted under the perspective that not single traits but suites of anatomical features are responsible for the adaptive capacity of plants in a specific environment.

2.2 Evolution of Plant Life Forms

The evolution of terrestrial vascular plants, from the primordial aquatic organisms to higher terrestrial plants, has been accompanied by increasing complexity in the structure and functions of their vegetative and reproductive organs. Many environmental factors drove the evolution of morpho-functional traits during the key 'moment' when organisms moved from the aquatic environment to colonise the lands. Indeed, some factors, including atmosphere composition and radiation, were changing during land plant evolution. On the other hand, gravity also played an important role during land colonisation, because plants, loosing hydrostatic pressure, perceived it with different intensity (Graham 1993; Rozema et al. 1997; Bateman et al. 1998). However, the major factor was water availability: the first land plants faced the problem of solving the conflict between water retention and the metabolic requirement to exchange gases with the atmosphere for photosynthesis (Niklas 1986).

The evolutionary history of plants is rich with examples of how specific traits arose to fulfil specific needs (e.g. for water translocation, for reproduction, to withstand mechanical stress, etc.) due to changing environmental conditions (Raven 1977) (Fig. 2.1). Although such traits involve any life organisation levels (from the development of multifaceted metabolic pathways up to organogenesis), there is common agreement that the development of specific forms and functions is controlled by mechanical and physiological constraints (Niklas 1986). A need for simplifying the variability of plant shape and size has lead to the application of plant functional types (Smith et al. 1997). Each of them might be considered the result of the interconnections between traits evolved at different levels (e.g. cytological, anatomical, phenological, physiological, biochemical and molecular) and designed to accomplish specific functions. Whilst single traits can be related to many functions, on the other hand, multiple traits can act in cooperation to achieve a specific aim through the development of definite adaptive strategies.

Focusing on the strategies to withstand water shortage, the adaptation mechanisms become more and more developed and complex moving from wet areas towards arid areas, where the regularity of drought events leads to the development of multifarious strategies (Monneveux and Belhassen 1996). Pursuing the need to avoid drying out of tissues, there is common agreement on the presence of an adaptive link between the leaf gas exchange capacity and hydraulic efficiency in above- and below-ground organs, as well as at the whole plant level (Brodribb 2010). The existence of a strong relation between leaf photosynthetic capacity and hydraulic supply of water to leaves has been shown in a group of conifers and angiosperm species, both vessel bearing and vessel less (Brodribb and Field 2000).

During evolution, two strategies of water use arose in plants leading to the two plant types known as desiccation tolerant (poikilohydric) and non-desiccation tolerant (homoiohydric) (Schulze et al. 2005). Poikilohydric organisms act as purely physical systems whose water content tends to reach the equilibrium with the humidity of the environment (e.g. lichens, some mosses). Such organisms do

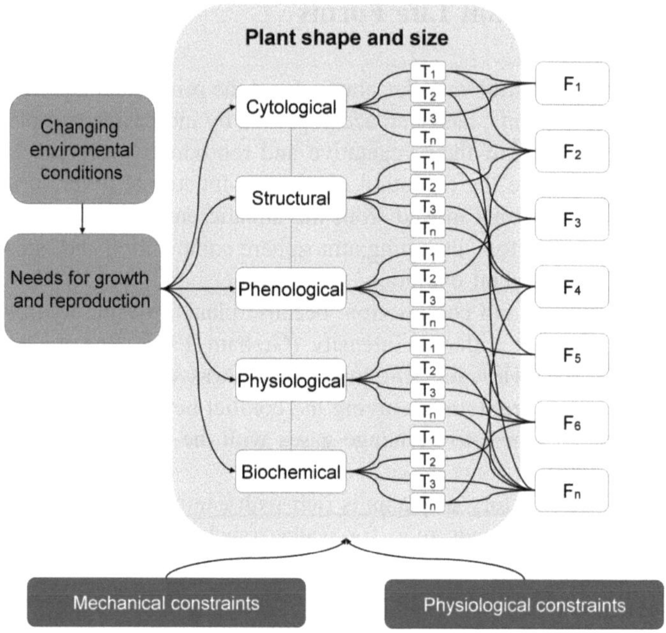

Fig. 2.1 Schematic view of how plant shape and size, depending on the combination of different traits (T_i) expressed at various organisation levels, are primed by the need to respond to environmental inputs and constrained by mechanical and physiological limitations. Single traits or combinations of them are responsible for specific adaptive strategies to fulfil precise functions (F_i)

not possess mechanisms to prevent desiccation: they consequently dry out and remain dormant when water availability decreases, but can rehydrate when water becomes available again. On the other hand, homoiohydric organisms (e.g. some ferns, flowering plants) are able to maintain high water content in their tissues independently on the conditions of the surrounding environment. Higher plants are homoiohydric. Among them, three types of resistance to drought are typically classified, namely escape, avoidance and tolerance strategies. The boundaries between these types are not clear-cut since they are not mutually exclusive (Levitt 1980). Escaping is a strategy based on the ability to complete the life cycle before the period of water shortage. In this case, plants do not experience water deficit because they are able to modulate their vegetative and reproductive phenology according to the most favourable seasons (Aronne and Wilcock 1997). Dehydration-avoidance strategies are based on the ability to maintain tissue water potential as high as possible through a variety of adaptive traits involving the minimisation of water loss and optimisation of water uptake. The reduction of transpiration occurs in water-saving plants, while water spenders rely on other means, such as osmotic adjustment, to prevent desiccation (Levitt 1980). Tolerance strategies occur in plants able to endure low tissue water potential through adaptive traits involving osmotic adjustment and the formation of more compact and stiff tissues.

Extreme desiccation tolerant plants, known as resurrection plants, can be found in angiosperms monocotyledonous and dicotyledonous plants (mainly in the Scrophulariaceae and Myrothamnaceae families). They are able to severely dehydrate, still fully recovering after rehydration as poikilohydric organisms.

Since escaping, avoidance and tolerance mechanisms can occur contemporarily in the same organism (Ludlow 1989), a definite classification of morpho-functional adaptive traits into the different strategies is not significant: the same traits can be beneficial for each of them. Such traits are generally constitutive rather than stress induced (Chaves et al. 2003).

2.3 Modulation of Growth: Phenology and Growth Forms

Apart from resurrection plants, some authors agree that the term drought resistant does not apply to many higher plants because they evolved different avoidance reactions based on the restriction of growth events into time windows when water supply is satisfactory (Grene et al. 2011). Such reactions include the shedding of aerial organs to reduce water loss due to transpiration or the regulation of plant phenology.

Most annual plants are a typical example of avoidance: the shortness of life cycle is accompanied by the completion of their reproductive cycle before the dry season. There is evidence that the transition between phenophases can be affected by increasing aridity depending on species-specific phenotypic plasticity, which is also influenced by the environment. According to Aronson et al. (1992), while desert plants respond to imposed drought by shortening their growth cycles with early seed maturation and senescence, Mediterranean annuals do not change phenophases promptly. More recently, Franks et al. (2007) demonstrated that flowering timing evolves rapidly (i.e. in a few generations) to escape drought.

For other biological types, including perennials and woody plants, adaptation to drought depends either on the evolutionary modification of specific organs, such as deep taproots or storage organs, or the onset of specific *habit* (e.g. shrub). In plants growing in deserts, morpho-anatomical traits to withstand long and severe periods of drought rely on: (1) the extreme reduction of leaves into thorns and (2) the development of succulent stems which perform photosynthesis and accumulate water in parenchyma tissues. In semiarid ecosystems, dominant forms of plant life include the evergreen sclerophyllous, drought deciduous and seasonally dimorphic shrubs which represent the main adaptive strategies of woody perennial species to water stress. The dominant role of shrubs in semiarid environments resides in the fact that they can grow in areas of high environmental stress where trees cannot survive (Rundel 1991; Wilson 1995).The different leaf shedding models derive from the diverse ability of plants to employ favourable periods to photosynthesise and allocate resources for growth, reproduction, or storage.

It is interesting that, although it is recognised that water shortage is the main limiting factor in Mediterranean-type ecosystems, here most woody perennials undertake highly resource demanding processes just during summer. As an example, Aronne and Wilcock (1997) found that less than one-fifth of the woody perennial species occurring in a Mediterranean shrubland avoided summer aridity by reproducing in spring; the remaining species either undertook most of the whole cycle in the summer or spent the arid period developing their fruits. In Mediterranean ecosystems, such timing of growth is possible because of the occurrence of specific morphological, anatomical, physiological and reproductive attributes. They concern all the organs, from leaves to the finest roots, and include sclerophylly, specialised xylem traits and root architecture, that enhance plant survival (Aronne and Wilcock 1994; Aronne and De Micco 2001; Lamont 1983; di Castri et al. 1981; Kummerow 1989; Rhizopoulou and Mitrakos 1990; Matosevic et al. 1997; De Micco and Aronne 2008; De Micco et al. 2008).

The timing of leaf shedding is also linked with periods of xylem growth in woody species. Indeed, the duration of cambial activity, as well as structural and chemical properties of the formed wood, is dependent on environmental factors, especially water availability. Summer deciduous plants, with drought-avoiding behaviour, are characterised by single summer dormancy in cambial activity (Cherubini et al. 2003), which results in the formation of annual rings as recorded in *Cistus ladanifer* L. (Patón et al. 1998). However, the occurrence of two periods of dormancy in cambial activity in seasonally dimorphic species leads to the formation of false rings as recorded in *Cistus incanus* L. (De Micco and Aronne 2009). False rings, often referred to as Intra-Annual Density Fluctuations (IADFs) or double rings, are frequent in woods of Mediterranean environments and are triggered by specific patterns of cambial activity. Here, IADFs are caused by the interruption of the normal course of growth due to a sudden drought event, thus determining a zone of wood characterised by an abrupt change in wood anatomical properties (Schulman 1938; De Micco et al. 2007, 2012; Battipaglia et al. 2010). Such specific anatomical properties arise in response to drought to enhance safer (i.e. less subject to cavitation or embolism), though slower, water transport (we will specify this in the paragraph 2.6). The analysis of the variation of anatomical properties and isotope composition along IADFs, being correlated to stomatal regulation, has proved to be useful to reconstruct phenological events from wood structure, thus unravelling how woody species vary their role of carbon sink in different environmental conditions triggered by climate change or regional modifications of land use (Battipaglia et al. 2010).

2.4 From the Resistant Dry Seed to the Fragile Seedling

Adaptive strategies against drought have been studied in adult plants of many arid and semiarid ecosystems. Much less information is available about the occurrence of defence strategies during the first phases of plant life after seed germination, although it is recognised that interspecific differences in seedling survival and juvenile growth performance along resource gradients are key factors controlling plant community structure and dynamics (Fenner and Kitajima 1999). Plant communities are first shaped by seed dispersal, which impose the habitat where the plants grow in, and then by the effect of environmental factors on seed survival, germination, seedling establishment and growth (Schupp 1995). Seedling establishment seems to be one of the most critical phases in the regeneration process by sexual reproduction. Once new gene combinations have been produced with the formation of seeds capable to germinate, survival and growth at early stages of plant development are major bottlenecks to successfully complete the reproductive cycle and to achieve canopy occupancy as demonstrated in tropical and cool-temperate forests (Fenner and Kitajima 1999; Sánchez-Gómez et al. 2006). Indeed, even if a species has a very specialised reproductive system that allows successful seed production and dispersal, other ecological factors may constrain seed germination and subsequent seedling establishment, growth and survival (Traveset et al. 2001). Seedling survival depends on the ability to cope with numerous environmental factors such as water availability, temperature, radiation, pathogens, herbivory and competitive interactions (Moles and Westoby 2004). However, the main reason for seedling mortality is drought which constraints the recruitment processes in time and/or in space (García-Fayos and Verdú 1998; Moles and Westoby 2004). Therefore, in arid regions recruitment occurs in restricted rainfall periods or in limited wet areas (Pugnaire et al. 2006; Padilla and Pugnaire 2007). Wherever dry periods, interspersed between rain events, are frequent, two aspects are fundamental in order to ensure rapid seedling establishment: prompt anchoring of juvenile seedlings to the substrate and immediate water absorption (Young and Martens 1991; Aronne and De Micco 2004). Given that the high desiccation tolerance of the embryo in the dry seed is rapidly lost during germination, survival highly depends on the rapidity in overtaking the early life stages.

Higher seedling survival is often linked to larger biomass allocation to roots. This allows a better water and nutrient uptake due to the possibility to explore larger volumes and deeper layers of soil (Davis 1989; Padilla et al. 2007). Survival of deep-rooted seedlings after drought is higher if compared to shallow-rooted ones, although this relation can be influenced by other species-specific factors (Padilla and Pugnaire 2007). Also the size of seeds seems to affect seedling survival during periods of water shortage because larger seeds are characterised by larger storage reserves which allow the development of larger seedlings with more expanded roots than smaller seeds (Fenner and Kitajima 1999). Nevertheless, by means of an experiment involving some Mediterranean woody species, Padilla et al. (2007) demonstrated the importance of water availability during the first stages of development of seedlings regardless of seed size. There is also evidence that root elongation is increased in drought-tolerant species and in species growing in dry sites, suggesting that root plasticity is under genetic control (Sharp et al. 2004). However, to what extent root plasticity is controlled by genetic principles or environmental constrains has not been clarified yet (Padilla et al. 2007). Actually, it is likely that many characters, such as seed size and vital strategy, act together in determining the capability of seedling establishment. As an example, rooting depth in seedlings of wild species has been shown to increase in response to drought especially in species that regenerate mainly from seeds after disturbance (seeders) (Reader et al. 1993). Moreover, growth reactions to water shortage can be influenced by the interaction with other factors, mainly with irradiance. More specifically, contrasting hypotheses attempt to explain the effects of drought on seedling performance along gradients of irradiance. According to the trade-off hypothesis, the effects of light limitation are more negative under drought conditions: the reduced carbohydrates produced by photosynthesis enhance the conflicts between the simultaneous demands for allocation of resources to develop both above-ground and below-ground biomass (Kubiske et al. 1996). On the other hand, many experiments support the facilitation hypothesis by which drought is less harmful under shadier conditions (Sánchez-Gómez et al. 2006). According to this hypothesis, so-called "nurse plants" facilitate growth and development of other plant species (target species) beneath their canopy. Indeed, they create micro-habitats that are more favourable for seed germination and/or seedling establishment. The nursing effect has been used to restore vegetation in arid and subarid zones in recent years (Ren et al. 2008).

Apart from biomass root/shoot ratio and root length, many other morphological parameters affect seedling establishment in Mediterranean environments. The development of specialised tissues, such as hypocotyl hairs, during seedling emergence has been shown to have composite positive effects involving facilitation of water uptake and prompt physical support (Aronne and De Micco 2004). Other anatomical traits favouring adaptation to drought, such as small diameter roots and specialised tissues to avoid water loss, are common also in adult plants and will be analysed in the following paragraphs.

2.5 Transpiration and Leaf Xeromorphy

Water absorbed by plant roots is lost in the atmosphere mostly through leaf transpiration. Transpiration rates vary widely among plant groups: from 10 to 0.1 g of water $dm^{-2}h^{-1}$ in hygrophytes and xerophytes respectively (Monneveux and Belhassen 1996). Plants in arid and semiarid environments show leaves with xeromorphic traits designed to reduce transpiration to a minimum under drought conditions. The reduction of transpiration can be reached through various means including leaf shedding (i.e. in drought deciduous species) as well as decrease of leaf number, leaf size and branching. In extreme conditions shedding of twigs, branches or larger portions of plants also occurs. In plants that appear leafless under drought conditions, photosynthesis is achieved thanks to green branches, stems or retained petioles (Fahn 1964). Sclerophylly of plants is also considered an adaption to drought: hard leaves do not suffer from permanent damage due to wilting and can completely recovery when favourable conditions are restored.

Sclerophylls are widely distributed in arid and semiarid environments together with seasonally dimorphic species. Sclerophylly has been interpreted as a phenomenon linked also to other functions as protection against pathogens or as response to scarce nutrient availability (Salleo and Nardini 2000). However, there is evidence that stiff, leathery leaves are widespread in species adapted to drought occurring in various environments throughout the world. Sclerophyllous leaves are characterised by reinforcing tissues (e.g. thick-walled epidermal cells, sclereids, etc.) which prevent the collapse of the whole structure when water availability is scarce (drought hypothesis), thus reducing the risk for mechanical damage. Under drought conditions, a sclerophyllous leaf slightly reduces its volume thanks to thick cuticle and thick-walled epidermal cells, but thin-walled mesophyll cells severely shrink resulting in an increase of intercellular spaces. This allows photosynthesis to remain active also in conditions of severe water stress when other leaf types wilt (Shields 1950).

Seasonally dimorphic species are characterised by a seasonal reduction in their transpiring surface: larger winter mesomorphic leaves growing on dolichoblasts (long twigs) are shed at the beginning of the arid season and are replaced by smaller summer xeromorphic leaves on new brachyblasts (short twigs) (Orshan 1964; Aronne and De Micco 2001).

In both sclerophyllous and summer leaves of seasonally dimorphic plants there are many similar traits allowing to withstand drought conditions. Such characteristics regard not only the morphological appearance of leaves on axes but also their anatomical properties both inside (e.g. features of palisade and spongy parenchyma tissues) and at surface levels (e.g. deposition of protective layers and features of stomata).

It is well accepted that phenomena such as paraheliotropism, steep leaf inclination and changes in the colour, due to alteration of pigment content, are frequent in dry habitats. They help reducing solar irradiation and consequently decrease leaf heating and transpiration rates as well as avoid damage to photosystems and phenomena of photo inhibition (Aronne and De Micco 2001; Arena et al. 2008). The lamina folding and the leaf rolling are valuable mechanisms to reduce transpiration (Fig. 2.2a). Leaf rolling, is frequent in grasses adapted to drought and is induced by turgor loss of bulliform cells occurring in the epidermis (Fig. 2.2b) or of other specialised mesophyll cells (Shields 1950).

Adaptation to drought conditions is achieved also through the low ratio of the external leaf surface to its volume. This is generally also accompanied by the occurrence of a compact structure (Fig. 2.2a, c) made of small mesophyll cells with thick cell walls, reduced intercellular spaces and a more compact network of veins. The increased leaf stiffness, although limiting gas exchange, might reduce water losses enhancing high leaf tissue density and dry mass per area (Fahn 1964; Mooney and Dunn 1970; Niinemets 2001). More recently, the decrease of total diffusion conductance throughout intercellular spaces from substomatal cavities to chloroplasts has been linked not only to reduced aquaporin conductance, but also to precise anatomical traits such as the reduction of chloroplast surface area exposed to intercellular space per unit leaf area (Miyazawa et al. 2008; Tosens et al. 2012). The presence of additional layers of palisade parenchyma at the expense of spongy tissue (Fig. 2.2 a, c, d) is also considered a way to increase the

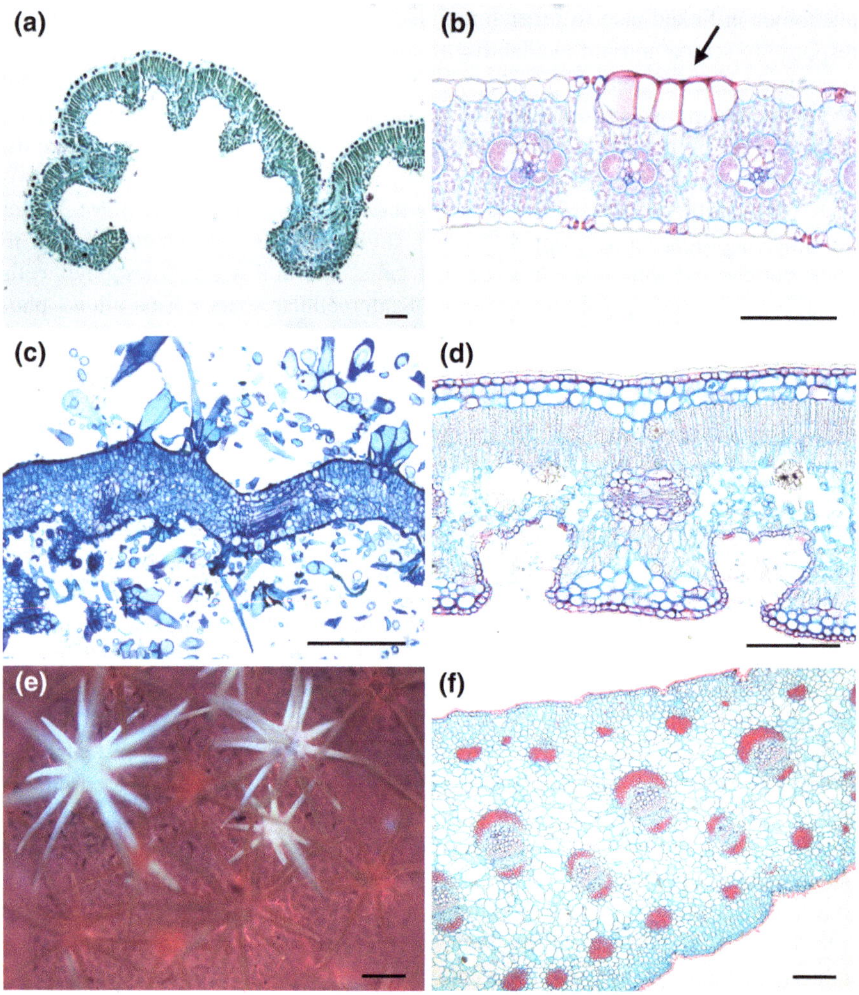

Fig. 2.2 Light (**a–d**, **f**) and epi-fluorescence (**e**) microscopy views of leaf cross-sections (**a–d**, **f**) and lamina surface (**e**) showing anatomical traits for resistance to drought: **a**. Folded lamina and compact structure in *Cistus monspeliensis* L.; **b**. Bulliform cells (*arrow*) in *Zea mais* L.; **c**. High leaf density and high frequency of trichomes in *Cistus incanus* L.; **d**. Additional layers of epidermis and palisade parenchyma, and stomata concentrated in crypts in *Nerium oleander* L.; **e**. High frequency of suberised trichomes in *C. incanus* L.; **f**. Thick cuticle, sunken stomata and water-storage parenchyma in *Yucca* sp. Bars correspond to 100 μm

path of water through intercellular spaces to reach stomata; this would be a strategy to increase water use efficiency (ratio of carbon dioxide fixed to water lost) (Lewis 1972). In xeric leaves, additional layers of palisade parenchyma can develop also adjacent to the lower epidermis leading to isobilateral anatomical structure which, together with steep leaf inclination, allows the optimisation of light interception in the early morning hours and in late afternoon. The increase of mesophyll thickness enhances the photosynthetic capacity if it is accompanied by an increase in the number of chloroplasts exposed near the surface area facing the intercellular spaces (Oguchi et al. 2005). However, other factors, including leaf developmental stage and light availability, are known to interact with drought in determining modifications of mesophyll and chloroplast differentiation, and ultimately of mesophyll diffusion conductance and photosynthetic capacity (Tosens et al. 2012).

Water losses are also severely affected by the characteristics of stomata. Adaptation to drought involves the decrease of stomata size, while stomata density shows a more plastic response to environmental changes. In a recent experiment, Xu and Zhou (2008) demonstrated that stomatal density increases under moderate water deficit, while it declines under severe drought. There is much information on how reduction of transpiration can be rapidly achieved through a physiological control of stomatal opening. However, stomatal closure is strongly controlled by hydraulic architecture of leaves and stems that determines the critical thresholds for vulnerability to cavitation (Sperry 2000). Plants of arid and semiarid environments show sunken stomata, often covered by resinous masses and wax layers or confined in deep crypts of the lamina (Fig. 2.2d) (Fahn 1964; Monneveux and Belhassen 1996). These crypts are often occluded by wax tubules or trichomes which might further reduce transpiration. However, the occurrence of stomata plugged with cuticular structures has been also shown as an adaptation to excess water in plants growing in rainforests and cloud forests: these plugs help maintaining photosynthetic activity by preventing the formation of a continuous water film that would impede diffusion of CO_2 into the leaf (Field et al. 1998). The presence of plugged and/or sunken stomata and hairy leaves is not exclusive of plants of arid environments and is also related to other functions such as protection against herbivory (Bongers 1973; Koster and Baas 1981). Although xeromorphic leaves are generally more hairy than mesomorphic ones (Fig. 2.2c, e), there is common agreement on the fact that dead trichomes participate in the reduction of transpiration when stomata are closed, while living trichomes might themselves increase water losses (Shields 1950).

Xeromorphic leaves are also characterised by the presence of a thick cuticle. The hydraulic permeability of the cuticle depends on its thickness, chemical composition and crystal forms of cuticular waxes embedded in the cuticle or deposited over it. Water permeability of cuticle is lower in xeromorphic than in mesomorphic leaves and it has been demonstrated that the properties of this barrier at leaf interface with atmosphere are genetically controlled (Riederer and Schreiber 2001). Cuticular water permeability also depends on relative humidity, decreasing as environment dries out thanks to chemical properties of cutin and wax domains (Bargel et al. 2004). Moreover, different cuticle composition can determine changes in reflectance thus indirectly affecting lamina heating (Monneveux and Belhassen 1996). Indeed, the occurrence of cuticular waxes, by increasing leaf reflectance at the visible and infrared light wavelengths, can reduce photoinhibition of photosynthesis and transpiration rates, having a positive effect on water use efficiency. Besides, UV radiation is also attenuated by the presence of flavonoids in the cuticle matrix or at the surface of epicuticular waxes (Bargel et al. 2004). Phenolic compounds can also filter excess radiation either in form of vacuolar accumulations in glandular leaf hairs, in epidermal and parenchyma cells, or linked to membranes throughout mesophyll cells as also shown in cortical cells of green photosynthesising twigs (Tattini et al. 2000; De Micco and Aronne 2007). Such phenolics also have an indirect role in adaptation to drought since, being feeding deterrents, they protect plants from permanent damage due to grazing and pathogen attacks. This is critical since development of spare organs would require an extra-additional energetic cost which is often unaffordable under drought conditions.

The above-reported traits are linked to the reduction of transpiration; however, xeromorphic leaves can be also characterised by the presence of multilayered epidermis or parenchyma tissues devoted to water storage (Fig. 2.2f). Such water-storage tissues show lower osmotic pressure than photosynthesising cells which, under conditions of low water availability, can obtain water from the water-storage cells. The latter are generally thin-walled cells which can shrink easily, though rapidly recover when water becomes available again (Fahn 1964). In xeromorphic leaves, the recovery of cell turgidity after shrinking without suffering mechanical damage is favoured by the occurrence of the so-called concertina cells whose involuted cell walls help rapid enlargement when water becomes available again (Aronne and De Micco 2001).

2.6 The Role of the Stem and Evolutionary Trends in Wood Anatomy

Symptoms of adaptation to drought in stems have to be searched in the presence of peripheral structures reducing water loss, in the formation of water-storage tissues and in the characteristics of the water transport system. Regarding water saving, it relies on the properties of thick cuticle and thick-walled epidermal cells as well as on the occurrence of suberized subepidermal layers in young organs or interxylary cork rings in older stems. Such structures constitute barriers arranged "in series" as hydraulic resistors regulating water exchanges at the plant–atmosphere interface (De Micco and Aronne 2011). Water storage is achieved through the presence of succulent stems, below-ground structures such as bulbs and rhizomes, and living wood fibres or living xylem parenchyma containing reserve materials (Fahn 1964).

Apart from the analysis of structures which reduce water loss or serve in water storage, most studies have focused on adaptive traits regulating the transport of water from roots to leaves. Considering the metabolic cost for the synthesis of xylem tissue, plants tailoring xylem formation to fit the expected evapo-transpirational demand of leaves should have an adaptive advantage during evolution (Sperry 2003). In the last few decades, there has been increasing interest in the study of the relationships between wood anatomy and environmental factors. In general, most wood anatomical traits can be interpreted based on their functional significance in the survival strategies of the species. They determine the hydraulic conductivity (efficiency) and vulnerability to cavitation (safety) of a given wood as well as the biomechanics of stems and branches (Zimmermann 1983; Carlquist 1975; Baas et al. 1983, 2004; Baas and Schweingruber 1987; Tyree and Sperry 1989; Niklas 1992; Tyree et al. 1994; Hacke and Sperry 2001). Adaptation to drought can be achieved through a sort of compromise between the need to maintain high conductivity when water is available, and to prevent embolism under conditions of aridity. The main ecological trends in wood anatomy highlight that, moving from mesic to xeric areas, woods tend to be less efficient in water flow, but more resistant to embolism and more robust (Fig. 2.3). In Mediterranean environments, this general tendency is often accompanied by a change of the plant habit from trees to evergreen/drought deciduous shrubs (De Micco et al. 2008). For many wood anatomical features, the existence of direct or indirect proportionality with the properties of efficiency/safety of water transport can be traced (Fig. 2.3). Such relations are sometimes ambiguous, considering that different combinations of various characteristics can tip the balance towards one extreme or the other. This balance between efficiency and safety is well achieved by many shrub species from Mediterranean semiarid ecosystems which have wood specialised for high conductivity when water is available (simple perforation plates, wide earlywood vessels, etc.), but also for safety during drought periods (narrow latewood vessels, vasicentric tracheids, etc.) (Carlquist 1988, 1989).

Xeric conditions apparently favoured selection for short elements with simple perforations while scalariform ones remained generally restricted to plant taxa with a mesic or boreal/alpine ecology (Baas 1986; Carlquist 1975). Simple perforation plates are a more evolved character than scalariform ones (Bailey and Tupper 1918; Wheeler and Baas 1991) and they offer the capability for conducting larger volumes of water per unit of time when water is available (Carlquist 1975; Christman and Sperry 2010). However, according to Sperry (2003), scalariform perforation plates would be advantageous under drought conditions because they favour refilling of embolised vessels in a passive way. More specifically, a scalariform perforation would divide large gas bubbles into smaller ones which may dissolve more rapidly than fewer, larger bubbles in refilling vessels with simple perforation plates.

Xeromorphic woods are generally characterised by an increased number of narrow vessels. Under dry conditions, the redundancy of conducting cells permit water transport despite deactivation of a part of the tissue (Carlquist 1975; Baas et al. 1983). Conductivity being proportional to the fourth power of the vessel radius (Van der Oever et al. 1981; Zimmermann 1983), narrow vessels only allow slow water flow rate. However, narrow vessels are valuable for safety since they guarantee water transport also when larger vessels are embolised (Carlquist 1975).

In arid and semiarid environments, vessel size may be limited also by the need to reduce intervessel pitting and embolism by air seeding which is a phenomenon described as gas being drawn through pit membrane pores (Wheeler et al. 2005;

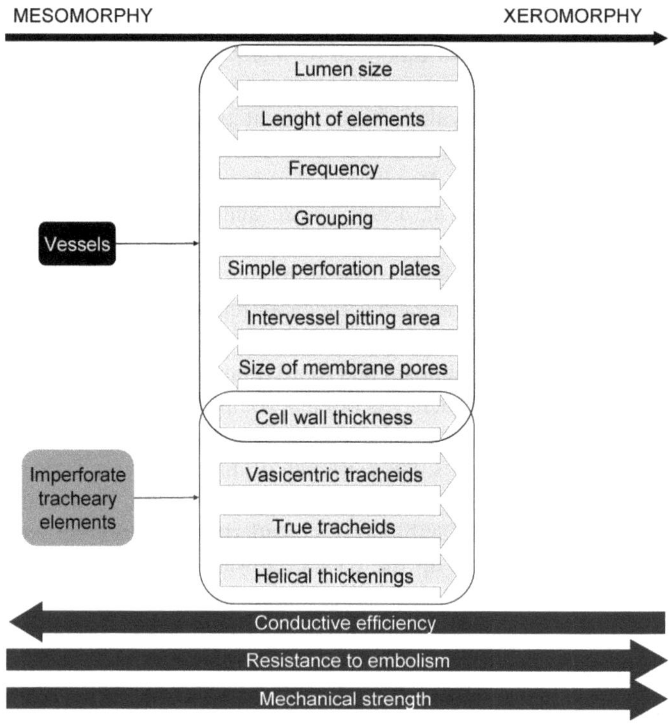

Fig. 2.3 Schematic view of the direct and inverse proportionality between the main attributes of xylem elements and functional traits of efficiency/safety against embolism and implosion. Different combinations of these traits might play a role in species geographical distribution in mesic and xeric areas

Sperry et al. 2006). The porosity of pit membranes between adjacent vessels might be designed to solve the conflict between the functional requirements to minimise vascular resistance, which favours thin, porous membranes, and to limit embolism spreading, which requires robust membranes and smaller pores. Indeed, a decrease in total pit area per vessel causes a decrease in the average size of membrane pore, and hence an increase in safety from cavitation (Wheeler et al. 2005). In the light of these considerations, traits such as pit membrane permeability, pit membrane area between vessels and pore size can be considered important adaptive traits potentially driving ecological differences between species (Jansen et al. 2009).

Vessel grouping is another phenomenon, favouring safety, common in the arid desert and Mediterranean flora (Carlquist 1989). Recently, the first empirical evidence of the positive relation between vessel grouping and cavitation resistance has been reported for seven *Acer* species (Lens et al. 2011). If a particular vessel in a group embolises, the surrounding active vessels maintain the three-dimensional conductive pathway. Almost the same role can be played by vasicentric tracheids which are distributed in sheaths around the vessels, especially in woods of semiarid ecosystems. According to Carlquist (1989), such tracheids have been "re-invented" in clades that have evolved more specialised woods where primitive tracheids have been supplanted by non-conducting fibre elements. The other mode of tracheid reinvention is represented by "vascular tracheids" that are formed only at the end of growth ring (Carlquist 1989). However, as in the case of very narrow vessels, the conductive rate through tracheids is extremely low but this is presumably not a problem since low transpiration and low conductive rates may be expected during periods of drought.

Another common property of woods from xeric floras is the occurrence of helical thickening in vessels.

Thickness and density of vessel sculpturing in *Acer* species has been shown to be correlated with resistance to cavitation (Lens et al. 2011). Apart from the increase of mechanical strength, helical sculpturing has a role in preventing the incidence and spreading of cavitation, because it increases wall surface and thereby water bonding to the surface (Carlquist 1989; Kohonen and Helland 2009). In woods of xeric environments, helical thickenings increase mechanical strength which can be also guaranteed by the presence of vessels and imperforate tracheary elements with very thick walls and narrow lumen (Sperry 2003).

Besides, other features, such as length of conduit elements, play an important role in defining efficiency of water transport and resistance to negative pressures, but, being correlated to vessel lumen diameter, are less studied.

In general, adaptation to drought can be achieved by means of various combinations of wood anatomical traits. The lack or the low expression of a specific trait can be compensated by the occurrence of other features addressed to the same goal; for example, greater wall thickness in any species may compensate for a lesser amount of imperforate tracheary elements when vessel frequency is very high as shown in *Cistus monspeliensis* L. by De Micco et al. (2008). Moreover, wood adaptive attributes can be considered additive (Carlquist 1989); such a property would explain species distribution along gradients of water availability in semiarid ecosystems such as Mediterranean types. In these ecosystems, moving from more mesic to arid areas there is a change in habit from deciduous to evergreen and seasonally dimorphic/drought deciduous that is also accompanied by an increasing occurrence of attributes in wood anatomy allowing adaptation to drought conditions (De Micco et al. 2008; De Micco and Aronne 2009).

Among the numerous wood anatomical attributes, two groups can be distinguished: those quite conserved and stable are mainly qualitative traits (e.g. porosity and morphology of xylem elements); the others regard quantitative attributes which are characterised by high plasticity under changing environmental conditions. Actually, even within the same plant, specific wood anatomical traits, such as vessel size and cell walls thickness, can quantitatively vary according to the season in a sort of seasonal dimorphism leading to the development of summer tree rings safer than winter tree rings as recently shown in the shrub *Cistus incanus* (De Micco and Aronne 2009). Hence, this high plasticity of specific wood anatomical traits is responsible for their intra-annual variability that can be linked with environmental fluctuations and used as proxy of growth cycles.

2.7 Some Adaptive Traits at the Root Level

A well-established response of plants to drought is the modification of the root/shoot dry matter ratio which tips in favour of the former. This determines an increase of root density over leaf area, a phenomenon bringing the advantages already discussed for seedlings. Adaptive strategies to drought are based also on many other traits which regard both root architecture and anatomy.

At the beginning of the twentieth century, there was a common idea that roots of arid environments had to grow very deep in the soil to withstand long periods of severe drought. Later it has been demonstrated that there is a variety of adaptive strategies. Among these, three types of root architecture are described for perennial species. The first type, typical of succulent species, is made of shallow roots, not growing deeper than about 20 cm. The second type comprises both long roots growing parallel to each other, a few metres deep, and shallow adventitious roots designed for quick uptake of water after brief precipitations. The third system is characterised by many lateral roots about 1 m long, accompanied by very long taproots which can develop tens of metres long (Phillips 1963; Kummerow 1981). Such taproots reach very deep layers of soil where phreatic water is permanently present, thus avoiding seasonal fluctuations of water availability. In some species such as *Agave deserti* Engelm, the root system is made of established and rain-induced roots which are produced on established ones within a few hours of rains and are shed when the soil dries (Hunt et al. 1987). Although representing an additional production cost for the plant, rain-induced roots confer an adaptive advantage in desert environments since they are characterised by higher hydraulic conductivity than established ones. In semiarid environments, both root systems characterised by shallow and deep roots coexist (Kummerow 1981). Moreover, in the Chilean matorral and Californian chaparral, some species have been reported to be strictly interconnected to each other (Kummerow 1981). Although main types of root architecture have been described, it is also commonly accepted that the distribution of roots throughout a soil is largely affected by moisture content of the superficial layers more than of deeper layers (Blum 1996).

As regards inner structure, the adaptive function of specific root attributes is commonly accepted even if consistent evolutionary trends among plant groups and along mesic–xeric gradients are less investigated than in stems. Xeromorphism at root level relies on the presence of traits linked to the regulation of water uptake, the avoidance of water loss and the formation of water-storage tissues (Fig. 2.4a).

The presence of small diameter roots under reduced water availability is considered as a strategy aimed to maximise absorptive surfaces, thus increasing rates of water and nutrient uptake (Eissenstat 1992). The control of water loss is also exerted by the presence of specialised tissues such as a rhizodermis with thickened outer cell walls, a well-developed suberised exodermis, often accompanied by many layers of thin- or thick-walled suberised cells (Fig. 2.4b–d). The presence of suberized layers of cells, at the periphery of the root, represents an important mechanism not only in the selection of nutrient uptake but especially because it regulates the inverse flux of water that, in extreme drought conditions, could pass from the root to the soil (Hose et al. 2001). It has been experimentally shown that the limitation imposed to root radial hydraulic conductivity by suberized layers increases during root development and during soil drying (North and Nobel 1995). After crossing epidermis and exodermis, water has to pass through the cortical parenchyma. The reduced number of cortical layers is considered of adaptive advantage under drought conditions because it shortens the way between the soil and the stele favouring quick radial water transport (Fahn 1964). However, considering the inverse flux of water under drought conditions, the rupture of cortical cells is considered a strategy to create cortical lacunae which interrupt the radial pathway for water movement from the stele to the soil (Fig. 2.4d) (Robards et al. 1979). In *Agave deserti* Engelm, the development of intercellular cortical lacunae has been shown to reduce radial conductivity across root cortex in response to soil drying (North and Nobel 1995). The opening of large fractures, especially in the outermost cortical layers, could also open a path for easy water release as found in *Opuntia ficus-indica* (L.) Miller under drying soil conditions (North and Nobel 1996). Recently, it has been also speculated that the presence of cortical lacunae is beneficial under drought conditions because it reduces root metabolic costs by transforming living cortical cells in air volume (Zhu et al. 2010). However, the formation of cortical lacunae can weaken root mechanical strength, making it more vulnerable in soils prone to swelling-shrinkage cycles consequent to frequent fluctuations in water availability (Striker et al. 2007). This weakening of root strength can be counterbalanced by the general finding of increased development of lignified tissues with cells characterised by thicker walls in various structures (e.g. additional schlerenchyma, increased vascular system, lignified pith) (Fig. 2.4d, h) (Mostajeran and Rahimi-Eichi 2008). The shrinkage of cortical parenchyma cells in roots growing under drought conditions is a common finding (Peña-Valdivia et al. 2010); however this phenomenon can be considered reversible especially when cell walls are reinforced by the deposition of suberin as shown in *Lygeum* spp. (Fig. 2.4e, f).

The development of endodermis with thicker cell walls and the formation of additional layers of cells with suberised walls around the stele (Fig. 2.4g, h) are considered means to prevent the desiccation of meristematic tissues such as the pericycle and other tissues inside the stele (North and Nobel 1992). Water deficit induces the formation of an higher number of endodermal cells with Casparian bands closer to root tip in *Opuntia ficus-indica* (North and Nobel 1996).This phenomenon is accompanied by other anatomical changes, including the formation of a higher number of suberised peridermal layers (phellem) which reduce water permeability of cortex as measured also in other species (Schönherr and Ziegler 1980). The relevance of the apoplastic barriers in controlling water flow depends on root developmental stage: in young unstressed roots, most of the radial hydraulic resistance is more evenly distributed across cortical layers, while in water-stressed old plants it is mainly due to exodermis and endodermis (Steudle 2000).

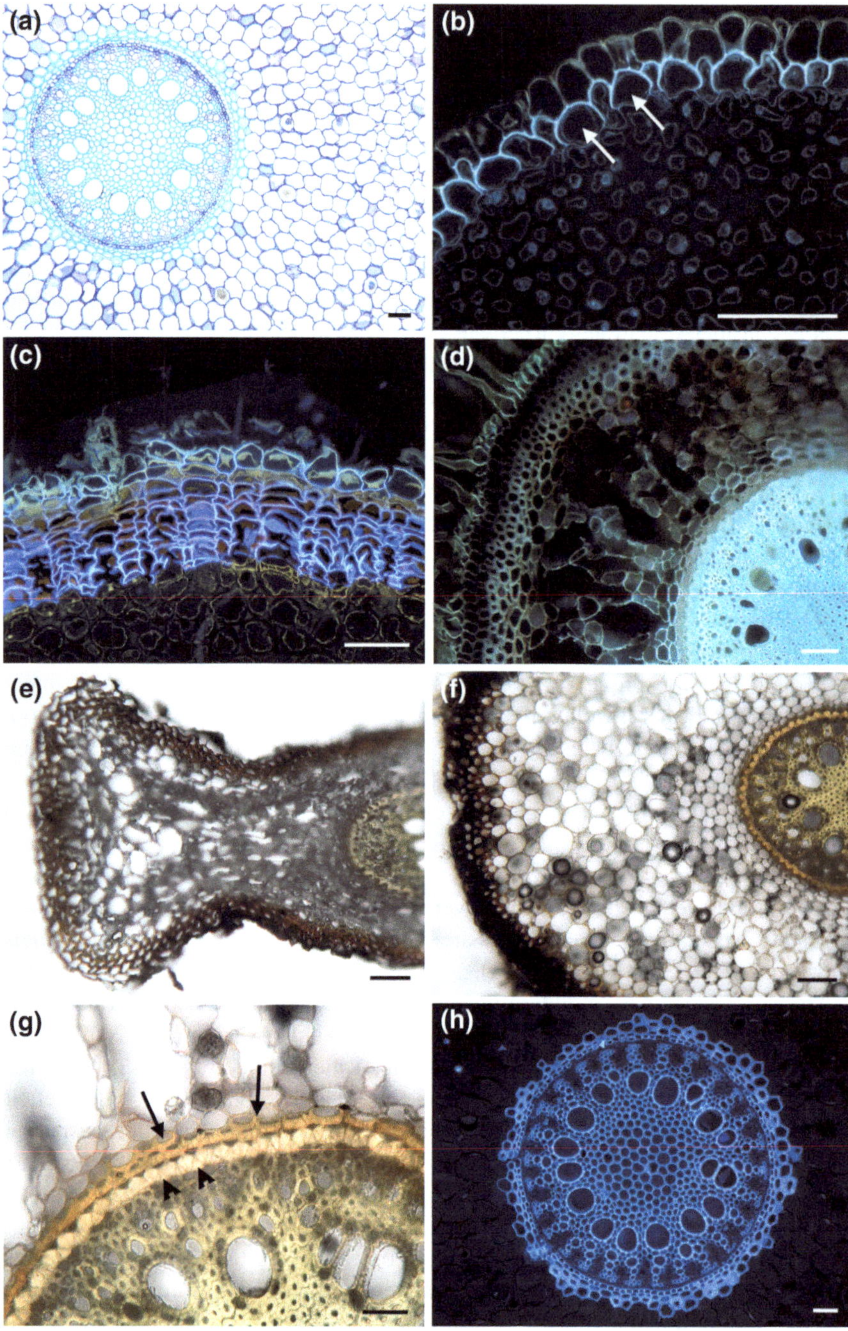
(a)
(b)
(c)
(d)
(e)
(f)
(g)
(h)

◀ **Fig. 2.4** Light (**a**, **e–g**) and epi-fluorescence (**b–d**, **h**) microscopy views of root cross-sections showing anatomical traits for resistance to drought: **a**. Water-storage parenchyma in *Asparagus acutifolius* L.; **b**. Suberised exodermis (*arrows*) in *Primula palinuri* Petagna; **c**. Suberised subepidermal cells in *Primula palinuri* Petagna; **d**. Suberised rhizodermis, thickened subepidermal cells and cortical lacunae in *Lygeum* spp.; **e**, **f**. Cortical parenchyma cells respectively shrunk and recovered after rewatering in *Lygeum* spp.; **g**. Endodermis (*arrowheads*) with thickened cell walls and additional layers of thick-walled cells (*arrows*) around the stele in *Lygeum* spp.; **h**. Additional layers of cells with thickened walls around the stele and lignified pith in *Asparagus acutifolius* L. Bars correspond to 100 μm

Once the water has entered the stele, it needs to be transported efficiently throughout the plant to maintain a continuous water supply to the leaves. In xeric conditions, plants have evolved morpho-functional traits that change root hydraulic conductance as a mechanism for regulating transpiration complementary to those played by aerial organs (Trubat et al. 2006). It is recognised that the efficiency of water transport and the hydraulic safety of vascular system are fundamental for the survival of plants in arid environments. Indeed, in the context of climate changes, considering an increase in the frequency of drought and of overall higher temperatures, vegetation of arid and semiarid environments have to cope with increased xylem vessel cavitation. Within this scenario, in these environments, the adaptive capacity of species is strictly linked to the characteristics of their xylem in the root as in the stem. Indeed, within the same plant, conductivity shows strong variability between different root types growing with different directions and at different depths, a phenomenon which should optimise water flow according to changing water availability (Fahn 1964; De Micco and Aronne 2010).

As final consideration, we may emphasise that plants showing different root architecture and various combinations of anatomical traits allowing adaptation to drought coexist in the same arid and semiarid environments. Moreover, it has been shown that water stress triggers a wide variety of root morphological and anatomical responses: various traits are adjusted with different intensities and trends in different species and even in different cultivars within species (Shao et al. 2008; Peña-Valdivia et al. 2010). A comprehensive understanding of the different mechanisms of root adaptation to water deficit remains a valuable goal because roots can be considered sensors that detect changes of water availability in soil and influence the resistance to drought at the whole plant level.

2.8 Conclusion

The importance of the strict relationship between structure and function in plants has been recently recalled, considering that major metabolic and physiological processes are ultimately regulated by the physics of the plant's structure (Brodribb 2009). The coordination between structure and function in plants is apparent when considering ecological trends in the evolution of morpho-anatomical traits. Such a coordination is especially strong under stressful conditions, when structural investment is regulated for reaching optimal carbon allocation to guarantee growth, survival and reproduction. A wide variation of morphological and anatomical traits accompanies plants of different ecosystems in a sort of harmonisation between plant response and environmental constraints to improve adaptation.

Adaptation to drought is based on many morpho-anatomical traits expressed in different organs at different levels. Since they are not mutually exclusive, different combinations of traits lead to different adaptive strategies. Moreover, various degrees of adaptation can coexist, allowing quite different biological forms to share the same environment. Survival against drought also relies on the capacity of plants to simultaneously cope with other stress factors, such as predation and excess radiation.

References

Arena C, Vitale L, De Santo Virzo A (2008) Paraheliotropism in *Robinia pseudoacacia* L.: an efficient strategy to optimise photosynthetic performance under natural environmental conditions. Plant Biol 10:194–201

Aronne G, De Micco V (2001) Seasonal dimorphism in the Mediterranean *Cistus incanus* L. subsp. *incanus.* Ann Bot-London 87(6):789–794

Aronne G, De Micco V (2004) Hypocotyl features of *Myrtus communis* L. (Myrtaceae): a manysided strategy for possible enhancement of seedling establishment in the Mediterranean environment. Bot J Linn Soc 145:195–202

Aronne G, Wilcock CC (1994) Reproductive characteristics and breeding system of shrubs of the Mediterranean region. Funct Ecol 8:69–76

Aronne G, Wilcock CC (1997) Reproductive phenology in Mediterranean macchia vegetation. Lagascalia 19(1–2):445–454

Aronson J, Kigel J, Shmida A, Klein J (1992) Adaptive phenology of desert and Mediterranean populations of annual plants grown with and without water stress. Oecologia 89:17–26

Atjay GL, Ketner P, Duvigneaud P (1979) Terrestrial primary production and phytomass. In: Bolin B, Degens ET, Kempe S, Ketner P (eds) The global carbon cycle, SCOPE Report 13. Wiley, UK, pp 129–181

Baas P (1986) Ecological patterns in xylem anatomy. In: Givnish TJ (ed) On the economy of plant form and function. Cambridge University Press, Cambridge, pp 327–352

Baas P, Schweingruber FH (1987) Ecological trends in the wood anatomy of trees, shrubs and climbers from Europe. IAWA Bull n s 8(3):245–274

Baas P, Werker E, Fahn A (1983) Some ecological trends in vessel characters. IAWA Bull n s 4(2–3):141–159

Baas P, Ewers FW, Davis SD, Wheeler EA (2004) Evolution of xylem physiology. In: Hemsley A, Poole I (eds) The evolution of plant physiology. Elsevier Scientific Publishing Company, Amsterdam, pp 273–295

Bailey IW, Tupper WW (1918) Size variation in tracheary cells. I. A comparison between the secondary xylem of vascular cryptogams, gymnosperms and angiosperms. In: Proceedings of the American Academy of Arts and Sciences 54:149–204

Bargel H, Barthlott W, Koch K, Schreiber L, Neinhuis C (2004) Plant cuticles: multifunctional interfaces between plant and environment. In: Hemsley A, Poole I (eds) The evolution of plant physiology. Elsevier Scientific Publishing Company, Amsterdam, pp 273–295

Bateman RM, Crane PR, Di Michele WA, Kenrick PR, Rowe NP, Speck T, Stein WE (1998) Early evolution of land plants. Annu Rev Ecol Syst 29:263–292
Battipaglia G, De Micco V, Brand WA, Linke P, Aronne G, Saurer M, Cherubini P (2010) Variations of vessel diameter and δ13C in false rings of *Arbutus unedo* L. reflect different environmental conditions. New Phytol 188(4):1099–1112
Blum A (1996) Crop responses to drought and the interpretation of adaptation. Plant Growth Regul 20:135–148
Bongers JM (1973) Epidermal leaf characters of the winteraceae. Blumea 21:381–411
Brodribb TJ (2009) Xylem hydraulic physiology: the functional backbone of terrestrial plant productivity. Plant Sci 177:245–251
Brodribb TJ, Field TS (2000) Stem hydraulic supply is linked to leaf photosynthetic capacity: evidence from new Caledonian and Tasmanian rainforests. Plant Cell Environ 23:1381–1388
Brodribb TJ, Field TS, Sack L (2010) Viewing leaf structure and evolution from a hydraulic perspective. Funct Plant Biol 37:488–498
Carlquist S (1975) Ecological strategies of xylem evolution. University of California Press, Berkeley
Carlquist S (1988) Comparative wood anatomy. Systematic, ecological, and evolutionary aspects of dicotyledon wood. Springer, Berlin
Carlquist S (1989) Adaptive wood anatomy of chaparral shrubs. In: Keely JE (ed) The California chaparral: paradigms re-examined. Los Angeles Country Museum of Natural History Contributions, Los Angeles, pp 25–35
Chaves MM, Maroco JP, Pereira JS (2003) Understanding plant responses to drought: from genes to the whole plant. Funct Plant Biol 30:239–264
Cherubini P, Gartner BL, Tognetti R, Bräker OU, Schoch W, Innes JL (2003) Identification, measurement and interpretation of tree rings in woody species from Mediterranean climates. Biol Rev 78:119–148
Christman MA, Sperry JS (2010) Single-vessel flow measurements indicate scalariform perforation plates confer higher flow resistance than previously estimated. Plant Cell Environ 33:431–443
Davis SD (1989) Patterns in mixed chaparral stands: differential water status and seedling survival during summer drought. In: Keeley SC (ed) The California chaparral: paradigms re-examined. Natural History Museum of Los Angeles County, Los Angeles, pp 97–105
De Micco V, Aronne G (2007) Anatomical features, monomer lignin composition and accumulation of phenolics in one-year-old branches of the Mediterranean *Cistus ladanifer* L. Bot J Linn Soc 155:361–371
De Micco V, Aronne G (2008) Twig morphology and anatomy of Mediterranean trees and shrubs related to water availability. Bot Helv 118:139–148
De Micco V, Aronne G (2009) Seasonal dimorphism in wood anatomy of the Mediterranean *Cistus incanus* L. subsp *incanus*. Trees-Struct Funct 23(5):981–989
De Micco V, Aronne G (2010) Root structure of *Rumex scutatus* L. growing on slopes. IAWA J 31(1):13–28
De Micco V, Aronne G (2011). Anatomy and lignin characterization of twigs in the chaparral shrub *Rhamnus californica*. IAWA J in press
De Micco V, Saurer M, Aronne G, Tognetti R, Cherubini P (2007) Variations of wood anatomy and δ13C within tree rings of coastal *Pinus pinaster* Ait. showing intra-annual density fluctuations. IAWA J 28(1):61–74
De Micco V, Aronne G, Baas P (2008) Wood anatomy and hydraulic architecture of stems and twigs of some Mediterranean trees and shrubs along a mesic-xeric gradient. Trees-Struct Funct 22:643–655
De Micco V, Battipaglia G, Brand WA, Linke P, Saurer M, Aronne G, Cherubini P (2012) Discrete versus continuous analysis of anatomical and δ13C variability in tree rings with intra-annual density fluctuations. Trees-Struct Funct 26:513–524
di Castri F, Goodall DW, Specht RL (eds) (1981) Ecosystems of the World 11, Mediterranean-Type Shrublands. Elsevier Scientific Publishing Company, Amsterdam, pp 309–315

Eissenstat DM (1992) Costs and benefits of constructing roots of small diameter. J Plant Nutr 15:763–782
Fahn A (1964) Some anatomical adaptations in desert plants. Phytomorphology 14:93–102
Field TS, Zwieniecki MA, Donoghue MJ, Holbrook NM (1998) Stomatal plugs of *Drimys winteri* (Winteraceae) protect leaves from mist but not drought. Proc Natl Acad Sci USA 95:14256–14259
Fenner M, Kitajima K (1999) Seed and seedling ecology. In: Pugnaire FI, Valladares F (eds) Handbook of functional plant ecology. Marcel-Dekker, New York, pp 589–621
Franks SJ, Sim S, Weis EA (2007) Rapid evolution of flowering time by an annual plant in response to a climate fluctuation. Proc Natl Acad Sci USA 104:1278–1282
García-Fayos P, Verdú M (1998) Soil seed bank, factors controlling germination and establishment of a Mediterranean shrub: *Pistacia lentiscus* L. Acta Oecol 19:357–366
Graham LE (1993) The origin of land plants. Wiley, New York
Grene R, Vasquez-Robinet C, Bohnert HJ (2011) Molecular biology and physiological genomics of dehydration stress. In: Lüttge U, Beck E, Bartels D (eds) Plant desiccation tolerance. Springer, Heidelberg, pp 255–288
Hacke UG, Sperry JS (2001) Functional and ecological xylem anatomy. Perspect Plant Ecol 4(2):97–115
Hose E, Clarkson DT, Steudle E, Hartung W (2001) The exodermis: a variable apoplastic barrier. J Exp Bot 52:2254–2264
Hulme M, Barrow EM, Arnell NW, Harrison PA, Johns TC, Downing TE (1999) Relative impacts of human-induced climate change and natural climate variability. Nature 397:688–691
Hunt ER, Zakir NJD, Nobel PS (1987) Water costs and water revenues for established and rain-induced roots of *Agave deserti*. Funct Ecol 1:125–130
Intergovernmental Panel on Climate Change (IPCC) (2007) Climate change 2007: mitigation. Contribution of working group III to the fourth assessment report of the intergovernmental panel on climate change. Cambridge University Press, Cambridge
Jansen S, Choat B, Pletsers A (2009) Morphological variation of intervessel pit membranes and implications to xylem function in angiosperms. Am J Bot 96:409–419
Kohonen MM, Helland A (2009) On the function of wall sculpturing in xylem conduits. J Bionic Eng 6:324–329
Koster J, Baas P (1981) Alveolar cuticular material in Myristicaceae. In: Cutler D (ed) The Plant Cuticle. Academic, London, pp 131–138
Kubiske ME, Abrams MD, Mostoller SA (1996) Stomatal and nonstomatal limitations of photosynthesis in relation to the drought and shade tolerance of tree species in open and understory environments. Trees-Struct Funct 11:76–82
Kummerow J (1981) Structure of roots and root systems. In: di Castri F, Goodall DW, Specht RL (eds) Ecosystems of the World 11, Mediterranean-type shrublands. Elsevier Scientific Publishing Company, Amsterdam, pp 269–288
Kummerow J. (1989) Structural aspects of shrubs in Mediterranean-type plant communities. Options Méditerranéennes—Série Séminaires 3: 5–11
Lamont BB (1983) Strategies for maximizing nutrient uptake in two Mediterranean ecosystems of low nutrient status. In: Kruger FJ, Mitchell DT, Jarvis JUM (eds) Mediterranean-type ecosystems: the role of nutrients. Springer, Berlin, pp 246–273
Lens F, Sperry JS, Christman MA, Choat B, Rabaey D, Jansen S (2011) Testing hypotheses that link wood anatomy to cavitation resistance and hydraulic conductivity in the genus *Acer*. New Phytol 190:709–723
Levitt J (1980) Response of plants to environmental stresses. Chilling, freezing, and high temperature stresses. Academic, New York
Lewis MC (1972) The physiological significance of variation in leaf structure. Sci Progr Oxford 60:25–51

Ludlow MM (1989) Strategies of response to water stress. In: Kreeb KH, Richter H, Hinckley TM (eds) Structural and functional responses to environmental stresses: water shortage. SPB Academic Publishing, The Hague, pp 269–281

Matosevic I, Costa G, Giovannetti M (1997) The mycorrhizal status of the woody Mediterranean shrub *Myrtus communis* L. Mycorrhiza 7(1):51–53

Maximov NA (1931) The physiological significance of the xeromorphic structure of plants. J Ecol 19:272–282

Mitrakos K (1980) A theory for Mediterranean plant life. Oecolog Plantar 15:245–252

Miyazawa SI, Yoshimura S, Shinzaki Y, Maeshima M, Miyake C (2008) Deactivation of aquaporins decreases internal conductance to CO_2 diffusion in tobacco leaves grown under long-term drought. Funct Plant Biol 35:556–564

Moles T, Westoby M (2004) What do seedlings die from and what are the implications for evolution of seed size? Oikos 106:193–199

Monneveux P, Belhassen E (1996) The diversity of drought adaptation in the wide. Plant Growth Regul 20:85–92

Moony HA, Dunn EL (1970) Photosynthetic systems of Mediterranean-climate shrubs and trees of California and Chile. Am Nat 104:447–453

Morison JIL, Morecroft MD (2006) Plant growth and climate change. Blackwell Publishing, Oxford

Mostajeran A, Rahimi-Eichi V (2008) Drought stress effects on root anatomical characteristics of rice cultivars (*Oryza sativa* L.). Pakistan J Biol Sci 11:2173–2183

Niinemets U (2001) Global-scale climatic controls of leaf dry mass per area, density, and thickness in trees and shrubs. Ecology 82:453–469

Niklas KJ (1986) Evolution of plant shape. Design constraints. Trends Ecol Evol 1(3):67–72

Niklas KJ (1992) Plant biomechanics: an engineering approach to plant form and function. University of Chicago Press, Chicago

North GB, Nobel PS (1992) Drought-induced changes in hydraulic conductivity and structure in roots of *Ferocactus acanthodes* and *Opuntia ficus-indica*. New Phytol 120:9–19

North GB, Nobel PS (1995) Hydraulic conductivity of concentric root tissues of *Agave deserti* Engelm. Under wet and drying conditions. New Phytol 130:47–57

North GB, Nobel PS (1996) Radial hydraulic conductivity of individual root tissues of *Opuntia ficus-indica* (L.) Miller as soil moisture varies. Ann Bot 77:133–142

Oguchi R, Hikosaka K, Hirose T (2005) Leaf anatomy as a constraint for photosynthetic acclimation: differential responses in leaf anatomy to increasing growth irradiance among three deciduous trees. Plant Cell Environ 28:916–927

Orshan G (1964) Seasonal dimorphism of desert and Mediterranean chamaephytes and its significance as a factor in their water economy. In: Rutter AJ, Whitehead FH (eds) The water relations of plants. Blackwell, Edinburgh, pp 206–222

Padilla FM, Pugnaire FI (2007) Rooting depth and soil moisture control Mediterranean woody seedling survival during drought. Funct Ecol 21:489–495

Padilla FM, Miranda JD, Pugnaire FI (2007) Early root growth plasticity in seedlings of three Mediterranean woody species. Plant Soil 296:103–113

Patón D, Azocar P, Tovar J (1998) Growth and productivity in forage biomass in relation to the age assessed by dendrochronology in the evergreen shrub *Cistus ladanifer* (L.) using different regression models. J Arid Environ 38:221–235

Peña-Valdivia CB, Sánchez-Urdaneta AB, Meza Rangel J, Juárez Muñoz J, García-Nava R, Celis Velázquez R (2010) Anatomical root variations in response to water deficit: wild and domesticated common bean (*Phaseolus vulgaris* L.). Biol Res 43:417–427

Phillips WS (1963) Depth of roots in soil. Ecology 44:424

Pugnaire FI, Luque MT, Armas C, Gutierrez L (2006) Colonization processes in semi-arid Mediterranean old-fields. J Arid Environ 65:591–603

Raven JA (1977) The evolution of vascular land plants in relation to supracellular transport processes. Adv Bot Res 5:153–219

Reader RJ, Jalili A, Grime JP, Spencer RE, Matthews NN (1993) A comparative-study of plasticity in seedling rooting depth in drying soil. J Ecol 81:543–550
Ren H, Long Y, Nan L (2008) Nurse plant theory and its application in ecological restoration in lower-subtropics of China. Prog Nat Sci 18(2):137–142
Rhizopoulou S, Mitrakos K (1990) Water relations of evergreen sclerophylls. Seasonal changes in water relations of eleven species from the same environment. Ann Bot-London 65:171–178
Riederer M, Schreiber L (2001) Effects of environmental factors on the water permeability of plant cuticles. J Exp Bot 52:2023–2033
Robards AWV, Clarkson DT, Sanderson J (1979) Structure and permeability of the epidermal/hypodermal layers of the sand sedge (*Carex arenzaria* L.). Protoplasma 101:331–347
Rozema J, Staaij J, Bjiorn LO, Caldwell M (1997) UV-B as an environmental factor in plant life: stress and regulation. Trends Ecol Evol 12(1):22–28
Rundel PW (1991) Shrub life forms. In: Chu E, Mooney HA, Winner WE, Pell EJ (eds) Responses of plants to multiple stresses. Academic, New York, pp 345–370
Salleo S, Nardini A (2000) Sclerophylly: evolutionary advantage or mere epiphenomenon? Plant Biosyst 134:247–259
Sánchez-Gómez D, Zavala MA, Valladares F (2006) Survival responses to irradiance are differentially influenced by drought in seedlings of forest tree species of the temperate-Mediterranean transition zone. Acta Oecol 30:322–332
Schönherr J, Ziegler H (1980) Water permeability of *Betula* periderm. Planta 147:345–354
Schulman E (1938) Classification of false annual rings in Monterey pine. Tree-ring Bull 4:4–7
Schulze ED, Beck E, Müller-Hohenstein K (2005) Plant ecology. Springer, Berlin
Schupp EW (1995) Seed seedling conflicts, habitat choice, and patterns of plant recruitment. Am J Bot 82:399–409
Schwartz MD (1999) Advancing to full bloom: planning phenological research for the twenty first century. Int J Biometeorol 42:113–118
Shao HB, Chu LY, Jaleel CA, Zhao CX (2008) Water-deficit stress-induced anatomical changes in higher plants. C R Biologies 33:215–225
Sharp RE, Poroyko V, Hejlek LG, Spollen WG, Springer GK, Bohnert HJ, Nguyen HT (2004) Root growth maintenance during water deficits: physiology to functional genomic. J Exp Bot 55:2343–2351
Shields LM (1950) Leaf xeromorphy as related to physiological and structural influences. Bot Rev 16:399–447
Smith TM, Shugart HH, Woodward FI (eds) (1997) Plant functional types. Cambridge University Press, Cambridge
Sperry JS (2000) Hydraulic constraints on plant gas exchange. Agric For Meterology 104(1):13–23
Sperry JS (2003) Evolution of water transport and xylem structure. Int J Plant Sci 164:S115–S127
Sperry JS, Hacke UG, Pittermann J (2006) Size and function in conifer tracheids and angiosperm vessels. Am J Bot 93:1490–1500
Steudle E (2000) Water uptake by roots: effects of water deficit. J Exp Bot 51:1531–1542
Striker GGP, Insausti P, Grimoldi AA, Vega AS (2007) Trade-off between root porosity and mechanical strength in species with different types of aerenchyma. Plant, Cell Environ 30:580–589
Tattini M, Gravano E, Pinelli P, Mulinacci N, Romani A (2000) Flavonoids accumulate in leaves and glandular trichomes of *Phillyrea latifolia* exposed to excess solar radiation. New Phytol 148:69–77
Tosens T, Niinemets U, Vislap V, Eichelmann H, Castro Díez P (2012) Developmental changes in mesophyll diffusion conductance and photosynthetic capacity under different light and water availabilities in Populul tremula: how structure constrains function. Plant Cell Environ 35:839–856
Traveset A, Riera N, Mas RE (2001) Ecology of fruit-colour polymorphism in *Myrtus communis* and differential effects of birds and mammals on seed germination and seedling growth. Funct Ecol 89:749–760

Trubat R, Cortina J, Vilagrosa A (2006) Plant morphology and root hydraulics are altered by nutrient deficiency in *Pistacia lentiscus* (L.) Trees-Struct Funct 20:334–339

Tyree MT, Sperry JS (1989) Vulnerability of xylem to cavitation and embolism. Annu Rev Plant Phys 40:19–38

Tyree MT, Davis SD, Cochard H (1994) Biophysical perspectives of xylem evolution: is there a tradeoff of hydraulic efficiency for vulnerability to dysfunction? IAWA J 15(4):335–360

Van den Oever L, Baas P, Zandee M (1981) Comparative wood anatomy of symplocos and latitude and altitude of provenance. IAWA Bull n s 2:3–24

Wheeler EA, Baas P (1991) A survey of the fossil record for dicotyledonous wood and its significance for evolutionary and ecological wood anatomy. IAWA J 12:275–332

Wheeler JK, Sperry JS, Hacke UG, HOANG N (2005) Intervessel pitting and cavitation in woody Rosaceae and other vesselled plants: a basis for a safety versus efficiency trade-off in xylem transport. Plant Cell Environ 28:800–812

Wilson BF (1995) Shrub stems: form and function. In: Gartner BL (ed) Plant stems: physiology and functional morphology. Academic, San Diego, pp 91–102

Xu Z, Zhou G (2008) Responses of leaf stomatal density to water status and its relationship with photosynthesis in a grass. J Exp Bot 59:3317–3325

Young JA, Martens E (1991) Importance of hypocotyl hairs in germination of *Artemisia* seeds. J. Range Manage 44:438–442

Zhu J, Brown KM, Lynch JP (2010) Root cortical aerenchyma improves the drought tolerance of maize (Zea mays L.). Plant Cell Environ 33:740–749

Zimmermann MH (1983) Xylem structure and the ascent of sap. Springer, Berlin

Chapter 3
Xylem Cavitation and Embolism in Plants Living in Water-Limited Ecosystems

A. Vilagrosa, E. Chirino, J.J. Peguero-Pina, T.S. Barigah, H. Cochard and E. Gil-Pelegrín

Abstract Water deficit is considered the main limiting factor for the establishment, survival, and growth of plants mainly in water-limited ecosystems. Plants have evolved a wide range of morphologic and functional mechanisms to adapt to arid environments. However, if the tension in the xylem conduits becomes too high, thus xylem cavitation can occur i.e., water column breakage. This results in the hydraulic disconnection of leaves and above-ground parts from roots because xylem conduits are filled with air and water vapor, and this phenomenon is called embolism. Therefore, the resistance of the xylem to cavitation and embolism is of paramount importance for plant functioning. In this chapter, we will review the role of plant hydraulics and xylem cavitation in the context of water-limited ecosystems and their relationship with other plant functional traits and with survival capacity. These topics will be analyzed and discussed on the basis of current knowledge and our research experiences.

A. Vilagrosa (✉) · E. Chirino
Fundación de la Comunidad Valenciana Centro de Estudios Ambientales del Mediterráneo (CEAM), Unidad Mixta de Investigación Universidad de Alicante-Fundación CEAM, PO Box 99 E-03080, Alicante, Spain
e-mail: a.vilagrosa@ua.es

J.J. Peguero-Pina
Grup de Recerca en Biologia de les Plantes en Condicions Mediterrànies, Departament de Biologia, Universitat de les Illes Balears, Carretera de Valldemossa, km 7.5 07071, Palma de Mallorca, Balears, Spain

T.S.Barigah · H. Cochard
INRA, UMR 547 PIAF, F-63100, Clermont-Ferrand, France

T.S.Barigah · H. Cochard
Clermont Université, Université Blaise Pascal, UMR 547 PIAF, BP 10448 F-63000, Clermont-Ferrand, France

E. Gil-Pelegrín
Unidad de Recursos Forestales, Centro de Investigación y Tecnología Agroalimentaria, Avenida de Montañana 930, E-50059, Zaragoza, Spain

R. Aroca (ed.), *Plant Responses to Drought Stress*,
DOI: 10.1007/978-3-642-32653-0_3, © Springer-Verlag Berlin Heidelberg 2012

3.1 Introduction

A major tradeoff of acquiring CO_2 for photosynthesis is the loss of water from plant leaves. Since water is essential for plant correct functioning, its limitation induces many dysfunctions in terrestrial plants. In this sense, soil water availability represents a significant environmental constraint in Mediterranean and other drylands in the world—a fact which may become even more accentuated in the future according to climate change projections (IPCC 2010). Several factors could promote frequent drought events: higher temperatures, earlier snowmelt, higher interannual variability, precipitation regimes with reduced rainfall amounts, and heat waves. Recent studies have linked widespread plant mortality to extreme drought events (Allen et al. 2010 and references herein). However, the effect of drought could be species specific since species from xeric sites can survive low soil water contents while other species from more mesic sites do not (Ryan 2011).

Drought leads to water deficit in plant tissues, which affects many physiological processes and can have mortal consequences for plant growth and survival (Levitt 1980; Larcher 1995). Among these processes, the xylem cavitation and loss of hydraulic conductivity have been recognized to play an important role in plant performance and drought resistance (Tyree and Sperry 1989). Cavitation is defined as the phenomenon of vapor-bubble formation of a flowing liquid in a region where the pressure of the liquid falls below its vapor pressure. This phenomenon occurs in the xylem of vascular plants when the tension of the sap within a conduit becomes high enough that dissolved air within the sap expands to fill vessels or tracheids. As a consequence, vapor-filled (embolized) conduits no longer hold sap, decrease xylem hydraulic conductance which, in its turn leads to stomatal closure, and even to the abscission of leaves, shoot dieback and finally to plant death (Tyree and Sperry 1989; Brodribb and Cochard 2009).

Recognition of the relevance of water transport in plant tissues began several centuries ago, around 1628, with the studies of Harvey and Malpigi (Richter and Cruiziat 2002). Later, Strasburger showed that woody stems immersed in different solutions will readily pull the solution up. In 1892, Joseph Böhm provided experimental proof that water could be lifted to considerable heights and that the force lifting the water had to be a purely physical one (Richter and Cruiziat 2002). Some years later, Dixon and Joly published their ideas and experiments about water transport (Dixon and Joly 1895), and Dixon (1914) wrote the book "Transpiration and the Ascent of Sap in Plants" (Steudle 2001). The cohesion-tension theory put forward in this book to explain the ascent of sap became universally accepted. In the middle of the twentieth century, John Milburn, Melvin Tyree, Martin H. Zimmermann and co-workers renewed and developed experimental approaches to comfort the cohesion-tension theory, hydraulic design, and xylem cavitation (Tyree and Zimmermann 2002; Pickard and Melcher 2005). From these preliminary studies to the present, a significant progress has been made in understanding plant hydraulic architecture including cavitation process and its impact on plant functioning and ecosystem dynamics. Therefore, the number of references in which "xylem cavitation" appears has recently shown a sharp increase: in contrast to the nine studies published between 1970 and 1990, there have been 827 publications on this subject from 1990 to 2011 (Source: Web of Science "xylem cavitation"). In addition to studies focused on explaining the functioning of xylem and cavitation processes, several current studies have also highlighted the importance of xylem cavitation and functional plant strategies in the context of climate change and recurrent drought episodes to provide an explanation for relevant mortality events occurring around the world (McDowell et al. 2008; Breshears et al. 2009; Allen et al. 2010).

3.1.1 Plant Functional Strategies in Water-Limited Ecosystems

Water absorption and management of water loss by transpiration are especially important to the survival and growth of species in water-limited environments (Kozlowski et al. 1991). For this reason, functional strategies of drought resistance developed by species in the context of multiple stress factors have been the subject of many studies (Levitt 1980; Larcher 1995). These drought strategies and functional mechanisms to resist water stress have a significant impact in regard to xylem cavitation. Drought resistance strategies fall into two main groups: avoidance mechanisms (*Drought avoidance*) and tolerance mechanisms (*Drought tolerance*) (Levitt 1980; Kozlowski et al. 1991; Jones 1992; Larcher 1995). These strategies rely on the prevalence of specific response mechanisms that allow plants to avoid or tolerate water stress by maintaining an active vegetative state (Table 3.1). Drought avoidance is based on maintaining appropriate hydration of tissues and cells under conditions of water deficiency. Species following this strategy can either reduce water loss by closing stomata early or increase water absorption (e.g. deep root systems). The drought tolerance strategy is promoted by physiological traits that permit continuous water transport, gas exchange or cell survival at low water content, and low water potentials (ψ), such as osmotic adjustment, increased structural reinforcement of the wood (Hacke et al. 2001), decreased vulnerability of xylem to embolism, and the ability of cells (especially meristems) to remain alive.

The third type of resistance would be to avoid or escape stress conditions (*Drought escape*). Many authors do not consider this strategy to be a true resistance mechanism since it occurs in plants that complete their life cycle before the stress affecting them. However, this group also includes vegetative species that exhibit partial dormancy during the summer, such as summer deciduous or semideciduous species (e.g., *Periploca sp, Withania sp*) or some rhizomatous grasses (*Stipa sp., Lygeum sp., Brachypodium sp.*) (Valladares et al. 2008). Despite the different mechanisms, there is no optimal strategy for water-limited ecosystems and plants develop different mechanisms according to their functional characteristics (Levitt 1980).

Table 3.1 Drought resistance strategies and main mechanisms involved (elaborated from Levitt 1980; Kozlowski et al. 1991; Jones 1992; Larcher 1995; Valladares et al. 2008)

Resistance strategy		Main mechanisms
Drought escape	Escape stress conditions	Vegetative dormancy Short life-span
Drought avoidance	Avoidance of stress conditions, isohydric response	Stomatal regulation Extensive root systems High capacity for water transport from roots to leaves. Small leaves, high leaf mass area ratio
Drought tolerance	Tolerance to decrease water potential, anisohydric response	Osmotic adjustment Cell-wall properties Antioxidant systems Compatible solutes High resistance to xylem cavitation

3.2 Hydraulic Architecture and Water Transport from Roots to Leaves

Since plants began to colonize the land, they have had to deal with the challenge of uptaking water and transporting it to distal tissues against gravitational forces. Although water transport is fundamental to the life of terrestrial plants, only some of the basic questions about the functionality of the transporting conduits have been extensively studied, while others still remain unclear (Zwieniecki and Holbrook 2009; Brodersen et al. 2010; Nardini et al. 2011).

3.2.1 The Ascent of Water

Water flow from the soil to the atmosphere is analyzed as a process based in the Soil–Plant–Atmosphere continuum, termed SPAC (Huber 1928; Van den Honert 1948) and the tension–cohesion mechanism performs the transport of water through the plant (Dixon and Joly 1895). This theory of water ascent in plants, also called cohesion–tension theory (C–T), was initially explained by Böhm (1893), followed by Dixon and Joly (1895). The C–T theory postulates that the cohesion of water molecules and the tensions generated in the water column are essential to the ascent of water. The rise of xylem sap during transpiration is due to capillary menisci of cell walls being evaporatively depleted by solar heating (Pickard and Melcher 2005). This theory constitutes a physical and analytical explanation of xylem transport in vascular plants (see Steudle 2001; Tyree and Zimmermann 2002).

According to the SPAC concept, both the cohesion of water molecules and the rigidity of xylem conduits allow a force exerted at one end of the conduits to be transmitted throughout the column to the other end. The C–T theory has been questioned several times because direct observations of the water potential of conduits, made with pressure probes (Steudle 1995), do not match the potential measured with a Scholander pressure chamber (Zimmermann et al. 1994; Canny 1995 and 2001) or other methodological approaches (Milburn 1996; Salleo et al. 1996; Wei et al. 1999; Cochard et al. 2000a; Kikuta et al. 2003). This led Canny to propose an alternate theory, "the compensating pressure theory" (Canny 1995). However, other studies have provided evidence to the contrary (Tyree 1997; Steudle 2001). This controversy has elapsed for several years (Stiller and Sperry 1999; Canny 2001; Zimmermann et al. 2004; Angeles et al. 2004). Other theories like the "root pressure" or "the capillary ascension" theories may explain the rise of water in small plants or herbs, but they are limited for explaining the ability of water transport to the top of the tall trees which is achieved by the mechanism explained by C–T theory (Steudle 1995).

The xylem is the tissue specialized in the transport of water from the roots to all parts of the plant. The transport of water through the xylem is produced by two factors: driving force and conductance—or its inverse, resistance (Boyer 1985). The driving force originates in the evaporation in the substomata cavities of the leaves. This evaporation allows the flow of water upward from the roots because it produces a negative hydrostatic pressure (suction tension), which is measured as water potential (Scholander et al. 1965). The other factor, conductance, is inversely proportional to the within-plant resistance to water transport in the xylem. This resistance is due to the friction of water molecules against conduit walls and the resistance of some other structures as perforation plates that connect the xylem conduits. As a result of these two factors (hydrostatic pressure and conductance), the water flow through the xylem conforms to Darcy's Law ($F = kh * \Delta\psi$), where water flow (F, kg s^{-1}) is a function of the hydraulic conductance (kh, kg s^{-1} MPa^{-1}) and the water potential gradient between the both extremes of the conduit ($\Delta\psi$ MPa).

Moreover, other theories analogous to Darcy's law have been used to describe water flow in plants, e.g., Ohm's law. In this case, the flow of water in the plant is considered to be a catenary process, where each catena component is viewed as a hydraulic conductance (analogous to an electrical conductance) across which the water flows (analogous to electric circuit; Tyree and Ewers 1991). In the same sense, Hagen–Poiseuille's law has also been applied to analyze the transport of water through xylem conduits, where $Kh = (\pi\rho/128\eta)\,\Sigma^n\,(d^4)$, assimilating the conducting elements to a capillary (Tyree and Ewers 1991; Tyree and Zimmermann 2002). This law states that hydraulic conductivity per unit pressure gradient (Kh) is proportional to the fourth power of the radius of the conducting element, where ρ is the fluid density, η is the coefficient of viscosity, d and n are, respectively, the diameter and the number of the conducting elements. It is worthy to note that pit membrane resistance connecting elements is neglected when referring to Poiseuille's law. At cell level, the passive movement of water is described by means of Newton's law ($v = (1/f)F$) of motion on the Earth where there is friction (Nobel 1999); where v is velocity of movement (m.s^{-1}), F is the force causing the movement (N), and f is the coefficient of friction (N s.m^{-1}). Water movement occurs when it moves from one point where it has high energy to another where energy is low.

In the plant, the loss of water vapor at leaf level is controlled by the opening and closing movement of the stomata. This mechanism determines the gas exchange in the leaf and the evaporative flux density (E). This flow from leaves is governed by Fick's law of gas diffusion in the air ($E = g_L\ (X_i{-}X_O)$; Tyree and Zimmermann 2002), where g_L is the diffusional conductance of the leaf (controlled by stomatal conductance, g_s), X_i is the mole fraction of water vapor at the evaporative surface of the palisade and mesophyll cells, and X_0 is the mole fraction of water vapor in the ambient air surrounding the leaf.

In summary, the hydraulic architecture of a plant can be regarded as a typical hydraulic system composed of: energy transfer (solar radiation on the leaves), a driving force (transpiration flux), pipes or conducting elements (vascular system), reservoirs (living cells), and regulating systems (essentially, the stomata), whose operation is based on the SPAC concept (see Tyree and Zimmermann 2002; Sperry 1995; Hoolbrook and Zwieniecki 2005).

3.2.2 Hydraulic Architecture and Water Flow Through the Plant

The most remarkable aspects of sap ascent are, on the one hand, that water moves through the xylem under tension (negative pressure) and, on the other hand, that all land plants potentially face the same fundamental tradeoff to obtain water efficiently and maintain hydraulic conductivity to protect photosynthesis, while running the risk of drought-induced cavitation. In the whole plant, long-distance transport of fluids occurs mainly through vascular tissues which do not show significant membrane barriers. Yet, living tissues can be the site of intense water flows during transpiration or expansion growth (Sperry 1995; Maseda and Fernandez 2006; Markesteijn et al. 2011). While water flows across leaves through the apoplastic, symplastic, and transmembrane pathways (Steudle and Peterson 1998; Maurel et al. 2008), the relative contributions of these pathways to hydraulic conductance are not well characterized in leaves, and also in roots (Voicu et al. 2009; Shatil-Cohen et al. 2011).

Because of the knowledge of hydraulic architecture has grown over the past years, it is now possible to carry out a better analysis of how hydraulic characteristics has allowed trees to cope with periods of drought in water limited ecosystems. One of the key features are plant segmentation and compartmentalization, i.e., any structural feature that confines xylem cavitation to small, distal, expendable organs in favor of larger and more important organs representing years

Fig. 3.1 Leaf and branch mortality in a cork oak tree (*Quercus suber*) after a severe summer drought in the Southeast of Portugal. Almost half of the tree was affected. (Photograph: Vilagrosa)

of growth and carbohydrate investment like trunk and main branches or root system (Tyree et al. 1993; Tyree and Zimmermann 2002; Hölttä et al. 2009). Furthermore, distal and expendable structures due to a large gradient of xylem water potential in shoots were found to be more vulnerable to cavitation in *Acer saccharum* and *Juglans regia* stems (Tyree et al. 1993; Tyree et al. 2003, Fig. 3.1). By analyzing the hydraulic architecture and loss of hydraulic conductivity in some seedlings of Mediterranean species under drought stress, hydraulic segmentation and compartmentalization of xylem cavitation and embolisms was found even in early stages of development. This segmentation matches leaf abscissions observed after severe drought conditions (Vilagrosa 2002; Vilagrosa et al. 2003, Fig. 3.2). This kind of plant segmentation is termed vulnerability segmentation. Nevertheless, under drought conditions where stomata remain closed all day long, transpiration will be very low and hydraulic segmentation will be ineffective because water potential will be nearly the same everywhere in the plant from the roots to the shoot tips (Tyree et al. 1993). Although the basic mechanisms of plant hydraulic architecture are fairly well understood, little is still known about how hydraulic properties are related to species' life-history strategies, such as drought and shade tolerance (Tyree and Ewers 1991; Markesteijn et al. 2011).

Plant species rely on different strategies to adapt to the water stress caused by drought spells. Among these strategies is the development of resilient xylem structures that support continuous water transport from roots to leaves (Hacke et al. 2001; Tyree and Zimmermann 2002; Maseda and Fernandez 2006), along

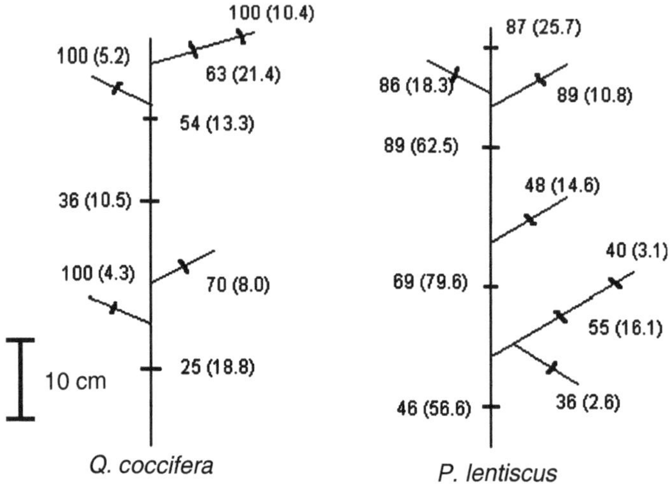

Fig. 3.2 Hydraulic diagrams for saplings of two Mediterranean species (kermes oak, *Quercus coccifera*, and mastic tree, *Pistacia lentiscus*) with percentage of cavitation (values outside brackets) and hydraulic conductivity (values inside brackets, LSC$\cdot 10^{-5}$, kg m^{-1} s^{-1} MPa^{-1}). Both species were in their early stages of development (3 years old and about 50 cm in height). In *Q. coccifera*, water potential was −7.0 MPa with an average of 75 % loss conductivity. For *P. lentiscus*, water potential was −5.9 MPa with an average of 65 % loss conductivity (Vilagrosa 2002)

with water channels at both ends of the water pathway. Indeed, Brodribb and Cochard (2009) demonstrated the role that stem and leaf hydraulics play in determining the drought tolerance of conifers, but they did not consider the role of aquaporins at both root and shoot level. Aquaporins are membrane channels that have a conserved structure and facilitate the transport of water and/or small neutral solutes (urea, boric acid, silicic acid) or gases (ammonia, carbon dioxide). Recently, (Cochard et al. 2007; Kaldenhoff et al. 2008) reported that aquaporins play a central role in plant water relations since they mediate the regulation of root water transport in response to a variety of environmental stimuli. Moreover, aquaporins are also shown to contribute significantly to water uptake by roots. For example, Javot and Maurel (2002) reported that mercury inhibits root hydraulic conductivity by 30–90 % in more than 10 plant species. Moreover, different approaches have established that aquaporins contribute significantly to the permeability of plant membranes in the leaves (Sack et al. 2004; Nardini et al. 2005; Cochard et al. 2007). Although, they facilitate water transport through internal leaf tissues during transpiration and in expanding tissues, their contribution to leaf water transport is still insufficiently understood (Maurel et al. 2008).

Impact of tree height on plant hydraulics remains a hanging question in this area of research (Spicer and Gartner 1998; Koch et al. 2004; Netting 2009). Spicer and Gartner (1998) studied mechanical properties of wood and they suggested that xylem formation represents a tradeoff between mechanical support and an efficient supply of water to leaves. In this sense, Spicer and Gartner (1998) analyzed the potential tradeoff between hydraulic and mechanical xylem functions by relating the branch hydraulic properties as consequence of changes in xylem anatomy due to the compression wood. Using branch segments of Douglas-fir (*Pseudotsuga menziesii*) they found that lower branch halves had significantly lower specific conductivity, higher specific gravity, lower water content and larger proportions of volume occupied by both cell wall and air than upper parts (halves) of the same branch. These differences in hydraulic properties between upper and lower halves suggested that compression wood, evidenced by the changes in the xylem anatomy, does reduce xylem transport efficiency (Spicer and Gartner 1998). However, other studies (Spicer and Gartner 2002) report that seedlings affected by compression wood maintained the same water potentials and stomatal conductance than controls seedlings. Finally, these authors conclude that reductions in stem specific conductivity, caused by compression wood, have a little impact on leaf-level processes when the entire soil–plant–atmosphere continuum is considered.

Later, some reports (Koch et al. 2004) indicate that tree height is limited by water transport constraints in taller trees which induce reductions in leaf net assimilation rate through stomatal regulation. Koch et al. (2004) observed that the tallest trees (i.e. higher than 120 m) are prone to an irreversible loss of hydraulic conductivity because of conduit embolisms when their xylem pressure decreases to values less than -1.9 MPa. They conclude that the maximum height is 122–130 m for *Sequoia sempervirens* originated from northern California. Moreover, they also observed an increase in stable carbon isotope composition ($\delta^{13}C$) and in leaf mass to area ratio for the higher parts of the trees as well as a decrease in leaf photosynthesis rates. Netting (2009) questioned these results because embolism-repair mechanisms may occur by -2.7 MPa and concludes that tree height may be limited by a long-term balance between dieback and sustained growth.

3.2.3 Changes in Hydraulic Architecture in Response to Different Processes

Previous sections defined hydraulic architecture in terms of the characteristics of the water-conducting structure of plants, such as design and hydraulic functions, and other related variables such as conductivity, porosity, and vulnerability (Tyree and Zimmermann 2002). Additional properties derived from these hydraulic variables, such as hydraulic segmentation and vulnerability segmentation, are also analyzed in hydraulic architecture studies. On the one hand, the aboveground shoot hydraulic architecture is associated with crown design: the vertical position of the branches within a crown affects both the morphologic and the physiological acclimation of the crown for light acquisition and hydraulic conductance (Yoshimura 2011). On the other hand, the root hydraulic architecture is also affected by the length, segmentation, and development of the root system among other variables. Considering that hydraulic architecture determines the capacity to supply all the plant parts with water, we will analyze the hydraulic architecture and its relationship to different plant processes such as regulation of transpiration, plant functional types, and how some abiotic conditions can change root hydraulic conductivity in this section.

3.2.3.1 Hydraulic Architecture, Water Availability, and Gas Exchange

Several studies related the hydraulic architecture of the xylem network to other plant processes like water availability or instantaneous gas exchange rates (Sperry 2000; Nardini and Salleo 2000; Hubbard et al. 2001). Changes in hydraulic architecture may affect water use efficiency (*WUE*) at leaf level and water potential gradient at the whole plant level (Brodribb and Holbrook 2003). Under conditions of high water demand, *P. lentiscus* seedlings with high root hydraulic conductance showed an increase in transpiration (*E*) without increasing the water potential gradient between soil and leaves (Hernández et al. 2009). In general, plants living in reduced water availability conditions produce changes in their xylem network such as increased hydraulic conductance to provide a better supply of water to leaves (Jacobsen et al. 2007a; Martínez-Vilalta et al. 2009).

Water transport is essential for carbon fixation and gas exchange in land-based plants. Coordination between hydraulic architecture and control of gas exchange at leaf level, expressed as transpiration (*E*), net CO_2 assimilation (*A*), and stomatal conductance (g_s), may also be related to the life-history features of a species. In this sense, the relationship between photosynthesis and leaf-specific hydraulic conductivity reflects a balance between carbon gain and water transport capacity. A study of 11 species in a preAlpine continental climate showed that shoot hydraulic conductance limited the maximum stomatal conductance (Nardini and Salleo 2000). These authors observed that species with lower maximum stomatal conductance had proportionally lower shoot hydraulic conductance than species with a higher water demand. Another study (Santiago et al. 2004) in a tropical area (annual rainfall 1800 mm) reported that the leaf-specific hydraulic conductivity (k_L) of upper branches was positively correlated with maximum rates of net CO_2 assimilation per unit leaf area (A_{area}) and stomatal conductance (gs) across 20 species of canopy trees. Moreover, maximum k_L showed a stronger relationship with A_{area} than with initial k_L suggesting that the assignment to photosynthetic capacity is proportional to maximum water transport capacity (Santiago et al. 2004). For seven co-occurring Mediterranean species (*Anthyllis cytisoides* L., *Genista scorpius* L. DC., *Myrtus communis* L., *Pistacia lentiscus* L., *Rosmarinus officinalis* L., *Spartium junceum* L. *and Ulex parviflorus* Pour.), Hernandez et al. (2010) found a positive correlation between leaf gas exchange variables and root hydraulic conductance scaled by stem cross-section area (K_{As}). These authors indicated that species with high specific root length (SRL) show high K_{As}. These patterns suggest that root system morphology determines both leaf gas exchange and root capacity for water transport. Adaptations in hydraulic design may differ considerably across species. Some of species' responses show a phylogenetic signal while others respond to functional syndromes or plant functional traits (Jacobsen et al. 2008). Indeed, Hernández et al. (2011) observed that Mediterranean species with different regenerative strategies after wildfires, i.e., seeder and resprouter species, showed contrasted patterns of hydraulic architecture and gas exchange, independent of phylogenetic affiliation patterns.

3.2.3.2 Effects of Nutrients and Light Availability

The general principle governing resource acquisition in response to stress is that plants allocate resources preferentially to functions that most strongly limit growth (Chapin 1991). Nitrogen, phosphorous and potassium fertilization significantly affects several components of hydraulic architecture, including cavitation resistance (Harvey and Van den Driessche 1997; Ewers et al. 2000; Trubat et al. 2006). Leaf-specific hydraulic conductance was reduced by 50 % in fertilized trees of *Pinus tadea* (in contrast to other treatments, including the control treatment), due to a higher leaf area per sapwood area and lower specific conductivity of plants and root xylem (Ewers et al. 2000). Moreover, specific conductivity of terminal stems increases with N supplies, whereas leaf-specific conductivity and wood density decreases (Bucci et al. 2006). Overall, fertilization treatments can change the root hydraulic conductance of seedlings, but the response is species-dependent. Hernandez et al. (2009) indicated that a high fertilization treatment increased the leaf area and root biomass of *P. lentiscus* seedlings, but reduced the values of root hydraulic conductance on a leaf area basis (K_{R-LA}). Seedlings cultivated under low fertilization showed values of K_{R-LA} around 34.0 ± 7.9 kg m^{-2} s^{-1} $MPa^{-1} \times 10^{-5}$ while under high fertilization values were around 9.2 ± 1.8 kg.m^{-2} s^{-1} $MPa^{-1} \times 10^{-5}$. Similar results to those previously reported for *P. lentiscus* were also observed for morphologic traits and root hydraulic conductance in *Pinus canariensis* (Luis et al. 2010). In this study, pine seedlings grown at full sunlight under high fertilization conditions showed lower K_{R-LA} than those treated with low fertilization doses (K_{R-LA} was 4.9 ± 1.0 and 1.3 ± 0.5 kg m^{-2} s^{-1} $MPa^{-1} \times 10^{-5}$ for low and high fertilization treatments, respectively). Hydraulic conductivity in roots of *Fraxinus pennsylvanica* was positively correlated with P concentration (Andersen et al. 1989), and an increased P supply reduced cavitation in the xylem of hybrid poplars as a result of conduit modifications (Harvey and Van den Driessche 1997).

Nutrient deficiency also affects plant morphology and root hydraulics (Ewers et al. 2000; Williamson et al. 2001, Bucci et al. 2006). In *P. lentiscus* (L.), Trubat et al. (2006) observed that nitrogen deficiency decreased aboveground biomass accumulation and leaf area, but did not affect below-ground biomass accumulation and root morphology. Phosphorus-deficient plants showed a decrease in leaf area and an increase in root length, root surface area, and *SRL*. Consequently, nitrogen and phosphorus deficiency reduced the root hydraulic conductance scaled by total root surface area (K_{R-RS}). K_{R-RS} in control seedlings was 1.3 ± 0.2 kg m^{-2} s^{-1} $MPa^{-1} \times 10^{-5}$

while in seedlings treated with deficiency of nitrogen and phosphorus was 0.8 ± 0.2 and 0.6 ± 0.0 kg m^{-2} s^{-1} MPa^{-1} × 10^{-5}, respectively. These authors also observed that root conductance per unit leaf area was not affected by treatments and concluded that low nutrient availability did not affect seedling capacity for maintaining water supply to leaves due to differences in root system development. Besides, Lovelock et al. (2006) found that plants grown on a P-deficient site showed lower leaf water potential, stomatal conductance and photosynthetic carbon-assimilation rates, and less conductive xylem, than plants on an N-limited site. These differences were markedly reduced when P fertilization was applied to the P-limited site. In contrast, fertilization with N on the N-limited site had little effect on either photosynthetic or hydraulic traits.

The concentration of cations in the xylem sap can alter the xylem water flow in angiosperm plants. Recent evidence of ion-mediated changes in pit membrane porosity suggests that plants should modulate hydraulic conductance (Cochard et al. 2010; Nardini et al. 2010) because the sap ionic composition has a strong effect on the xylem hydraulic conductance of a large number of species (Zwieniecki et al. 2001; Cochard et al. 2010). The current interpretation of this phenomenon is attributed to both the porous structure and the high pectin content of interconduits pit membranes (Nardini et al. 2007; Jansen et al. 2011) while other studies invoke the shrinking of some king of polyelectrolyte polymers as the result of electrostatic events (van Doorn et al. 2011). Herbette and Cochard (2010) brought up the regulating role of calcium in cavitation resistance of trees. For these authors, removing calcium from the stem-cell wall leads to an increase in xylem vulnerability to cavitation without impairing the xylem hydraulic conductance. As vulnerability to cavitation relied on pit membranes conduits, their structure made of hemicelluloses, and pectins should play a major role (Hafren et al. 2000; Willats et al. 2001).

Light availability is known to affect several morphological and functional plant traits (Valladares and Pearcy 1998). The effect of sunlight on hydraulic characteristics and vulnerability to cavitation was found to vary with light availability and environmental conditions. Low light availability altered the hydraulic design and produced higher vulnerability to cavitation in branches of *Fagus sylvatica* trees (Cochard et al. 1999; Barigah et al. 2006; Herbette et al. 2010). In contrast, in some Mediterranean species, shading treatments did not produce significant differences at root level e.g., changes in root hydraulic conductance in some Mediterranean species (Hernández et al. 2009; Luis et al. 2010). Light availability produce changes in stem hydraulic conductance (K_{stem}) which can be related with changes in the ionic concentration in the xylem sap (Nardini et al. 2010). Field measurements on *Laurus nobilis* revealed that sun-exposed branches had 60 % higher K_{stem} with respect to shade-exposed branches. Nardini et al. (2010) linked these results to higher xylem sap potassium concentration in the sun-exposed branches. The authors indicated that this ionic effect (high potassium concentration) is apparently a mechanism of plants to decrease the hydraulic resistance in the illuminated branches, thus diverting water flows toward sun-exposed foliage, preventing leaf dehydration, and favoring stomatal aperture and high rates of photosynthesis.

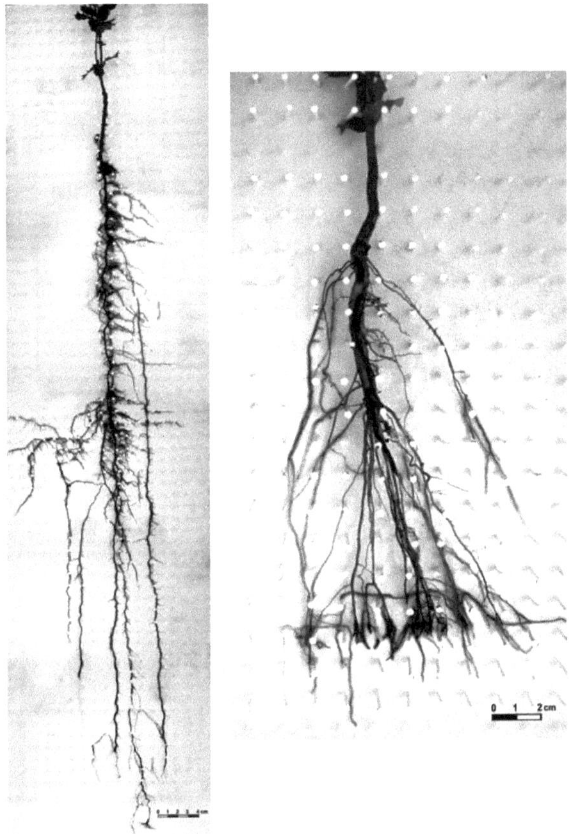

Fig. 3.3 Morphologic variability in the root systems of *Quercus ilex* L. of the same age subjected to different rooting environments. *Left*, seedling grown in a 100 cm deep container; *right*, seedling grown in a 20 cm deep container. (Photographs: Pemán)

3.2.3.3 Effects of Root Development

A plant's capacity to transport water from soil to leaves is highly dependent on its root architecture. Morphologic root features vary among species and are significantly correlated with root hydraulic conductance and leaf gas exchange variables, such as transpiration, net CO_2 assimilation and stomatal conductance (Addington et al. 2006; Peman et al. 2006; Chirino et al. 2008). A study carried out by Chirino et al. (2008) showed that the nursery cultivation of *Q. suber* seedlings in deep containers CCL-30 (depth: 30 cm, diameter: 5 cm; volume: 589 cm^3) improves the morpho-functional attributes related to root growth and functionality as compared to seedlings cultivated in standard containers. Similar results were reported by Pemán et al. (2006) who compared the functionality of root systems differing in size and shape (Fig. 3.3). The use of deep containers led to produce seedlings with a long tap root able to reach deeper soil horizons by means of higher growth in biomass and number of new roots. In this sense, deep containers

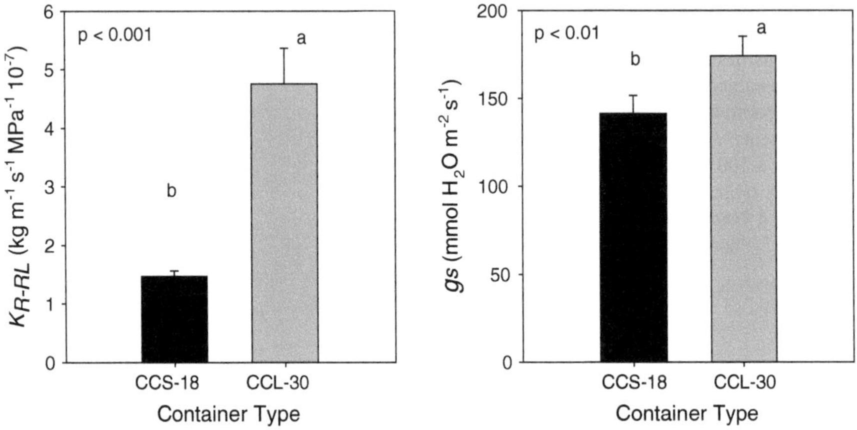

Fig. 3.4 Root hydraulic conductance per root unit length (K_{R-RL}) and maximum stomatal conductance (gs) in seedlings cultivated in standard containers (*black bars*, CCS-18; depth: 18 cm, diameter: 5 cm; volume: 300 cm^3) and in deep containers (*gray bars*, CCL-30; depth: 30 cm, diameter: 5 cm; volume: 589 cm^3). Different letters indicate significant differences at $P < 0.05$

Fig. 3.5 Mean vessel diameter for seedlings of *Quercus coccifera* seedlings grown in both types of containers (the same as in Fig. 3.4). Differences between the two groups of seedlings were statistically significant (P < 0.05)

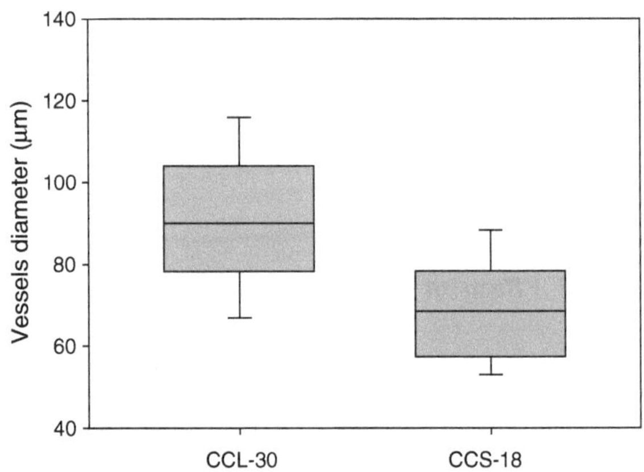

favored a high water transport capacity, expressed as root hydraulic conductance scaled by leaf area or root length (Fig. 3.4 Left) or also accounted for changes in vessel size (Fig. 3.5). These morpho-functional advantages contributed to a better water status under drought stress conditions as brought up by in higher stomatal conductance values (Chirino et al. 2009; Fig. 3.4 Right). Therefore, root characteristics as rooting depth, biomass, and xylem conduits can increased the plant hydraulic efficiency to supply water to leaves.

3.3 Xylem Cavitation and Embolism

As pointed out above, sap is transported under tension (i.e. negative pressure) according to the tension-cohesion theory. When the water tension inside the conduits is too negative, the breakage of the water column can occur, a phenomenon known as xylem cavitation (Tyree and Sperry 1989; Tyree and Zimmermann 2002). Embolism results from this phenomenon and rises in plants as a consequence of either by water stress (Tyree and Sperry 1989) or freeze–thaw cycles during winter (Sperry and Sullivan 1992). This phenomenon can cause a substantial reduction in xylem water transport and thus exerts a debilitating influence on plant water status by decreasing its hydraulic conductance. Thus, plants can no longer provide water properly to leaves and other tissues.

In relation to water stress induced cavitation, several mechanisms have been postulated for the formation or entry of air bubbles in the xylem conduits (see Tyree et al. 1994 for a review of all mechanisms), but the most commonly accepted mechanism is the "air seeding" postulated by Zimmermann and Crombie (Jarbeau et al. 1995; Salleo et al. 1996; Tyree and Zimmermann 2002). According to that mechanism, cavitation occurs when an air bubble outside a water-filled conduit is aspirated into the xylem element through pit pores as a consequence of the tension. In addition, Sperry and Saliendra (1994) showed that the vulnerability of the xylem to cavitation increased with enhancing the permeability of the pores after altering the surface tension of the xylem sap. The small size of the pores in the pit membranes prevents the air entry from an embolized xylem element to another functional conduit until the pressure gradient exceeds a certain threshold. This threshold depends on the radius of the pore. When the tension rises, the water potential becomes more negative, and suction from the air embolized xylem conduits occurs, creating a meniscus. If the meniscus radius equals the pore radius, air bubbles will be suctioned in. This is expressed by the formula:

$$\psi_x = -2\tau \cos a / r_m$$

where ψ_x is the difference between the critical pressure potential of xylem and air pressure to move the gas meniscus in the pore and thus match the radius of the meniscus (r_m) to the radius of the pores in the pit membrane. τ is the surface tension of water (0.072 Pa m at 25 °C) and "a" is the contact angle between the meniscus and the pore (Tyree and Sperry 1989).

Therefore, cavitation resistance depends on the size of the largest pores in the pit membranes in angiosperms. In gymnosperms, inter-tracheid pits are quite different in structure and function from inter-vessel angiosperm pits (Tyree and Zimmermann 2002). The mechanism of "air seeding" in gymnosperms would be somewhat different due to special characteristics of pits (Tyree and Zimmermann 2002). Pit membrane porosity and characteristics of torus and margo structures are involved in cavitation resistance (Sperry and Tyree 1990; Tyree and Zimmermann 2002). Three mechanisms have been proposed in conifers: (1) a rupture of pit margo microfibrils, (2) a displacement of the pit torus from its normal sealing position over the pit aperture, and (3) a rupture of an air–water menisci in a pore of the pit margo (Sperry and Tyree 1990; Cochard 2006; Pittermann et al. 2006; Cochard et al. 2009). The first studies pointed out that cavitation could then occur by elastic stretching of the margo or by rupture of the microfibril strands (Sperry and Tyree 1990). Pittermann et al. (2010) examined pit structure in 15 species of *Cupressaceae* exhibiting a broad range of cavitation resistances. These authors observed that cavitation resistance was most closely correlated to the ratio of the torus to pit aperture diameter and they suggested that impermeable torus has the capacity to exert a valve effect by sealing the pit aperture as consequence of xylem tensions during drought. Other studies suggested that cavitation in conifers could also be provoked by the capillary failure of an air–water meniscus located between the torus and the pit wall aperture because of they found a linear relation between sample vulnerability to cavitation and fluid surface tension (Cochard 2006, 2009). This mechanism could rather be similar to the mechanism proposed for angiosperms. Recently, Delzon et al. (2010) observed in 40 conifer tree species that high cavitation resistance exhibited both a high flexibility of the margo and a large overlap between the torus and the pit aperture, allowing the torus to tightly seal the pit aperture. These results support the hypothesis of seal capillary-seeding as the most likely mode of air-seeding, and suggest that the adhesion of the torus to the pit border may be the main determinant of cavitation resistance in conifers (Delzon et al. 2010).

3.3.1 Methodologies for Measuring Xylem Cavitation and Embolism

One of the oldest methods for detecting cavitation is the acoustic registration of the cavitation event, both in the audible (Low-Frequency) range (Milburn and Johnson 1966; Crombie et al. 1985) and in the ultrasonic range (Tyree et al. 1984). The rapid relaxation of the tension in the liquid (xylem water) or in the solid components of the conductive tissues would explain the audio and ultrasonic emission respectively (Tyree and Sperry 1989; Lo Gullo and Salleo 1991). This method has been used in many studies for recording cavitation events in stems (Tyree and Sperry 1989; Mayr and Sperry 2010) or leaves (Salleo et al. 2001; Trifilo et al. 2003). A critical analysis of the advantages and disadvantages of registering acoustic emissions as a technique for detecting cavitation and interpreting its effects in plant hydraulics can be found in Jackson and Grace (1996). As a main conclusion, these authors consider the method useful for defining water potential threshold values for cavitation in order to explain differences in drought tolerance among individuals or species. They also considered that it has only a limited use while calculating the amount of hydraulic conductivity loss registered at the different water potential values.

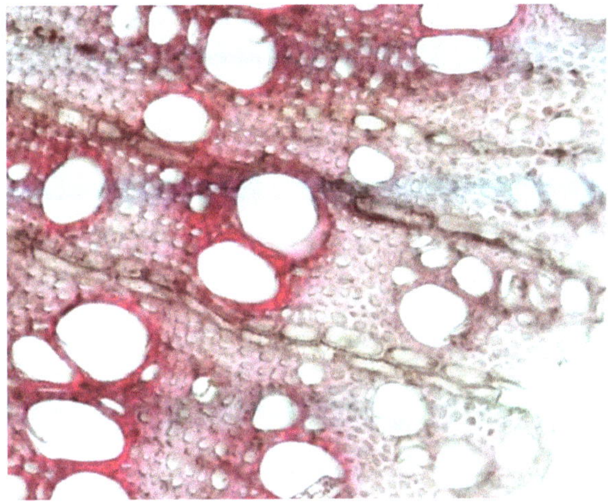

Fig. 3.6 Transverse section of *Pistacia lentiscus* stem stained with basic fuchsine. Functional vessels were stained in red while non-functional vessels remained unstained (*right*). Photograph: Vilagrosa

Anatomic methods have also been useful for the study of xylem cavitation. The oldest procedure is based on the perfusion of a sample at low pressure with a dye like safranin, basic fuchsin, alcian blue, phloxine B (Lo Gullo and Salleo 1991; Peguero-Pina et al. 2011a). This method implies a microscopic observation of a previously perfused segment for an accurate distinction between embolized and non-embolized conduits (Corcuera et al. 2006). This allows visualizing the water conducting cross-sectional area of the sample (Fig. 3.6). The use of cryo-scanning electron microscopy enables highly accurate distinctions between water- and air-filled conduits (Cochard et al. 2000) or tracheids (Mayr and Cochard 2003; Cochard et al. 2004). This method brings up directly the presence of water in the conduits is directly confirmed. However, the relative complexity of the sampling procedure and the cost of the required equipment represent a clearly a limitation for its extensive use.

The ‘hydraulic’ method relies on a conductivity apparatus devised by Sperry et al. (1988). It constitutes the main reference for quantifying xylem embolism and the method consists in estimating the hydraulic conductance of a stem segment before and after successive perfusions under pressure with degassed and filtered (0.10–0.22 μm) water solution to measure the initial conductivity, assuming that the possible embolism in open vessels or tracheids has not been displaced. The perfusion evacuates or dissolves air contained in the embolized xylem conduits. The initial to final conductivity ratio gives a quantitative value of embolism level i.e., the related percent loss conductivity (PLC). The percentage of initial hydraulic conductivity related to the maximum one is an estimation of the amount of embolism.

The apparatus allows to assess the hydraulic conductivity in stem (Lo Gullo and Salleo 1991), root (Alder et al. 1996) and petiole segments (Cochard et al. 2000a). This technique (Fig. 3.7) is now widespread but remains laborious and delicate since it requires the use of a precision balance (it is unusable in the field) and the focused plant material requires being connected to a water-filled tubing system to

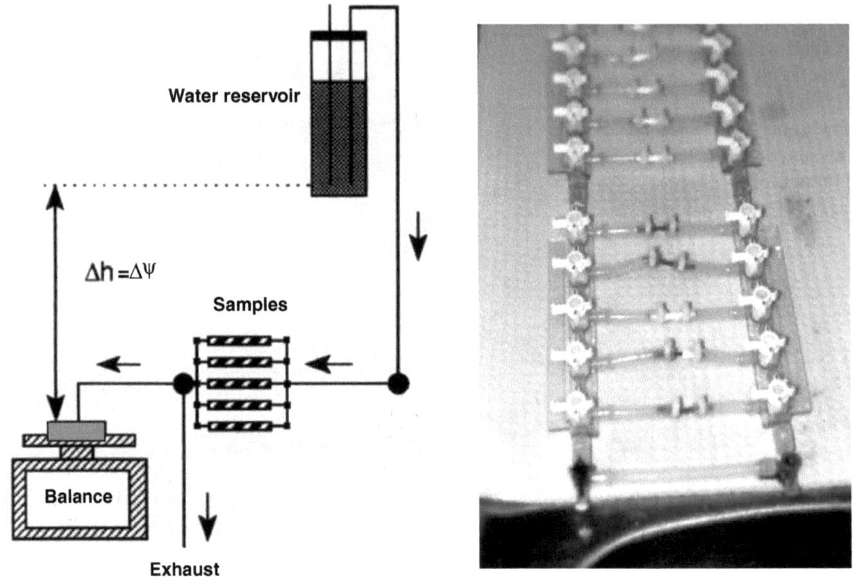

Fig. 3.7 Hydraulic method for embolism determination as designed by Sperry et al. (1988). *Left*, schematic representation of the hydraulic system. The water is allowed to flow at low pressure from the water reservoir to the balance through a water-filled tubing system and each plant material. The pressure gradient is determined by the difference in height between the balance and the water reservoir. After the initial measurement, the whole hydraulic system is flushed at high pressure to remove any embolisms. *Right*, a photograph of the plant material (stem segments) connected to the tubing system through which water flows from the reservoir to the balance

be perfused at low pressure (up to 10 kPa) and high pressure (100–175 kPa). Recently, an improved version of the Sperry's apparatus (Xyl'EM, *xyl*em *e*mbolism *m*eter) was designed (Cochard et al. 2000b). It measures the water flows with a high precision liquid flowmeter (Liquiflow, Instrutec, France) instead of an accurate balance. Thus, easy and fast measurements can be performed in the laboratory as well as in field with the Xyl'EM. Another variation of the method proposed by Sperry et al. (1988) can be found in Spicer and Gartner (1998), Mayr and Cochard (2003). In this procedure, a 'Micro-Sperry apparatus' was used, which allows the perfusion of small areas of the xylem by using steel cannulas inserted in the wood of a branch segment.

Different solutions have been used to prevent the microbial proliferation that induces a long-term decline in conductivity due to the clogging of the conduits (Sperry et al. 1988). Among these are oxalic (Sperry et al. 1988; Spicer and Gartner 1998), citric (Jarbeau et al. 1995) or hydrochloric acid (Kolb et al. 1996; Vilagrosa et al. 2003), KCl solutions (Lo Gullo and Salleo 1991) and, more recently, commercial silver biocides (Mayr and Cochard 2003; Peguero-Pina et al. 2011b). The low flows through the stem have to be registered as a mass flow with a digital balance (LoGullo and Salleo 1991; Vilagrosa et al. 2003) or a high precision liquid mass flowmeter (Cochard et al. 2000b). In more complex structures, the methods for measuring embolism have to be adapted to overcome the difficulties associated with the features of the sample, although the basic principle—perfusion before and after embolism dissolution with carefully prepared solutions—is maintained. Kolb et al. (1996) designed a setup for branched systems, such as whole shoots and roots. For this purpose, the shoot and most root tips are previously cut off before being enclosed in a vacuum chamber. The use of vacuum allows the establishment of pressure differences, inducing a flow of the solution through the complex system while minimizing the risk of refilling the potentially embolized conduits. The hydraulic conductance is then determined from the slope of the linear relationship between the flow rate and the different vacuum pressures applied.

The hydraulic method is the most widely developed technique for measuring the different degrees of water stress in plant, shoot, or root segments. Initially, to reach different levels of water stress, dehydration was applied by means of drying out the samples. The water potential of samples was determined by psychrometric or pressure chamber techniques. Later, negative water potentials were reached by applying plants organs to positive potentials in a Scholander chamber (Cochard et al. 1992). The reduction in hydraulic conductivity in relation to the applied pressure showed similar results to those found with the dehydration method (Cochard et al. 1992; Sperry and Saliendra 1994). From these results, the air injection methodology was developed (Sperry and Saliendra 1994). In this method, the branch or root segment is introduced into a pressure chamber with both ends open. This allows access to both ends of the branch to be measured. The hydraulic conductivity is measured with the segments exposed to increases in pressure within the chamber.

Nowadays, the centrifugation method is one of the most used which generates negative water potentials in shoot and root segments (Alder et al. 1996). The main advantage of this method is its ability to handle more samples per unit of time, while the drawbacks are the requirements of an ultracentrifuge and a customized rotor. Currently, this method has undergone several transformations in order to make determinations in shorter periods of time (see Alder et al. 1996; Cochard et al. 2005; Li et al. 2008 for detailed information), including ultracentrifuges in which the hydraulic conductivity of stem segments is directly measured (Cavitron method; Cochard et al. 2005). However, certain inconsistencies in the results, especially in species with long vessels, suggest the use of caution while determining vulnerability curve (VC)s to cavitation with these methods (Cochard et al. 2010; Choat et al. 2010; Sperry et al. 2011).

The xylem vulnerability to drought stress-induced cavitation of a tree species is assessed by relating cavitation events to water potential (ψ). It typically consists in plotting PLC values versus ψ (see Fig. 3.10a for examples of vulnerability curves in some *Quercus* species). The resulting curve is a vulnerability curve . Loss hydraulic conductivity is expressed as percent of maximum stem-segment conductivity. PLC is determined by collecting stem segments previously cut under water and measuring the flow of water through the segment for a given low

Fig. 3.8 Bubbles coming out from a cavitated stem segment when it is injected with degassed water at high pressure (0.1 MPa). Photograph: Vilagrosa

pressure (K_i). After this first measure, stem segments are perfused with water to high pressure (typically, at 0.1 MPa) to remove air from any embolized conduit (Fig. 3.8). Then, the water flow is determined again (K_{max}). PLC is calculated as $100(1 - K_i/K_{max})$ (Sperry and Tyree 1990).

3.3.2 Xylem Cavitation in the Context of Plant Functionality: Leaf, Stem, and Root Vulnerability

The concept of (SPAC) allows us to study the flow of water through plants from the source (soil) to the final sink (atmosphere). Among all the well-defined resistances to water flow (see Black 1979), stomata have traditionally been regarded as the main variable in the regulation of the flow from the plant leaf (Weatherley 1976; Sack and Holbrook 2006). The study of whole plant hydraulic has shown that: (1) stomata are not the only way to regulate the water movement through the SPAC (Luu and Maurel 2005) and (2) the break in the continuum due to embolism in the xylem should induce a shift in the whole resistance to water flow throughout the plant which can result in a critical dysfunction of the vascular transport pathway (Tyree and Cochard 1996; Tsuda and Tyree 2000). The paramount importance of maintaining the values of the different resistances to the movement of water through the xylem has been the theme of many papers since the first decades of the twentieth century (Tyree and Sperry 1989).

The movement through the xylem starts as soon as water has reached the root stele and constitutes one of the most complex mechanisms in the physiology of plant water relations. Effectively, the complexity of the water flow through the root is a reflection of the inherent anatomic complexity of this plant organ (Steudle and Peterson 1998). This affirmation can be better appreciated if an upper scale, the architecture of the entire root system, is also computed as an essential feature contributing to the complexity (Peman et al. 2006). In spite of this complexity, significant progress has been made toward understanding and modeling the biophysics and biochemistry of water movement in the roots, both across the non-vascular pathway, i.e., radially from the root surface to the xylem, and across the vascular pathway, i.e., through the xylem toward the demanding areas of the plant (Tyree et al. 1994; North et al. 2004).

The relative importance of root xylem cavitation and embolism in the physiology of the plant is mediated by the contribution, in terms of resistance to the whole plant water, of the non-vascular and vascular pathways, respectively (Huang and Nobel 1993). Frensch and Steudle (1989) proposed that the flow through the non-vascular pathway in small roots is the main resistance element in the entire process. Different studies have shown that this resistance increases when plants are subjected to drought stress, as a consequence of deep anatomic changes (North and Nobel 1992) and/or biochemical, aquaporin-mediated effects (North et al. 2004). Lo Gullo et al. (1998) showed how severe drought stress caused anatomic changes in roots of *Olea oleaster*, which formed a two-layered exodermis with thicker suberized walls, and a three- to four-layered endodermis with completely suberized tangential walls. In terms of seasonal water relations in Mediterranean plants, such an effect implies that the plant, once its root system is affected, has to recover from these changes by resuming the growth of root tips and producing new lateral roots. The need for reinvestment in the root system after a severe drought should be taken into consideration when analyzing the recovery of physiological activity in Mediterranean woody plants after the summer period (Corcuera et al. 2005).

In spite of the above considerations, the study of vascular pathway resistance in root hydraulics and how it increases as a result of xylem cavitation has been the subject of many papers due to the great importance of this process in the whole plant physiological response to drought. Root xylem properties generally differ from stem ones (McElrone et al. 2004), a fact that has led many researchers to consider that the former may be more tightly linked to the control of the entire water movement in the SPAC (Jackson et al. 2000). Among other differences between root and stem xylem features, roots are generally more vulnerable to xylem cavitation than stems (Sperry and Saliendra 1994, Alder et al. 1996, Hacke and Sauter 1996; Jackson et al. 2000). These results suggest that they may be the weakest link along the soil-to-leaf water transport pathway (Martínez-Vilalta et al. 2002; McElrone et al. 2004, Fig. 3.9). Therefore, incorporating information on the hydraulic properties of roots in addition to that of stems in comparative studies may help identify adaptive relationships between xylem function and other physiological traits (Maherali et al. 2006).

Since the first papers on xylem cavitation (Tyree and Sperry 1989) , the studies of the integrity of stem conductivity at different water stress levels have served as a way of explaining niche segregation among closely related (Tyree and Cochard 1996; Hao et al. 2008) or congeneric species (Tognetti et al. 1998; Vilagrosa et al. 2003). Figure 3.10 shows the different performances in terms of vulnerability to water stress induced cavitation among different *Quercus* species covering a wide

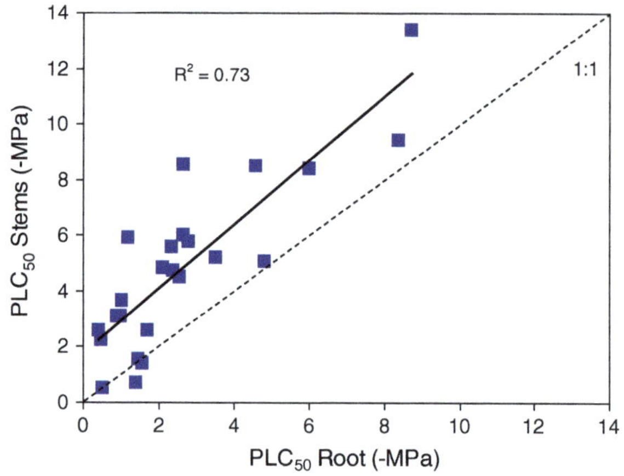

Fig. 3.9 Comparison of the water potential at PLC_{50} (50 % loss of hydraulic conductivity) between roots and shoots for 24 species. Methods for induced cavitation were dehydration, air injection, and centrifugation. (From: Martínez-Vilalta et al. 2002; Oliveras et al. 2003; Stout and Sala 2003, McElrone et al. 2004; Pratt et al. 2007; Froux et al. 2005; Hukin et al. 2005; Domec et al. 2006)

range of habitats—from mesic to xeric—in Europe. The different thresholds derived from the establishment of cavitation curves have been used for determining the distribution of plant species within (Maherali et al. 2006) and among vegetation types (Maherali et al. 2004; Bhaskar et al. 2007). Moreover, the collapse of the hydraulic conductivity of the trunk or branches has been associated with some processes of forest decline (Corcuera et al. 2006; Hoffmann et al. 2011, Peguero-Pina et al. 2011a) or with the lack of seedling recruitment in Mediterranean areas (Esteso-Martínez et al. 2006). These examples affirm the outstanding contribution of vulnerability to drought induced cavitation in woody plants in the ecology of ecosystems.

Different papers (Tyree et al. 1993; Nardini and Salleo 2000; Barigah et al. 2006) have shown that leaf hydraulic resistance constitutes the substantial component of the whole shoot or plant hydraulic. However, how much the vascular pathway contributes to this resistance should be computed before discussing the importance of leaf xylem cavitation in the SPAC dynamics. For Nardini et al. (2001), the main hydraulic restriction in the leaf lies in the non-vascular part of the pathway. The complexity of this mechanism is as high as in roots, in as far as many processes are involved (Sack et al. 2004; Sack and Holbrook 2006), including changes in aquaporin activity (Martre et al. 2002). The partitioning of the hydraulic resistance of the whole leaf (R_{leaf}) into the vascular resistance (R_{xylem}) and the non-vascular resistance ($R_{outside\ xylem}$) has been a matter of study with quite controversial results. Sack and Holbrook (2006) offered a good revision of this matter, concluding that R_{xylem} and $R_{outside\ xylem}$ are quite similar in magnitude, although some important variation in partitioning has been found when species from different habitats have been compared.

Changes in whole leaf hydraulic resistance due to cavitation in the complex xylem network of the leaf are well documented (Sack et al. 2004), and include discussion on its importance as an ecophysiological process and explanations of biogeographic trends in leaf size (Scoffoni et al. 2011). This type of study encompasses cavitation in the petiole (e.g., Cochard et al. 2000), midrib (Nardini

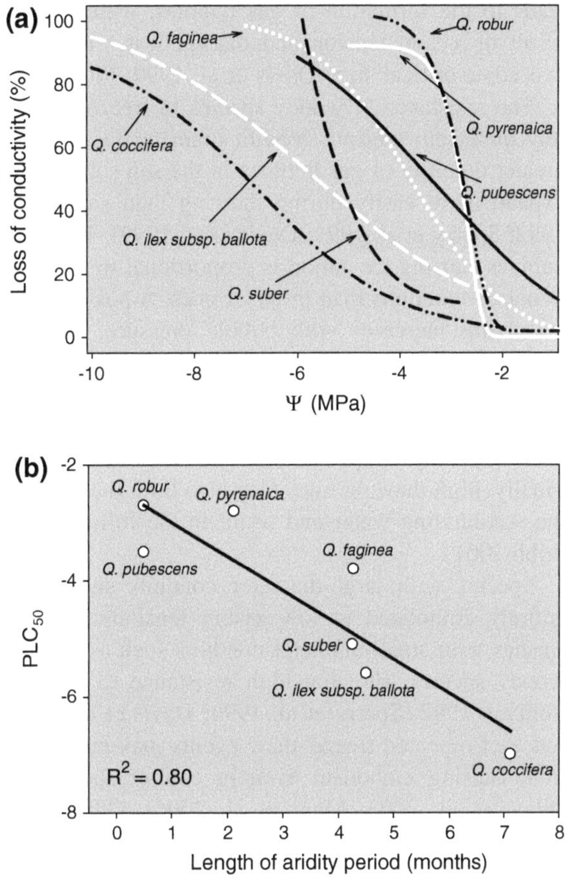

Fig. 3.10 **a** Vulnerability curves to drought-induced cavitation in several European *Quercus* species from different habitats. In this figure, Loss of conductivity or PLC is plotted as function of water potential (ψ). Redrawn from Tognetti et al. (1998) for *Q. pubescens*, Tyree and Cochard (1996) for *Q. robur* and *Q. suber*, Vilagrosa et al. (2003) for *Q. coccifera*, Corcuera et al. (2005) for *Q. ilex* subsp. *ballota*, Esteso-Martínez et al. (2006) for *Q. faginea* and Corcuera et al. (2006) for *Q. pyrenaica*. **b** Relationships between the length of the aridity period, i.e., the time-span during which the curve of monthly mean temperature values lies above the monthly precipitation curve in an ombro-thermic Gaussen-type graph, and the water potential at 50 % loss of hydraulic conductivity (PLC_{50})

et al. 2001) and veins (Salleo et al. 2003). Nevertheless, the quantitative importance of cavitation in the vascular pathway of the whole leaf hydraulics is still under discussion (Trifilo et al. 2003). Effectively, Zufferey et al. (2011) conclude that cavitation in the petiole in grapevines can serve as a "hydraulic fuse", preserving the integrity of other organs by limiting transpiration during water stress.

3.3.3 *Other Factors Affecting Xylem Cavitation: Low Temperature and Pathogens*

Low temperatures are a major factor limiting the transport of water in the xylem (Sperry et al. 1994; Cochard et al. 2000; Feild and Brodribb 2001). The occurrence of freeze–thaw events can cause long-term effects in the hydraulic functions of the tree by inducing embolism (Mayr et al. 2006). Freezing of the conducting elements leads to the formation of gas bubbles, which may expand during thawing as a result of xylem tension and disrupt water transport when the bubble diameter exceeds a critical size (Davis et al. 1999; Pittermann and Sperry 2003).

The resistance of woody species to freeze–thaw embolism has been shown to rely on xylem tension, conduit diameter i.e., the larger the conduit volume, the greater quantity of gas bubbles in the sap solution, resulting in large bubbles that expand more easily during thawing than smaller bubbles (Sperry and Sullivan 1992; Sperry et al. 1994; Davis et al. 1999). This suggests that the diameter of the bubbles causing cavitation is proportional to the conduits diameter, being smaller in narrow conduits than in wider ones. A possible reason for this is that the rate of dissolving increases with bubble pressure, which is inversely proportional to bubble diameter (La Place's law; Pittermann and Sperry 2006). The degree of xylem embolism is also sensitive to the lowest temperature experienced during freezing (Pockman and Sperry 1997), although the effect of decreasing temperature may only be pronounced in some species (Pittermann and Sperry 2003). Finally, high thawing rates may also limit the rate that gas bubbles can dissolve in the surrounding water and result in the formation of embolisms (Feild and Brodribb 2001).

Species with large-diameter conduits such as ring-porous trees are almost entirely embolized at low xylem tensions after a freeze–thaw event whereas species with small-diameter conduits such as tracheid-bearing and diffuse-porous woody species, showing high resistance to freeze–thaw embolism (Sperry and Sullivan 1992; Sperry et al. 1994; Davis et al. 1999). In spite of this, the occurrence of repeated freeze–thaw events may reduce the resistance to xylem cavitation, causing embolism even in conifer species with small-diameter tracheids (Mayr et al. 2003; Mayr et al. 2006). Other factors that could cause further decreases of water potential—low soil temperatures (which limit water uptake by roots) and high radiation events during winter (which increase the loss of water by transpiration)—may induce embolism in a more vulnerable xylem due to the occurrence of repeated freeze–thaw cycles (Peguero-Pina et al. 2011a). Thus, the combined stress of low water potentials and a high number of freeze–thaw events is sufficient to induce embolism in conifers (Mayr et al. 2006).

Xylem cavitation can also be promoted by vascular diseases, which may reduce the hydraulic conductivity of the xylem by air seeding (Tyree and Sperry 1989). However, recent studies have shown that plant pathogens inhabiting the water flow under tension induce xylem cavitation through the clogging or obstruction of the xylem conduits. Thus, Martín et al. (2005) showed that the pathogenic fungi *Ophiostoma novo-ulmi* caused xylem vessel blocking and cavitation in *Ulmus minor*. In this way, McElrone et al. (2008) observed that the major effect of infection by the xylem-limited bacterial pathogen *Xylella fastidiosa* occurred due to reduced hydraulic conductance caused by clogging of the conduits, and not by increased cavitation of xylem elements. More recently, Raimondo et al. (2010) stated that the mitosporic fungus *Phoma tracheiphila* caused xylem impairment and led to leaf shedding and plant dieback in *Citrus aurantium*. Finally, it should be noted that the combined effect of pathogen and water stress induced reductions in hydraulic conductivity could promote extensive xylem dysfunction in plants subjected concurrently to both stresses (McElrone et al. 2008).

3.3.4 Functional Implications of Xylem Cavitation: Flow Limitations to Gas Exchange, Recovery, and Refilling

Each day, leaves of plants have to allow the evaporation of a large amount of water to uptake carbon dioxide in the stomatal cavities and also regulate their surface temperature. The machinery behind the process of water transport has to be remarkably efficient and reliable, and any dysfunction may impair tree hydration (Zwieniecki and Holbrook 1998; Salleo et al. 2000; Cochard 2006). Any drop in hydraulic conductance as a consequence of xylem cavitation leads to an impairment of many physiological processes (Nardini and Salleo 2000; Pratt et al. 2005; Jacobsen et al. 2007b). For trees, the occurrences of embolisms are commonplace as this phenomenon takes place even in well-watered plants (Milburn 1991; Brodribb and Holbrook 2004). A substantial loss of free water will result in desiccation or dehydration which increases sap tension and leads to runaway embolism (Tyree and Ewers 1991). Plants respond to drought stress by means of physiological, biochemical, and metabolic adjustments that take place throughout all plant organs (Nardini et al. 2011). For example, vascular plants in Mediterranean regions face a predictable and marked annual drought period (Jacobsen et al. 2007b) and cannot avoid water stress-induced cavitation entirely as long as they transport water under tension (Wheeler and Stroock 2008). These species should preserve xylem function if they are to maintain water transport throughout the dry and hot summer months when water shortage produces extremely negative water potentials in the xylem (Davis et al. 2002; Ackerly 2004). Not all plant species are able to withstand such negative xylem pressures and water stress-induced cavitation since embolism have been related to shoot dieback and whole-plant death (Davis et al. 2002, Martínez-Vilalta et al. 2002, Vilagrosa et al. 2003, Jacobsen et al. 2007b). According to Zwieniecki and Holbrook (1998), Domec et al. (2006) and Zufferey et al. (2011) some species can recover from their decreased hydraulic conductivity. To regain hydraulic conductivity that has been impaired by cavitation, some plants must refill their embolized conduits rapidly because of metabolic and gene expression processes which may occur within a few seconds, minutes, or hours. These processes depend on the species and genotype, the extent and severity of the water loss, the age and stage of development, the organ and cell type, and the subcellular compartment (for review, see Thapa et al. 2011). Furthermore, plant response depends on the nature of the water shortage, inducing physiological responses to short-term changes, acclimation to a certain amount of water availability and adaptation to drought (Levitt 1980; Larcher 1995).

Xylem recovery from embolism is necessary to maintain the transpiration of new leaves produced after drought resumption. Two recovery mechanisms seem to have evolved: building up new functional conduits and refilling of embolized ones through an active mechanism. The first mechanism is common to almost terrestrial plant species with secondary cambium for which the production of new conduits increases overall xylem functionality. Recovery from embolism is based solely on this mechanism in several tree species (Cochard et al. 2001). The second mechanism implies a refilling of air-filled conduits when the xylem water potential is near or above the atmospheric value (Tyree and Yang 1992; Tyree et al. 1999). However, recent observations indicate the occurrence of embolism removal even in the presence of negative pressures in the xylem (Holbrook and Zwieniecki 1999; Salleo et al. 2006) associated with osmosis rather than the mechanically expected potential in intact woody plants, a none-straightforward mechanism which seems to violate the laws of thermodynamics (Clearwater and Goldstein 2005; Tyree et al. 1999). Moreover, for the former authors, many of the major biophysical questions regarding refilling under tension remain unanswered.

Since understanding how embolism repair under tension occurs is critical for evaluating the costs and constraints of transporting water in a metastable state, Zwieniecki and Holbrook (2009) devoted a conceptual framework as a guideline for future studies to understand how xylem refilling under tension takes place. Recently, Brodersen et al. (2010) reported a breakthrough in xylem embolism repair under tension based on experimental observations and theoretical modeling. Using the potential of 3D high-resolution computed tomography for in vivo monitoring of embolism repair, they visualized water droplets apparently exiting from vasicentric cells and entering embolized vessels of *Vitis vinifera*. The droplets expand over time and refill vessels while forcing the dissolution of entrapped gases at substantially negative stem water potentials ranging between −0.45 and −0.75 MPa (see Nardini et al. 2011 for review). Yet, embolism repair under negative pressure continues to be an unanswered question since it is restricted to a few number of species investigated.

3.3.5 Cavitation Plasticity: Variations Within the Tree, Populations, Ecotypes, or Cultivars

Cavitation resistance, at the interspecific level, has been widely documented across species and varies considerably (Maherali et al. 2004; Bréda et al. 2006). Comparisons of more or less distantly related taxa growing in contrasting environments suggest that increased cavitation resistance is a key component of species growing in drought-prone habitats (Tyree and Ewers 1991; Hacke et al. 2000; Pockman and Sperry 2000). To date, Machado and Tyree (1994) report similar levels of drought-induced cavitation in deciduous and evergreen plants, but higher levels in tropical plants than in temperate forest species. Even, reports are much scarcer at the intraspecific level, resistance to cavitation can vary significantly between populations of the same species (Alder et al. 1996; Herbette et al. 2010; Wortemann et al. 2011) or between closely related genotypes (Pammenter and Vander Willigen 1998; Sangsing et al. 2004; Cochard et al. 2007; Lamy et al. 2011), and between organs of the same plant (Tyree et al. 1993; Sperry and Saliendra 1994; Hacke and Sauter 1996; Cochard et al. 1999). Furthermore, Variation in resistance to drought-induced xylem cavitation has also been demonstrated among cultivars of the same species (Neufeld et al. 1992; Alsina et al. 2007).

Cochard et al. (1999) and Herbette et al. (2010) assessed the within-tree variation in vulnerability to cavitation due to light and seasonal effects. They found that light-exposed branches are less vulnerable than shaded ones. Furthermore, Lemoine et al. (2002) reported that the apical parts of beech trees experience both the lowest leaf water potential and the lowest vulnerability to cavitation. Moreover, in beech (*F. sylvatica* L.) the number of leaf primordia preformed in the buds determines the length and the type (long versus short) of annual growth units, and thus, branch growth and architecture (Cochard et al. 2005). Actually, different plant organs have different vulnerabilities to cavitation, with roots being the most susceptible and woody stems the least susceptible (Mencuccini and Comstock 1997). Accordingly, low water potentials are most likely to cause runaway cavitation in the roots but not in the other organs. For these authors, green twigs might be adapted to the favorable water conditions of the growing season, while woody stems would be prone to endure prolonged periods of drought stress. Within an organ, wider conduits are generally more susceptible to drought-induced cavitation, but across organs or species this trend is very weak (Hacke et al. 2000). This is probably because the cavitation threshold is not determined by conduit diameter but by the pore diameter in conduit walls (Tyree and Zimmermann 2002). Moreover, a positive relationship is found between conduit wall reinforcement and stem cavitation resistance (Hacke et al. 2001). Cavitation features at leaf scale has been little investigated. Nevertheless, there is some evidence that water-stress induced reductions in leaf hydraulic conductance (K_{leaf}) result from leaf xylem cavitation (Nardini et al. 2001; Johnson et al. 2009). Recently, Johnson et al. (2009) measured leaf hydraulic conductance concurrently with ultrasonic acoustic emission (UAE) in the dehydrating leaves of two vessel-bearing and two tracheid-bearing species to determine whether declining K_{leaf} was associated with an accumulation of cavitation events. In addition, they captured images of leaf internal structure with a cryo-scanning electron microscopy which enabled the detection of empty versus full as well as deformed conduits. They reported values of water potential inducing 50 % loss conductivity ranging from −1.5 to −2.4 MPa across their study species. Contrary to previous observations (Cochard et al. 2004), they did not report evidence of deformation or collapse in either *Pinus nigra* or *Pinus ponderosa* xylem tracheids (or vessels of the two angiosperm species) due to reductions in K_{leaf}.

At the population level, Mencuccini and Comstock (1997) showed differences in vulnerability to cavitation related to differences in the environment of origin among populations of the desert shrub *Ambrosia dumosa* (Gray) Payne when grown in a common garden. Differences in vulnerability to cavitation among populations were evident in *A. dumosa,* but not in *Hymenoclea salsola.* Several studies have analyzed for variations in vulnerability to cavitation among populations in forestry species (Tognetti et al. 1998; Martínez-Vilalta et al. 2009; Herbette et al. 2010; Corcuera et al. 2011). The results of these works are contrasted and reflect a high intraspecific phenotypic plasticity. Furthermore, Corcuera et al. (2011) found a good convergence between hydraulic traits, xylem embolism, and drought tolerance of the populations. Therefore, the adjustment of hydraulic properties at leaf and shoot level can play an important role in avoiding cavitation risks (Martínez-Vilalta et al. 2009).

At the cultivar level, few studies have focused on the variability within a single species found over a broad range of environmental conditions. However, Alsina et al. 2007 compared eight 10-year-old grapevine cultivars and found large differences between them based on their sensitivity to drought-induced cavitation. Franks et al. (1995) reported that seedlings of *Eucalyptus camadulensis* from drier environments in Australia exhibited higher resistances to drought-induced xylem cavitation compared with seedlings from wetter environments.

3.3.6 Coordination of Processes Between Cavitation and Cell and Leaf Function

The vulnerability of xylem to cavitation (i.e. apoplastic resistance) is a key parameter for understanding the role of hydraulic architecture in leaf and whole-plant function. This parameter has been associated with variations in any plant functional traits that controls cell and plant functionality (Vilagrosa et al. 2010). Specifically, photosystem II (PSII) functionality, photosynthetic pigment composition, cell membrane integrity, and protein stability play an important role in cell resistance to drought stress at symplastic level. Thus, under intense drought and temperature stresses, energy dissipation mechanisms could be promoted, mediated by changes in the de-epoxidation state of the xanthophyll cycle (Morales et al. 2006) and/or photoinhibition in the photosynthetic apparatus of water-stressed leaves (Demmig-Adams and Adams 2006). Moreover, many phospholipids of biological membranes undergo phase transitions and membrane fusions that are disruptive to membrane structure and function. For these reasons, cell membranes are also considered as one of the first targets of many stresses, and maintenance of their integrity and stability under water stress conditions is a major component of drought tolerance in plants (Earnshaw 1993).

The production of a xylem with a great resistance to cavitation, which demands high investments in construction costs (Pratt et al. 2007; McCulloh et al. 2011), would be accompanied by a high tolerance at leaf level to drought conditions by means of some adaptive processes, multigene expression or regulatory networks (Valliyodan and Nguyen 2006). Thus, the functional coordination between the resistance to xylem cavitation and the response of stomatal processes has been studied by several authors, establishing the concept of "safety margin" i.e., the water potential at stomatal closure and the onset of cavitation events (Tyree and Sperry 1988; Meinzer et al. 2009). Wide safety margins are common in plants living in environments with severe water limitations which are able to withstand severe drought periods with leaf water potentials lower than those inducing stomatal closure but much higher than those producing the onset of cavitation events (Vilagrosa et al. 2003; Peguero-Pina et al. 2008; 2009). In this way, Vilagrosa et al. (2010) showed another safety margin between symplast resistance at leaf level and xylem cavitation which would reflect a mechanism to maintain high structural and functional stability at leaf level. Thus, the capacity to maintain costly structures (cell membranes, metabolic and photosynthetic machinery, etc.) may be an important factor for survival and competitiveness in water-limited environments.

3.4 New Frontiers and Challenges with Xylem Cavitation as a Limiting Factor

3.4.1 Cavitation as the Limiting Factor for Plant Survival: Cavitation vs Starvation

Climatic models predict a reduction in the amount of precipitation and changes in rainfall seasonality in drylands (IPCC 2010), with important effects on drought severity and duration, and/or the frequency of heat waves (Hertig and Jacobeit 2008). This phenomenon, recently referred to as "global change-type droughts" (Breshears et al. 2009), may affect species survival, distribution, and productivity (McDowell et al. 2008, van Mantgem et al. 2009; Allen et al. 2010; Zeppel et al. 2011). Plant species must adapt to new climatic conditions or become locally extinct and, in this process, natural communities may undergo substantial changes in their composition and structure (Fig. 3.11). Increasing aridity as projected by climate change models could threaten the survival of not only young pine forests, but also mature stands in some areas. In fact, this occurs right now worldwide (Allen et al. 2010; Dietze and Moorcroft 2011).

As drought affects several physiological processes at plant and cell level, numerous studies have pointed out that xylem vulnerability to drought-induced cavitation plays a key role in drought resistance (Tyree and Sperry 1989). Indeed, vulnerability to cavitation differs widely among species (Pockman and Sperry 2000), and a low vulnerability leads to a higher drought tolerance (Tyree and Ewers 1991; Maherali et al. 2004). As already pointed out, xylem cavitation as a consequence of drought stress reduces hydraulic conductivity in the root to leaf pathway, preventing plants from supplying their leaves with water (Vilagrosa et al. 2003; Bréda et al. 2006). However, this effect can be species specific since different plant functional strategies can modulate the degree of stress suffered and the extent of plant damage (Miranda et al. 2010; Pratt et al. 2010; Quero et al. 2011).

Fig. 3.11 Drought impacts on Mediterranean macchia ecosystem, Eastern of Spain in summer 2003. Several species were affected by leaf shedding during the severe drought conditions of the summer (Photograph: Vilagrosa)

In this sense, Ogaya et al. (2003) reported a differential drought effects on several coexisting species representing different functional types (shrubs and trees) in a *Quercus ilex* forest in Eastern Spain. Within the same ecosystem, Hernández et al. (2009); Luis et al. (2010); Pratt et al. (2010) find that inter- and intraspecies differences in root system development, rooting strategy, and xylem hydraulic traits can determine the different levels of drought stress suffered by plants, regardless of the xylem resistance to cavitation. In a study of seedling response to reforestation in a Mediterranean ecosystem (Vilagrosa 2002; Vilagrosa et al. 2003), it was observed that species most resistant to xylem cavitation (e.g., *Q. coccifera*) consistently displayed the most negative water potential for similar conditions of water availability (Fig. 3.12a) , and this was consequence of the lowest root development and, therefore, the least access to soil water availability (Fig. 3.12b). In contrast, the species less resistant to cavitation, like *P. lentiscus,* developed a larger root system than *Q. coccifera* (Vilagrosa et al. 1997; Vilagrosa, 2002). This allowed *P. lentiscus* to get the ability to explore large and deep amount of soil and sustained hydration and less negative water potential (Fig. 3.12). In spite of its high cavitation resistance, *Q. coccifera* have always shown higher mortality rates than *P. lentiscus* within the same environmental conditions. However, Maherali et al. (2004), Pratt et al. (2010), and Meinzer et al. (2009) analyzed a wide range of functional groups and different ecosystems and pointed out that, in general, the lower the summer water potential the higher the resistance to xylem cavitation.

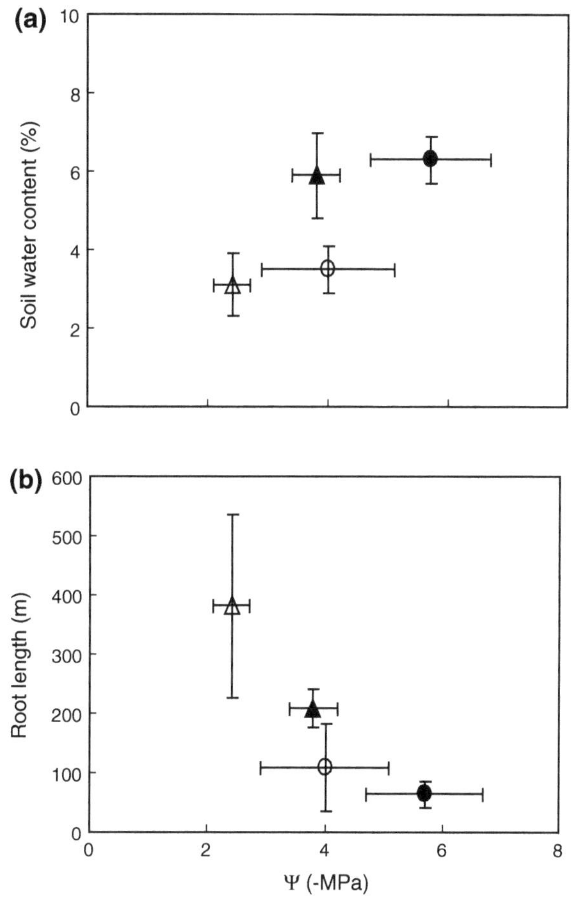

Fig. 3.12 **a** Relationships between leaf water potential at predrawn (ψ) and soil water content experienced by seedlings of *Q. coccifera* and *P. lentiscus* in the same plot. **b** Relationships between leaf water potential (ψ) and root development measured as length of fine roots growing in the soil for the same seedlings. Circles and triangles stand for *Quercus coccifera* and *Pistacia lentiscus*,, respectively. Close and open symbols represent different periods along the year, concretely three (*black symbols*) and 6 months (*white symbols*) after outplanting. All seedlings had the same age. (Redraw form Fonseca 1999)

Cavitation, however, may not be the only or even the most decisive factor behind drought-induced mortality. Multiple hypotheses on the mechanisms of drought-induced mortality have been proposed. A competing hypothesis, i.e., the carbon-starvation hypothesis, has recently been forwarded (McDowell et al. 2008; Allen et al. 2010). For these authors, carbon starvation would occur in trees that regulate stomatal conductance to prevent cavitation (Fig. 3.13). This hypothesis states that mortality may be the result of decreases in C fixation under prolonged non-lethal drought, which would produce a debilitating effect on the trees and predispose them to indirect damages like insect or fungal attacks (McDowell et al. 2008). Thus, plants will die as consequence of starvation or diseases. Some studies in Mediterranean Spain attributed the mortality of *Q. suber* and *Q. ilex* trees to *Phytophtora* sp. attacks on trees that were previously damaged by drought (Brasier 1996). These results are in line with the observed increase in insect attacks in European forests attributable to warmer conditions (Netherer and Schopf 2010). Nevertheless, the hypothesis of starvation has been questioned due to a lack of direct evidence of low C in dying plants and to different patterns of carbon

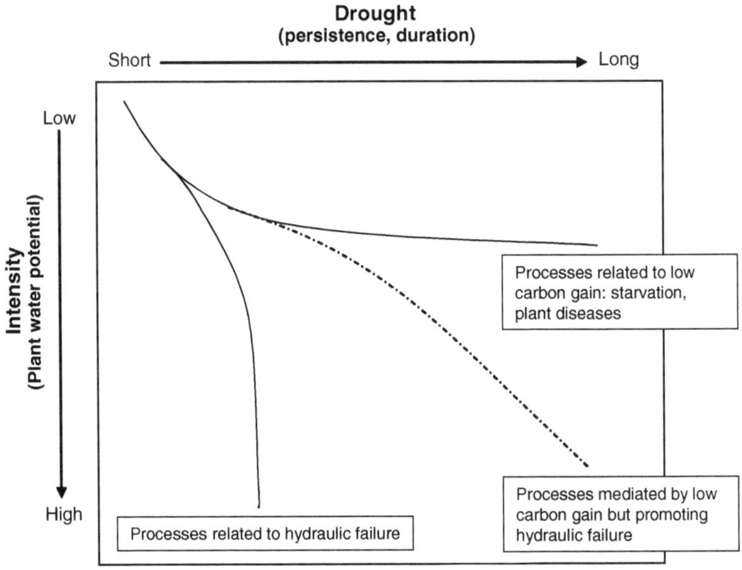

Fig. 3.13 Conceptual model relating hydraulic failure and carbon starvation as a function of drought stress duration and intensity. (From: McDowell et al. 2008 with modifications)

translocation in trees (Sala 2009; Zeppel et al. 2011; Galvez et al. 2011, but see: Galiano et al. 2011).

An intermediate pathway (Fig. 3.13, dashed line) would be the case of plants with low carbon fixation as a consequence of a high stomatal regulation which will produce a reduced capacity to growth roots and leaves. These plants would impair their ability to survive intense drought events. Therefore, lower capacity of roots to growth and colonize the soil as showed in Fig. 3.12 in *Q. coccifera* could be an example of such intermediate pathway, with higher rates of drought stress and mortality than *P. lentiscus* (Vilagrosa et al. 1997 and 2003). In fact, Landhäusser and Lieffers (2012) reported that reductions in carbon fixation produced loss of roots consequence of decreases in root reserves in aspen trees. This response was related to hydraulic failures and dying of mature trees. In this context, accumulated non-structural carbohydrates can play an additional key role in determining sprouting capacity after drought (Paula and Ojeda 2009). In addition to the amount of carbon reserves, the capacity to recover and regrowth new plant structures after rewatering can be a determinant factor to survive after drought resumption. Below-ground organs like those present in several Mediterranean and dryland species (such as lignotubers and burls) allow resprouter species to persist by growing new organs after disturbance (Naveh and Lieberman 1993). Vilagrosa et al. (2003) showed that Mediterranean shrubs as *Q. coccifera* and *Pistacia lentiscus* were able to resprout until they reached a full hydraulic failure, i.e., 100 % of the xylem conduits were embolized. These authors found that leaf shedding depends on the species and the drought severity (Fig. 3.14). When drought was severe, plants

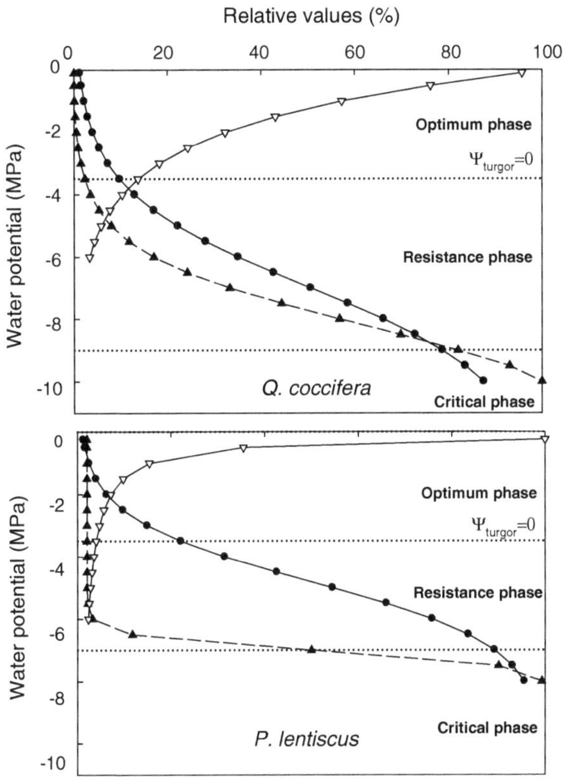

Fig. 3.14 Curves showing three different phases during an extreme drought stress period in two drought resistant Mediterranean shrubs: *Quercus coccifera* and *Pistacia lentiscus*. We identified three phases before plant death: the optimum phase coincided with positive turgor which produced only a stomatal regulation to avoid dehydration. No damages were observed. After this phase, the resistance phase corresponded to the first damages which lead to a few losses of leaves and a moderate loss of conductivity. All the plants recovered after irrigation. The critical phase corresponded to the extreme stress conditions, i.e., loss of conductivity higher than 80 %, which produced a noteworthy leaf shedding with important mortality for target plants. However, some plants were able to resprout from the base of their stems after rewatering. Other plants died probably because they experienced almost a complete loss of conductivity (i.e., 100 % xylem cavitation) (From Vilagrosa 2002; Vilagrosa et al. 2003 and 2010). Open and close triangles and close circles stand for stomatal conductance, leaf shedding, and xylem cavitation, respectively. To facilitate comparisons between species, the units are expressed as a percentage of their maximum values

should need to rebuild the entire canopy (branches and leaves), while in other cases, only new leaves need to be grown. Vulnerability to drought should affect the investment needed to recover after drought episode, i.e., the amount of carbon needed to rebuild the canopy. Thus, a plant's ability to preserve functional meristems under drought should substantially affect its capacity to withstand intense and recurrent droughts (Vilagrosa et al. 2003; Brodribb and Cochard 2009; Vilagrosa et al. 2010; Hoffmann et al. 2011).

3.4.2 Combination of Factors Affecting Xylem Cavitation in the Context of Global Warming: Summer Drought or Winter Frost?

In Mediterranean areas, summer drought and winter cold are the main factors limiting plant growth (Mitrakos 1980). The existence of two potential stress periods must be considered when explaining the ecological functioning of Mediterranean vegetation (Corcuera et al. 2005). On the one hand, substantial negative water potentials can be found in plants experiencing periods of intense summer drought, which can drive the xylem tension high enough to induce water stress-mediated xylem cavitation. On the other hand, winter cold is prone to limits the transport of water in the xylem and the occurrence of freeze–thaw events can cause long-term effects on the hydraulic functions of the tree because of embolism.

The survival of plant species in climates with water limitations should be related to the resistance of their xylem to cavitation (Pockman and Sperry 2000; Maherali et al. 2004). Martínez-Vilalta et al. (2009) detected an adjustment between hydraulic traits and environmental conditions, resulting in greater hydraulic capacity on dry sites. However, the same study underlies the fact that vulnerability to drought-induced cavitation was unrelated to climate dryness. For Addington et al. (2006) a greater hydraulic sufficiency of stems to supply water to leaves may be required to prevent excessive diurnal and seasonal declines in leaf water potential under high evaporative demand. Besides, Bhaskar et al. (2007) compared pairs of species and they found that a shift in hydraulic traits was a common response in species from Mediterranean-type climates. This shift seemed to be a strategy to cope with the high evaporative demand caused by atmospheric dryness.

One way to improve the xylem hydraulic capacity is to increase xylem conduit diameters, which would decrease xylem hydraulic resistivity (Pittermann et al. 2006). However, this functional advantage may be a risky solution. Different studies have shown that the large early-wood vessels of ring-porous trees, which are responsible for the majority of the water flow in this kind of xylem (Corcuera et al. 2004), are more vulnerable to cavitation than small vessels (Cochard and Tyree 1990; Corcuera et al. 2006). Recently, this fact has been mechanistically explained by the "rare pit" hypothesis (Christman et al. 2012). This hypothesis predicts that vessels with a large area of inter-vessel pits have a greater chance of presenting a pit with relatively low air-seeding pressure, reducing the threshold pressure for cavitation of the whole conduit (Christman et al. 2009). These mechanisms could also explain the accumulation of evidences that seems to indicate that the larger the conduit the more vulnerable to water stress induced cavitation when the comparison is done among species (Wheeler et al. 2005; Hacke et al. 2006).

Another risk associated to the formation of wide conduits is the well-known increased vulnerability to freezing–thawing induced cavitation during winter that species with large-diameter conduits can experience (Cochard and Tyree 1990;

Sperry and Sullivan 1992; Mayr and Sperry 2010). Therefore, a new ecological tradeoff arises, i.e., an improved hydraulic efficiency—which would help to cope with summer drought—could prevent the plant from occupying areas where the temperatures decrease below zero during winter.

Recently, Peguero-Pina et al. (2011b) studied this tradeoff by comparing under different environmental conditions the hydraulic traits of *Abies alba* and *A. pinsapo,* two closely related fir species occurring in the Iberian Peninsula. In fact, *A. alba* proliferates in humid European mountains (including the Spanish Pyrenees) while *A. pinsapo,* a relict species, grows only in a few restricted areas of Mediterranean mountains in Spain and Morocco, where intense summer drought periods are commonplace. Although both species does not differ in the resistance to cavitation, these authors find that *A. pinsapo* shows a higher efficiency of water transport in terms of specific and leaf hydraulic conductance. These results should contribute to maintaining its xylem tension below the threshold for rapidly increasing cavitation. Thus, the improvement of xylem sufficiency for supplying water to the transpiring needles in *A. pinsapo* can be considered a necessary and sufficient condition for coping with the high atmospheric evaporative demand, a characteristic of the Mediterranean summer. For Mayr and Sperry (2010), it is noteworthy that the higher K_s in *A. pinsapo* is largely a result of its wider tracheids, suggesting that this species may be much more vulnerable to freeze–thaw induced cavitation than *A. alba*. Nowadays, the expansion of *A. pinsapo* in the Iberian Peninsula toward Northern sites occupied by *A. alba* can be limited by the great occurrence of freeze–thaw events in these areas. However, the predicted increase in temperature and aridity due to global climate change (Barnett et al. 2005) can allow the geographic displacement of *A. pinsapo* in latitude and/or altitude.

3.5 Concluding Remarks

During last decades, plant hydraulic architecture and xylem cavitation have evolved as one of the major trends in plant water-relations studies. However, our understanding of the plant hydraulic architecture is far from being completely understood. Particularly, the two ends of the plant hydraulic pathway i.e., roots and leaves, remain poorly studied even new advances are currently in progress. In addition, aquaporins and their relative contribution to water flow within apoplastic, symplastic, and transmembrane pathways are not well characterized yet mainly in leaves and roots.

Hydraulic efficiency and vulnerability to cavitation determine many aspects of plant functioning. Several studies have reported that adaptation of hydraulic characteristics is integrally linked to ecosystem properties as light, nutrients, or water. A breakthrough in this area would allow us to predict plant responses at individual and community level in the context of changing environmental conditions. In this sense, there are currently uncertainties about climate-induced plant mortality and particularly, the mechanisms that drive forests dieback worldwide as well as the physiological thresholds priming the tree death. Research with large databases on functional features including those relating plant hydraulics and vulnerability to cavitation will allow an integrated analysis providing common response patterns that can assist to ecosystem management.

Plant functional traits and biochemical processes involved in water transport, gas exchange regulation, and plant damage or cell death are clearly noteworthy to relationships between species and adaptation to the environment, especially in the light of the new perspectives on climate projections. Although the mechanisms of xylem cavitation are relatively well established, there is still a need for settling and sharing a common protocol to improve our ability to compare species and biomes for their functional traits and drought-resistance strategies. The relative importance of root and/or stem xylem cavitation and their relative contribution to whole plant water flow and their capacity for recovery after drought should be taken into consideration when analyzing physiological activity and plant response to drought episodes. Furthermore, the role of genomics in the control and expression of genes involved in the regulations and mechanisms affecting the water transport system may deserve much more attention. In addition, the analysis of combinations of stress factors i.e., drought, freezing–thawing and their influence on xylem cavitation, and plant hydraulics should improve the capacity to integrate different processes in overall plant and ecosystem functioning.

Acknowledgements This chapter reviews and summarizes some of the outputs generated from research projects funded by the European Commission (*CREOAK project QLRT-2001-01594; FUME Grant Agreement n° 243888*), the Spanish Ministry of Science and Innovation (*PERSIST CGL2006-07126*; GRACCIE-CSD2007-00067 *CONSOLIDER-INGENIO Program; SUMIDER-OS SUM2008-00004-C03-03; CASISFOR AGL2010-21153-C02-02; SURVIVE CGL2011-30531-C02-02*) and the Spanish Ministry of the Environment (*ESTRES 063/SGTB/2007/7.1*), as well as by the Regional Government of Valencia (*XYLREFOR GRUPOS 03/155; APLITEC GV 05/208; INNOVA GVPRE/2008/085; FEEDBACKS-PROMETEO/2009/006*).

We greatly thank the Regional Forest Service (Consellería de Infraestructures, Territori i Medi Ambient, Regional Government of Valencia) for funding our research goals. We also sincerely thank Jacqueline Scheiding for the English proofreading of the text and two anonymous reviewers who provided excellent feedback for the manuscript improvement. Fundación CEAM is supported by Generalitat Valenciana.

References

Ackerly DD (2004) Functional strategies of chaparral shrubs in relation to seasonal water deficit and disturbance. Ecol Monogr 74:25–44

Addington RN, Donovan LA, Mitchell RJ, Vose JM, Pecot SD, Jack SB, Hacke UG, Sperry JS, Oren R (2006) Adjustments in hydraulic architecture of *Pinus palustris* maintain similar stomatal conductances in xeric and mesic habitats. Plant Cell Environ 29:535–545

Alder NN, Sperry JS, Pockman WT (1996) Root and stem xylem embolism, stomatal conductance and leaf turgor in *Acer grandidentatum* populations along a soil moisture gradient. Oecologia 105:293–301

Allen CD, Macalady AK, Chenchouni H, Bachelet D, Mcdowell N, Vennetier M, Kitzberger T, Rigling A, Breshears DD, Hogg EH, Gonzalez P, Fensham R, Zhang Z, Castro J, Demidova

N, Lim JH, Allard G, Running SW, Semerci A, Cobb N (2010) A global overview of drought and heat-induced tree mortality reveals emerging climate change risks for forests. Forest Ecol Manag 259:660–684

Alsina MM, De Herralde F, Aranda X, Save R. Biel C (2007) Water relations and vulnerability to embolism are not related: experiments with eight grapevine cultivars. Vitis 46:1–6

Andersen CP, Bussler BH, Chaney WR, Pope PE, Byrnes WR (1989) Concurrent establishment of ground cover and hardwood trees on reclaimed mined land and un mined reference sites. Forest Ecol Manag 28:81–99

Angeles G, Bond B, Boyer JS, Brodribb T, Brooks JR, Burns MJ, Cavender-Bares J, Clearwater M, Cochard H, Comstock J, Davis SD, Domec J-C, Donovan L, Ewers F, Gartner B, Hacke U, Hinckley T, Holbrook NM, Jones HG, Kavanagh K, Law B, López-Portillo J, Lovisolo C, Martin T, Martínez-Vilalta J, Mayr S, Meinzer FC, Melcher P, Mencuccini M, Mulkey S, Nardini A, Neufeld HS, Passioura J, Pockman WT, Pratt RB, Rambal S, Richter H, Sack L, Salleo S, Schubert A, Schulte P, Sparks JP, Sperry J, Teskey R, Tyree M (2004) The cohesion-tension theory. New Phytol 163:451–452

Barigah TS, Ibrahim T, Bogard A, Faivre-Vuillin B, Lagneau LA, Montpied P, Dreyer E (2006) Irradiance-induced plasticity in the hydraulic properties of saplings of different temperate broad-leaved forest tree species. Tree Physiol 26:1505–1516

Barnett TP, Adam JC, Lettenmaier DP (2005) Potential impacts of a warming climate on water availability in snow-dominated regions. Nature 438:303–309

Bhaskar R, Valiente-Banuet A, Ackerly DD (2007) Evolution of hydraulic traits in closely related species pairs from mediterranean and nonmediterranean environments of North America. New Phytol 176:718–726

Black CR (1979). The relationship between transpiration rate, water potential, and resistances to water movement in sunflower (*Helianthus annuus* L). J Exp Bot 30:235–243

Böhm J (1893) Capillarität und Saftsteigen. Ber Dtsch Bot Ges 11:203–212

Boyer JS (1985) Water transport. Ann Rev Plant Physiol 36:473–516

Brasier CM (1996) Phytophthora cinnamomi and oak decline in southern Europe. Environmental constraints including climate change. Ann Sci For 53:347–358

Bréda N, Huc R, Granier A, Dreyer E (2006) Temperate forest trees and stands under severe drought: a review of ecophysiological responses, adaptation processes and long-term consequences. Ann For Sci 63:625–644

Breshears DD, Myers OB, Meyer CW, Barnes FJ, Zou CB, Allen CD, Mcdowell NG, Pockman WT (2009) Tree die-off in response to global change-type drought: mortality insights from a decade of plant water potential measurements. Front Ecol Environ 7:185–189

Brodersen CR, McElrone AJ, Choat B, Matthews MA, Shackel KA (2010) The dynamics of embolism repair in xylem: in vivo visualizations using high-resolution computed tomography. Plant Physiol 154:1088–1095

Brodribb TJ, Cochard H (2009) Hydraulic failure defines the recovery and point of death in water-stressed conifers. Plant Physiol 149:575–584

Brodribb TJ, Holbrook NM (2003) Stomatal closure during leaf dehydration, correlation with other leaf physiological traits. Plant Physiol 132:2166–2173

Brodribb TJ, Holbrook NM (2004) Stomatal protection against hydraulic failure: a comparison of coexisting ferns and angiosperms. New Phytol 162:663–670

Bucci SJ, Scholz FG, Goldstein G, Meinzer FC, Franco AC, Campanello PI, Villalobos-Vega R, Bustamante M, Miralles-Wilhelm F (2006) Nutrient availability constrains the hydraulic architecture and water relations of savannah trees. Plant Cell Environ 29:2153–2167

Canny MJ (1995a) Potassium cycling in *Helianthus*: ions of the xylem sap and secondary vessel formation. Phil Trans R Soc Lond B 348:457–469

Canny MJ (2001) Contributions to the debate on water transport. Am J Bot 88:43–46

Chapin FS (1991) Effects of multiple stresses on nutrient availability and use. In: Mooney HA, Winner WE, Pell EJ, Chu E (eds), Response of plants to multiple stresses. Academic Press, San Diego, pp 67–88

Chirino E, Vilagrosa A, Cortina J, Valdecantos A, Fuentes D, Trubat R, Luis VC, Puértolas J, Bautista S, Baeza J, Penuelas JL, Vallejo VR (2009) Ecological restoration in degraded drylands: the need to improve the seedling quality and site conditions in the field. In: Grossberg SP (ed), Forest management, Nova Science Publishers, Inc., New York, pp 85–158

Chirino E, Vilagrosa A, Hernández EI, Matos A, Vallejo VR (2008) Effects of a deep container on morpho-functional characteristics and root colonization in *Quercus suber* L. seedlings for reforestation in mediterrranean climate. Forest Ecol Manag 256:779–785

Choat B, Drayton WM, Brodersen C, Matthews MA, Shackel KA, Wada H, McElrone AJ (2010) Measurement of vulnerability to water stress-induced cavitation in grapevine: a comparison of four techniques applied to a long-vesseled species. Plant Cell Environ 33:1502–1512

Christman MA, Sperry JS, Adler FR (2009) Testing the 'rare pit' hypothesis for xylem cavitation resistance in three species of *Acer*. New Phytol 182:664–674

Christman MA, Sperry JS, Smith DD (2012) Rare pits, large vessels, and extreme vulnerability to cavitation in a ring-porous tree species. New Phytol 193:713–720

Clearwater MJ, Goldstein G (2005) Embolism repair and long distance water transport. In: Holbrook NM, Zwieneicki MA. (eds) Vascular transport in plants, Elsevier Academic Press, Burlington, pp 375–400

Cochard H (2006) Cavitation in trees. C R Physique 7:1018–1026

Cochard H, Bodet C, Ameglio T, Cruiziat P (2000). Cryo-scanning electron microscopy observations of vessel content during transpiration in walnut petioles. Facts or artifacts? Plant Physiol 124:1191–1202

Cochard H, Cruiziat P, Tyree MT (1992) Use of positive pressures to establish vulnerability curves: further support for the air-seeding hypothesis and implications for pressure-volume analysis. Plant Physiol 100:205–209

Cochard H, Damour G, Bodet C, Tharwat I, Poirier M, Ameglio T (2005) Evaluation of a new centrifuge technique for rapid generation of xylem vulnerability curves. Physiol Plant 124:410–418

Cochard H, Forestier S, Améglio T (2001) A new validation of the Scholander pressure chamber technique based on stem diameter variations. J Exp Bot 52:1361–1365

Cochard H, Froux F. Mayr S, Coutand C (2004) Xylem wall collapse in water-stressed pine needles. Plant Physiol 134:401–408

Cochard H, Herbette S, Barigah T, Badel E, Ennajeh M, Vilagrosa A (2010) Does sample length influence the shape of xylem embolism vulnerability curves? A test with the cavitron spinning technique. Plant Cell Environ 33:1543–1552

Cochard H, Herbette S, Hernandez E, Hölttä T, Mencuccini M (2010) The effects of sap ionic composition on xylem vulnerability to cavitation. J Exp Bot 61:275–285

Cochard H, Lemoine D, Dreyer E (1999) The effects of acclimation to sunlight on the xylem vulnerability to embolism in *Fagus sylvatica* L. Plant Cell Environ 22:101–108

Cochard H, Martin R, Gross P, Bogeat-Triboulot MB (2000a) Temperature effects on hydraulic conductance and water relations of *Quercus robur* L. J Exp Bot 51:1255–1259

Cochard H, Bodet C, Ameglio T, Cruiziat P (2000b) Cryo-scanning electron microscopy observations of vessel content during transpiration in walnut petioles. Facts or artifacts? Plant Physiol 124:1191–1202

Cochard H, Venisse JS, Barigah TS, Brunel N, Herbette S, Guilliot A, Tyree MT, Sakr S (2007) Putative role of aquaporins in variable hydraulic conductance of leaves in response to light. Plant Physiol 143:122–130

Cochard H. (2006) Cavitation in trees. C R Phys 7:1018–1026

Cochard H, Holtta T, Herbette S, Delzon S, Mencuccini M (2009) New insights into the mechanisms of water-stress-induced cavitation in conifers. Plant Physiol 151:949–954

Corcuera L, Camarero JJ, Gil-Pelegrín E (2004) Effects of a severe drought on growth and wood anatomical properties of *Quercus faginea*. IAWA Jour 25:185–204

Corcuera L, Camarero JJ, Sisó S, Gil-Pelegrin E (2006) Radial-growth and wood-anatomical changes in overaged *Quercus pyrenaica* coppice stands: functional responses in a new Mediterranean landscape 20:91–98

Corcuera L, Cochard H, Gil-Pelegrin E, Notivol E (2011) Phenotypic plasticity in mesic populations of *Pinus pinaster* improves resistance to xylem embolism (P50) under severe drought. Trees-Struct Funct 25:1033–1042

Corcuera L, Morales F, Abadia A, Gil-Pelegrin E (2005) Seasonal changes in photosynthesis and photoprotection in a *Quercus ilex* subsp *ballota* woodland located in its upper altitudinal extreme in the Iberian Peninsula. Tree Physiol 25:599–608

Crombie D S, Hipkins M F, Milburn A (1985) Gas penetration of pit membranes in the xylem of rhododendron as the cause of acoustically detectable sap cavitation. Aust J Plant Physiol 12:445–53

Davis SD, Ewers FW, Sperry JS, Portwood KA, Crocker MC, Adams GC (2002) Shoot dieback during prolonged drought in *Ceanothus* chaparral of California: a possible cause of hydraulic failure. Am J Bot 89:820–828

Davis SD, Sperry JS, Hacke UG (1999) The relationship between xylem conduit diameter and cavitation caused by freezing. Am J Bot 86:1367–1372

Delzon S, Douthe C, Sala A, Cochard H (2010) Mechanism of water-stress induced cavitation in conifers: bordered pit structure and function support the hypothesis of seal capillary-seeding. Plant Cell Environ 33:2101–2111

Demmig-Adams B, Adams III WW (2006) Photoprotection in an ecological context: the remarkable complexity of thermal energy dissipation. New Phytol 172:11–21

Dietze M, Moorcroft PR (2011) Tree mortality in the eastern and central United States: patterns and drivers. Glob Change Biol 17:3312–3326

Dixon HH, Joly J (1895) On the ascent of sap. Philos Trans R Soc Lond 186:563–576

Domec J-C, Scholz FG, Meinzer FC, Goldstein G, Villalobos-Vega R (2006) Diurnal and seasonal variation in root xylem embolism in neotropical savanna woody species: impact on stomatal control of plant water status. Plant Cell Environ 29:26–35

van Doorn WG, Hiemstra T, Fanourakis D (2011) Hydrogel regulation of xylem water flow: an alternative hypothesis. Plant Physiol 157:1642–1649

Earnshaw MJ (1993) Stress indicators: electrolyte leakage. In: Hendry GAF, Grime JP (eds) Methods in comparative plant ecology. Chapman and Hall, London, pp 152–154

Esteso-Martinez J, Camarero JJ, Gil-Pelegrin E (2006) Competitive effects of herbs on *Quercus faginea* seedlings inferred from vulnerability curves and spatial-pattern analyses in a Mediterranean stand (Iberian System, northeast Spain). Ecoscience 13:378–387

Ewers B, Oren R, Sperry J (2000) Influence of nutrient versus water supply on hydraulic architecture and water balance in *Pinus taeda*. Plant Cell Environ 23:1055–1066

Feild TS, Brodribb T (2001) Stem water transport and freeze-thaw xylem embolism in conifers and angiosperms in a Tasmanian treeline heath. Oecologia 127:314–320

Fonseca DE (1999) Manipulaciones de las características morfoestructurales de plantonesde especies forestales mediterráneas producidas en vivero. Zaragoza M.Sc Thesis. CIHEAM-IAMZ, Zaragoza

Franks PJ, Gibson A, Bachelard EP (1995) Xylem permeability and embolism susceptibility in seedlings of *Eucalyptus camaldulensis* Dehnh from two different climatic zones. Aust J Plant Physiol 22:15–21

Frensch J, Steudle E (1989) Axial and radial hydraulic resistance to roots of maize (*Zea mays* L). Plant Physiol 91:719–726

Froux F, Ducrey M, Dreyer E, Huc R (2005) Vulnerability to embolism differs in roots and shoots and among three mediterranean conifers: consequences for stomatal regulation of water loss? Trees-Struct Funct 19:137–144

Galiano L, Martínez-Vilalta J, Lloret F (2011) Carbon reserves and canopy defoliation determine the recovery of Scots pine 4 yr after a drought episode. New Phytol 190:750–759

Galvez DA, Landhäusser SM, Tyree MT (2011) Root carbon reserve dynamics in aspen seedlings: does simulated drought induce reserve limitation? Tree Physiol 31:250–257

Hacke U, Sauter JJ (1996) Drought-induced xylem dysfunction in petioles, branches and roots of *Populus balsamifera* L. and *Alnus glutinosa* (L) Gaertn. Plant Physiol 111:413–417

Hacke UG, Sperry JS, Pittermann J (2000) Drought experience and cavitation resistance in six shrubs from the great basin, Utah. Basic Appl Ecol 1:31–41

Hacke UG, Sperry JS, Pockman WT, Davis SD, McCulloh KA (2001) Trends in wood density and structure are linked to prevention of xylem implosion by negative pressure. Oecologia 126:457–461

Hacke UG, Sperry JS, Wheeler JK, Castro L (2006) Scaling of angiosperm xylem structure with safety and efficiency. Tree Physiol 26:689–701

Hafren J, Daniel G, Westermark U (2000) The distribution of acidic and esterified pectin in cambium, developing xylem and mature xylem of *Pinus sylvestris*. IAWA J 21:157–168

Hao GY, Hoffmann WA, Scholz FG, Bucci SJ, Meinzer FC, Franco AC, Cao KF Goldstein G (2008) Stem and leaf hydraulics of congeneric tree species from adjacent tropical savanna and forest ecosystems. Oecologia 155:405–415

Harvey HP, Van den Driessche R (1997) Nutrition, xylem cavitation and drought resistance in hybrid poplar. Tree Physiol 17:647–654

Herbette S, Wortemann R, Awad H, Huc R, Cochard H, Barigah TS (2010) Insights into xylem vulnerability to cavitation in *Fagus sylvatica* L: phenotypic and environmental sources of variability. Tree Physiol 30:1448–1455

Herbette S. Cochard H (2010) Calcium is a major determinant of xylem vulnerability to cavitation. Plant Physiol 153:1932–1939

Hernández EI, Pausas JG, Vilagrosa A (2011) Leaf physiological traits in relation to resprouter ability in the Mediterranean Basin. Plant Ecol 212:1959–1966

Hernández EI, Vilagrosa A, Luis VC, Llorca M, Chirino E, Vallejo VR (2009) Root hydraulic conductance, gas exchange and leaf water potential in seedlings of *Pistacia lentiscus* L and *Quercus suber* L grown under different conditions of fertilization and light regimes. Environ Exp Bot 67:269–276

Hernández EI, Vilagrosa A, Pausas JG, Bellot J (2010) Morphological traits and water use strategies in seedlings of Mediterranean coexisting species. Plant Ecol 207:233–244

Hertig E, Jacobeit J (2008) Downscaling future climate change: temperature scenarios for the Mediterranean area. Global Planet Change 63:127–131

Hoffmann WA, Marchin RM, Abit P, Lau OL (2011) Hydraulic failure and tree dieback are associated with high wood density in a temperate forest under extreme drought. Glob Change Biol 17:2731–2742

Holbrook NM, Zwieniecki MA (1999) Embolism repair and xylem tension. Do we need a miracle? Plant Physiol 120:7–10

Holbrook NM, Zwieniecki MA (2005) Vascular trasnport in plants. Elsevier Academic Press, USA

Hölttä T, Cochard H, Nikinmaa E, Mencuccini M (2009) Capacitive effect of cavitation in xylem conduits: results from a dynamic model. Plant Cell Environ 32:10–21

Huang B, Nobel PS (1993) Hydraulic condctivity and anatomy along lateral roots of cacti—changes with soil water status. New Phytol 123:449–507

Hubbard RM, Ryan MG, Stiller V, Sperry JS (2001) Stomatal conductance and photosynthesis vary linearly with plant hydraulic conductance in ponderosa pine. Plant Cell Environ 24:113–121

Huber B (1928) Weitere quantitative Untersuchungen über das Wasserleitungs system der Pflanzen. Jahrb Wiss Bot 67:877–959

Hukin D, Cochard H, Dreyer E, Le Thiec D, Bogeat-Triboulot MB (2005) Cavitation vulnerability in roots and shoots: does *Populus euphratica* Oliv., a popular from arid areas of Central Asia, differ from other poplar species?. J Exp Bot 53:2003–2010

IPCC (2010) Meeting report of the intergornmental panel on climate change, expert meeting on detection and attribution related to anthropogenic climate change. IPCC Working Group I Technical support unit, university of Bern, Switzerland, pp 55

Jackson GE, Grace J (1996) Field measurements of xylem cavitation: are acoustic emissions useful? J Exp Bot 47:1643–1650

Jackson RB, Sperry JS, Dawson TE (2000) Root water uptake and transport: using physiological processes in global predictions. Trends Plant Sci 5:482–488

Jacobsen AL, Agenbag L, Esler KJ, Pratt RB, Ewers FW, Davis SD. 2007a. Xylem density, biomechanics, and anatomical traits correlate with water stress in seventeen evergreen shrub species of the Mediterranean-type climate region of South Africa. J Ecol 95:171–183

Jacobsen AL, Pratt RB, Davis SD, Ewers FW (2008) Comparative community physiology: non-convergence in water relations among three semi-arid shrub communities. New Phytol 180:100–113

Jacobsen AL, Pratt RB, Ewers F, Davis SD (2007) Cavitation resistance among 26 chaparral species of southern California. Ecol Monogr 77:99–115

Jansen S, Gortan E, Lens F, Lo Gullo MA, Salleo S, Scholz A, Stein A, Trifilò P, Nardini A (2011) Do quantitative vessel and pit characters account for ion-mediated changes in the hydraulic conductance of angiosperm xylem? New Phytol 189:218–228

Jarbeau JA, Ewers FW, Davis SD (1995) The mechanism of water-stress-induced embolism in two species of chaparral shrubs. Plant Cell Environ 18:189–196

Javot H, Maurel C (2002) The role of aquaporins in root water uptake. Ann Bot 90:301–13

Johnson DM, Woodruff, DR McCulloh KA, Meinzer FC (2009) Leaf hydraulic conductance, measured in situ, declines and recovers daily: leaf hydraulics, water potential and stomatal conductance in four temperate and three tropical tree species. Tree Physiol 29:879–887

Jones HG (1992) Plants and Microclimate. A quantitative approach to environmental plant physiology. Cambridge University Press, Cambridge

Kaldenhoff R, Ribas-Carbó M, Flexas-Sans J, Lovisolo C, Heckwolf M, Uehlein N (2008) Aquaporins and plant water balance Plant Cell Environ 31:658–666

Kikuta SB, Hietz P, Richter H (2003) Vulnerability curves from conifer sapwood sections exposed over solutions with known water potentials. J Exp Bot 54:2149–2155

Koch GW, Sillett SC, Jennings GM, Davis SD (2004) The limits to tree height. Nature 428: 851–854

Kolb KJ, Sperry JS, Lamont BB. (1996) A method for measuring xylem hydraulic conductance and embolism in entire root and shoot systems. J Exp Bot 47:1805–1810

Kozlowski T T, Kramer P J, Pallardy S G (1991) The physiological ecology of woody plants. Academic Press, Toronto

Lamy JB, Bouffier L, Burlett R, Plomion C, Cochard H, Delzon S (2011) Uniform selection as a primary force reducing population genetic differentiation of cavitation resistance across a species range. PLoS ONE 6: e23476. doi:10.1371/journal.pone.0023476

Landhäusser SM, Lieffers V (2012) Defolitaion increases risk of carbon starvation in root systems of mature aspen. Trees-Struct Funct 26:653–661

Larcher W (1995) Physiological plant ecology. Ecophysiology and stress physiology of functional groups. Springer, Berlin

Lemoine D, Cochard H, Granier A (2002) Within crown variation in hydraulic architecture in beech (*Fagus sylvatica* L.): evidence for a stomatal control of embolism. Ann For Sci 59: 19–27

Levitt J (1980) Responses of plants to environmental stresses. Academic Press, New York

Li Y, Sperry JS, Taneda H, Bush SE, Hacke UG (2008) Evaluation of centrifugal methods for measuring xylem cavitation in conifers, diffuse- and ring-porous angiosperms. New Phytol 177:558–568

Lo Gullo MA, Nardini A, Salleo S, Tyree MT (1998) Changes in root hydraulic conductance of *Olea oleaster* seedlings following drought stress and irrigation. New Phytol 140:25–31

Lo Gullo MA, Salleo S (1991) Three different methods for measuring xylem cavitation and embolism: a comparison. Ann Bot London 67 (5):417–424

Lovelock C, Feller I, Ball M, Engelbrecht B, Ewe (2006) Differences in plant function in phosphorus- and nitrogen limited mangrove ecosystems. New Phytol 72:514–522

Luis VC, Llorca M, Chirino E, Hernández E, Vilagrosa A (2010) Differences in morphology, gas exchange and root hydraulic conductance before planting in *Pinus canariensis* seedlings growing under different fertilization and light regimes. Trees-Struct Funct 24:1143–1150

Luu DT, Maurel C (2005) Aquaporins in a challenging environment: molecular gears for adjusting plant water status. Plant Cell Environ 28:85–96

Machado JL, Tyree MT (1994) Patterns of hydraulic architecture and water relations of two tropical canopy trees with contrasting leaf phonologies: *Ochroma pyramidale* and *Pseudobombax septenatum* Tree Physiol 14:219–240
Maherali H, Moura CF, Caldeira MC, Willson CJ, Jackson RB (2006) Functional coordination between leaf gas exchange and vulnerability to xylem cavitation in temperate forest trees. Plant Cell Environ 29:571–538
Maherali H, Pockman WT, Jackson RB (2004) Adaptative variation in the vulnerability of woody plants to xylem cavitation. Ecology 85:2184–2199
Markesteijn L, Poorter L, Bongers F, Paz H, Sack L (2011) Hydraulics and life history of tropical dry forest tree species: coordination of species' drought and shade tolerance. New Phytol 191:480–495
Martín JA, Solla A, Coimbra MA, Gil L (2005) Metabolic distinction of *Ulmus minor* xylem tissues after inoculation with *Ophiostoma novo-ulmi*. Phytochemistry 66:2458–2467
Martínez-Vilalta J, Cochard H, Mencuccini M, Sterck F, Herrero A, Korhonen JFJ, Llorens P, Nikinmaa E, Nolè A, Ripullone F, Sass-Klaassen U, Zweifel R (2009) Hydraulic adjustments of Scots pine across Europe. New Phytol 184:353–364
Martínez-Vilalta J, Prat E, Oliveras I, Piñol J (2002) Xylem hydraulic properties of roots and stems of nine Mediterranean woody species. Oecologia 133:19–29
Martre P, Morillon R, Barrieu F, North GB, Nobel PS, Chrispeels MJ (2002) Plasma membrane aquaporins play a significant role during recovery from water deficit. Plant Physiol 130:2101–2110
Maseda PH, Fernandez RJ (2006) Stay wet or else: three ways in which plants can adjust hydraulically to their environment. J Exp Bot 57:3963–3977
Maurel C, Verdoucq L, Luu DT, Santoni V (2008) Plant aquaporins: membrane channels with multiple integrated functions. Ann Rev Plant Biol. 59:595–624
Mayr S, Cochard H (2003) A new method for vulnerability analysis of small xylem areas reveals that compression wood of Norway spruce has lower hydraulic safety than opposite wood. Plant Cell Environ 26:1365–1371
Mayr S, Sperry JS (2010) Freeze-thaw-induced embolism in *Pinus contorta*: centrifuge experiments validate the 'thaw-expansion hypothesis' but conflict with ultrasonic emission data. New Phytol 185:1016–1024
Mayr S, Gruber A, Bauer H (2003) Repeated freeze-thaw cycles induce embolism in drought stressed conifers (Norway spruce, stone pine). Planta 217:436–41
Mayr S, Wieser G, Bauer H (2006) Xylem temperatures during winter in conifers at the alpine timberline. Agr Forest Meteorol 137:81–88
McCulloh K, Meinzer F, Sperry JS, Lachenbruch B, Voelker S, Woodruff JC (2011) Comparative hydraulic architecture of tropical tree species representing a range of successional stages and wood density. Oecologia 167:27–37
McDowell N, Pockman W, Allen D, Breshears D, Cobb N, Kolb T, Plaut J, Sperry J, West A, Williams D, Yepez EA (2008) Mechanisms of plant survival and mortality during drought: why do some plants survive while others succumb to drought? New Phytol 178:719–739
McElrone AJ, Jackson S, Habdas P (2008) Hydraulic disruption and passive migration by a bacterial pathogen in oak tree xylem. J Exp Bot 59:2649–2657
McElrone AJ, Pockman WT, Martínez-Vilalta J, Jackson RB (2004) Variation in xylem structure and function in stems and roots of trees to 20 m depth. New Phytol 163:507–517
Meinzer FC, Johnson DM, Lachenbruch B, McCulloh KA, Woodruff DR (2009) Xylem hydraulic safety margins in woody plants: coordination of stomatal control of xylem tension with hydraulic capacitance. Funct Ecol 23:922–930
Mencuccini, M, Comstock J (1997) Vulnerability to cavitation in populations of two desert species, *Hymenoclea salsola* and *Ambrosia dumosa*, from different climatic regions. J Exp Bot 48:1323–1334
Milburn JA. (1991) Cavitation and emboli, development and significance. In: Raghavendra AS (ed) Physiology of trees. Wiley, New York, pp 163–173
Milburn JA (1996) Sap ascent in vascular plants: challengers to the cohesion theory ignore the significance of immature xylem and the recycling of Münch water. Ann Bot 78:399–407

Milburn JA, Johnson RPC (1966) The conduction of sap. II. Detection of vibrations produced by sap cavitation in *Ricinus* xylem. Planta 66:4–52

Miranda JD, Padilla FM, Martínez-Vilalta J, Pugnaire FI (2010) Woody species of a semiarid community are only moderately resistant to cavitation. Funct Plant Biol 37:828–839

Mitrakos K (1980) A theory for Mediterranean plant life. Acta Oecol-Oec Plant 1245–252

Morales F, Abadía A, Abadía J (2006) Photoinhibition and photoprotection under nutrient deficiencies, drought, and salinity. In: Demmig-Adams B, Adams III WW, Mattoo AK (eds) Photoprotection, photoinhibition, gene regulation and environment. Springer, Dordrecht, pp 65–85

Nardini A, Gascó A, Trifilo P, Lo Gullo MA, Salleo S (2007) Ion-mediated enhancement of xylem hydraulic conductivity is not always suppressed by the presence of Ca^{2+} in the sap. J Exp Bot 58:2609–2615

Nardini A, Grego F, Trifilò P, Salleo S (2010) Changes of xylem sap ionic content and stem hydraulics in response to irradiance in *Laurus nobilis*. Tree Physiol 30:628–635

Nardini A, Lo Gullo MA, Salleo S (2011) Refilling embolized xylem conduits: Is it a matter of phloem unloading?. Plant Sci 180: 604–611

Nardini A, Salleo S (2000) Limitation of stomatal conductance by hydraulic traits: sensing or preventing xylem cavitation? Trees-Struct Funct 15:14–24

Nardini A, Salleo S, Andri S (2005) Circadian regulation of leaf hydraulic conductance in sunflower (*Helianthus annuus* L. cv. Margot). Plant Cell Environ 28:750–759

Nardini A, Tyree MT, Salleo S (2001) Xylem cavitation in the leaf of *Prunus laurocerasus* and its impact on leaf hydraulics. Plant Physiol 125:1700–1709

Naveh Z and Lieberman AS (1993) Landscape ecology theory and applications. (2nd edn), Springer, New York

Netherer S, Schopf A (2010) Potential effects of climate change on insect herbivores general aspects and a specific example (Pine processionary moth, *Thaumetopoea pityocampa*). Forest Ecol Manag 259:831–838

Netting AG (2009) Limitations within "the limits to tree height". Brief communication. Am J Bot 96:542–544

Neufeld HS, Grantz DA, Meinzer FC, Goldstein G, Crisosto GM, Crisosto C (1992) Genotypic variability in vulnerability of leaf xylem to cavitation in water-stressed and well-irrigated sugarcane. Plant Physiol 100:1020–1028

Nobel PS (1999). Physicochemical and environmental plant physiology. Academic Press, San Diego, London

North GB, Martre P, Nobel PS (2004) Aquaporins account for variations in hydraulic conductance for metabolically active root regions of *Agave deserti* in wet, dry, and rewetted soil. Plant Cell Environ 27: 219–228

North GB, Nobel PS (1992) Drought-induced changes in hydraulic conductivity and structure in roots of *Ferocactus acanthodes* and *Opuntia ficus-indica*. New Phytologist 120:9–19

Ogaya R, Peñuelas J, Martínez-Vilalta J, Mangirón M (2003). Effect of drought on diameter increment of *Quercus ilex*, *Phillyrea latifolia*, and *Arbutus unedo* in a holm oak forest of NE Spain. Forest Ecol Manage 180:175–184

Oliveras I, Martínez-Vilalta J, Jimenez-Ortiz T, Lledó MJ, Escarré A, Piñol J (2003) Hydraulic properties of *Pinus pinea* and *Tetraclinis articulata* in a dune ecosystem of Eastern Spain. Plant Ecol 169:131–141

Pammenter NW, Vander Willigen C (1998) A mathematical and statistical analysis of the curves illustrating vulnerability of xylem to cavitation. Tree Physiol 18:589–593

Paula S, Ojeda F (2009) Belowground starch consumption after recurrent severe disturbance in three resprouter species of the genus *Erica*. Botany 87: 253–259

Peguero-Pina JJ, Alquézar-Alquézar JM, Mayr S, Cochard H, Gil-Pelegrín E (2011a) Embolism induced by winter drought may be critical for the survival of *Pinus sylvestris* L. near its southern distribution limit. Ann Sci 68:565–574

Peguero-Pina JJ, Morales F, Flexas J, Gil-Pelegrín E, Moya I (2008) Photochemistry, remotely sensed physiological reflectance index and de-epoxidation state of the xanthophyll cycle in *Quercus coccifera* under intense drought. Oecologia 156:1–11

Peguero-Pina JJ, Sancho-Knapik D, Cochard H, Barredo G, Villarroya D, Gil-Pelegrín (2011b) Hydraulic traits are associated with the distribution range of two closely related Mediterranean firs, *Abies alba* Mill. and *Abies pinsapo* Boiss. Tree Physiol 31:1067–1075

Peguero-Pina JJ, Sancho-Knapik D, Morales F, Flexas J, Gil-Pelegrín E (2009) Differential photosynthetic performance and photoprotection mechanisms of three Mediterranean evergreen oaks under severe drought stress. Funct Plant Biol 36:453–462

Peman J, Voltas J, Gil-Pelegrin E (2006) Morphological and functional variability in the root system of *Quercus ilex* L. subject to confinement: consequences for afforestation. Ann For Sci 63:425–430

Pickard WF, Melcher PJ (2005) Perspectives on the biophysics of xylem transport. In: Holbrook NM and Zwieniecki MA (eds) Vascular transport in plants. Elsevier Academic Press, Burlington, San Diego, pp 4–18

Pittermann J, Choat B, Jansen S, Stuart SA, Lynn L, Dawson TE (2010) The Relationships between xylem safety and hydraulic efficiency in the Cupressaceae: The evolution of pit membrane form and function. Plant Physiol 153:1919–1931

Pittermann J, Sperry JS (2003) Tracheid diameter is the key trait determining the extent of freezing-induced embolism in conifers. Tree Physiol 23:907–914

Pittermann J, Sperry JS (2006) Analysis of freeze-thaw embolism in conifers: the interaction between cavitation pressure and tracheid size. Plant Physiol 140:374–382

Pittermann J, Sperry JS, Hacke U, Wheeler JK, Sikkema EH. 2006. Inter-tracheid pitting and the hydraulic efficiency of conifer wood: the role of tracheid allometry and cavitation protection. Am J Bot 93:1265–1273

Pockman WT, Sperry JS (1997) Freezing-induced xylem cavitation and the northern limit of *Larrea tridentata*. Oecologia 109:19–27

Pockman WT, Sperry JS (2000) Vulnerability to xylem cavitation and the distribution of Sonoran desert vegetation. Am J Bot 87:1287–1299

Pratt RB, Ewers FW, Lawson MC, Jacobsen AL, Brediger M, Davis SD (2005) Mechanisms for tolerating freeze-thaw stress of two evergreen chaparral species: *Rhus ovata* and *Malosma laurine* (Anacardiaceae). Am J Bot 92:1102–1113

Pratt RB, Jacobsen AL, Ewers FW, Davis SD (2007) Relationships among xylem transport, biomechanics and storage in stems and roots of nine *Rhamnaceae* species of the California chaparral. New Phytol 174:787–798

Pratt RB, North GB, Jacobsen AL, Ewers FW, Davis SD (2010) Xylem root and shoot hydraulics is linked to life history type in chaparral seedlings. Funct Ecol 24:70–81

Quero JL, Sterck FJ, Martínez-Vilalta J, Villar R (2011) Water-use strategies of six co-existing Mediterranean woody species during a summer drought. Oecologia 166:45–57

Raimondo F, Nardini A, Salleo S, Cacciola SA, LoGullo MA. 2010. A tracheomycosis as a tool for studying the impact of stem xylem dysfunction on leaf water status and gas exchange in *Citrus aurantium* L. Trees 24:327–333

Richter H, Cruiziat P (2002) A brief history of the study of water movement in the xylem. Online document. Essay 4.1. Available via: http://5e.plantphys.net/article.php?ch=4&id=98

Ryan MG (2011) Tree responses to drought. Tree Physiol 31:237–239

Sack L, Holbrook NM (2006). Leaf Hydraulics. Annu. Rev. Plant Biol 57:361–81

Sack L; Streeter CM, Holbrook NM (2004) Hydraulic analysis of water flow through leaves of sugar maple and red oak. Plant Physiol 134:1824–1833

Sangsing K, Kasemsap P, Thanisawanyangkura S, Sangkhasila K, Gohet E, Thaler P, Cochard H (2004) Xylem embolism and stomatal regulation in two rubber clones (*Hevea brasiliensis* Muell. Arg.). Trees-Struct Funct 18:109–114

Sala A. (2009) Lack of direct evidence for the carbon-starvation hypothesis to explain drought-induced mortality in trees. PNAS 106:E68

Salleo S, Lo Gullo MA, De Paoli D, Zippo M (1996) Xylem recovery from cavitation induced embolism in young plants of *Laurus nobilis*, a possible mechanism. New Phytol 132:47–56

Salleo S, Lo Gullo MA, Raimondo F, Nardini A (2001) Vulnerability to cavitation of leaf minor veins: any impact on leaf gas exchange? Plant Cell Environ 24:851–859

Salleo S, Nardini A, Pitt F, Lo Gullo MA (2000) Xylem cavitation and hydraulic control of stomatal conductance in Laurel (*Laurus nobilis* L.). Plant Cell Environ 23:71–79
Salleo S, Raimondo F, Trifilo P, Nardini A (2003) Axial-to-radial water permeability of leaf major veins: a possible determinant of the impact of vein embolism on leaf hydraulics? Plant Cell Environ 26:1749–1758
Salleo S, Trifilo P, Lo Gullo M (2006) Phloem as a possible major determinant of rapid cavitation reversal in stems of *Laurus nobilis* (laurel). Funct Plant Biol 33:1063–1074
Santiago LS, Goldstein G, Meinzer FC, Fisher JB, Machado K, Woodruff D, Jones T (2004) Leaf photosynthetic traits scale with hydraulic conductivity and wood density in Panamanian forest canopy trees. Oecologia 140:543–550
Shatil-Cohen A, Attia Z, Moshelion M (2011) Bundle-sheath cell regulation of xylem-mesophyll water transport via aquaporins under drought stress: a target of xylem-borne ABA? Plant J 67:72–80
Scholander PF, Hammel HT, Bradstreet ED, Hemmingsen EA (1965) Sap pressures in vascular plants. Science 148:339–346
Scoffoni C, Rawls M, McKown A, Cochard H, Sack L (2011) Decline of leaf hydraulic conductance with dehydration: relationship to leaf size and venation architecture. Plant Physiol 156:823–843
Sperry JS (1995) Limitations on stem water transport and their consequences. In: Gartner BL (ed) Plant stems: Physiology and functional morphology, Academy Press Limited, London, pp 105–124
Sperry JS (2000) Hydraulic constraints on plant gas exchange. Agric For Meteorol 104:13–23
Sperry JS, Christman MA, Torrez-Ruiz JM, Taneda H, Smith DD (2011) Vulnerability curves by centrifugation: is there an open vessel artifact, and are "r" shaped curves necessarily invalid? Plant Cell Environ 35:601–610
Sperry JS, Donnelly JR, Tyree MT (1988) A method for measuring hydraulic conductivity and embolism in xylem. Plant Cell and Environ 83:414–417
Sperry JS, Nichols KL, Sullivan JEM, Eastlack SE (1994) Xylem embolism in ring-porous, diffuse-porous, and coniferous trees of Northern Utah and Interior Alaska. Ecology 75: 1736–1752
Sperry JS, Saliendra NZ (1994) Intra- and inter-plant variation in xylem cavitation in *Betula occidentalis*. Plant Cell Environ 17:1233–1241
Sperry JS, Sullivan JEM (1992) Xylem embolism in response to freeze-thaw cycles and water-stress in ring-porous, diffuse-porous, and conifer species. Plant Physiol 100:605–613
Sperry JS, Tyree MT (1990) Water-stress-induced xylem embolism in three species of conifers. Plant Cell Environ 13:427–436
Spicer R, Gartner B (1998) Hydraulic properties of Douglas-fir (*Pseudotsuga menziesii*) branches and branch halves with reference to compression wood. Tree Physiol 18:777–784
Spicer R, Gartner B (2002) Compression wood has little impact on the water relations of Douglas-fir (*Pseudotsuga menziesii*) seedlings despite a large effect on shoot hydraulic properties. New Phytol 154:633–640
Steudle E (1995) Botany—trees under tension. Nature 378:663–664
Steudle E (2001) The cohesion-tension mechanism and the acquisition of water by plants root. Ann Rev Plant Physiol. Plant Mol Biol 52:847–875
Steudle E, Peterson CA (1998) How does water get through roots? J Exp Bot 49:775–788
Stiller V, Sperry JS (1999) Canny's compensating pressure theory fails a test. Amer J Bot 86:1082–1086
Stout DL, Sala A (2003) Xylem vulnerability to cavitation in *Pseudotsuga menziesii* and *Pinus ponderosa* from contrasting habitats. Tree Physiol 23:43–50
Thapa G, Dey M, Sahoo L, Panda SK (2011) An insight into the drought stress induced alterations in plants. Biol Plantarum 55:603–613
Tognetti R, Longobucco A, Raschi A (1998) Vulnerability of xylem to embolism in relation to plant hydraulic resistance in *Quercus pubescens* and *Quercus ilex* co-occurring in a Mediterranean coppice stand in central Italy. New Phytol 139:437–447

Trifilo P, Nardini A, Lo Gullo MA, Salleo S (2003) Vein cavitation and stomatal behaviour of sunflower (*Helianthus annuus*) leaves under water limitation. Physiol Plant 119: 409–417

Trubat R, Cortina J, Vilagrosa A (2006) Plant morphology and root hydraulics are altered by nutrient deficiency in *Pistacia lentiscus*. Trees-Struct Funct 20:334–339

Tsuda M, Tyree MT (2000) Hydraulic conductance measured by the high pressure flow meter in crop plants. J Exp Bot 51:823–828

Tyree MT, Salleo S, Nardini A, Lo Gullo MA, Mosca R (1999) Refilling of embolized vessels in young stems of laurel. Do we need a new paradigm? Plant Physiol 120:11–21

Tyree MT (1997) The cohesion-tension theory of sap ascent: current controversies J Exp Bot 48:1753–1765

Tyree MT, Cochard H (1996) Summer and winter embolism in oaks: impact on water relations. Ann For Sci 53:173-180

Tyree MT, Cochard H, Cruiziat P, Sinclair B, Ameglio T (1993) Drought-induced leaf shedding in walnut: evidence for vulnerability segmentation. Plant Cell Environ 16:879–882

Tyree MT, Davis SD, Cochard H (1994) Biophysical perspectives of xylem evolution: is there a tradeoff of hydraulic efficiency for vulnerability to dysfunction? IAWA J 15:335–360

Tyree MT, Dixon MA, Thompson RG (1984) Ultrasonic acoustic emissions from the sapwood of *Thuja occidentalis* measured inside a pressure bomb. Plant Physiol 74:1046–1049

Tyree MT, Engelbrecht BMJ, Vargas G, Kursar TA (2003) Desiccation tolerance of five tropical seedlings in Panama. Relationship to a field assessment of drought performance. Plant Physiol 132:1439–1447

Tyree MT, Ewers FW (1991) The hydraulic architecture of trees and other woody plants. New Phytologist, 119:345–360

Tyree MT, Salleo S, Nardini A, Lo Gullo MA, Mosca R (1999) Refilling of embolized vessels in young stems of laurel. Do we need a new paradigm? Plant Physiol 120:11–21

Tyree MT, Sperry JS (1988) Do woody plants operate near the point of catastrophic xylem dysfunction caused by dynamic water stress? Plant Physiol 88:574–580

Tyree MT, Sperry JS (1989) Vulnerability of xylem cavitation and embolism. Ann Rev Plant Physol Mol Bio 40:19–38

Tyree MT, Yang S (1992) Hydraulic conductivity recovery versus water pressure in xylem of *Acer saccharum*. Plant Physiol 100:669–676

Tyree MT, Yang S, Cruiziat P, Sinclair B (1994) Novel methods of measuring hydraulic conductivity of tree root systems and interpretation using AMAIZED. A maize-root dynamic model for water and solute transport. Plant Physiol 104:189–199

Tyree MT, Zimmermann MH (2002) Xylem structure and the ascent of sap, 2nd edition. Springer-Verlag, Berlin, Germany

Valladares F, Pearcy RW (1998) The functional ecology of shoot architecture in sun and shade plants of *Heteromeles arbutifolia* M. Roem., a Californian chaparral shrub. Oecologia 114:1–10

Valladares F, Vilagrosa A, Penuelas J, Ogaya R, Camarero JJ, Corchera L, Sisó S, Gil-Pelegrín E (2008) Estrés hídrico: ecofisiología y escalas de sequía. In: Valladares F (ed) Ecología del bosque mediterráneo en un mundo cambiante, 2nd edn. Organismo autónomo de parques nacionales. Ministerio de medio ambiente, Madrid, Spain, pp 165–192

Valliyodan B, Nguyen HT (2006) Understanding regulatory networks and engineering for enhanced drought tolerance in plants. Curr Opin Plant Biol 9:189–195

Van den Honert TH (1948) Water transport implants as a catenary process. Discussions of the Faraday Society 3:146–153

van Mantgem PJ, Stephenson NL, Byrne JC, Daniels LD, Franklin JF, Fulé PZ, Harmon ME, Larson AJ, Smith JM, Taylor AH et al. (2009) Widespread increase of tree mortality rates in the Western United States. Science 323:521–524

Vilagrosa A (2002). Estrategias de resistencia al déficit hídrico en *Pistacia lentiscus* L. y *Quercus coccifera* L. PhD Thesis. Universidad de Alicante, Alicante (Spain)

Vilagrosa A, Bellot J, Vallejo VR, Gil-Pelegrín E (2003) Cavitation, stomatal conductance, and leaf dieback in seedlings of two co-occurring Mediterranean shrubs during an intense drought. J Exp Bot 54:2015–2024

Vilagrosa A, Morales F, Abadía A, Bellot J, Cochard H, Gil-Pelegrín E (2010) Are symplast tolerance to intense drought conditions and xylem vulnerability to cavitation coordinated? An integrated analysis of photosynthetic, hydraulic and leaf level processes in two Mediterranean drought-resistant species. Environ Exp Bot 69:233–242

Vilagrosa A, Seva JP, Valcecantos A, Cortina J, Alloza JA, Serrasolses I, Diego V, Abril M, Ferran A, Bellot J, Vallejo VR (1997) Plantaciones para la restauraci6n forestal en la comunidad Valenciana. In: La restauraci6n de la cubierta vegetal en la comunidad Valenciana. V. R. Vallejo (Ed.). Pp: 435–598

Voicu M, Cooke JEK, Zwiazek JJ (2009) Aquaporin gene expression and apoplastic water flow in bur oak (*Quercus macrocarpa*) leaves in relation to the light response of leaf hydraulic conductance. J Exp Bot 60:4063–4075

Weatherley PE (1976) Introduction: water movement through plants. Phil Trans R Soc Lond B series 273:435–444

Wei C, Tyree MT, Steudle E (1999) direct measurement of xylem pressure in leaves of intact maize plants. A test of the cohesion-tension theory taking hydraulic architecture into consideration. Plant Physiol 121:1191–1205

Wheeler JK, Sperry JS, Hacke UG, Hoang N (2005) Inter-vessel pitting and cavitation in woody rosaceae and other vesselled plants: a basis for a safety vs. Efficiency trade-off in xylem transport. Plant Cell Environ 28:800–812

Wheeler TD, Stroock AD (2008) The transpiration of water at negative pressures in a synthetic tree. Nature, 455:208–212

Willats WG, McCartney L, Mackie W, Knox JP (2001) Pectin: cell biology and prospects for functional analysis. Plant Mol Biol 47:9–27

Williamson L, Ribrioux S, Fitter AH, Leyser MO 2001) Phosphate availability regulates root system architecture in *Arabidopsis thaliana*. Plant Physiol 126:875–882

Wortemann R, Herbette S, Barigah TS, Fumanal B, Alia R, Ducousso A, Gomory D, Roeckel-Drevet P, Cochard H (2011) Genotypic variability and phenotypic plasticity of cavitation resistance in *Fagus sylvatica* L. across Europe. Tree Physiol 31:1–8

Yoshimura K (2011) Hydraulic function contributes to the variation in shoot morphology within the crown in *Quercus crispula*. Tree Physiol 31 774–781

Zeppel MJB, Henry D, Adams HD, Anderegg WRL (2011) Mechanistic causes of tree drought mortality: recent results, unresolved questions and future research needs. New Phytol 192:800–803

Zimmermann U, Meizer FC, Benkert R, Zhu JJ, Schneider H, Goldstein G, Kuchenbrod E, Haase A (1994) Xylem water transport: is the availabe evidence consistent with the cohesion theory? Plant Cell Environ 17:1169–1181

Zimmermann U, Schneider H, Wegner LH, Haase A. 2004. Water ascent in tall trees: does evolution of land plants rely on a highly metastable state? New Phytologist 162:575–615

Zufferey V, Cochard H, Ameglio T; Spring JL, Viret, O (2011). Diurnal cycles of embolism formation and repair in petioles of grapevine (*Vitis vinifera* cv. Chasselas). J Exp Bot 62:3885–3894

Zwieniecki MA, Holbrook NM (1998) Diurnal variation in xylem hydraulic conductivity in white ash (*Fraxinus americana* L.) red maple (*Acer rubrum* L.) and red spruce (*Picea rubens* Sarg.). Plant Cell Environ 21:1173–1180

Zwieniecki MA, Holbrook NM (2009) Confronting Maxwell's demon: biophysics of xylem embolism repair. Trends Plant Sci 14:530–534

Zwieniecki MA, Melcher PJ, Holbrook NM (2001) Hydrogel control of xylem hydraulic resistance in plants. Science 291:1059–1062

Part II
Physiological Responses

Chapter 4
Regulation of Root Water Uptake Under Drought Stress Conditions

Ricardo Aroca and Juan Manuel Ruiz-Lozano

Abstract Drought is one of the most stressful conditions limiting plant yield around the world. Most efforts have been made in studying how aerial parts contribute to plant drought tolerance, being the role of roots less investigated. However, there are studies where a correlation between root water uptake capacity under drought conditions and drought tolerance has been found. Root water uptake capacity depends on morphological, anatomical, and molecular features of roots. A correlation between investment in root biomass during drought and tolerance to drought is hard to establish fromthe literature data. A better correlation between drought tolerance and root length density has been found. However, the capacity of absorbing water varies along a given root. Also, apoplastic barriers develop under drought conditions, limiting root water transport to some extent. Here new findings that question this assumption are presented. Finally, the exact role of aquaporin in the regulation of root hydraulic properties under drought conditions is far from being understood. This lack of knowledge is mainly caused by the large number of aquaporin isoforms present in the genomes of plants (up to 70 in cotton). Sub-cellular localization and knockout studies for each kind of aquaporins are needed in order to clarify their role in the regulation of root water uptake under drought conditions.

R. Aroca (✉) · J. M. Ruiz-Lozano
Departamento de Microbiología del Suelo y Sistemas Simbióticos, Estación Experimental del Zaidín (CSIC), Profesor Albareda 1, 18008 Granada, Spain
e-mail: raroca@eez.csic.es

R. Aroca (ed.), *Plant Responses to Drought Stress*,
DOI: 10.1007/978-3-642-32653-0_4, © Springer-Verlag Berlin Heidelberg 2012

4.1 Introduction

Drought stress is one of the main environmental factors causing restrictions of plant production in the world (Khan et al. 2010). The descent of the soil water potential is the environmental factor that initiates plant responses to drought. In fact, such descent induces the generation of several signals such as the production of abscisic cid (ABA) in root tissues (Wang et al. 1999) or some hydraulic signals transmitted along the root axis (Ionenko et al. 2012). At the same time, since water always moves following a water potential gradient, roots need to decrease their hydraulic conductivity(L) and their osmotic potential in order to avoid dehydration in the root cells (North and Nobel 1992). Although stomatal closure is the most evident symptom of drought stress (Farooq et al. 2009), changes in root hydraulic properties are also common responses to drought conditions (North and Nobel 1992; Silva et al. 2004). Indeed, dehydration avoidance during drought stress is a consequence of a tight balance between stomatal movements and root water uptake capacity (Aroca et al. 2008). Thus, Matsuo et al. (2009) found a positive correlation between shoot dry weight and root hydraulic conductance values among three rice genotypes under drought conditions. Hence, a conciliation between avoiding root tissue dehydration and taking enough water to supply shoot demands under drought conditions has to be reached (Ionenko et al. 2012).

Root water uptake capacity is the consequence of intrinsic root hydraulic properties of a given plant (Knipfer et al. 2011). Water can flow from soil to xylem vessels following three different pathways: apoplastic, symplastic, and transcellular ones. The apoplastic path includes the water circulating along cell walls and intercellular spaces. The symplastic path comprises the water circulating inside the cells through plasmodesmata. Finally, the transcellular path consists of the water flowing across the cell membranes (plasmalemma and tonoplast). Symplastic and transcellular paths cannot be distinguish experimentally, so the sum of both pathways is called cell-to-cell pathway. At the same time, there is an exchange of water among the three pathways depending on the environmental conditions. Thus, under non-stressful conditions water circulates mainly through the apoplastic path due to hydrostatic forces. Conversely, when transpiration is restricted at night or under stressful conditions, water flows predominantly through the cell-to-cell path because of the osmotic forces. Under some circumstances, it has been found that water flows only via the cell-to-cell path (Knipfer and Fricke 2010). For revisions about general concepts of root water uptake readers are referred to Zhao et al. (2004) and Aroca et al. (2012).

Some roots develop exo- and endodermis, which restrict the water flow through the apoplastic path, and therefore water has to cross cell membranes to reach the xylem vessels (Steudle and Peterson 1998). There are many situations in which water has to cross at least one membrane before reaching xylem vessels and consequently it has to flow through aquaporins.

Aquaporins are membrane intrinsic proteins that allow the passage of water and other small molecules through them always following an osmotic gradient (Maurel et al. 2008). In plants, aquaporins are divided into five subfamilies: plasma membrane intrinsic proteins (PIPs), tonoplast intrinsic proteins (TIPs), nodulin-26-like intrinsic proteins (NIPs), small and basic intrinsic proteins (SIPs), and only in dicots, uncharacterized intrinsic proteins (XIPs) (Maurel et al. 2008; Bienert et al. 2011; Lopez et al. 2012). Aquaporins specially PIPs and TIPs should be involved in the regulation of root water uptake under drought conditions (Aroca et al. 2012). In this chapter, we reviewed the current knowledge about how drought stress affects root morphology, anatomy, and aquaporins regulation in plants and how these changes modify root water uptake capacity.

4.2 Drought Effects on Root Morphology and Anatomy

The most common effect of drought stress is an increment of the root:shoot ratio (Wu and Cosgrove 2000). Such a change in the proportion of root biomass with respect to shoot biomass can be caused by a reduction in shoot growth without any change in root growth (Zollinger et al. 2006), by a direct investment in root growth (McMichael and Quisenberry 1991), or because root growth is less restricted than shoot growth (Li 1998). The increase of root: shoot ratio under drought conditions has been seen as a mechanism to explore more soil volume in order to absorb water from deeper soil layer, which is not available for less developed roots. However, no clear relationship between root:shoot ratio and drought tolerance has been always observed (McMichael and Quisenberry 1991; Li 1998; Liu et al. 2000; Matsui and Singh 2003; Zollinger et al. 2006). This lack of correlation can be explained because there are differences in water uptake capacity in the root system. Thus, Camposeo and Rubino (2003) observed that under drought conditions cowpea roots increased the capacity of absorbing water only of the deeper younger roots. Similar results were found by Kondo et al. (2000) in maize plants. At the same time, a better correlation between root:shoot ratio and drought tolerance has been observed when the shoot part was expressed in terms of area instead of weight (Matsui and Singh 2003). In addition, root length or root length density under drought conditions have been correlated with water stress tolerance. Thus, Grzesiak et al. (1999) found that a drought tolerant variety of maize had longer nodal roots than a sensitive variety under drought conditions. Furthermore, Jongrungklang et al. (2011) observed a correlation between root density in deeper soil layers and yield in different peanut genotypes with contrasting tolerance to pre-flowering drought stress. However, although it could be a relationship between root growth under drought conditions and drought tolerance (Grezsiak et al. 1999; Jongrungklang et al. 2011), it is clear that this parameter alone cannot explain the higher capacity of water absorption of such roots. Thus, when comparing maize versus rice, Kondo et al. (2000) found that maize plants were able to absorb more water from deeper soil layers under severe drought conditions than rice, and this was not only explained by a better root development of maize plants, but also by a higher water uptake capacity per unit of root surface. So, there must be intrinsic

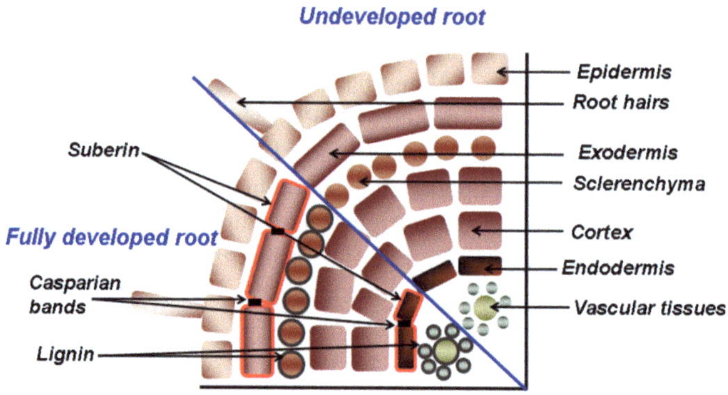

Fig. 1 Scheme of inner parts of the undeveloped (*right portion*) or fully developed (*left portion*) roots. Suberin layers are drawn in *red*, casparian bands in *black* and lignin in *grey*

properties of root tissues that make a particular root more efficient in terms of water absorption under drought conditions than others.

Roots may develop apoplastic barriers to water and solutes from just 10 mm from their root apexes to their basal zones. These barriers may consist of suberin layers deposited in the inner parts of the cell walls of endo and/or exodermis, lignin depositions in sclerenchyma and stellar cells, and casparian bands between exo and/or exodermis cells (Schreiber 2010; Ranathunge et al. 2011, Fig. 4.1). It is well known that these barriers increase under drought conditions (Huang and Nobel 1992) and can diminish root water uptake capacity estimated as root hydraulic conductivity(L) (Huang and Nobel 1992; Steudle 2000). However, most recently it has been found that these barriers do not always decrease L. Thus, Ranathunge et al. (2011) found that the development of casparian bands in both exo and endodermis, the increase in suberin layers, and the lignification of stellar tissues and the sclerenchyma in rice roots subjected to oxygen deprivation had no effect on L values. Moreover, Ranathunge and Schreiber (2011) described that Arabidopsis mutant plants accumulating less amounts of aliphatic suberin had indeed higher L values than wild-type plants, while Arabidopsis mutant plants accumulating higher amounts of aliphatic suberin did not present lower L values than wild type plants. These two studies question the relationship between depositions of apoplastic barriers and root water uptake capacity. Such discrepancies can be explained by the contribution of passage cells in the endodermis to water flux, by the formation of pores big enough to permit water flow or by different polymerization of suberin (Ranathunge and Schreiber 2011; Ranathunge et al. 2011).

Besides the studies cited above, it has been reported that suberization (Huang and Nobel 1992), lignification (North and Nobel 2000), and development of casparian bands (Stasovski and Peterson 1993) occur also under drought conditions, and that these barriers could impair root water absorption (Steudle 2000; Aroca et al. 2012). Many studies dealing with the relationship between apoplastic barriers and L values have been done in desert succulent plants. Thus, North and Nobel (2000) subjected different portions of *Agave deserti* roots to different water regimes and observed an inverse correlation between the development of suberin layers and lignification of endodermal cells and L. Upon rewatering, water uptake is restored via the proximal portions of roots, which develop fewer suberin layers than the distal ones (North and Baker 2007). In fact, Lo Gullo et al. (1998) found that roots that developed a well-formed suberin layer in both exo- and endodermis were unable to recover their capacity to absorb water upon rewatering. Similar results were found by Meyer et al. (2011) in *Iris germanica* plants. *I. germanica* roots posses a multiseriate exodermis whose suberization and formation of casparian bands are increased under drought conditions. By puncturing application it was demonstrated that this suberized exodermis impaired L values.

The development of these apoplastic barriers can be also seen as a mechanism to diminish the loss of water from roots to soil and therefore avoiding cellular death. Taleisnik et al. (1999) found a good correlation between the presence of suberized exodermal layers and a diminution of water lost from root to soil. However, in *Chloris gayana* plants it was observed that soil sheaths around the root formed under drought conditions were more determinant than exodermis development to restrict water loss from roots to soil (Taleisnik et al. 1999). It has also been reported that chickpea roots lost their epidermal and cortical cells upon exposure to drought, but the remaining endodermal cells and stele were able to resume growth upon soil rehydration, therefore, suberized endodermal cells prevented collapse of stellar cells (Spaeth and Cortes 1995).

Besides apoplastic barriers, some plants such as maize and rice develop an aerenchyma in their roots in response to drought (Zhu et al. 2010; Yang et al. 2012). This aerenchyma formation restricts root water uptake capacity and, depending of its intensity, it may improve or reduce plant drought tolerance (Zhu et al. 2010; Yang et al. 2012).

So, although it seems that apoplastic barriers may diminish root water uptake capacity under drought conditions, the most recent findings (Ranathunge and Schreiber 2011; Ranathunge et al. 2011) indicate that the amount or even the composition of suberin layers cannot totally explain the differences in L. Therefore, the type of polymerization of the apoplastic barrier, the spaces inside suberized layers, and the role of passage cells should be all taken into account to explain L responses to drought. Also, how different Arabidopsis suberin mutants behave under drought conditions should be tested. Simultaneously, since water can flow through two different paths in its radial way to the xylem vessels, the participation of aquaporins in the cell-to-cell path must also be considered.

4.3 Drought Effects on Root Hydraulic Conductivity (L)

As previously commented, L is very sensitive to drought (Aroca et al. 2012). In most of the studies, L decreased upon exposure to drought (North and Nobel 1991, 1998; Silva et al. 2004; Trifilo et al. 2004; Aroca et al. 2006). However, under some special environmental conditions or in specific genotypes, an increase in L was observed during drought exposure (Wikbergi and Ogreni 2007; Alsina et al. 2011). In these two studies woody plants tolerant to drought (specific willow and grape clones) were used. Alsina et al. (2011) found that a root clone of grapevine tolerant to drought increased its L during summer, and such an increase was correlated with a greater production. Similarly, Wikbergi and Ogreni (2007) found that the higher the ratio between shoot hydraulic conductance and L during moderate drought, the higher the biomass production in willow. So, these woody plants enhanced their L values during drought to overcome the loss of water by leaves in order to keep growing. How this increment in L takes place is still unknown. However, McLean et al. (2011) found that in a riparian tree (*Melaleuca argentea*) one half of the roots increased the capacity of absorbing water in a wet zone when the other half of the roots was deprived of water. Such an increase of water absorption capacity in wetted roots was accompanied by an increase in L and in the abundance of PIP1 aquaporin proteins. Hence, this study shows up the importance of aquaporins in controlling L, and the existence of signal communication between dehydrated and wetted roots.

However, drought stress causes L diminution in most of the cases (Aroca et al. 2012), most probably in order to avoid the flow of water outside the roots. Several factors may be causing the inhibition of L under drought conditions. Anatomical and morphological changes must be involved as commented above. Another likely cause could be a descent in the hydraulic conductivity of root cells (Lc). However, Vandeleur et al. (2009) found that L diminished in Chardonnay grapevine roots during prolonged drought stress while Lc of cortical cells increased. In fact, these authors proposed that the lower inhibition of L caused by drought in Chardonnay grapevine plants when compared to Grenache ones was due to the increase in Lc of cortical cells in the former cultivar, since both types developed similar apoplastic barriers. Most recently, Hachez et al. (2012) found an increase in Lc of cortical cells after 2 h of exposure to an osmotic shock (simulated by 10 % polyethylene glycol) in maize roots without any further change in L or even a decrease. Consequently these authors postulated that other cells beside cortical ones should be checkpoints in L regulation, since there was no time enough to build new apoplastic barriers. Nevertheless, the increases in Lc caused by drought could be seen as a mechanism to tolerate water stress when an osmotic adjustment occurs at the same time, and water could not flow back to the soil. Gong et al. (2010) found a higher osmotic potential gradient between roots and shoots in a drought-tolerant oat genotype than in a sensitive one under drought conditions. So, this higher osmotic potential gradient could facilitate the transport of water from roots to shoots in the tolerant oat genotype.

Other cause of L inhibition during drought can be the formation of root xylem embolism. When leaf transpiration exceeds root water absorption under drought conditions the water continuum of some xylem vessels can be disrupted and air bubbles can be formed. Although stem xylem embolism under drought conditions has been extensively studied (Tyree and Ewers 1991; Martínez-Vilalta et al. 2004), root xylem embolism also occurs (Martínez-Vilalta et al. 2002; Domec et al. 2004; Limousin et al. 2010). Indeed, Martínez-Vilalta et al. (2002) studying nine woody species from Mediterranean oak forest found that all of them presented higher xylem vulnerability in the roots than in the stems. Also, these authors showed that roots were working closer to this water potential point of loss of conductivity than stems. Similar results were found by Limousin et al. (2010) in *Quercus ilex* trees. So, root xylem embolism was more probably to occur in roots than in stems. These results are important since root xylem embolism has been seldom taken into account in studies about embolism caused by drought. More studies dealing with this specific field are absolutely required.

4.3.1 Hydraulic Lift

One of the mechanisms that plants have to overcome during drought stress is the redistribution of water from wetted to dried zones of the soil. This mechanism is called hydraulic lift (Liste and White 2008). In fact, a correlation between the capacity of plants to transfer water from deeper wet zones of the soil to dryer ones and tolerance to drought has been observed in different maize genotypes (Wan et al. 2000). This hydraulic lift not only achieves better water status of drought-tolerant plants, but also increases the transpiration rate and therefore, can increase both CO_2 assimilation rate and plant growth (Wan et al. 2000). In the same way, increase of hydraulic lift under drought conditions has been observed in several crop species and such an increase was related with the maintenance of photosynthesis rate unchanged under drought conditions (Zegada-Lizarazu and Iijima 2004). Hydraulic lift not only contributes to absorption of water by the same plant that is moving the water, but also by adjacent plants. Hence, the use of plants with deeper roots systems, reaching groundwater planting aside of the interested crops may delay adverse drought effects in such crops (Sekiya and Yano 2004). Also, this agronomical technique can be improved by shadowing of the plant doing the hydraulic lift (Sekiya and Yano 2004). Finally, water redistribution from wet to dryer soil layers can also operate from top to bottom soil layers after raining events (Oliveira et al. 2005). In spite of the great importance of hydraulic lift for drought stress tolerance, the molecular mechanisms behind this process are totally unknown. Hydraulic lift depends on soil and root conductivities, water potential gradients, and the intensity of transpiration. Up-to-date, how these factors are interconnected is not completely understood (Neumann and Cardon 2012).

4.4 Drought Effects on Aquaporins

One of the causes of the regulation of root water uptake under drought conditions can be the regulation of aquaporins activity (Aroca et al. 2012). Thus, in some cases a correlation between changes induced by drought stress in L and PIPs protein abundance has been found (Aroca et al. 2006, 2007), but in other cases

Table 1 Plant species, drought treatment, effects on L, PIPs protein abundance and bibliographic references

Plant species	Drought treatment	Effects on L	PIPs protein amount	Reference
Phaseolus vulgaris	4 d without watering	Decreased	PIP2 descent PIP1 unchanged	Aroca et al. (2006)
Phaseolus vulgaris	4 d without watering	Decreased	PIP2 descent PIP1 increased	Aroca et al. (2007)
Jatropha curcas	24 h at 40 % PEG[a]	Decreased	PIP2 increased	Zhang et al. (2007)
Oryza sativa	10 h 20 PEG	Unchanged	PIP1 descent	Lian et al. (2004)
Zea mays	4 d without watering	Unchanged	PIP2;1 and PIP2;5 descent PIP1;2 unchanged	Ruiz-Lozano et al. (2009)

[a] PEG: Polyethyleneglycol

such correlation is lacking (Lian et al. 2004; Zhang et al. 2007; Ruiz-Lozano et al. 2009, Table 4.1). Curiously, the correlation was always found when L was compared with PIP2 protein abundance and not when it was compared with PIP1 proteins (Table 4.1). This fact could be explained since PIP2 aquaporins always showed more water transport capacity than PIP1 when expressed in heterologous systems (Azad et al. 2008; Otto et al. 2010). However, PIP1 aquaporins can modulate the activity of PIP2 aquaporins by forming heterotetramers (Fetter et al. 2004). On the other hand, the correlation between PIP2 protein abundance and L was also found in plants subjected to water deficit during several days and by suppressing irrigation, but there was no relationship in plants subjected to short-term drought treatments (few hours) or in plants submitted to polyethylene glycol (PEG) as an osmotic agent (Table 4.1). It is curious that Zhang et al. (2007) found a descent of L and at the same time an increase in PIP2 protein amount. This discrepancy could be caused by localization of PIP2 proteins in internal membranes where they cannot contribute to water transport from external medium. Similar findings have been extensively described for salt stress (Boursiac et al. 2005, 2008; Li et al. 2011; Luu et al. 2012), but regarding drought treatment, it is just a theory that needs to be confirmed experimentally.

Most of the antibodies used in the works summarized in Table 4.1 were non-specific for a single aquaporin PIP isoform, recognizing several isoforms of PIP1s or PIP2s proteins (Aroca et al. 2006). Thus, although these antibodies show a general pattern of PIPs protein responses to drought, each PIP isoform may respond in a different way to drought stress. 10 out the 19 analyzed PIP genes shown in Table 4.2 were downregulated, 5 upregulated and 4 unchanged after drought stress treatment. At the same time, in most of these studies L was not measured or the expression of other PIP aquaporin genes remained unanalyzed. Therefore, it is difficult to find a correlation between PIP aquaporin gene expression and L.

Another approach to ascertain the role of aquaporins in the regulation of L under drought conditions is the use of transgenic plants overexpressing or lacking

Table 2 Plant species, drought treatment, L response, PIP gene expression, and bibliographic references

Plant species	Drought treatment	L response	PIP expression	Reference
Phaseolus vulgaris	4 d without watering	Decreases	*PvPIP*2;1 and *PvPIP*1;1 up-regulated	Aroca et al. (2006)
Nicotiana tabacum	PEG treatment (−0.35 MPa) 24 h	Decreases	*NtPIP*1;1 and *NtPIP*2;1 down-regulated, *NtAQP*1 up-regulated	Mahdieh et al. (2008)
Glycine max	10 d at 70 % of soil water holding capacity	Not measured	*GmPIP*1 and *GmPIP*2 down-regulated	Porcel et al. (2006)
Lactuca sativa	10 d at 70 % of soil water holding capacity	Not measured	*LsPIP*1 and *LsPIP*2 down-regulated	Porcel et al. (2006)
Populus nigra clones Poli and 58–861	3 weeks without watering plus 2 weeks at 25 % of field capacity	Not measured	*PIP*1;2 down-regulated in Poli and up-regulated in 58–861	Cocozza et al. (2010)
Lactuca sativa	Grown at −0.3 MPa	Not measured	*LsPIP*2 down-regulated	Alguacil et al. (2009)
Zea mays	4 d without watering	Unchanged	*ZmPIP*1;1 up-regulated, *ZmPIP*1;2, 1;5, 2;1, 2;2 unchanged and *ZmPIP*2;5 and 2;6 down-regulated	Ruiz-Lozano et al. (2009)

specific aquaporin genes. However, there are very few studies investigating the effects of altering both aquaporin expression and L in plant drought tolerance. Among them, Lian et al. (2004) found that drought-sensitive rice genotype overexpressing a PIP1 gene showed an enhancement of L under drought conditions and a better water status than untransformed plants. However, several researchers evaluated how the overexpression of aquaporin genes modifies drought tolerance of a given plant (Table 4.3). Only three out the ten examples described in Table 4.3 show an increase of drought tolerance in the transformed plants, all of them involving a PIP1 aquaporin gene. In other three cases, overexpression of PIP1 gene induced drought sensitivity (Table 4.3). The fact that a decrease in drought tolerance was achieved in the two studies in which a TIP gene was overexpressed is also remarkable (Peng et al. 2007; Wang et al. 2011). It was postulated that TIP aquaporins could be involved in the osmotic adjustment under osmotic stress conditions such as drought stress, using the transport of water through the tonoplast as a buffering system (Li et al. 2008). Consequently, a plant would lose its osmotic adjustment capacity when forced to transport too much water through the tonoplast.

Table 3 Plant species, overexpressed gene, drought treatment, traits analyses, and source of a compilation of researches where an aquaporin gene was overexpressed

Plant species	Overexpressed gene	Drought treatment	Traits analyses (drought tolerance)	Reference
Arabidopsis thaliana	*VfPIP*1	25 d without watering	More root growth, less transpiration rate and mortality (increased)	Cui et al. (2008)
Arabidopsis thaliana	*PgTIP*1	10 d without watering	More transpiration and dehydration (decreased)	Peng et al. (2007)
Nicotiana tabacum	*BjPIP*1	10 % PEG 6,000 2–30 h	Less transpiration and dehydration (increase)	Zhang et al. (2008)
Nicotiana tabacum	*AtPIP*1*b*	Stop watering (no data on duration)	More wilting (decreased)	Aharon et al. (2003)
Nicotiana tabacum	*BnPIP*1	10 d without watering	Less wilting (increased)	Yu et al. (2005)
Arabidopsis thaliana	*AtPIP*1;4 or *AtPIP*2;5	10 d without watering	Less survival (decreased)	Jang et al. (2007a)
Arabidopsis thaliana	*GsTIP*2;1	Grown at 200–300 mM Manitol	Less survival (decreased)	Wang et al. (2011)
Arabidopsis thaliana	*CsPIP*1;1	Grown at 4d at 300 mM Manitol	Less survival (decreased)	Jang et al. (2007b)
Arabidopsis thaliana	*CfPIP*2;1	Grown at 4d at 300 mM Manitol	More survival (increased)	Jang et al. (2007b)

Knockout studies suggest that PIP aquaporins could regulate L (Javot et al. 2003; Postaire et al. 2010), although their role during drought stress is far to be understood. The high diversity of PIP aquaporins in plants (up to 28 depending on the species; Park et al. 2010) contributes to the difficulty of reaching conclusions. Also, PIP proteins are not always located in the plasma membrane, being able to migrate to other cell membranes depending on several environmental stimuli (Boursiac et al. 2005, 2008; Luu et al. 2012). Hence, more research is needed to clarify the role of aquaporins in the response of L to drought stress.

4.5 Conclusion

Traditionally stomatal function during drought stress has been considered as a key point in the regulation of plant tolerance to drought. However, root water uptake capacity under drought conditions seems to be crucial when coping with water deficit events (Matsuo et al. 2009). The different water uptake capacity under drought conditions among different plant species/genotypes is not related with a higher investment in root biomass, but with specific hydraulic characteristics of some roots (Kondo et al. 2000). The involvement of apoplastic barriers in restricting root water uptake (Ranathunge and Schreiber 2011; Ranathunge et al. 2011), and the involvement of root embolism should be further explored. At the same time, the frequently observed lack of correlation between PIP aquaporin abundance or expression and L under drought conditions deserves future studies in order to be clarified. Moreover, the specific subcellular localization of each PIP isoform during drought has to be studied and a complete set of knockout lines for each PIP isoform have to be tested in relation to their drought tolerance and L response to water stress.

References

Aharon R, Shahak Y, Wininger S, Bendov R, Kapulnik Y, Galili G (2003) Overexpression of a plasma membrane aquaporin in transgenic tobacco improves plant vigor under favourable growth conditions but not under drought or salt stress. Plant Cell 15:439–447

Alguacil MM, Kohler J, Caravaca F, Roldán A (2009) Differential effects of *Pseudomonas mendoica* and *Glomus intraradices* on lettuce plants physiological response and aquaporin PIP2 gene expression under elevated atmospheric CO_2 and drought. Microb Ecol 58:942–951

Alsina MM, Smart DR, Bauerle T, de Herralde F, Biel C, Stockert C, Negron C, Savé R (2011) Seasonal changes of whole root conductance by a drought-tolerant grape root system. J Exp Bot 62:99–109

Aroca R, Ferrante A, Vernieri P, Chrispeels MJ (2006) Drought, abscisic acid and transpiration rate effects on the regulation of PIP aquaporin gene expression and abundance in *Phaseolus vulagaris* plants. Ann Bot 98:1301–1310

Aroca R, Porcel R, Ruiz-Lozano JM (2007) How does arbuscular mycorrhizal symbiosis regulate root hydraulic properties and plasma membrane aquaporins in *Phaseolus vulgaris* under drought, cold or salinity stresses? New Phytol 173:808–816

Aroca R, Vernieri P, Ruiz-Lozano JM (2008) Mycorrhizal and non-mycorrhizal *Lactuca sativa* plants exhibit contrasting responses to exogenous ABA during drought stress and recovery. J Exp Bot 59:2029–2041

Aroca R, Porcel R, Ruiz-Lozano JM (2012) Regulation of root water uptake under abiotic stress conditions. J Exp Bot 63:42–57

Azad AK, Katsuhara M, Sawa Y, Ishikawa T, Shibata H (2008) Characterization of four plasma membrane aquaporins in tulip petals: a putative homolog is regulated by phosphorylation. Plant Cell Physiol 49:1196–1208

Bienert GP, Bienert MD, Jahn TP, Boutry M, Chaumont F (2011) Solanaceae XIPs are plasma membrane aquaporins that facilitate the transport of many uncharged substrates. Plant J 66:306–317

Boursiac Y, Chen S, Luu DT, Sorieul M, van den Dries N, Maurel C (2005) Early effects of salinity on water transport in Arabidopsis roots. Molecular and cellular features of aquaporin expression. Plant Physiol 139:790–805

Boursiac Y, Boudet J, Postaire O, Luu DT, Tournaire-Roux C, Maurel C (2008) Stimulus-induced downregulation of root water transport involves reactive oxygen species-activated cell signalling and plasma membrane intrinsic protein internalization. Plant J 56:207–218

Camposeo S, Rubino P (2003) Effect of irrigation frequency on root water uptake in sugar beet. Plant Soil 253:301–309

Cocozza C, Cherubini P, Reiger N, Saurer M, Frey B, Tognetti R (2010) Early effects of water deficit on two parental clones of *Populus nigra* grown under different environmental conditions. Funct Plant Biol 37:244–254

Cui XH, Hao FS, Chen H, Chen J, Wang XC (2008) Expression of the *Vicia faba* VfPIP1 gene in Arabidopsis thaliana plants improves their drought resistance. J Plant Res 121:207–214

Domec JC, Warren JM, Meinzer FC, Brooks JR, Coulombe R (2004) Native root xylem embolism and stomatal closure in stands of Douglas-fir and ponderosa pine: mitigation by hydraulic redistribution. Oecologia 141:7–16

Farooq M, Wahid A, Kobayashi N, Fujita D, Basra SMA (2009) Plant drought stress: effects, mechanisms and management. Agron Sustain Dev 29:185–212

Fetter K, Van Wilder V, Moshelion M, Chaumont F (2004) Interactions between plasma membrane aquaporins modulate their water channel activity. Plant Cell 16:215–228

Gong DS, Xiong YC, Ma BL, Wang TM, Ge JP, Qin XL, Li PF, Kong HY, Li ZZ, Li FM (2010) Early activation of plasma membrane H^+-ATPase and its relation to drought adaptation in two contrasting oat (*Avena sativa* L.) genotypes. Environ Exp Bot 69:1–8

Grzesiak S, Hura T, Grzesiak MT, Pienkowski S (1999) The impact of limited soil moisture and waterlogging stress conditions on morphological and anatomical root traits in maize (Zea mays L.) hybrids of different drought tolerance. Acta Physiol Plant 21:305–315

Hachez C, Veselov D, Ye Q, Reinhardt H, Knipfer T, Fricke W, Chaumont F (2012) Short-term control of maize cell and root water permeability through plasma membrane aquaporin isoforms. Plant Cell Environ 35:185–198

Huang BR, Nobel PS (1992) Hydraulic conductivity and anatomy for lateral roots of Agave deserti during root-growth and drought-induced abscission. J Exp Bot 43:1441–1449

Ionenko IF, Dautova NR, Anisimov AV (2012) Early changes of water diffusional transfer in maize roots under the influence of water stress. Environ Exp Bot 76:16–23

Jang JY, Lee SH, Rhee JY, Chung GC, Ahn SJ, Kang H (2007a) Transgenic Arabidopsis and tobacco plants overexpressing an aquaporin respond differently to various abiotic stresses. Plant Mol Biol 64:621–632

Jang JY, Rhee JY, Kim DG, Chung GC, Lee JH, Kang H (2007b) Ectopic expression of a foreign aquaporin disrupts the natural expression patterns of endogenous aquaporin genes and alters plant responses to different stress conditions. Plant Cell Physiol 48:1331–1339

Javot H, Lauvergeat V, Santoni V, Martin-Laurent F, Guclu J, Vinh J, heyes J, Franck KI, Schaffner AR, Bouchez D, Maurel C (2003) Role of a single aquaporin isoform in root water uptake. Plant Cell 15:509–522

Jongrungklang N, Toomsan B, Vorasoot N, Jogloy S, Boote KJ, Hoogenboom G, Patanothai A (2011) Rooting traits of peanut genotypes with different yield responses to pre-flowering drought stress. Field Crop Res 120:262–270

Khan HR, Paull JG, Siddique KHM, Stoddard FL (2010) Faba bean breeding for drought-affected environments: a physiological and agronomic perspective. Field Crop Res 115:279–286

Knipfer T, Fricke W (2010) Root pressure and solute reflection coefficient close to unit exclude a purely apoplastic pathway of radial water transport in barley (*Hordeum vulgare*). New Phytol 187:159–170

Knipfer T, Besse M, Verdeil JL, Fricke W (2011) Aquaporin-facilitated water uptake in barley (*Hordeum vulgare* L.) roots. J Exp Bot 62:4115–4126

Kondo M, Murty MVR, Aragones DV (2000) Characteristics of root growth and water uptake from soil in upland rice and maize under water stress. Soil Sci Plant Nutr 46:721–732

Li C (1998) Variation of seedling traits of *Eucalyptus microtheca* origins in different water regimes. Silvae Genetica 47:132–136

Li GW, Peng YH, Yu X, Zhang MH, Cai WM, Sun WN, Su WA (2008) Transport functions and expression analysis of vacuolar membrane aquaporins in response to various stresses in rice. J Plant Physiol 165:1879–1888

Li XJ, Wang XH, Yang Y, Li RL, He QH, Fang XH, Luu DT, Maurel C, Lin XJ (2011) Single-molecule analysis of PIP2;1 dynamics and partitioning reveals multiple modes of Arabidopsis plasma membrane aquaporin regulation. Plant Cell 23:3780–3797

Lian HL, Yu X, Ye Q, Ding XS, Kitagawa Y, Kwak SS, Su WA, Tang ZC (2004) The role of aquaporin RWC3 in drought avoidance in rice. Plant Cell Physiol 45:481–489

Limousin JM, Longepierre D, Huc R, Rambal S (2010) Change in hydraulic traits of Mediterranean *Quercus ilex* subjected to long-term through fall exclusion. Tree Physiol 30:1026–1036

Liste HH, White JC (2008) Plant hydraulic lift of soil water: implications for crop production and land restoration. Plant Soil 313:1–17

Liu ZM, Thompson K, Spencer RE, Reader RJ (2000) A comparative study of morphological responses of seedling roots to drying soil in 20 species from different habitats. Acta Bot Sin 42:628–635

Lo Gullo MA, Nardini A, Salleo S, Tyree MT (1998) Changes in root hydraulic conductance (K-R) of *Olea oleaster* seedlings following drought stress and irrigation. New Phytol 140:25–31

Lopez D, Bronner G, Brunel N, Auguin D, Bourgerie S, Brignolas F, Carpin S, Tournaire-Roux C, Maurel C, Fumanal B, Martin F, Sakr S, Label P, Julien JL, Gousset-Dupont A, Venisse JS (2012) Insight into *Populus* XIP aquaporins: evolutionary expansion, protein functionality, and environmental regulation. J Exp Bot 63:2217–2230

Luu DT, Martiniere A, Sorieul M, Runions J, Maurel C (2012) Fluorescence recovery after photobleaching reveals high cycling dynamics of plasma membrane aquaporins in Arabidopsis roots under salt stress. Plant J 69:894–905

Mahdieh M, Mostajeran A, Horie T, Katsuhara M (2008) Drought stress alters water relations and expression of PIP-type aquaporin genes in *Nicotiana tabacum* plants. Plant Cell Physiol 49:801–813

Martínez-Vilalta J, Prat E, Oliveras I, Piñol J (2002) Xylem hydraulic properties of roots and stems of nine Mediterranean woody species. Oecologia 133:19–29

Martínez-Vilalta J, Sala A, Piñol J (2004) The hydraulic architecture of Pinaceae: a review. Plant Ecol 171:3–13

Matsui T, Singh BB (2003) Root characteristics in cowpea related to drought tolerance at the seedling stage. Exp Agric 39:29–38

Matsuo N, Ozawa K, Mochizuki T (2009) Genotypic differences in root hydraulic conductance of rice (*Oryza sativa* L.) in response to water regimes. Plant Soil 316:25–34

Maurel C, Verdoucq L, Luu DT, Santoni V (2008) Plant aquaporins: membrane channels with multiple integrated functions. Annu Rev Plant Biol 59:595–624

McLean EH, Ludwig M, Grierson PF (2011) Root hydraulic conductance and aquaporin abundance respond rapidly to partial root-zone drying events in a riparian *Melaleuca* species. New Phytol 192:664–675

McMichael BL, Quisenberry JE (1991) Genetic variation for root-shoot relationships among cotton germplasm. Environ Exp Bot 31:461–470

Meyer CJ, Peterson CA, Steudle E (2011) Permeability of *Iris germanica's* multiseriate exodermis to water, NaCl, and ethanol. J Exp Bot 62:1911–1926

Newmann RB, Cardon ZG (2012) The magnitude of hydraulic redistribution by plant roots: a review and synthesis of empirical and modelling studies. New Phytol 194:337–352

North GB, Baker EA (2007) Water uptake by older roots: Evidence from desert succulents. HortScience 42:1103–1106

North GB, Nobel PS (1991) Changes in hydraulic conductivity and anatomy caused by drying and rewetting roots of *Agave deserti* (Agavaceace). Am J Bot 78:906–915

North GB, Nobel PS (1992) Drought-induced changes in hydraulic conductivity and structure in roots of *Ferocactus acanthodes* and *Opuntia ficus-indica*. New Phytol 120:9–19

North GB, Nobel PS (1998) Water uptake and structural plasticity along roots of a desert succulent during prolonged drought. Plant Cell Environ 21:705–713

North GB, Nobel PS (2000) Heterogeneity in water availability alters cellular development and hydraulic conductivity along roots of a desert succulent. Ann Bot 85:247–255

Oliveira RS, Dawson TE, Burgess SSO, Nepstad DC (2005) Hydraulic redistribution in three Amazonian trees. Oecologia 145:354–363

Otto B, Uehlein N, Sdorra S, Fischer M, Ayaz M, Belasategui-Macadam X, Heckwolf M, Lachnit M, Pede N, Priem N, Reinhard A, Siegfart S, Urban M, Kaldenhoff R (2010) Aquaporin tetramer composition modifies the function of tobacco aquaporins. J Biol Chem 285:31253–31260

Park W, Scheffler BE, Bauer PJ, Campbell BT (2010) Identification of the family of aquaporin genes and their expression in upland cotton (*Gossypium hirsutum* L.). BMC Plant Biol 10:142
Peng YH, Lin WL, Cai WM, Arora R (2007) Overexpression of a *Panax ginseng* tonoplast aquaporin alters salt tolerance, drought tolerance and cold acclimation ability in transgenic Arabidopsis plants. Planta 226:729–740
Porcel R, Aroza R, Azcón R, Ruiz-Lozano JM (2006) PIP aquaporin gene expression in arbuscular mycorrhizal *Glycine max* and *Lactuca sativa* plants in relation to drought stress tolerance. Plant Mol Biol 60:389–404
Postaire O, Tournaire-Roux C, Grondin A, Boursiac Y, Morillon R, Schaffner AR, Maurel C (2010) A PIP1 aquaporin contributes to hydrostatic pressure-induced water transport in both the root and rosette of Arabidopsis. Plant Physol 152:1418–1430
Ranathunge K, Schreiber L (2011) Water and solute permeabilities of Arabidopsis roots in relation to the amount and composition of aliphatic suberin. J Exp Bot 62:1961–1974
Ranathunge K, Lin JX, Steudle E, Schreiber L (2011) Stagnat deoxygenated growth enhances root suberization and lignifications, but differentially affects water and NaCl permeabilities in rice (*Oryza sativa* L.) roots. Plant Cell Environ 34:1223–1240
Ruiz-Lozano JM, Alguacil MM, Bárzana G, Vernieri P, Aroca R (2009) Exogenous ABA accentuates the differences in root hydraulic properties between mycorrhizal and non mycorrhizal maize plants through regulation of PIP aquaporins. Plant Mol Biol 70:565–579
Schreiber L (2010) Transport barriers made of cutin, suberin and associated waxes. Trends Plant Sci 15:546–553
Sekiya N, Yano K (2004) Do pigeon pea and sesbania supply groundwater to intercropped maize through hydraulic lift? Hydrogen stable isotope investigation of xylem waters. Field Crop Res 86:167–173
Silva FCE, Shaleva A, Maroco JP, Almeida MH, Chaves MM, Pereira JS (2004) Responses to water stress in two *Eucalyptus globules* clones differing in drought tolerance. Tree Physiol 24:1165–1172
Spaeth SC, Cortes PM (1995) Root cortex death and subsequent initiation and growth of lateral roots from bare steles of chickpeas. Can J Bot 73:253–261
Stasovski E, Peterson CA (1993) Effects of drought and subsequent rehydration on the structure, vitality, and permeability of *Allium cepa* adventitious roots. Can J Bot 71:700–707
Steudle E (2000) Water uptake by root: effects of water deficit. J Exp Bot 51:1531–1542
Steudle E, Peterson CA (1998) How does water gets through roots? J Exp Bot 49:775–788
Taleisnik E, Peyrano G, Cordoba A, Arias C (1999) Water retention capacity in root segments differing in the degree of exodermis development. Ann Bot 83:19–27
Trifilo P, Raimondo F, Nardini A, Lo Gullo MA, Salleo S (2004) Drought resistance of Ailanthus altissima: root hydraulics and water relations. Tree Physiol 24:107–114
Tyree MT, Ewers FW (1991) The hydraulic architecture of tress and other woody-plants. New Phytol 119:345–360
Vandeleur RK, Mayo G, Shelden MC, Gilliham M, Kaiser BN, Tyerman SD (2009) The role of plasma membrane intrinsic protein aquaporins in water transport through roots: diurnal and drought stress responses reveal different strategies between isohydric and anisohydric cultivars of grapevine. Plant Physiol 149:445–460
Wan CG, Xu WW, Sosebee RE, Machado S, Archer T (2000) Hydraulic lift in drought-tolerant and -susceptible maize hybrids. Plant Soil 219:117–126
Wang HF, Zhang JH, Liang JH, Yin WL (1999) Root and xylem ABA changes in response to soil drying in two woody plants. Chin Sci Bull 44:2236–2241
Wang X, Li Y, Ji W, Bai X, Cai H, Zhu D, Sun XL, Chen LJ, Zhu YM (2011) A novel *Glycine soja* tonoplast intrinsic protein gene responds to abiotic stress and depresses salt and dehydration tolerance in transgenic *Arabidopsis thaliana*. J Plant Physiol 168:1241–1248
Wikbergi J, Ogreni E (2007) Variation in drought resistance, drought acclimation and water conservation in four willow cultivars used for biomass production. Tree Physiol 27:1339–1346
Wu YJ, Cosgrove DJ (2000) Adaptation of roots to low water potentials by changes in cell wall extensibility and cell wall proteins. J Exp Bot 51:1543–1553

Yang X, Li Y, Ren B, Ding L, Gao C, Shen Q, Guo S (2012) Drought-induced root aerenchyma formation restricts water uptake in rice seedlings supplied with nitrate. Plant Cell Physiol 53:495–504

Yu QJ, Hu YL, Li JF, Wu Q, Lin ZP (2005) Sense and antisense expression of plasma membrane aquaporin *BnPIP1* from *Brassica napus* in tobacco and its effects on plant drought tolerance. Plant Sci 169:647–656

Zegada-Lizarazu W, Iijima M (2004) Hydrogen stable isotope analysis of water acquisition ability of deep roots and hydraulic lift in sixteen food crop species. Plant Prod Sci 7:427–434

Zhang Y, Wang YX, Jiang LD, Xu Y, wang YC, Lu DH, Chan F (2007) Aquaporin JcPIP2 is involved in drought responses in *Jatropha curcas*. Acta Biochim Biophys Sin 39:787–794

Zhang YX, Wang Z, Chai TY, Wen ZS, Zhang HM (2008) Indian mustard aquaporin improves drought and heavy-metal resistance in tobacco. Mol Biotechnol 40:280–292

Zhao CX, Deng XP, Zhang SQ, Ye Q, Steudle E, Shan L (2004) Advances in the studies on water uptake by plant roots. Acta Bot Sin 46:505–514

Zhu JM, Brown KM, Lynch JP (2010) Root cortical aerenchyma improves the drought tolerance of maize (*Zea mays* L.). Plant Cell Environ 33:740–749

Zollinger N, Kjelgren R, Cerny-Koenig T, Kopp K, Koenig R (2006) Drought responses of six ornamental herbaceous perennials. Sci Hortic 109:267–274

Chapter 5
The Response of Photosynthesis to Soil Water Stress

Jaume Flexas, Alexander Gallé, Jeroni Galmés, Miquel Ribas-Carbo and Hipólito Medrano

Abstract The physiological and molecular basis of photosynthetic responses to limited soil water availability (water stress) has been intensively examined over the last decade(s). Therefore, this chapter highlights the major achievements of the underlying processes of photosynthetic limitation under drought, an increasingly important issue within the context of climate change. Restricted CO_2 diffusion to the sites of carboxylation inside the chloroplast has been demonstrated to be the main limiting factor for photosynthesis, particularly during the early phases of stress. Stomatal (g_s) and mesophyll conductance (g_m), the two leaf diffusion components, contribute differently to this limitation, being largely influenced by the degree of water deficit. Thus, photosynthetic acclimation to drought and its recovery from drought depend primarily on the capacity to adjust g_m and g_s rapidly. The basis of g_m and g_s regulation is not fully understood, but several genetic, metabolic, and structural factors involved have been recently described. Secondary stress factors such as excessive light and elevated temperatures affect photosynthetic performance too, implying efficient photoprotection a necessary feature for stress-resistant plants.

J. Flexas (✉) · A. Gallé · J. Galmés · M. Ribas-Carbo · H. Medrano
Universitat de les Illes Balears, Grup de Recerca en Biologia de les Plantes en Condicions Mediterrànies, Carretera de Valldemossa Km 7.5, 07122 Palma de Mallorca, Spain
e-mail: jaume.flexas@uib.es

A. Gallé
e-mail: alexander.galle@uib.es

J. Galmés
e-mail: jeroni.galmes@uib.es

M. Ribas-Carbo
e-mail: mribas@uib.es

H. Medrano
e-mail: hipolito.medrano@uib.es

R. Aroca (ed.), *Plant Responses to Drought Stress*,
DOI: 10.1007/978-3-642-32653-0_5,

5.1 Introduction

Low water availability is considered as the main environmental factor limiting plant growth and yield worldwide, especially in semi-arid areas. It is well documented that one of the primary physiological impacts of drought is on photosynthesis (Flexas et al. 2004a). Presently, not only in the arid and semiarid zones drought is a constraint, but it is also increasingly affecting temperate regions occasionally subjected to severe drought events (Giorgi and Lionello 2008; Battisti and Naylor 2009). Improving the knowledge on photosynthesis responses to water stress is essential for the development of deficit irrigation programs, as well as for improving the accuracy of ecosystem productivity predictions from climate data.

There has been some controversy regarding the main physiological targets responsible for photosynthetic impairment under drought and/or salinity (Chaves 1991; Flexas and Medrano 2002; Lawlor and Cornic 2002). As a consequence of stomatal closure, CO_2 diffusion from the atmosphere to the site of carboxylation is reduced, and this is often regarded as the main cause for decreased photosynthesis under drought and salinity (Centritto et al. 2003; Loreto et al. 2003; Chaves and Oliveira 2004; Flexas et al. 2004a, b, 2009; Grassi and Magnani 2005; Chaves et al. 2009; Peeva and Cornic 2009). On the other hand, some authors have suggested that metabolic impairment due to water stressinduced decrease of leaf water content and increased ion concentration is more limiting for photosynthesis than stomatal closure (Tezara et al. 1999, 2002; Tang et al. 2002). More recently, decreased leaf internal diffusion of CO_2 (i.e., decreased mesophyll conductance, g_m) has been identified as another potential cause for photosynthesis impairment under drought (Flexas et al. 2002; Galmés et al. 2007a) and salinity (Centritto et al. 2003). No general consensus exists on the relative importance of each of these factors in limiting photosynthesis under drought, which could in fact depend on the prevailing light conditions (Zhou et al. 2007a; Flexas et al. 2009; Gallé et al. 2009), leaf and plant age (Varone et al. 2012), and differ between different drought intensities, as well as between drought imposition, acclimation, and recovery upon rewatering (Flexas et al. 2006a). The aim of this chapter is to review the current state-of-knowledge on photosynthetic responses of plants to water stress. We will focus mostly on C_3 plants, for which sufficient knowledge has been gained as to build up some generalizations. Studies are scarcer for C_4 plants (Ghannoum 2009) and, especially, for CAM plants, which should be an important research priority for the near future.

5.2 Diffusional and Biochemical Limitations During Drought Imposition in C_3 Plants

Because stomatal closure is among the earliest physiological events occurring in response to drought, and because stomatal conductance (g_s) often correlates strongly with net CO_2 assimilation (A_N), it has been frequently assumed that stomatal closure reduces CO_2 uptake in drought leaves. On the other hand, applying large CO_2 concentrations around leaves overcomes diffusion limitations to CO_2 and several reports have shown that very high CO_2 fully restores maximum photosynthesis in water stressed leaves (Kaiser 1987; Cornic et al. 1989). Moreover, by stripping the epidermis—where stomata reside—from leaves, stomatal limitation is removed, allowing CO_2 to freely diffuse into the leaf. Using this approach, some studies have shown that photosynthesis can be fully restored in water stressed leaves, supporting the idea of a stomatal limitation (Dietz and Heber 1983; Schwab et al. 1989). In other studies, however, it was not possible to restore A_N in water stressed leaves by either applying high CO_2 (Graan and Boyer 1990; Tezara et al. 1999) or stripping leaf epidermis (Tang et al. 2002). These discrepancies in the results obtained in similar experiments have fed the controversy as to whether stomatal or non-stomatal limitations—the latter often used as synonymous of biochemical limitations—are the main cause for decreased photosynthesis under water stress. One study by Tezara et al. (1999), in particular, had a strong influence in this debate. These authors suggested that the main limiting factor for photosynthesis under water stress was impaired photophosphorylation due to decreased chloroplast ATPase activity. Despite the important influence of this paper, the data in Tezara's study were relatively scarce and not fully conclusive. The experiment was performed by inducing a progressive water stress to sunflower plants, and leaf water potential was used as the indicator for water stress severity. While leaf ATP content decreased progressively along the entire gradient of water potential used during the experiment, this could be due to decreased ATP synthesis in the chloroplast or in other cell compartments. Data for chloroplast ATP content were available only at very low water potentials, when photosynthesis was already close to zero. Almost immediately after its publication, some authors claimed against the validity of the conclusions of this study (Cornic 2000), while others showed that impaired Rubisco activity, RuBP regeneration and even photoinhibition occur at similar levels of water stress as those causing impaired photophosphorylation (Flexas and Medrano 2002). Indeed, other reports have suggested that inactivation of Rubisco is the main cause for photosynthesis limitation under drought (Parry et al. 2002), including another study by Tezara et al. (2002) in drought stressed sunflower.

In an attempt to solve for the occurrence of this apparent controversy, Lawlor and Cornic (2002) compiled literature data and reanalyzed them using leaf relative water content as a common indicator for the intensity of water stress. Unfortunately, such analysis did not solve the controversy, as two clearly opposed patterns were found. The so-called 'Type I' response (see Fig. 5.1a) consisted in progressive decreases of A_N as RWC decreased, mainly as a consequence of stomatal closure, since the photosynthetic capacity (i.e., after overcoming stomatal limitations) was unaffected until very low RWC was reached. The 'Type II' response, in contrast, consisted in parallel decreases of A_N and photosynthetic capacity as RWC declined, supporting a predominant role for non-stomatal limitations (Fig. 5.1a). The occurrence of 'Type I' or 'Type II' responses was not found to be associated to specific genotypes or conditions during the experiments, so that even a single species could display one or another response depending on the study. In other words, there is no clear evidence on what does each type depends on, so that the 'types' may most likely reflect differences in the definition/determination method of photosynthetic capacity. In parallel, Flexas et al. (2002) showed in a study with field-grown grapevines that using daily maximum g_s as the common indicator for water stress intensity homogenized the responses between genotypes and growing conditions, which were different when using RWC as a reference. Moreover, it was shown that mesophyll conductance to CO_2 (g_m)—i.e., the facility for internal CO_2 diffusion inside leaves—, which had been typically considered large and constant, progressively decreased as water stress intensified, just as it does in salt-stressed plants (Centritto et al. 2003). This finding implied that not only stomatal closure and impaired photosynthetic capacity could be responsible for drought-induced photosynthesis, but also decreased g_m could play a major role. A limited leaf mesophyll conductance to CO_2 implies that removing leaf epidermis and/or increasing CO_2 concentration around leaves may not always result in fully saturating Rubisco. In this sense, many of the data used to conclude the occurrence of biochemical impairment of photosynthesis could not be valid.

Flexas and Medrano (2002) extended the use of g_s as the common basis for assessing the degree of drought to literature data on many species, and found general relationships between g_s and A_N regardless of the species and experimental conditions. Flexas et al. (2004b) specifically showed a common response between Type I, Type II, and even Type III (consisting in very isohydric plants, whose RWC does not change appreciably during drought imposition, see Fig. 5.1a), when using g_s instead of RWC as the reference parameter indicative of water stress. More recent studies have further re-analyzed data on a g_s basis, from which a large consensus has been reached as to how progressive water stress sequentially induces downregulation of the different components associated to photosynthesis (Flexas et al. 2004a,b, 2006a,b; de Souza et al. 2005; Grassi and Magnani 2005; Jiang et al. 2006; Galmés et al. 2007a; Zhou et al. 2007a; Liu et al. 2010). The common pattern consists in three phases, as follows (see Fig. 5.1b):

Phase 1: Mild water stress ($g_s > 0.15$ mol H_2O m^{-2} s^{-1}). When g_s declines from a maximum to about 0.15 mol H_2O m^{-2} s^{-1}, as a consequence of progressively increasing water shortage, decreased g_s is the only cause for reduced A_N. The rate of linear electron transport (ETR), g_m, the activities of photosynthetic enzymes, the maximum quantum efficiency of PSII (Fv/Fm), the maximum velocity of carboxylation ($V_{c,max}$), etc., remain constant during this phase. The rate of photorespiration (PR) progressively increases, as a consequence of decreased substomatal (C_i) and chloroplast (C_c) CO_2 concentrations. It is remarkable that this threshold is very common among plants, so that even plants whose maximum g_s is below 0.15 mol H_2O m^{-2} s^{-1}—for instance, slcerophyll species—accomplish this rule, being never found on phase 1 but starting at phase 2 from field capacity. This is recognizable by the fact that these plants show linear rather than curvilinear responses of ETR or g_m to g_s (Galmés et al. 2007a), i.e., these parameters do not remain constant at any gs, unlike what happens in phase 1.

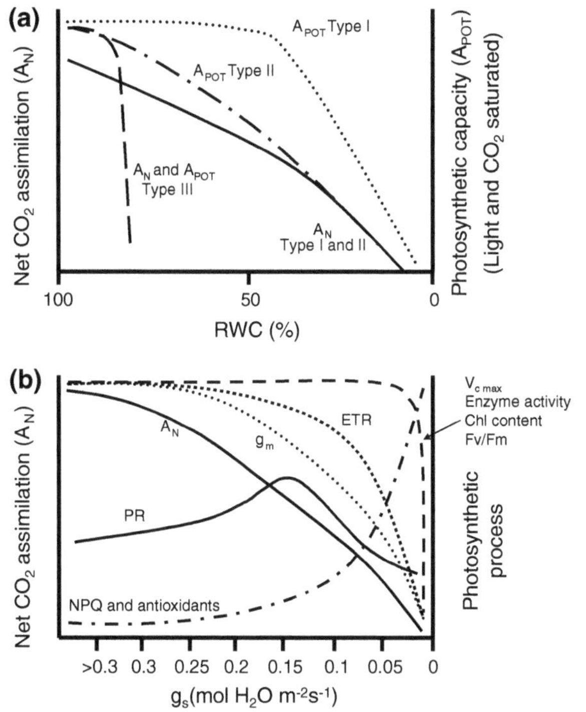

Fig. 5.1 **a** The response of net photosynthesis (A_N) and photosynthetic capacity (A_{POT}) to water stress when using leaf relative water (RWC) content as the reference for stress intensity. Three different Types of response appear, depending on the experiment. Modified after Lawlor and Cornic (2002) and Flexas et al. (2004b). **b** Generalized response of net photosynthesis (A_N) and several parameters related to photosynthetic capacity (see text for the meaning of abbreviations) to water stress when using daily maximum leaf stomatal conductance (g_s) content as the reference for stress intensity

Phase 2: Moderate water stress ($0.15 > g_s > 0.05$ mol H_2O m^{-2} s^{-1}). During this phase, further decreases of g_s are accompanied by large decreases of g_m. Some studies suggest that the decrease in g_m is associated with changes in the abundance of some aquaporins (Flexas et al. 2006c; Miyazawa et al. 2008), although further studies are required to fully elucidate how internal diffusion of CO_2 in leaves is regulated. Still during this phase, small but significant decreases of ETR occur, which are accompanied by subtle increases of non-photochemical quenching of chlorophyll fluorescence (NPQ), the activity of antioxidant enzymes, and the amounts of non-enzymatic antioxidants. These responses suggest that the leaves are preparing for conditions of severe stress, which can lead to secondary oxidative stress. The use of traditional gas exchange analysis (i.e., A_N-C_i curves) to evaluate the presence of non-stomatal limitations in this phase leads to the erroneous conclusion that $V_{c,max}$ is decreased. In fact, these apparent decreases in $V_{c,max}$ are simply an effect of decreased g_m, and the analysis of A_N -C_c curves shows almost constant $V_{c,max}$ during this phase. Neglecting g_m also results in erroneous estimates of the maximum rate of ETR (J_{max}, Centritto et al. 2003; Aganchich et al. 2009).

Phase 3: Severe water stress ($g_s < 0.05$ mol H_2O m^{-2} s^{-1}). When g_s drops below this threshold value the photosynthetic capacity is impaired in many studies, although not in others. Interestingly, the metabolic impairment generally consists of the simultaneous inhibition of all photosynthetic enzymes, decreases in chlorophyll and protein contents, and the appearance of permanent damage to the photosystems, which suggest that the leaves are enduring oxidative stress and/or inducing leaf senescence and remobilization of leaf nutrients (Munne-Bosch et al. 2001). In the studies where metabolic impairment is not apparent, it remains unclear whether it is indicative of highly resistant genotypes or because water stress was ameliorated by, e.g., moderate to low light and/or mild temperature. Metabolic impairment, when it occurs, follows a tightly regulated system, consisting in an impaired Rubisco activity at low g_s due to reduced C_c (Flexas et al. 2006b; Galmés et al. 2011a, b). However, whether the impairment was due to reduced activation state of Rubisco, reduced enzyme concentration and/or increased concentration of tightly bound inhibitors seems to be dependent on the species and the technique used to impose water stress (Bota et al. 2004; Flexas et al. 2006c). On the other hand, Zhou et al. (2007a) found that Rubisco impairment was strongly correlated to the presence of H_2O_2 regardless of the light conditions during water stress, suggesting either a role of H_2O_2 as a secondary messenger in the regulation of Rubisco activity or as causing factor of oxidative stress.

This response pattern to progressively declining g_s is similar to that often observed to progressively declining fraction of transpirable soil water (FTSW, e.g. Centritto et al. 2011), simply because g_s responds directly to FTSW rather than to leaf water potential or RWC (Turner et al. 1985). For this reason, the use of FTSW instead of RWC or even g_s as a tool to assess the general responses of photosynthetic capacity-related parameters seems an easy and promising way of comparing results from different experiments, species, etc.

A similar pattern of response has been described for C_4 plants (Hura et al. 2006, 2007; Ghannoun, 2009), except that leaf photosynthetic metabolism of C_4 plants—particularly Rubisco in bundle sheath cells—seems somewhat more sensitive to water stress than it is in C_3 plants (Carmo-Silva et al. 2007, 2008a,b). The response of photosynthesis to water stress in C_4 plants may be to some extent different to that of C_3 plants. Although generally stomatal limitations are also present, and may be dominant at the very early stages of water stress imposition (Marques da Silva and Arrabaça 2004), biochemical limitations often appear also at early stages of drought, immediately after stomatal limitations (Du et al. 1996; Lal and Edwards 1996; Saccardy et al. 1996; Ghannoun et al. 2002, 2003; Carmo-Silva et al. 2007, 2008a, b). In particular, Rubisco seems very sensitive to water stress, declining linearly with stress intensity in many C_4 species (Du et al. 1996; Carmo-Silva et al. 2007) but not in some others (Lal and Edwards 1996). By contrast, PEPC and C_4 acid decarboxylases are more resistant to water stress, declining only eventually under severe stress conditions (Saccardy et al. 1996; Carmo-Silva et al. 2008a). Although g_m is thought to be not limiting in C_4 plants because of their CO_2-concentrating mechanisms, it has been suggested that bundle sheath CO_2 leakiness could increase under water stress, leading to reduced CO_2 availability and a contributing to a less efficient fixation (Carmo-Silva et al. 2008b). Potential differences in the response to stress among different C_4 subtypes remain elusive (Carmo-Silva et al. 2007, 2008a, b).

In summary, there is a quite general response of photosynthesis to progressive water stress imposition across C_3 plants when g_s is used as the normalizing reference for the level of water stress intensity. Although g_s can respond very differently to leaf water potential, soil water content, etc., depending on the species or genotypes and on environmental conditions, the factors limiting photosynthesis at any given g_s can be accurately predicted. Despite the universality of the pattern, some of the limitations can differ depending on the prevailing light conditions during drought. For instance, under low light conditions g_m is not so highly depressed by water stress (Flexas et al. 2009, Galle et al. 2009). Moreover, the extent of the different limitations appear to vary during acclimation in some cases (Jiang et al. 2006), although not in others (Flexas et al. 2009), and they certainly vary significantly depending on the prevailing air humidity conditions (Perez-Martin et al. 2009) and on the age of leaves and plants subjected to water stress (Varone et al. 2012). Further studies are required in which photosynthesis responses to water stress are evaluated under different conditions, to improve our capacity for predicting photosynthesis limitations under water stress.

5.3 Secondary Oxidative Stress Effects on Photosynthesis Under Combined Excess Light and Water Stress

When CO_2 availability in the chloroplasts is drastically reduced under drought, the use of electrons for CO_2 assimilation in the Calvin cycle decreases. Excess of electrons can be diverted to other processes, such as PR or thermal dissipation; the latter being considered a major process in plant photoprotection under stress conditions (Demmig-Adams and Adams 2006). However, when these processes are saturated, ETR components become overreduced, resulting in electrons being transferred to oxygen at PSI or via the Mehler reaction. This generates reactive oxygen species (ROS), such as superoxide (O_2^-), hydrogen peroxide (H_2O_2), and the hydroxyl radical that can cause oxidative damage to the photosynthetic apparatus if the plant is not efficient in scavenging these molecules.

Depending on the duration and severity of the stress, the plant species, the specific tissue/organ and the phase of development, ROS can elicit antioxidant responses, typically observed under mild stress, or can lead to accelerated senescence, programmed cell death or necrosis (Levine 1999; Munne-Bosch and Alegre 2002). Acclimation to stress is generally associated with enhanced activity of the antioxidant molecules, which are able to remove oxygen radicals or repair the damage, thus keeping ROS concentration relatively low (Smirnoff 1998; Dat et al. 2000). Antioxidant activity include enzymatic and non-enzymatic mechanisms such as (a) the superoxide dismutases (SODs), which catalyze the dismutation of O_2^- to H_2O_2, (b) the catalases (CATs), which are responsible for the removal of H_2O_2, and (c) the enzymes and metabolites of the ascorbate-glutathione cycle, which are also involved in the removal of H_2O_2 (Foyer and Noctor 2003). Increased activities of plant antioxidant systems under stress have been considered characteristic of drought-resistant species (e.g., Gao et al. 2009). In a recent work by Rivero et al. (2007), the expression of isopentenyltransferase (IPT) that catalyzes the rate-limiting step in cytokinin (CK) synthesis led to increased concentrations of enzymes associated with the glutathione-ascorbate cycle in transgenic plants, and resulted in the suppression of drought-induced leaf senescence and in an outstanding drought-tolerance. In transgenic plants with elevated CK production that were subjected to drought, a 20 % of the upregulated transcripts were related to ROS metabolism. The efficient scavenging of ROS protects the photosynthetic apparatus during drought stress, leading to improved water use efficiency of the transgenic plants during and after stress. The presence of high light intensity and high ROS levels during water stress have been shown to exacerbate drought effects on g_m (Flexas et al. 2009; Galle et al. 2009), leaf photochemistry and Rubisco activity (Zhou et al. 2007a). It has been also hypothesized that oxidative damage to the chloroplast ATPase, which was caused by ROS under conditions of low CO_2 and excess light, resulted in a water stress-induced decrease of the photosynthetic capacity (Lawlor and Tezara 2009).

In addition to the deleterious effects, ROS can also serve as secondary messengers in the signaling for the activation of defence responses (Dat et al. 2000). This dual function of ROS, first described in responses to pathogens and later demonstrated in response to several abiotic stresses, presumably plays an important role on the acclimation processes (Dat et al. 2000), which may deserve better attention in future studies.

5.4 Photosynthesis Limitations During Rewatering After Drought Stress

The carbon balance of a plant following a complete period of water stress and recovery may depend as much on the velocity and degree of the recovery of photosynthesis after stress relief, as it certainly depends on the degree and velocity of photosynthesis decline during stress imposition (Flexas et al. 2006a; Centritto et al. 2011). In general, plants subject to severe water stress recover only 40–60 % of the maximum photosynthesis rate during the day after rewatering, and recovery continues during the next few days, although maximum photosynthesis rates are not fully recovered in some cases (Kirschbaum 1988; Sofo et al. 2004; Grzesiak et al. 2006; Bogeat-Triboulot et al. 2007; Gallé et al. 2007). The extent and velocity of recovery may depend on the severity of the stress endured prior to rewatering. The influence of previous water stress episodes was illustrated by Miyashita et al. (2005) and Grzesiak et al. (2006). Here, we grouped data available from the literature in three intensity levels of previous stress episodes to obtain a more general picture on how this affects the velocity of recovery after rewatering (Table 5.1). For the three groups, average g_s prior to rewatering was ≤ 0.05 mol H_2O m^{-2} s^{-1}, i.e., they were all at Phase 3 described in Sect. 5.2. However, the three groups differed in the endured inhibition of photosynthesis. When A_N was, on average, 36 % of maximum values observed in control plants, total recovery of A_N occurred in 4 days. When A_N was only 23 % of control, then full recovery took up

Table 5.1 Time required for achieving full recovery of photosynthesis upon rewatering after water stress. Water stress intensities (reflected by g_s and A_N before rewatering) were classified into 3 different ranges

g_s before re-watering ($mol\ H_2O\ m^{-2}\ s^{-1}$)	A_N before re-watering (% of control values)	Time to achieve full recovery of A_N (days)
0.018 ± 0.004	3.0 ± 1.2	18.2 ± 6.7
0.046 ± 0.010	22.9 ± 3.2	6.2 ± 2.5
0.048 ± 0.012	36.1 ± 1.8	4.3 ± 1.0

Data from: de Souza et al. (2004), Miyashita et al. (2005), dos Santos et al. (2006), Hura et al. (2006), Bougeat-Triboulot et al. (2007), Cai et al. (2007), Galle et al. (2007), Montanaro et al. (2007), Perez-Perez et al. (2007), Pou et al. (2008) and Galle et al. (2009)

to 6 days, and when it was as low as 3 % of controls full recovery took up to 18 days (Table 5.1).

Concerning the physiological mechanisms limiting recovery, Kirschbaum (1988) showed that recovery after a severe dehydration was a two-stage process: the first stage occurs during the first days upon rewatering, and is associated to recovery of water status and stomata reopening; the second stage lasts several days and likely requires de novo synthesis of photosynthetic proteins. Concerning the first of these two phases, however, in some species a sustained downregulation of g_s after rewatering imposes substantial limitations to photosynthesis, while increasing intrinsic water-use efficiency (Bogeat-Triboulot et al. 2007; Gallé and Feller 2007; Gallé et al. 2007, 2009; Galmés et al. 2007a; Flexas et al. 2009; Xu et al. 2009). Orange trees that endured severe water stress do not fully recover g_s even after two months of rewatering (Fereres et al. 1979). In some of these cases, limited recovery of leaf-specific hydraulic conductivity is the likely cause for the long-term downregulation of g_s after rewatering (Galmés et al. 2007b; Pou et al. 2008). Resco et al. (2009) have demonstrated specifically that the number of days necessary to reach maximum photosynthesis after a pulse of rain (i.e., the inverse of velocity of recovery) depends on antecedent, drought-induced percentage loss of hydraulic conductance, and Brodribb and Cochard (2009) reached a similar conclusion in conifers. Alternatively, in beech stomatal occlusions formed during water stress restrained recovery of g_s after rewatering (Gallé and Feller 2007). During the first days after rewatering, g_m remains low in some species (Galmés et al. 2007a) but it reverses fast in many others (Flexas et al. 2009; Galle et al. 2009, 2011; Varone et al. 2012), for which it is not regarded as a major limiting factor for photosynthesis recovery.

As for the second phase, Bogeat-Triboulot et al. (2007) showed recently that recovery after water stress, determined 10 days after rewatering, was accompanied by increases in some photosynthetic proteins, particularly Rubisco activase and proteins of the water splitting complex, although increased proteins transcripts were not detected. In the cases where photosynthesis recovery is slow and/or incomplete, photoinhibition and/or oxidative stress have been suggested as possible causes (Sofo et al. 2004; Gallé et al. 2007).

In summary, restricted velocity of photosynthesis recovery after rewatering causes significant losses of carbon gains in plants, and the precise understanding of the mechanisms leading to such slow recovery requires further investigation.

5.5 Photosynthesis Acclimation: Genomics and Proteomics

The response of photosynthesis to water stress may differ between acclimated and non-acclimated plants. Acclimation to water stress may lead to homeostatic compensation for the initial negative effects of water stress on photosynthesis, and involves gene expression and modification of plant physiology and morphology, taking place in days to weeks. Osmotic adjustment is perhaps the best example of physiological acclimation to water stress (Chaves and Oliveira 2004; see also Chap. 7). Very few evidences for photosynthetic acclimation to water stress have been described. For instance, in some studies, leaves unfolded under water stress show somewhat higher photosynthesis rates than non-acclimated leaves, associated with morphological adaptations and higher ETR rates (Maury et al. 1996; Kitao et al. 2003; Galmés et al. 2006). Recently, Galle et al. (2011) observed a permanent downregulation of g_s in *Cistus* plants subject to periodical cycles of water stress and recovery, appearing after the first cycle, which can be considered an acclimation response. However, *Quercus* plants showed almost identical photosynthetic responses to each repeated drought cycle, i.e., showing no evidence for acclimation. Similarly, in many studies with different species in which a constant level of water stress was kept for a week or more, in order to allow for acclimation, no clear evidence for photosynthetic acclimation was found (Flexas et al. 2009; Galle et al. 2009; Varone et al. 2012).

Because acclimation involves gene expression, an alternative approach to evaluate the mechanisms leading to acclimation may be looking at changes in gene expression and protein contents, i.e., by genomic and proteomic approaches. In such studies, photosynthetic pathways are in general not among the most altered by the stress (Killian et al. 2007; Chaves et al. 2009). For example, in *Thellungiella* (a stress tolerant relative of *Arabidopsis*), only 15 % of all genes downregulated are involved in photosynthesis (Wong et al. 2006). In rice, alterations in photosynthesis related genes are mostly associated with stress recovery but not with stress imposition (Zhou et al. 2007b). Even in those photosynthetic genes responding to stress, the most common trend is a downregulation, i.e., they would not contribute to acclimation of photosynthesis, but rather to its further decline. In addition, the alterations found at transcriptomic level are larger (5–10 %) than at protein level (usually less than 1 %). However, transcriptomic analysis also in *Vitis* have shown that some photosynthetic genes, like those of the Rubisco activase, some Calvin cycle enzymes and some PSI- and PSII-related genes are instead upregulated during acclimation to water stress (Cramer et al. 2007). Although proteomic analysis showed that some photosynthetic proteins were downregulated during water stress, it also confirmed that some—notably

Rubisco and sedoheptulose-1,5-bisphosphatase— were indeed upregulated (Vincent et al. 2007). Moreover, the number of stress responsive genes or proteins seems to be related with the stress intensity (Cramer et al. 2007) and with tissue origin (Zhou et al. 2007b). On the other hand, a recent proteomics study in peanut cultivars has suggested that the response can also differ in drought-sensitive and drought-tolerant genotypes (Kottapalli et al. 2009). For instance, while Rubisco large and small subunits were decreased in both types of cultivars, some PSII proteins were decreased and ATP synthase increased only in the tolerant genotype (Kottapalli et al. 2009). Still, precaution has to be taken when analyzing gene expression data, since they may not necessarily have a reflection in the protein contents and physiological properties, due to post-transcriptional regulation. For instance, Bogeat-Triboulot et al. (2007) did not observe any correlation between the abundance of transcripts and proteins in an experiment with *Populus euphratica* subjected to gradual soil water depletion.

In summary, there is little evidence for acclimation of photosynthesis to water stress conditions in the short term, while the genetic basis of photosynthetic acclimation needs still to be determined.

5.6 Concluding Remarks

Major progress in the understanding of the physiological and molecular limitations and the signaling events underlying photosynthetic responses to drought has been achieved over the last decade(s), which is reviewed in the present chapter. The available evidence suggests that restricted CO_2 diffusion to the carboxylation sites is the main limiting factor for photosynthesis, especially during the early phases of stress. This is initially due to reduced stomatal conductance (g_s), but it also involves reduced mesophyll conductance (g_m). Under more severe drought conditions, which occur mostly under high irradiance, metabolic impairment (a consequence of oxidative stress) corresponds invariably to, and seems to be caused by, low g_s. Both physiological and molecular approaches suggest that very little acclimation of photosynthesis to prolonged water stress occurs among most plants. The extent and velocity of photosynthesis recovery upon rewatering depends on the intensity of water stress endured, and specifically on the degree of photosynthetic inhibition. The mechanisms limiting rapid recovery appear to diverge between plants and environmental conditions, but slow recovery of stomatal conductance seems to be the most common cause.

References

Aganchich B, Wahbi S, Loreto F, Centritto M (2009) Partial root zone drying: regulation of photosynthetic limitations and antioxidant enzymatic activities in young olive (*Olea europaea*) saplings. Tree Physiol 29:685–696

Battisti DS, Naylor RL (2009) Historical warnings of future food insecurity with unprecedented seasonal heat. Science 323:240–244

Bogeat-Triboulot M-B, Brosche M, Renaut J, Jouve L, Le Thiec D, Fayyaz P, Vinocur B, Witters E, Laukens K, Teichmann T, Altman A, Hausman J-F, Polle A, Kangasjarvi J, Dreyer E (2007) Gradual soil water depletion results in reversible changes of gene expression, protein profiles, ecophysiology, and growth performance in *Populus euphratica*, a poplar growing in arid regions. Plant Physiol 143:876–892

Bota J, Medrano H, Flexas J (2004) Is photosynthesis limited by decreased Rubisco activity and RuBP content under progressive water stress? New Phytol 162:671–681

Brodribb TJ, Cochard H (2009) Hydraulic failure defines the recovery and point of death in water-stressed conifers. Plant Physiol 149:575–584

Cai H, Biswas DK, Shang AQ, Zhao LJ, Li WD (2007) Photosynthetic response to water stress and changes in metabolites in *Jasminum sambac*. Photosynthetica 45:503–509

Carmo-Silva AE, Soares AS, da Silva JM, da Silva AB, Keys AJ, Arrabaça MC (2007) Photosynthetic responses of three C_4 grasses of different metabolic subtypes to water deficit. Funct Plant Biol 34:204–213

Carmo-Silva A, Bernardes da Silva A, Keys A, Parry M, Arrabaça MC (2008a) The activities of PEP carboxylase and the C4 acid decarboxylases are little changed by drought stress in three C4 grasses of different subtypes. Photosynth Res 97:223–233

Carmo-Silva AE, Powers SJ, Keys AJ, Arrabaça MC, Parry MAJ (2008b) Photorespiration in C4 grasses remains slow under drought conditions. Plant, Cell Environ 31:925–940

Chaves MM (1991) Effects of water deficits on carbon assimilation. J Exp Bot 42:1–16

Chaves MM, Flexas J, Pinheiro C (2009) Photosynthesis under drought and salt stress: regulation mechanisms from whole plant to cell. Ann Bot 103:551–560

Chaves MM, Oliveira MM (2004) Mechanisms underlying plant resilience to water deficits: prospects for water-saving agriculture. J Exp Bot 55:2365–2384

Centritto M, Loreto F, Chartzoulakis K (2003) The use of low [CO_2] to estimate diffusional and non-diffusional limitation of photosynthetic capacity of salt-stressed olive saplings. Plant, Cell Environ 26:585–594

CentrittoM Brilli F, Fodale R, Loreto F (2011) Different sensitivity of isoprene emission, respiration and photosynthesis to high growth temperature coupled with drought stress in black poplar (*Populus nigra*) saplings. Tree Physiol 31:275–286

Cornic G, Le Gouallec JL, Briantais JM, Hodges M (1989) Effect of dehydration and high light on photosynthesis of two C_3 plants. *Phaseolus vulgaris L. and Elastostema repens* (hour.) Hall f.). Planta 177:84–90

Cornic G (2000) Drought stress inhibits photosynthesis by decreasing stomatal aperture—not by affecting ATP synthesis. Trends Plant Sci 5:187–188

Cramer GR, Ergül A, Grimplet J, Tillett RL, Tattersall EAR, Bohlman MC, Vincent D, Sonderegger J, Evans J, Osborne C, Quilici D, Schlauch KA, Schooley DA, Cushman JC (2007) Water and salinity stress in grapevines: early and late changes in transcript and metabolite profiles. Funct Integr Genomics 7:111–134

Dat J, Vandenabeele S, Vranova E, Montagu MV, Inze D, Breusegem FV (2000) Dual action of the active oxygen species during plant stress responses. Cellular and molecular life sciences: CMLS 57:779–795

Demmig-Adams B, Adams WW (2006) Photoprotection in an ecological context: the remarkable complexity of thermal energy dissipation. New Phytol 172:11–21

Dietz KJ, Heber U (1983) Carbon dioxide gas exchange and the energy status of leaves of *Primula palinuri* under water stress. Planta 158:349–356

Dos Santos MG, Vasconcelos Ribeiro R, Ferraz de Oliveira R, Machado EC, Pimentel C (2006) The role of inorganic phosphate on photosynthesis recovery of common bean after a mild water deficit. Plant Sci 170:659–664

Du YC, Kawamitsu Y, Nose A, Hiyane S, Murayama S, Wasano K, Uchida Y (1996) Effects of water stress on carbon exchange rate and activities of of photosynthetic enzymes in leaves of sugarcane (*Saccharum* sp.). Aust J Plant Physiol 23:719–726

Fereres E, Cruz-Romero G, Hoffman GJ, Rawlins SL (1979) Recovery of orange trees following severe water stress. J Appl Ecol 16:833–842

Flexas J, Medrano H (2002) Drought-inhibition of photosynthesis in C_3 plants: stomatal and non-stomatal limitations revisited. Ann Bot 89:183–189

Flexas J, Bota J, Escalona JM, Sampol B, Medrano H (2002) Effects of drought on photosynthesis in grapevines under field conditions: an evaluation of stomatal and mesophyll limitations. Funct Plant Biol 29:461–471

Flexas J, Bota J, Loreto F, Cornic G, Sharkey TD (2004a) Diffusive and metabolic limitations to photosynthesis under drought and salinity in C_3 plants. Plant Biol 6:269–279

Flexas J, Bota J, Cifre J, Escalona JM, Galmés J, Gulias J, Lefi EK, Martinez-Canellas SF, Moreno MT, Ribas-Carbo M, Riera D, Sampol B, Medrano H (2004b) Understanding down-regulation of photosynthesis under water stress: future prospects and searching for physiological tools for irrigation management. Ann Appl Biol 144:273–283

Flexas J, Bota J, Galmés J, Medrano H, Ribas-Carbo M (2006a) Keeping a positive carbon balance under adverse conditions: responses of photosynthesis and respiration to water stress. Physiol Plant 127:343–352

Flexas J, Ribas-Carbo M, Bota J, Galmés J, Henkle M, Martinez-Canellas S, Medrano H (2006b) Decreased rubisco activity during water stress is not induced by decreased relative water content but related to conditions of low stomatal conductance and chloroplast CO_2 concentration. New Phytol 172:73–82

Flexas J, Ribas-Carbó M, Hanson DT, Bota J, Otto B, Cifre J, McDowell N, Medrano H, Kaldenhoff R (2006c) Tobacco aquaporin NtAQP1 is involved in mesophyll conductance to CO_2 in vivo. Plant J 48:427–439

Flexas J, Baron M, Bota J, Ducruet J-M, Galle A, Galmés J, Jimenez M, Pou A, Ribas-Carbo M, Sajnani C, Tomas M, Medrano H (2009) Photosynthesis limitations during water stress acclimation and recovery in the drought-adapted Vitis hybrid Richter-110 (*V. berlandieri*x*V. rupestris*). J Exp Bot 60:2361–2377

Foyer CH, Noctor G (2003) Redox sensing and signalling associated with reactive oxygen in chloroplasts, peroxisomes and mitochondria. Physiol Plant 119:355–364

Galle A, Feller U (2007) Changes of photosynthetic traits in beech saplings (*Fagus sylvatica* L.) under severe drought stress and during recovery. Physiol Plant 131:412–421

Galle A, Haldimann P, Feller U (2007) Photosynthetic performance and water relations in young pubescent oak (*Quercus pubescens*) trees during drought stress and recovery. New Phytol 174:799–810

Galle A, Florez-Sarasa I, Tomas M, Pou A, Medrano H, Ribas-Carbo M, Flexas J (2009) The role of mesophyll conductance during water stress and recovery in tobacco (*Nicotiana sylvestris*): acclimation or limitation? J Exp Bot 60:2379–2390

Galle A, Florez-Sarasa I, Aououad HE, Flexas J (2011) The Mediterranean evergreen *Quercus ilex* and the semi-deciduous *Cistus albidus* differ in their leaf gas exchange regulation and acclimation to repeated drought and re-watering cycles. J Exp Bot 62:5207–5216

Galmés J, Medrano H, Flexas J (2006) Acclimation of rubisco specificity factor to drought in tobacco: discrepancies between in vitro and in vivo estimations. J Exp Bot 57:3659–3667

Galmés J, Medrano H, Flexas J (2007a) Photosynthetic limitations in response to water stress and recovery in Mediterranean plants with different growth forms. New Phytol 175:81–93

Galmés J, Flexas J, Save R, Medrano H (2007b) Water relations and stomatal characteristics of Mediterranean plants with different growth forms and leaf habits: responses to water stress and recovery. Plant Soil 290:139–155

Galmés J, Conesa MA, Ochogavia JM, Perdomo JA, Francis DM, Ribas-Carbo M, Save R, Flexas J, Medrano H, Cifre J (2011a) Physiological and morphological adaptations in relation to water use efficiency in Mediterranean accessions of *Solanum lycopersicum*. Plant, Cell Environ 34:245–260

Galmés J, Ribas-Carbo M, Medrano H, Flexas J (2011b) Rubisco activity in Mediterranean species is regulated by the chloroplastic CO_2 concentration under water stress. J Exp Bot 62:653–665

Gao D, Gao Q, Xu H-Y, Ma F, Zhao C-M, Liu JQ (2009) Physiological responses to gradual drought stress in the diploid hybrid *Pinus densata* and its two parental species. Trees 213:717–728

Ghannoum O (2009) C4 photosynthesis and water stress. Ann Bot 103:635–644

Ghannoun O, von Caemmerer S, Conroy JP (2002) The effect of drought on plant water use efficiency on nine NAD-ME and nine NADP-ME Australian C_4 grasses. Funct Plant Biol 29:1337–1348

Ghannoum O, Conroy JP, Driscoll SP, Paul MJ, Foyer CH, Lawlor DW (2003) Nonstomatal limitations are responsible for drought-induced photosynthetic inhibition in four C_4 grasses. New Phytol 159:599–608

Giorgi F, Lionello P (2008) Climate change projections for the Mediterranean region. Global Planet Change 63:90–104

Graan T, Boyer JS (1990) Very high CO_2 partially restores photosynthesis in sunflower at low water potentials. Planta 181:378–384

Grassi G, Magnani F (2005) Stomatal, mesophyll conductance and biochemical limitations to photosynthesis as affected by drought and leaf ontogeny in ash and oak trees. Plant, Cell Environ 28:834–849

Grzesiak MT, Grzesiak S, Skoczowski A (2006) Changes of leaf water potential and gas exchange during and after drought in triticale and maize genotypes differing in drought tolerance. Photosynthetica 44:561–568

Hura T, Grzesiak S, Hura K, Grzesiak M, Rzepka A (2006) Differences in the physiological state between triticale and maize plants during drought stress and followed rehydration expressed by the leaf gas exchange and spectrofluorimetric methods. Acta Physiologiae Plantarum 28:433–443

Hura T, Hura K, Grzesiak M, Rzepka A (2007) Effect of long-term drought stress on leaf gas exchange and fluorescence parameters in C_3 and C_4 plants. Acta Physiologiae Plantarum 29:103–113

Jiang Q, Roche D, Monaco TA, Hole D (2006) Stomatal conductance is a key parameter to assess limitations to photosynthesis and growth potential in barley genotypes. Plant Biology 8:515–521

Kaiser WM (1987) Effects of water deficit on photosynthetic capacity. Physiol Plant 71:142–149

Kilian J, Whitehead D, Horak J, Wanke D, Weinl S, Batistic O, D'Angelo C, Bornberg-Bauer E, Kudla J, Harter K (2007) The AtGenExpress global stress expression data set: protocols, evaluation and model data analysis of UV-B light, drought and cold stress responses. Plant J 50:347–363

Kirschbaum MUF (1988) Recovery of photosynthesis from water-stress in *Eucalyptus pauciflora*—A process in two stages. Plant, Cell Environ 11:685–694

Kitao M, Lei TT, Koike T, Tobita H, Maruyama Y (2003) Higher electron transport rate observed at low intercellular CO_2 concentration in long-term drought-acclimated leaves of Japanese mountain birch (*Betula ermanii*). Physiol Plant 118:406–413

Kottapalli KR, Rakwal R, Shibato J, Burow G, Tissue D, Burke J, Puppala N, Burow M, Payton P (2009) Physiology and proteomics of the water-deficit stress response in three contrasting peanut genotypes. Plant, Cell Environ 32:380–407

Lal A, Edwards GE (1996) Analysis of inhibition of photosynthesis under water stress in the C4 species *Amaranthus cruentus* and *Zea mays*: electron transport, CO_2 fixation and carboxylation capacity. Funct Plant Biol 23:403–412

Lawlor DW, Cornic G (2002) Photosynthetic carbon assimilation and associated metabolism in relation to water deficits in higher plants. Plant, Cell Environ 25:275–294

Lawlor DW, Tezara W (2009) Causes of decreased photosynthetic rate and metabolic capacity in water-deficient leaf cells: a critical evaluation of mechanisms and integration of processes. Ann Bot 103:561–579

Levine A (1999) Oxidative stress as a regulator of environmental responses in plants. In: Lerner HR (ed) Plant responses to environmental stresses: from phytohormones to genome reorganization. CRC Press, ISBN 0824700449, 9780824700447, pp 248–266

Liu C-C, Liu Y-G, Guo K, Zheng Y-R, Li G-Q, Yu L-F, Yang R (2010) Influence of drought intensity on the response of six woody karst species subjected to successive cycles of drought and rewatering. Physiol Plant 139:39–54

Loreto F, Centritto M, Chartzoulakis K (2003) Photosynthetic limitations in olive cultivars with different sensitivity to salt stress. Plant, Cell Environ 26:595–601

Marques da Silva J, Arrabaça MC (2004) Photosynthesis in the water-stressed C_4 grass *Setaria sphacelata* is mainly limited by stomata with both rapidly and slowly imposed water deficits. Physiol Plant 121:409–420

Maury P, Mojayad F, Berger M, Planchon C (1996) Photochemical response to drought acclimation in two sunflower genotypes. Physiol Plant 98:57–66

Miyashita K, Tanakamaru S, Maitani T, Kimura K (2005) Recovery responses of photosynthesis, transpiration, and stomatal conductance in kidney bean following drought stress. Environ Exp Bot 53:205–214

Miyazawa S-I, Yoshimura S, Shinzaki Y, Maeshima M, Miyake C (2008) Deactivation of aquaporins decreases internal conductance to CO_2 diffusion in tobacco leaves grown under long-term drought. Funct Plant Biol 35:553–564

Montanaro G, Dichio B, Xiloyannis C (2007) Response of photosynthetic machinery of field-grown kiwifruit under Mediterranean conditions during drought and re-watering. Photosynthetica 45:533–540

Munne-Bosch S, Alegre L (2002) Plant aging increases oxidative stress in chloroplasts. Planta 214:608–615

Munne-Bosch S, Mueller M, Schwarz K, Alegre L (2001) Diterpenes and antioxidative protection in drought-stressed *Salvia officinalis* plants. J Plant Physiol 158:1431–1437

Parry MAJ, Andralojc PJ, Khan S, Lea PJ, Keys AJ (2002) Rubisco activity: effects of drought stress. Ann Bot 89:833–839

Peeva V, Cornic G (2009) Leaf photosynthesis of *Haberlea rhodopensis* before and during drought. Environ Exp Bot 65:310–318

Perez-Martin A, Flexas J, Ribas-Carbo M, Bota J, Tomas M, Infante JM, Diaz-Espejo A (2009) Interactive effects of soil water deficit and air vapour pressure deficit on mesophyll conductance to CO_2 in *Vitis vinifera* and *Olea europaea*. J Exp Bot 60:2391–2405

Pérez-Pérez JG, Syvertsen JP, Botía P, García-Sánchez F (2007) Leaf water relations and net gas exchange responses of salinized Carrizo citrange seedlings during drought stress and recovery. Ann Bot 100:335–345

Pou A, Flexas J, Alsina MD, Bota J, Carambula C, de Herralde F, Galmés J, Lovisolo C, Jimenez M, Ribas-Carbo M, Rusjan D, Secchi F, Tomas M, Zsofi Z, Medrano H (2008) Adjustments of water use efficiency by stomatal regulation during drought and recovery in the drought-adapted Vitis hybrid Richter-110 (*V. berlandieri* x *V. rupestris*). Physiol Plant 134:313–323

Resco V, Ewers BE, Sun W, Huxman TE, Weltzin JF, Williams DG (2009) Drought-induced hydraulic limitations constrain leaf gas exchange recovery after precipitation pulses in the C_3 woody legume, *Prosopis velutina*. New Phytol 181:672–682

Rivero RM, Kojima M, Gepstein A, Sakakibara H, Mittler R, Gepstein S, Blumwald E (2007) Delayed leaf senescence induces extreme drought tolerance in a flowering plant. Proc Nat Acad Sci 104(49):19631–19636

Saccardy K, Cornic G, Brulfert J, Reyss A (1996) Effect of drought stress on net CO_2 uptake by *Zea* leaves. Planta 199:589–595

Schwab KB, Schreiber U, Heber U (1989) Response of photosynthesis and respiration of resurrection plants to desiccation and rehydration. Planta 177:217–227

Smirnoff N (1998) Plant resistance to environmental stress. Curr Opin Biotechnol 9:214–219

Sofo A, Dichio B, Xiloyannis C, Masia A (2004) Effects of different irradiance levels on some antioxidant enzymes and on malondialdehyde content during rewatering in olive tree. Plant Sci 166:293–302

de Souza RP, Machado EC, Silva JAB, Lagoa A, Silveira JAG (2004) Photosynthetic gas exchange, chlorophyll fluorescence and some associated metabolic changes in cowpea (*Vigna unguiculata*) during water stress and recovery. Environ Exp Bot 51:45–56 de Souza CR, de Maroco JP, dos Santos TP, Rodrigues ML, Lopes CM, Pereira JS, Chaves MM (2005) Control of stomatal aperture and carbon uptake by deficit irrigation in two grapevine cultivars. Agric Ecosyst Environ 106:261–274

Tang AC, Kawamitsu Y, Kanechi M, Boyer JS (2002) Photosynthetic oxygen evolution at low water potential in leaf discs lacking an epidermis. Ann Bot 89:861–870

Tezara W, Mitchell VJ, Driscoll SD, Lawlor DW (1999) Water stress inhibits plant photosynthesis by decreasing coupling factor and ATP. Nature 401:914–917

Tezara W, Mitchell V, Driscoll SP, Lawlor DW (2002) Effects of water deficit and its interaction with CO_2 supply on the biochemistry and physiology of photosynthesis in sunflower. J Exp Bot 53:1781–1791

Turner NC, Schulze E-D, Gollan T (1985) The responses of stomata and leaf gas exchange to vapour pressure deficits and soil water content. Oecologia 65:348–355

Varone L, Ribas-Carbo M, Cardona C, Galle A, Hl Medrano, Gratani L, Flexas J (2012) Stomatal and non-stomatal limitations to photosynthesis in seedlings and saplings of Mediterranean species pre-conditioned and aged in nurseries: different response to water stress. Environ Exp Bot 75:235–247

Vincent D, Ergul A, Bohlman MC, Tattersall EAR, Tillett RL, Wheatley MD, Woolsey R, Quilici DR, Joets J, Schlauch K, Schooley DA, Cushman JC, Cramer GR (2007) Proteomic analysis reveals differences between *Vitis vinifera* L. cv. Chardonnay and cv. Cabernet Sauvignon and their responses to water deficit and salinity. J Exp Bot 58:1873–1892

Wong CE, Li Y, Labbe A, Guevara D, Nuin P, Whitty B, Diaz C, Golding GB, Gray GR, Weretilnyk EA, Griffith M, Moffatt BA (2006) Transcriptional profiling implicates novel interactions between abiotic stress and hormonal responses in *Thellungiella*, a close relative of Arabidopsis. Plant Physiol 140:1437–1450

Xu Z, Zhou G, Shimizu H (2009) Are plant growth and photosynthesis limited by pre-drought following rewatering in grass? J Exp Bot 60:3737–3749

Zhou YH, Lam HM, Zhang JH (2007a) Inhibition of photosynthesis and energy dissipation induced by water and high light stresses in rice. J Exp Bot 58:1207–1217

Zhou J, Wang X, Jiao Y, Qin Y, Liu X, He K, Chen C, Li Ma, Wang J, Xiong L, Zhang Q, Fan L, Deng X (2007b) Global genome expression analysis of rice in response to drought and high-salinity stresses in shoot, flag leaf, and panicle. Plant Mol Biol 63:591–608

Chapter 6
Water Use Strategies of Plants Under Drought Conditions

Eunice L. V. A. Bacelar, José M. Moutinho-Pereira, Berta M. C. Gonçalves, Cátia V. Q. Brito, José Gomes-Laranjo, Helena M. F. Ferreira and Carlos M. Correia

Abstract The growing concerns about water scarcity have focused more attention on water management in agriculture and promotion of water conservation through improved water use efficiency (WUE). Depending on the main purpose of the study, WUE can be estimated at multiple scales, from leaf to whole plant, crop, yield, and ecosystem levels. Drought resistance and WUE are not synonymous and their association is often misunderstood. Effectively, two water use strategies may be employed by woody plants under drought conditions. The prodigal water use behavior is beneficial in conditions where water supply is interrupted for short periods only. The conservative water use is favorable in conditions where a long dry period prevails and is associated with high capacity for drought resistance and slow growth rates. In this chapter, we also examine how human manipulation such as breeding and agricultural management techniques will offer new opportunities to improve plant water use under drought conditions.

6.1 Introduction

Climate change has emerged in recent years as one of the most critical phenomena affecting our planet. The latest scientific findings reported by the Intergovernmental Panel on Climate Change (IPCC), show changes in some extreme events, particularly daily temperature extremes and heat waves. Precipitation has

E. L. V. A. Bacelar (✉) · J. M. Moutinho-Pereira · B. M. C. Gonçalves · C. V. Q. Brito · J. Gomes-Laranjo · H. M. F. Ferreira · C. M. Correia
Centre for the Research and Technology of Agro-Environmental and Biological Sciences (CITAB)/Department of Biology and Environment, Universidade de Trás-os-Montes e Alto Douro (UTAD), Apartado 1013, 5001-801 Vila Real, Portugal
e-mail: areale@utad.pt

R. Aroca (ed.), *Plant Responses to Drought Stress*,
DOI: 10.1007/978-3-642-32653-0_6, © Springer-Verlag Berlin Heidelberg 2012

generally increased over land north of 30°N from 1900 to 2005, but there is evidence that droughts will intensify over the coming century in southern Europe and the Mediterranean region, central Europe, central North America, Central America and Mexico, northeast Brazil, and southern Africa (IPCC 2011). Given these facts, the question is which are and will be their impacts on plants, animals, and ecosystems.

Climate change can have both beneficial and detrimental impacts on plants. Researchers believe that climate change will increase productivity in certain crops and regions, and reduce productivity in others (Karl et al. 2009). Many species show positive responses to elevated carbon dioxide and low levels of warming, but higher levels of warming often negatively affect growth and yields. Nevertheless, it is expected that heat, drought, climate-associated pests and diseases, flooding, and extreme weather events will all contribute to a decline in plant productivity.

As a result of increasing populations and unsustainable practices in water usage, severe strains on freshwater resources are being felt around the world, especially in the rapidly developing countries. Water is an important component of the metabolism of all living organisms, facilitating many vital biological reactions by being a solvent, a transport medium, and an evaporative coolant (Bohnert et al. 1995). In plants and other photoautotrophs, water plays the additional role of providing the energy necessary to drive photosynthesis. Thus, any limitation in the availability of water has a great influence on plant life. The growing concerns about water scarcity have focused more attention on water management in agriculture and promotion of water conservation through improved water use efficiency (WUE). The way to increase WUE in plants is through human manipulation such as breeding or crop management techniques. Manipulating plant drought resistance mechanisms will offer new opportunities to improve WUE in water-scarce environments.

6.2 Water Use Strategies and Drought Resistance Mechanisms

Many environmental conditions can lead to water deficit in plants. Periods of little or no rainfall can lead to drought episodes. The period of time without rainfall actually needed to produce a drought episode depends mainly on the water holding capacity of the soil and the rate of evapotranspiration (ET) (Jones 1992; Larcher 1995; Kozlowski and Pallardy 1997). Water deficit in plants can also occur in environments in which water is not limiting (Bray et al. 2000). In saline habitats, the presence of high salt concentrations makes it more difficult for plant roots to extract water from the environment. Soil temperatures between 0 and 15 °C reduce soil and plant hydraulic conductance and eventually water uptake through an increase in the viscosity of water. However, the decrease in water viscosity cannot fully explain the decrease in the root water uptake rate (Bloom et al. 2004) and the reduction of root hydraulic conductivity has also biological–metabolic causes (Aroca et al. 2012). It has been suggested that the decrease in root hydraulic conductivity upon exposure to low temperatures could be caused by an inhibition of aquaporin activity (Wan et al. 2001; Murai-Hatano et al. 2008; Ionenko et al. 2010; Aroca et al. 2012). Exposure to temperatures below 0 °C, known as freezing temperatures, can lead to cellular dehydration as water leaves the cells and forms ice crystals in the intercellular spaces. Occasionally, well-watered plants will show periodic signs of water stress such as transient loss of turgor during some periods of the day. In this case, wilting indicates that the transpirational water loss has exceeded the rate of water absorption.

The ability of plants to survive the consequences of drought is termed drought resistance. It is important not to confound the significance of drought resistance and WUE. In fact, these terms are not synonymous, although this is often the case in the literature (Blum 2005; Morison et al. 2008). Both features are often conferred by plant traits that are mutually exclusive (Blum 2005). In some situations, higher drought resistance is linked to identical (Nelson et al. 2007) or even lower (Moutinho-Pereira et al. 2007; Bacelar et al. 2009) WUE, whereas in others enhanced WUE does not necessarily mean an improved drought resistance (Masle et al. 2005). In other cases, however, increased drought resistance can be associated with higher WUE (Bacelar et al. 2007a; Moutinho-Pereira et al. 2007; Rivero et al. 2007). Differences in the time of the day and season when they are determined and on the genotypes, plant age, and the experimental environmental conditions may explain these discrepancies.

Drought adaptation depends on the severity of water deficit and two different water use strategies may be employed by woody plants (Passioura 1982). The prodigal water use behavior is beneficial in conditions where water supply is interrupted for short periods only. In this situation there is little danger of serious desiccation despite rapid water use, and it enables a plant to grow quickly. In this strategy, plants present high stomatal conductance, high C_i/C_a ratio, and low WUE_{leaf} that are positively correlated with high photosynthetic rate (Bacelar et al. 2009). The conservative water use behavior is beneficial in conditions where a long dry period prevails, enabling the plant to use the available water efficiently. Conservative water use strategies are associated with high WUE_{leaf}, high capacity for drought resistance, and slow growth rates.

There is no universal way by which drought resistance can be achieved and in consequence the different components of such capacity have been classified in different ways by different authors. Classically, plant resistance to drought has been divided into escape, avoidance, and tolerance strategies (Levitt 1972; Turner 1986). Nevertheless, these strategies are not mutually exclusive and, in practice, plants may combine a range of response types (Ludlow 1989).

6.2.1 Drought Escape

Plants that escape drought, like desert ephemerals and annual crop and pasture plants, exhibit a high degree of developmental plasticity, being able to complete their life cycle before water deficit occurs. Escape strategies rely on successful reproduction before the onset of severe stress. Improved reproductive success also includes better partitioning of assimilates to developing fruits and seeds. This is associated with the plant's ability to store reserves in some organs (stems and roots) and to mobilize them for fruit production, a response well documented in crop plants, such as cereals (Bruce et al. 2002) and some legumes (Chaves et al. 2002).

6.2.2 Drought Avoidance

Plants that tend to avoid drought generally have tissues that are very sensitive to dehydration, and thus they have to avoid water deficits whenever water shortage occurs (Ludlow 1989). Dehydration avoidance is common to both annuals and perennials and is associated with a variety of adaptive traits. The most effective protection against drought is a deep, extensively branched root system that can absorb water from a large volume of soil (Arndt 2000). Moreover, roots with low hydraulic conductance or few but long roots can permit a slow but sustainable supply of water to the plant (Passioura 1983).

Water movement from the roots to the atmosphere is controlled by the conductance of the components of the water pathway (Lovisolo and Schubert 1998). Traditionally, stomatal conductance and root conductivity have been considered the main controlling factors of water flow in the plant (Jones 1992). However, the efficiency of water transport depends on constraints to water movement that affect hydraulic conductivity (Tyree and Ewers 1991). It has been reported that water stress affects shoot conductivity by inducing embolism in xylem vessels (Schultz and Matthews 1988; Tyree and Sperry 1989; Tognetti et al. 1996) or by a reduction in the vessel diameter (Lovisolo and Schubert 1998). Adaptation requires a fine balance because features that reduce vulnerability to cavitation, narrow conduits, and many inter-conduit connections, also reduce hydraulic conductivity that generates the high xylem tensions that trigger embolism (Sperry 2003). In general, vessels with narrow diameters are less susceptible to embolism (Lovisolo and Schubert 1998). However, variations in xylem conduit diameter can radically affect the function of the conducting system because of the fourth-power relationship between radius and flow through a capillary tube, as described by the Hagen-Poiseuille law (Fahn et al. 1986). Under water stress conditions, some plants are able to increase or maintain the level of protection against cavitation and turn more efficient the water transport through the xylem, revealing a trade-off between safety (low vulnerability to cavitation) and efficiency (large hydraulic conductivity) (Bacelar et al. 2007a).

Leaves growing under water stress usually develop smaller but more densely distributed stomata. This modification makes a leaf able to reduce transpiration by a quicker onset of stomatal regulation (Larcher 1995). Stomata are mainly confined to the abaxial surface and are often hidden beneath dense hairs (peltate scales or trichomes) or in depressions (grooves or crypts). In this way the boundary layer resistance is increased and the air outside the stomata becomes moister (Larcher 1995). Other morpho-anatomical traits that help to minimize water loss during drought include leaf rolling (Schwabe and Lionakis 1996), dense leaf pubescence (Karabourniotis and Bornman 1999; Liakoura et al. 1999; Bacelar et al. 2004), a thick cuticle and epicuticular wax layer (Leon and Bukovac 1978; Liakoura et al. 1999; Richardson and Berlyn 2002; Bacelar et al. 2004), heavily lignified tissue (Richardson and Berlyn 2002), smaller mesophyll cells and less intercellular spaces (Bongi et al. 1987; Mediavilla et al. 2001). Moreover, leaf movements, such as paraheliotropism, can also prevent damage by high temperatures, dehydration, and photoinhibition (Ludlow 1989). Another common feature in water stressed plants is the reduction of the canopy leaf area through reduced growth and shedding of older leaves (Bacelar et al. 2007a). This usually begins with the oldest leaves and progresses toward stem tips. Although the loss of leaves results also in a reduction of the photosynthetic surface it consequently reduces water loss and prolongs survival (Kozlowski et al. 1991).

Because of their ability to postpone or avoid tissue water deficits, plants that avoid drought could be said to have good short-term survival. However, they have poor long-term survival, because the avoidance mechanisms eventually fail to prevent dehydration of tissues that are relatively sensitive to desiccation (Ludlow 1980).

6.2.3 Drought Tolerance

Drought tolerance is the ultimate drought strategy (Connor 2005) and it refers to adaptations that allow the plant to withstand the stress, with or without a reduction in performance.

Osmotic adjustment is generally regarded as an important drought tolerance mechanism. Most of the adjustment can usually be accounted for by increases in concentration of a variety of common solutes including sugars, organic acids, and ions, especially K^+ (Taiz and Zieger 1998). Enzymes extracted from the cytosol of plant cells have been shown to be severely inhibited by high concentrations of ions. These accumulations during osmotic adjustment appears to occur mainly in the vacuoles, that are kept out of contact with enzymes in the cytosol or subcellular organelles (Taiz and Zieger 1998). Because of this compartmentation of ions, other solutes must accumulate in the cytosol to maintain water potential equilibrium within the cell. These other solutes, called compatible solutes or compatible osmolytes, are organic compounds that do not interfere with enzyme functions. These compatible solutes may protect the cell membrane and the metabolic machinery under dehydration (Chaves et al. 2003).

The osmotic compounds synthesized in response to water stress include amino acids like proline, aspartic acid, and glutamic acid (Samuel et al. 2000; Hamilton and Heckathorn 2001; Bacelar et al. 2009), methylated quaternary ammonium compounds (Rathinasabapathi et al. 2001), carbohydrates (Vijn and Smeekens 1999), polyols (Smirnoff 1998) and low-molecular-weight proteins (Ingram and Bartels 1996).

One of the most studied compatible solutes is the amino acid proline. Accumulation of high proline content in cells has also been associated with prevention of protein denaturation, preservation of enzyme structure and activity (Rajendrakumar et al. 1994; Samuel et al. 2000), and protection of membranes from damage by reactive oxygen species (ROS) produced under drought and high light conditions (Hamilton and Heckathorn 2001).

Betaines are quaternary ammonium compounds in which the nitrogen atom is fully methylated (Chen and Murata 2002). The most common betaines in plants include glycine betaine (GB, the most widely studied betaine), as well as proline betaine, β-alanine betaine, choline-O-sulfate, and 3-dimethylsulfoniopropionate (Rhodes and Hanson 1993; McNeil et al. 1999). GB is widely distributed in higher plants and is synthesized in many species at elevated rates in response to environmental stresses, appearing to be a critical determinant of stress tolerance (Rhodes and Hanson 1993). In addition, studies in vitro have shown that GB is effective as a compatible solute in stabilizing the quaternary structures of enzymes and complex proteins, as well as in maintaining the highly ordered state of membranes (Papageorgiou and Murata 1995).

In several species, polyols accumulate in leaves in response to water stress (Noiraud et al. 2000). The most common polyols in plants include acyclic forms, mannitol, glycerol, sorbitol, and cyclic forms (cyclitols), ononitol, and pinitol (Ashraf and Harris 2004). In addition to their role in osmoregulation, polyols also function as oxygen radical scavengers. For example, mannitol was found in vitro to act as a scavenger of hydroxyl radical (Halliwell and Gutteridge 1999), thereby protecting the proteins from oxidative damage in drought stressed plants (Moran et al. 1994). Recent studies support that species like olive tree cope with drought and salinity by coordinating mannitol transport with intracellular metabolism (Conde et al. 2011).

In general, soluble sugar content tends to be maintained in the leaves of droughted plants, in spite of lower rates of carbon assimilation (Chaves et al. 2003). This is achieved at the expenses of starch, which drastically declines (Chaves 1991). This response favors osmoregulation, enhances desiccation tolerance and allows metabolic activity to be regained more quickly in young leaves (Chaves et al. 2003).

During drought periods, plants undergo many physiological changes and induce a large number of genes for adaptation (Ingram and Bartels 1996). A typical change in gene expression is the induction of genes involved in the synthesis of low-molecular-weight proteins, e.g., dehydrins and late embryogenesis abundant (LEA) proteins (Ingram and Bartels 1996). The functions of LEA proteins are multiple as they are involved not only in seed maturation and desiccation but also in vegetative organs, where they can stabilize enzyme complexes and membrane structures and promote ion sequestration (Close 1996; Garay-Arroyo et al. 2000). Dehydrins are involved in a range of responses to abiotic stress that share a dehydration component. However, the mechanisms by which dehydrin synthesis is enhanced and functions in vivo remain poorly understood (Cellier et al. 2000). There is increasing evidence that another class of proteins, the small heat-shock proteins (sHsps), may play a role in cellular protection during dehydration (Alamillo et al. 1995; Alpert and Oliver 2002). Strong evidence supports that sHsps function as molecular chaperones (Hendrick and Hartl 1995; Hong and Vierling 2000) that bind to partially folded or denatured substrate proteins and thereby prevent irreversible aggregation or promote correct substrate folding (Sun et al. 2002). They are also described as being developmentally regulated, being abundant in dry mature seeds, and disappearing during germination (Wehmeyer et al. 1996).

Membrane-associated carriers or transporters are probably involved in differentially distributing osmolytes within the cell and may also participate in regulating the transport and distribution of these solutes throughout the plant (Bray et al. 2000). Aquaporins have been implicated as the major facilitators for the movement of water across membranes (Chrispeels and Agre 1994; Bray et al. 2000). In addition, the capacity of aquaporins to transport small neutral solutes and/or gases has raised the intriguing possibility that aquaporins may work as membrane channels with multiple functions (Tyerman et al. 2002). Aquaporins may facilitate water movement in drought stressed tissues and promote the rapid recovery of turgor on watering (Bray et al. 2000).

Changes in the elastic properties of plant cell walls can also contribute to turgor maintenance under drought (Patakas and Noitsakis 1999). An inelastic cell wall should act to increase the uptake of water from drying soil because more negative water potentials would result from a relatively small decrease in water content (Schulte 1993). On the other hand, a more flexible cell should allow shrinkage during water loss, thus maintaining high turgor for a given change in water content (Patakas and Noitsakis 1999). In a study with olive tree (Bacelar et al. 2009) it was observed that cultivars that employ a prodigal water use strategy revealed high tissue elasticity, whereas conservative water use cultivars revealed high tissue rigidity.

An extreme desiccation tolerance is found in the resurrection plants as well as in some ferns, non-vascular plants, algae, and lichens (Chaves et al. 2003). The leaves of resurrection plants, like *Craterostigma plantagineum*, can equilibrate with air close to 0–2 % (v/v) relative humidity and are still able to fully recover their physiological activity upon rehydration (Bartels et al. 1990). When water is withheld from mature individuals of *C. plantagineum*, changes rapidly occur at the messenger RNA and protein levels, eventually leading to the tolerant state (Ingram and Bartels 1996).

The disadvantage of tolerance of water deficits is that the continued carbon fixation is inevitably accompanied by water loss. Thus, plant water status must ultimately fall, what can result in some cases in death (Arndt 2000).

When the use of absorbed light in either photosynthesis or photorespiration and the thermal dissipation are not enough to cope with excess energy, the production of highly reactive molecules is exacerbated (Chaves et al. 2003). The excess of reducing power determines a redirection of photon energy into processes that favor the production of ROS, mainly in the photosynthetic (Asada 1999) and mitochondrial electron transport chains (Møller 2001). To mitigate the oxidative damage initiated by ROS, plants have developed a complex antioxidative defence system that includes enzymatic and non-enzymatic antioxidants. These compounds and enzymes are not distributed uniformly, so defence systems vary among specific subcellular compartments. The degree to which the activities of antioxidant enzymes and the amount of antioxidants increase under drought stress are extremely variable among several plant species and even between cultivars of the same species (Bacelar et al. 2006; Bacelar et al. 2007b).

6.3 Estimation of WUE at Multiple Scales

WUE could be defined in many ways, depending on the scale of measurement and the main purpose of the studies. For plant physiologists, the basic unit of production could be moles of carbon gained in photosynthesis (A) in relation to water used in transpiration (E) or to stomatal conductance (g_s), which permits the calculation of instantaneous WUE_{leaf} (A/E) or intrinsic WUE_{leaf} (A/g_s), also found as WUE_i in the literature. For farmers and agronomists, the unit of production is much more likely to be the accumulated plant biomass or the yield of harvested product achieved from the water made available to the plants through precipitation or irrigation. With this in mind, in this chapter we present the concepts of whole plant WUE (WUE_{WP}), crop WUE (WUE_C), and yield WUE (WUE_Y). Finally, mainly for forestry and environmental issues the concept of ecosystem WUE (WUE_E) is important, defined as the net carbon uptake per amount of water lost from the ecosystem.

Intrinsic WUE_{leaf} presents substantial intra and interspecific variation and is also dependent on environmental conditions, mainly drought and co-occurring stresses such as elevated temperature, high vapor pressure deficit, and elevated photosynthetic photon flux density. For instance, A/g_s was shown to range between around 20 and 80 μmol CO_2/mol H_2O during a typical summer day, with low values generally in the afternoon, when comparing three Portuguese grapevine cultivars (Moutinho-Pereira et al. 2007). On the other hand, in Touriga Nacional grapevine cultivar WUE_{leaf} decreased during the day in the more stressful conditions of Port wine country, while it increased more than 100 % in other sites of the Douro Valley, to values higher than 100 μmol CO_2/mol H_2O (Moutinho-Pereira et al. 2004). Meanwhile, Bacelar et al. (2007a) found in olive tree intrinsic WUE_{leaf} values in the range of 80–130 μmol CO_2/mol H_2O, being the higher values under regulated deficit irrigation (RDI), at midday and during the afternoon, whereas in droughted trees it decreased from 110 μmol CO_2/mol H_2O in the morning to 80 μmol CO_2/mol H_2O at midday. Meanwhile, in full irrigated trees, WUE_{leaf} increased from 80 to 90 μmol CO_2/mol H_2O between these periods. Thus, although A/g_s typically increases under water stress conditions (Bota et al. 2001;

Flexas et al. 2002), under some environmental conditions this may not occur, such as in some regions of the inland Iberian Peninsula or other regions with Mediterranean weather conditions. Thus, these results support the possibility of improving WUE_{leaf} by genetic engineering. WUE_{leaf} could be ameliorated by the increment of net carbon assimilation at leaf level (i.e., the difference between carbon gain in photosynthesis and carbon loss in respiration), or by the reduction in stomatal conductance or transpiration rate.

The carbon isotope discrimination ($\Delta^{13}C$), a measure of the $^{13}C/^{12}C$ ratio in plant material relative to the value of the same ratio in the air on which plants feed, can be used as a long-term indicator of WUE_{leaf}. Negative correlations between $\Delta^{13}C$ and water use efficiency have been demonstrated in many species, as predicted by theory (reviewed in Farquhar et al. 1989). These observations point to $\Delta^{13}C$ as a potential candidate for use in breeding for greater agronomic water use efficiency (Condon et al. 2004). In addition, the measurements of the O isotope enrichment of plant organic material ($\Delta^{18}O$) can provide complementary information to that inferred from $\Delta^{13}C$ in analyses of plant water use efficiency. Specifically, $\Delta^{18}O$ could provide information about the ratio of ambient to intercellular vapour pressures, and thus about the leaf to air vapor pressure difference, during photosynthesis (Cernusak et al. 2008).

Whole plant WUE represents the whole plant carbon and biomass acquisition per amount of transpired water. Thus, higher WUE_{WP} could be obtained by the drop of plant water losses, the decrease of growth respiration and, mainly, the decrease of maintenance respiration, that uses a substantial proportion of the total carbon assimilated (Amthor 2000), or the increase of photosynthesis by either increasing total leaf surface or increasing leaf photosynthetic capacity. Water losses depend on leaf transpiration rate, which is a function of leaf conductance (including cuticular and especially stomatal conductances) to water vapor, and on dimension and structure of canopy, such as leaf area index and leaf orientation. Leaf conductance depends on well-identified morpho-anatomical and physiological traits that may be targeted for genetic manipulation. Meanwhile, the dimension and structure of the canopy could be regulated by agronomic management techniques.

Closely related with WUE_{WP} is WUE_C, although in this case the water consumed by the crop includes the transpired water plus the water lost by physical processes, such as soil evaporation and runoff. Thus, in addition to the aspects reported before to WUE_{WP}, higher WUE_C could be achieved by the implementation of agronomic practices that reduce the amount of water lost without being used by the plants.

WUE_Y is a variant of WUE_{WP} that represents the carbon and biomass allocated to yield components (mainly fruits and seeds, but also to other plant organs, depending on plant species and the aim of the production). Therefore, higher partitioning to yield organs could increase WUE_Y.

It is important to emphasize that improving WUE_{leaf} may not necessarily result in improving WUE_{WP},, WUE_C ,and WUE_Y, because of the interference of canopy and ambient processes (Flexas et al. 2010). The absence of a significant association between A/g_s and WUE_{WP} was found by our group in olive tree (Bacelar et al.

2007a) reflecting the difference in timescale of both processes (from seconds to months) and non-accounted energy expenses in growth and maintenance in the long-term water use (Maroco et al. 2000). The study of Bacelar et al. (2007a) also shows that WUE_{WP} was enhanced by 64 % in Cobrançosa cultivar under drought conditions, while in the rest of the genotypes it was either unaffected ('Verdeal Transmontana') or considerably decreased ('Madural'). Thus, for production on relatively dry sites, cultivars that are characterized by high values of WUE_{WP}, such as 'Cobrançosa', appear to be the most promising.

For the determination of WUE_E, two different methods based on Bowen ratio and eddy covariance systems could be used to evaluate the H_2O and CO_2 fluxes (Emmerich 2007). One method uses the ratio of net ecosystem exchange (NEE) of carbon dioxide to ET by the ecosystem for the growing season or some other time period (Tubiello et al. 1990). A more instantaneous evaluation method of WUE_E is the regression of daily daytime NEE versus ET with the slope value of the regression line, a measure of WUE_E (Baldocchi et al. 2001). Ecosystem differences in CO_2 and H_2O fluxes have important management implications including primary productivity, carbon sequestration, and rangeland health (Emmerich 2007).

6.4 Selection and Breeding Programs to Improve WUE

Breeding for drought resistance has long been part of the breeding practice in most crops that have been grown under dryland conditions (Blum 2011). Plant breeders use a step-wise selection procedure to identify the best performing genotypes in early generations (Bänzinger et al. 2000; Atlin 2003). First, many genotypes are evaluated with few replicates and at few sites. Later, the most successful genotypes or their descendants are evaluated with more replicates and at more sites. Finally, the breeder reduces both the number of genotypes and the variation among genotypes, mainly by eliminating the poor-performing fraction. According to Bänzinger et al. (2000), the key to breeding for drought tolerance is to manage stress. This is done by conducting experiments partly or entirely in the dry season and managing the stress through irrigation.

In the past, breeding efforts to improve drought tolerance have been hindered by its quantitative genetic basis and our poor understanding of the physiological basis of yield in water-limited conditions (Blum 1998; Passioura 2002). Breeders and physiologists addressed the question of how physiological approaches in plant breeding could have greater impact, namely through (Jackson et al. 1996; Reynolds et al. 2001):

- Focusing physiological work on a suitable range of germplasm;
- working with larger populations to enable extrapolation of findings to breeding processes;

- identifying traits for use as indirect selection criteria, in addition to those already used in core breeding programs;
- identifying traits for use as selection criteria in introgression programs;
- conducting selection trials in more representative environments;
- developing tools that could be quickly and easily applied to large numbers of segregating lines.

According to Tuberosa and Salvi (2006), drought is the most devastating stress and the most recalcitrant to breeder's efforts of all the abiotic stresses that curtail crop productivity. Moreover, breeding for specific physiological traits that are expected to impart a yield advantage in dry environments has been notoriously difficult and unsuccessful (Richards et al. 2001). In fact, breeding for drought tolerance is further complicated by the fact that several types of environmental stresses can challenge plants simultaneously. High temperatures, high irradiance, scarcity of water, and nutrient deficiencies are commonly encountered under normal growing conditions but may not be amenable to management through traditional farm practices (Mittler 2006; Fleury et al. 2010). In general, it is crucially important, according to Bacon (2004), to select genotypes able to optimize water harvest and WUE. Flexas et al. (2010) suggested the enhancement of WUE by biotechnology and genetic engineering in grapevine, mainly by: (1) increasing CO_2 diffusion to the sites of carboxylation without increasing stomatal conductance, thanks to increase in mesophyll conductance to CO_2, which partly depends on aquaporins; and (2) improving the carboxylation efficiency of Rubisco that could be achieved by replacing Rubisco from grapevine with Rubiscos from other C_3 species with higher specificity for CO_2.

On the other hand, indiscriminant selection for higher WUE with the assumption that it equates with improved drought resistance or improved yield under stress might bring about serious negative consequences (Blum 2005). If low water use is the breeder's target it is highly probable that selection for this end can be achieved by directly selecting simple traits, such as reduced leaf area, without measuring WUE. Therefore, Blum (2009) recommended that plant breeders targeting water-limited environments should consider skipping the use and reference to WUE and consider plant constitutive and adaptive traits which drive the effective use of water (EUW) and the resultant dehydration avoidance as major traits for yield improvement in drought prone environments.

At the molecular level, pathways and gene networks between abiotic stresses overlap (Shinozaki and Yamaguchi-Shinozaki 2007). Moreover, some biochemical mechanisms may have opposing effects under different stresses; therefore tackling tolerance to one stress may lead to sensitivity to another (Fleury et al. 2010). Although the reductionist approach of studying isolated stress has considerably increased our knowledge of tolerance mechanisms, interaction between multiple stresses and stress combinations should be studied so as to make progress relevant to field conditions (Fleury et al. 2010). In addition to these confounding environmental factors, a drought research program should also consider plant phenology (Fukai and Cooper 1995; Fleury et al. 2010). For example, breeding for a shortened crop life cycle has been a very successful strategy in C_3 cereals in Mediterranean conditions (Araus et al. 2002). However, in well-developed agricultural regions, crop flowering time has already been optimized by breeders so that the plant's phenology matches its environment (Passioura 2007).

Pang et al. (2011) found evidences for the development of new perennial legume forages for a wide range of drought prone agricultural regions in breeding/selection programs. The inter- and intraspecific variability of responses to drought stress in the plants examined by those authors suggests a wide range of strategies in perennial herbaceous legumes to cope with drying conditions. For example, leaf osmotic adjustment was observed in *Bituminaria bituminosa* var. *albomarginata* accessions 6 and 10, both accessions of *Cullen australasicum* and *Kennedia prostrata* and leaf water potential was maintained at a high level in *B. bituminosa* var. *albomarginata* accessions 6 and 10 and *Macroptilium atropurpureum*.

Campos et al. (2004) refereed that the association between genotype and phenotype must be better understood and quantified to improve the ability to predict phenotypic performance based on genetic information. Tinker (2002) specifies that genomics and high throughput laboratory-based analysis of DNA are indispensable to comprehending gene-phenotype associations. This will be critically important for drought tolerance, where performance is regulated by many loci and subject to multiple genotype x environment and gene x gene interactions (epistasis) (Campos et al. 2004). Also, crop simulation models can assist investigation of the physiological and genetic basis of genotype variation (Shorter et al. 1991) and help selection decisions in plant breeding (Cooper et al. 1993).

6.5 Improving WUE by Agricultural Management Techniques

It is widely accepted that water is the most universally limiting factor in agricultural production systems (Centritto 2005). Natural factors related with the environment and genetic characteristics are basically responsible by the definition of the plant water requirements and, in the last circumstance, by their WUE. However, the farmer's decision, through the crop managements that ensure a better equilibrium between plants and environment, cannot be minimized. This role is much crucial as the local aridity index is higher. Improving of WUE_C is essential for agricultural sustainability under higher potential ET- associated soil moisture deficits exacerbated by the global climatic change (Flexas et al. 2010). To achieve higher WUE_C the increment/validation of scientific monitoring tools is necessary, such as infrared thermometry (Grant et al. 2007), trunk diameter sensors (Conejero et al. 2007), passive chlorophyll fluorescence (Flexas et al. 2002), hyperspectral reflectance indices (Rodríguez-Pérez et al. 2007), or sap flow meters (Fernández et al. 2008). The impact of agricultural management tools on WUE_C has been studied by many researchers. From the social and economic points of view, this subject is of crucial importance for many world regions where the lack of food is a direct consequence of the insufficient and unreliable rainfall and changing rainfall patterns (FAO 2010). This section intends to refer some management techniques, either at soil or plant level, that can optimize crop water use efficiency and crop yield and its quality attributes.

6.5.1 At Soil Level

In those arid and semi-arid areas where the highest crop water requirements coincide with periods without rain, some soil properties such as texture, effective depth, and organic matter content are important to ensure both a high soil water-holding capacity and a good root development. The soil volume that can be explored by the roots is a key factor that affects the plant–water relations and mineral nutrition and therefore its vigor and yield (Keller 2005). Generally, plants are less affected by severe drought when the root system is deeper. This situation should be enhanced by greater soil depth and adequate fertility (Rodrigues et al. 2011).

Soil texture affects the erodibility and the availability of water, air, and mineral nutrients (Northcote 1988). In general, sandy soils show poor water availability while more argillaceous textures lead to soils more compact and oxygen-poor, impairing the development of the root system. In an increasingly mechanized agriculture, this negative aspect is exacerbated by excessive traffic of heavy farm machinery.

Soil management has great influence on crop–water relations. Cover crops keep the soil protected against erosion, reduce soil compaction, increase water infiltration, introduce organic carbon, and may prevent nitrate leaching. In periods of greater water scarcity, however, this practice affects crops in a manner comparable to deficit irrigation, because the covers use water and reduce nutrient availability for the crops (Lipecki and Berbeć 1997; Rodrigues et al. 2011). Conversely, the critical situation of water resources in most arid regions makes the soil surface management one of the most critical aspects of the agriculture practices applied to rainfed crops (Rodrigues et al. 2005). Thus, it is desirable that cover vegetation dries out naturally (using species that employ a drought escape strategy) or artificially before the drought period, creating a mulch of dead material on the ground. Therefore, the soil rest permanently protected, the water infiltration increases, and the root system develops freely close to the soil surface. In addition, the mulch reduces soil temperature and water losses by evaporation in summer (Van Huyssteen et al. 1984; Possigham 1992; Rodrigues et al. 2011).

The mycorrhizal associations have been studied due to their advantages in arid environments. These associations vary widely in structure and functions, but arbuscular mycorrhizas (AM) are by far the most prevalent of all mycorrhizal categories. It is estimated that 80 % of all terrestrial plants form this type of association, including many agriculturally and horticulturally important crop species (Peterson 2004; Mohammadi Goltapeh et al. 2008). Inoculate soils with appropriate AM fungi may directly enhance root water uptake, contributing to preserve physiological activity in plants under severe drought conditions. Mycorrhizal fungi improved WUE in species such as safflower and wheat (Bryla and Duniway 1997) and watermelon (Kaya et al. 2003). Among the mechanisms involved, many are related to the physiology of the host plant, but some are related to the extraradical mycelia (Liu et al. 2007).

6.5.2 At Plant Level

At plant level, assuming that the plant breeding programs already took into account the greater intrinsic water use capacity, there are numerous cultural practices that can be implemented from the beginning of crop cycle to adult stage, to improve this attribute. Normally, these cultural strategies have in common the achievement of high environmental resources efficiency (light, temperature, CO_2, water and nutrients), with significant benefit to crop water economy. First, the management of the crop training system has a fundamental role to reach this goal (Smart 1974). This subject includes a set of cultural techniques selected by the farmers to train and guide the structure of the plant to its best performance. Some of these cultural practices have permanent effects for the crop life, such as the plantation density, the row orientation, and the plant shape, while others are susceptible to annual changes, e.g., all operations performed on the size of the branches and total leaf area during dormancy and the growing season (Carbonneau et al. 1981; Huglin and Schneider 1998). The row orientation has an important influence on WUE_C, both because of the light and thermal canopy microclimate and the thermal soil microclimate effects. Specifically for vineyards in dry and hot regions, some authors (Carbonneau et al. 1981; Intrieri et al. 1998) have proposed an N–S orientation rather than an E–W one, to improve light interception early in the morning (east side) and late in the afternoon (west side). This leads to higher photosynthetic rates, mostly in the morning, when the environmental conditions are relatively more favorable. On the other hand, at midday, the total leaf surface exposed to light is reduced (just the top receives sunlight), which may be helpful in reducing transpiration and preventing the leaf water potential dropping below the threshold values for runaway embolism (Moutinho-Pereira et al. 2001).

As refereed, the plant density has a crucial role for soil moisture exploitation. Usually, for rainfed fruit tree orchards and vineyards grown in the driest areas, the plant density must be lowered for each plant to have a greater rootzone and thus increasing the available water for each plant. However, it may increase the water loss by ET because the total leaf area and the soil area exposed to direct solar radiation are higher (Yuste 2007).

Pruning and removal of shaded leaves influence plant vigor, affecting the ratio between the total leaf area and sun exposed leaf area (Baeza et al. 2005). When this ratio increases, WUE decreases. In very hot and dry regions, given the weak vigor that plants usually present, the maintenance of the leaf surface in good health guarantees greater photosynthetic productivity and allows lower water loss because the integrity of protective tissues, the epidermis and cuticle, is better safeguarded.

In arid regions, irrigation is undoubtedly a cultural practice with enormous benefits for the plant growth and development (Fereres and Evans 2006). However, under limited water resources, the technical, economic, and cultural irrigation managements require careful decisions. The efficient management of limited water resources requires irrigation scheduling techniques based on deficit supplies with respect to crop water needs (Tognetti et al. 2005). Deficit irrigation, defined as the application of water below full crop water requirements, is an important tool to achieve the goal of reducing irrigation water use (Fereres and Soriano 2007). Techniques such as RDI, where a water deficit is applied in a closely controlled way in low water stress sensitivity periods (Chalmers et al. 1981; Kriedemann and Goodwin 2003), and Partial Root-zone Drying (PRD), in which half of the root-zone is irrigated alternatively in scheduled irrigation events (Dry and Loveys 1998), are being increasingly valued because they allow to reconcile the advantages of a moderate water stress for a more efficient water use and an adequate development of the plants (Santos et al. 2007). In Australia, PRD has resulted in water savings of up to 50 % compared with the conventional irrigation of grapevines. An additional response of crops to PRD appears to be a shift of root growth to deeper soil layers, whereas under standard drip irrigation roots are often concentrated in the surface soil, which makes crops less drought resistant (Keller 2005).

Without disregarding the importance that irrigation can have for improving crop performance, it is crucial to develop mitigation alternatives, not only in economic terms but also in terms of environmental sustainability. Among these mitigation alternatives, there has been a major effort by the scientific community to study the effect of inorganic substances on the improvement of light microclimate and water relations of leaves. Reflective antitranspirants (e.g., kaolin, Bordeaux mixture), have interesting properties because they can lower leaf temperature by increasing reflection of infrared radiation (Moutinho-Pereira et al. 2001) (Fig. 6.1). Lowered leaf temperature reduces the vapor pressure gradient between the leaf and the bulk air. Since this is the driving force behind transpiration, the plant water consumption is reduced (Glenn and Puterka 2005). Jifon and Syvertsen (2003) found that midday WUE of grapefruit kaolin-sprayed leaves was 25 % higher than that of control leaves, indicating improved WUE under high light intensity. However, the results of kaolin application require further demonstration since they have often been contradictory, ranging from increased to decreased yields and photosynthetic rates.

6.6 Water Use Strategies of Native and Invasive Plants

In xeric environments, WUE is likely to be a target of natural selection (Heschel et al. 2002). As previously discussed in this chapter, plants under natural conditions need to greatly improve their WUE to increase their chances of survival in

Fig. 6.1 Kaolin and Bordeaux mixture application in grapevine: **a** and **b** field trials in the Douro Demarcated Region; treatments: **c** control; **d** kaolin and **e** Bordeaux mixture

water -scarce environments. Due to climate change, biological invasions may become a serious problem under natural conditions. Competitive strategies are important for the successful establishment and persistence of invasive plants (Vilá and Weiner 2004). Tolerance strategies may also help in the successful establishment or dominance of invasive species (Allred et al. 2010). An important trait is the ability of seedlings to develop root systems quickly, before the dry summer begins (Roché et al. 1994). Greater competitive ability could include greater use of limiting resources, such as water, thereby inhibiting establishment, survival, and reproduction of native species (Blicker et al. 2003). According to McDowell (2002) invasive species increase the efficiency of resource capture by maintaining a high rate of photosynthesis per unit water loss.

To identify the factors that make some plant species troublesome invaders, many studies have compared native and invasive plant performance (Daehler 2003). Hill et al. (2006) found that invasive species had lower diurnal variation in leaf water potential than native ones despite the high transpiration. This capacity may be due to the ability of water uptake trough deeper roots (Hill et al. 2006).

Increased WUE may also be an important contribution to the success of invasive species in areas where water is limiting (Hill et al. 2006). The efficiency of water use can be caused by variations in the environment or leaf functional traits (Seibt et al. 2008). Invasive *Rubus* plants had higher instantaneous WUE than native ones, probably due to their thicker leaves (McDowell 2002). However, Hill et al. (2006) found that WUE was higher in invasive species than in natives only under the most unusually dry conditions. Leaf carbon isotope composition allowed McAlpine et al. (2008) to conclude that the higher WUE of invasive plants/species resulted from increased carbon gain and not from strategies that limit water loss. Blicker et al. (2003) found that mature invasive plants may prolong their growing season by increasing WUE during mid- and late summer, when neighboring semi-arid plants are quiescent. On the other hand, Funk and Vitousek (2007) and Cavaleri and Sack (2010) reported similar values of WUE among invasive and native species.

Some studies support the hypothesis that high photosynthetic rate is an important strategy of invasive ability (Kloeppel and Abrams 1995; Baruch and Goldstein 1999; McDowell 2002; Hill et al. 2006; Jiang et al. 2009; Allred et al. 2010). Invasive species can maintain higher photosynthetic rates by modifying leaf anatomy and structure. Specific leaf area (SLA) represents light-capturing area deployed per dry mass allocated (Hamilton et al. 2005). A high SLA is a trait often associated with invasiveness (Baruch and Goldstein 1999; Lake and Leishman 2004; Hamilton et al. 2005; Burns and Winn 2006; Grotkopp and Rajmánek 2007; Feng et al. 2008; Allred et al. 2010). A high SLA reflects the occurrence of thinner leaves, which are less costly to produce and to maintain (Lee and Graham 1986) and, due to the associated rapid leaf production, allows an opportunistic capture of solar energy to invasive plants (Grotkopp and Rajmánek 2007). Thus, species with a high SLA have a shorter investment return rate and greater potential for fast growth (Lake and Leishman 2004). However, some authors defend that a high SLA favors invasive species only under some conditions, such as when in new environments with non-limiting resources (Burns 2006; Burns and Winn 2006; Leishman et al. 2007).

In general, invasive species had greater WUE and adaptations at root level that allow a more efficient water uptake. The most relevant attributes that favor the invaders are the greater photosynthetic rates associated to a high SLA. However, these attributes may vary according to the species and the environment in which they are established. This subject is still controversial, and the problem of what makes a species invasive is far from solved.

6.7 Conclusions

The available freshwater is limited in many parts of the planet due to multiple factors, such as climate change, population increase, and rapid economic development. The greatest potential savings can be made in agriculture. In fact, small improvements in WUE by human manipulation such as breeding and agricultural management techniques can have a significant effect on water availability. However, the concept of WUE is of limited application when discussing the fitness of plants for water-limited environments (Jones 2004). Intrinsic WUE_{leaf} presents substantial intra- and interspecific variation, and depends on environmental conditions. Moreover, improving WUE_{leaf} may not necessarily result in improved WUE_{WP}, WUE_C and WUE_Y. For natural ecosystems, improvements in WUE are only valuable if there is little competition. In fact is not useful for a species to have a conservative water use, with a high WUE, if the competitors will use all the water available first.

References

Alamillo J, Almoguera C, Bartels D, Jordano J (1995) Constitutive expression of small heat shock proteins in vegetative tissues of the resurrection plant *C. plantagineum*. Plant Mol Biol 29(5):1093–1099

Allred BW, Fuhlendorf SD, Monaco TA, Will RE (2010) Morphological and physiological traits in the success of the invasive plant *Lespedeza cuneata*. Biol Invasions 12:739–749

Alpert P, Oliver MJ (2002) Drying without dying. In: Black M, Pritchard H (eds) Desiccation and survival in plants: drying without dying. CAB International, Wallingford, pp 4–31

Amthor JS (2000) The mccree-de wit-penning de vries-thornley respiration paradigms: 30 years later. Ann Botany 86:1–20

Araus JL, Slafer GA, Reynolds MP, Royo C (2002) Plant breeding and drought in C3 cereals: what should we breed for? Ann Botany 89:925–940

Arndt SK (2000) Mechanisms of drought resistance in the tropical fruit tree *Ziziphus*. PhD Thesis, Vienna

Aroca R, Porcel R, Ruiz-Lozano JM (2012) Regulation of root water uptake under abiotic stress conditions. J Exp Bot 63(1):43–57

Asada K (1999) The water–water cycle in chloroplasts scavenging of active oxygens and dissipation of excess photons. Annu Rev Plant Physiol Plant Mol Biol 50:601–639

Ashraf M, Harris PJC (2004) Potential biochemical indicators of salinity tolerance in plants. Plant Sci 166:3–16

Atlin G (2003) Improving drought tolerance by selecting for yield. In: Fischer KS, Lafitte R, Fukai S, Atlin G, Hardy B (eds) Breeding rice for drought-prone environments. International Rice Research Institute, Los Baños (Philippines), pp 14–22

Bacelar EA, Correia CM, Moutinho-Pereira JM, Gonçalves B, Lopes JI, Torres-Pereira JM (2004) Sclerophylly and leaf anatomical traits of five field-grown olive cultivars growing under drought conditions. Tree Physiol 24:233–239

Bacelar EA, Moutinho-Pereira JM, Gonçalves BC, Ferreira HF, Correia CM (2007a) Changes in growth, gas exchange, xylem hydraulic properties and water use efficiency of three olive cultivars under contrasting water availability regimes. Environ Exp Bot 60:183–192

Bacelar EA, Moutinho-Pereira JM, Goncalves BC, Lopes JI, Correia CM (2009) Physiological responses of different olive genotypes to drought conditions. Acta Physiol Plant 31(3):611–621

Bacelar EA, Santos DL, Moutinho-Pereira JM, Gonçalves BC, Ferreira HF, Correia CM (2006) Immediate responses and adaptative strategies of three olive cultivars under contrasting water availability regimes: changes on structure and chemical composition of foliage and oxidative damage. Plant Sci 170(3):596–605

Bacelar EA, Santos DL, Moutinho-Pereira JM, Lopes JI, Gonçalves BC, Ferreira TC, Correia CM (2007b) Physiological behavior, oxidative damage and antioxidative protection of olive trees grown under different irrigation regimes. Plant Soil 292:1–12

Bacon MA (2004) Water use efficiency in plant biology. Blackwell Publishing CRC Press, Oxford

Baeza P, Ruiz C, Cuevas E, Sotés V, Lissarrague J (2005) Ecophysiological and agronomic response of Tempranillo grapevines to four training systems. Am J Enol Viticul 56:129–138

Baldocchi D, Falge E, Gu L, Olson R, Hollinger D, Running S, Anthoni P, Bernhofer Ch, Davis K, Evans R, Fuentes J, Goldstein A, Katul G, Law B, Lee X, Malhi Y, Meyers T, Munger W, Oechel W, Paw KT, Pilegaard K, Schmid HP, Valentini R, Verma S, Vesala T, Willson K, Wosfsy S (2001) Fluxnet: a tool to study the temporal and spatial variability of ecosystem scale carbon dioxide, water vapor, and energy flux densities. B Am Meteorol Soc 82:2415–2434

Bänzinger M, Edmeades GO, Beck D, Bellon M (2000) Breeding for drought and nitrogen stress tolerance in maize: from theory to practice. D.F., Cimmyt, Mexico

Bartels D, Schneider K, Terstappen G, Piatkowski D, Salamini F (1990) Molecular cloning of abscisic acid–modulated genes which are induced during desiccation of the resurrection plant *C. plantagineum*. Planta 181:27–34

Baruch Z, Goldstein G (1999) Leaf construction cost, nutrient concentration, and net CO_2 assimilation of native and invasive species in Hawaii. Oecologia 121:183–192

Blicker PS, Olson BE, Wraith JM (2003) Water use and water-use efficiency of the invasive *Centaurea maculosa* and three native grasses. Plant Soil 254:371–381

Bloom AJ, Zwieniecki MA, Passioura JB, Randall LB, Holbrook NM, St.Crair DA (2004) Water relations under root chilling in a sensitive and tolerant tomato species. Plant, Cell Environ 27:971–979

Blum A (1998) Plant breeding for stress environments. CRC Press, Boca Raton

Blum A (2005) Drought resistance, water-use efficiency, and yield potential—are they compatible, dissonant, or mutually exclusive? Aust J Agr Res 56:1159–1168

Blum A (2009) Effective use of water (EUW) and not water-use efficiency (WUE) is the target of crop yield improvement under drought stress. Field Crop Res 112:119–123

Blum A (2011) Plant breeding for water-limited environments. Springer Science+Business Media, New York

Bohnert HJ, Nelson DE, Jensen RG (1995) Adaptations to environmental stresses. Plant Cell 7:1099–1111

Bongi G, Mencuccini M, Fontanazza G (1987) Photosynthesis of olive leaves: effect of light flux density, leaf age, temperature, peltates and H_2O vapour pressure deficit on gas exchange. J Am Soc Hortic Sci 112:143–148

Bota J, Flexas J, Medrano H (2001) Genetic variability of photosynthesis and water use in Balearic grapevine cultivars. Ann Appl Biol 138:353–365

Bray EA, Bailey-Serres J, Weretilnyk E (2000) Response to abiotic stresses. In: Buchanan BB, Gruissem W, Jones RL (eds) Biochemistry and molecular biology of plants. American Society of Plant Physiologists Press, Rockville, pp 1158–1203

Bruce WB, Edmeades GO, Barker TC (2002) Molecular and physiological approaches to maize improvement for drought tolerance. J Exp Bot 53:13–25

Bryla DR, Duniway JM (1997) Effects of mycorrhizal infection on drought tolerance and recovery in safflower and wheat. Plant Soil 197:95–103

Burns JH (2006) Relatedness and environment affect traits associated with invasive and noninvasive introduced Commelinaceae. Ecol Appl 16(4):1367–1376

Burns JH, Winn AA (2006) A comparison of plastic responses to competition by invasive and non-invasive congeners in the Commelinaceae. Biol Invasions 8:797–807

Campos H, Cooper M, Habben JE, Edmeades GO, Schussler JR (2004) Improving drought tolerance in maize: a view from industry. Field Crop Res 90:19–34

Carbonneau A, Casteran P, Leclair Ph (1981) Principles de choix de systèmes de conduite pour des vignobles temperes et definitions pratiques utilisables en reglementation. Connaiss Vigne Vin 15(2):97–124

Cavaleri MA, Sack L (2010) Comparative water use of native and invasive plants at multiple scales: a global meta-analysis. Ecology 91(9):2705–2715

Cellier F, Conejero G, Casse F (2000) Dehydrin transcript fluctuations during a day/night cycle in drought-stressed sunflower. J Exp Bot 51:299–304

Centritto M (2005) Photosynthetic limitations and carbon partitioning in cherry in response to water deficit and elevated [CO_2]. Agricult Ecosys Environ 106:233–242

Cernusak LA, Winter K, Aranda J, Turner BL (2008) Conifers, angiosperm trees, and lianas: growth, whole-plant water and nitrogen use efficiency, and stable isotope composition (δ^{13}C and δ^{18}O) of seedlings grown in a tropical environment. Plant Physiol 148:642–659

Chalmers DJ, Mitchell PD, van Heek L (1981) Control of peach tree growth and productivity by regulated water supply, tree density, and summer pruning. J Am Soc Hort Sci 106:307–312

Chaves MM (1991) Effects of water deficits on carbon assimilation. J Exp Bot 42:1–16

Chaves MM, Maroco JP, Pereira JS (2003) Understanding plant response to drought: from genes to the whole plant. Funct Plant Biol 30:239–264

Chaves MM, Pereira JS, Maroco J, Rodrigues ML, Ricardo CP, Osório ML, Carvalho I, Faria T, Pinheiro C (2002) How plants cope with water stress in the field. Photosynthesis and growth. Ann Bot 89:907–916

Chen THH, Murata N (2002) Enhancement of tolerance of abiotic stress by metabolic engineering of betaines and other compatible solutes. Curr Opin Plant Biol 5:250–257

Chrispeels MJ, Agre P (1994) Aquaporins: water channel proteins of plant and animal cells. Trends Biochem Sci 19:421–425

Close T (1996) Dehydrins: emergence of a biochemical role of a family of plant dehydration proteins. Physiol Plantarum 97:795–803

Conde A, Silva P, Agasse A, Conde C, Gerós H (2011) Mannitol transport and mannitol dehydrogenase activities are coordinated in *Olea europaea* under salt and osmotic stresses. Plant Cell Physiol 52(10):1766–1775

Condon AG, Richards RA, Rebetzke GJ, Farquhar GD (2004) Breeding for high water-use efficiency. J Exp Botany 55:2447–2460

Conejero W, Alarcón JJ, García-Orellana Y, Nicolás E, Torrecillas A (2007) Evaluation of sap flow and trunk diameter sensors used for irrigation scheduling in early maturing peach trees. Tree Physiol 27:1753–1759

Connor DJ (2005) Adaptation of olive (*Olea europaea* L.) to water-limited environments. Aust J Agric Res 56:1181–1189

Cooper M, Delacy IH, Eisemann RL (1993) Recent advances in the study of genotype × environment interactions and their application to plant breeding. In: Imrie BC, Hacker JB (eds) Focused plant improvement: towards responsible and sustainable agriculture. In: Proceedings

tenth Australian plant breeding conference, Vol 1. Organising Committee, Australian Convention and Travel Service, Canberra, pp 116–131
Daehler CC (2003) Performance comparisons of co-occurring native and alien invasive plants: implications for conservation and restoration. Annu Rev Ecol Evol Syst 34:183–211
Dry PR, Loveys BR (1998) Factors influencing grapevine vigour and the potential for control with partial rootzone drying. Aust J Grape Wine Res 4:140–148
Emmerich WE (2007) Ecosystem water use efficiency in a semiarid shrubland and grassland community. Rangeland Ecol Manage 60:464–470
Fahn A, Werker E, Baas P (1986) Wood anatomy and identification of trees and shrubs from Israel and adjacent regions. Isr Acad Sci Hum, Jerusalem
FAO (2010) Climate change implications for food security and natural resources management in Africa. Twenty-sixth regional conferences for Africa. Luanda, Angola
Farquhar GD, Ehleringer JR, Hubick KT (1989) Carbon isotope discrimination and photosynthesis. Annu Rev Plant Physiol Plant Mol Biol 40:503–537
Feng Y-L, Fu G-L, Zheng Y-L (2008) Specific leaf area relates to the differences in leaf construction cost, photosynthesis, nitrogen allocation, and use efficiencies between invasive and noninvasive alien congeners. Planta 228:383–390
Fereres E, Evans RG (2006) Irrigation of fruit trees and vines: an introduction. Irrigation Sci 24:55–57
Fereres MA Soriano (2007) Deficit irrigation for reducing agricultural water use. J Exp Bot 58:147–159
Fernández JE, Green SR, Caspari HW, Diaz-Espejo A, Cuevas MV (2008) The use of sap flow measurements for scheduling irrigation in olive, apple and Asian pear trees and in grapevines. Plant Soil 305:91–104
Fleury D, Jefferies S, Kuchel H, Langridge P (2010) Genetic and genomic tools to improve drought tolerance in wheat. J Exp Bot 61(12):3211–3222
Flexas J, Bota J, Escalona JM, Sampol B, Medrano H (2002) Effects of drought on photosynthesis in grapevines under field conditions: an evaluation of stomatal and mesophyll limitations. Funct Plant Biol 29:461–471
Flexas J, Galmes J, Galle A, Gulias J, Pou A, Ribas-Carbo M, Tomas M, Medrano H (2010) Improving water use efficiency in grapevines: potential physiological targets for biotechnological improvement. Aust J Grape Wine Res 16:106–121
Fukai S, Cooper M (1995) Development of drought-resistant cultivars using physio-morphological traits in rice. Field Crop Res 40:67–86
Funk JL, Vitousek PM (2007) Resource-use efficiency and plant invasion in low-resource systems. Nature 446:1079–1081
Garay-Arroyo A, Colmenero-Flores JM, Garciarrubio A, Covarrubias AA (2000) Highly hydrophylic proteins in prokaryotes and eukaryotes are common during conditions of water deficit. J Biol Chem 275:5668–5674
Glenn DM, Puterka GJ (2005) Particle films: a new technology for agriculture. Hortic Rev 31:1–45
Grant OM, Tronina L, Jones HG, Chaves MM (2007) Exploring thermal imaging variables for the detection of stress responses in grapevine under different irrigation regimes. J Exp Bot 58:815–825
Grotkopp E, Rejmánek M (2007) High seedling relative growth rate and specific leaf area are traits of invasive species: phylogenetically independent contrasts of woody angiosperms. Am J Bot 94(4):526–532
Halliwell B, Gutteridge JMC (1999) Free radicals in biology and medicine, 3rd edn. Oxford University Press, Oxford
Hamilton EW, Heckathorn SA (2001) Mitochondrial adaptations to sodium chloride: complex I is protected by anti-oxidants and small heat-shock proteins, while complex II is protected by proline and betaine. Plant Physiol 126:166–1274

Hamilton MA, Murray BR, Cadotte M, Hose GC, Baker AC, Harris CJ, Licari D (2005) Life-history correlates of plant invasiveness at regional and continental scales. Ecol Lett 8:1066–1074

Hendrick JP, Hartl FU (1995) The role of molecular chaperones in protein folding. Faseb J 9:1559–1569

Heschel MS, Donohue K, Hausmann N, Schmitt J (2002) Population differentiation and natural selection for water-use efficiency in *Impatiens capensis* (Balsaminaceae). Int J Plant Sci 163(6):907–912

Hill JP, Germino MJ, Wraith JM, Olson BE, Swan MB (2006) Advantages in water relations contribute to greater photosynthesis in *C. maculosa* compared with established grasses. Int J Plant Sci 167(2):269–277

Hong S-W, Vierling E (2000) Mutants of *Arabidopsis thaliana* defective in the acquisition of tolerance to high temperature stress. P Natl Acad Sci U S A 97:4392–4397

Huglin P, Schneider C (1998) Biologie et écologie de la vigne. Technique and Documentation, 2e éd. Paris

Ingram J Bartels D (1996) The molecular basis of dehydration tolerance in plants. Annu Rev Plant Physiol Plant Mol Biol 47:377–403

Intrieri C, Poni S, Rebucci B, Magnanini E (1998) Row orientation effects on whole-canopy gas exchange of potted and field-grown grapevines. Vitis 37(4):147–154

Ionenko IF, Anisimov AV, Dautova NR (2010) Effect of temperature on water transport through aquaporins. Biol Plant 54:488–494

IPCC (2011) The special report on managing the risks of extreme events and disasters to advance climate change adaptation. http://www.ipcc.ch/ Cited 1 Dez 2011

Jackson P, Robertson M, Cooper M, Hammer G (1996) The role of physiological understanding in plant breeding: from a breeding perspective. Field Crop Res 49:11–37

Jiang L-F, Luo Y-Q, Chen J-K, Li B (2009) Ecophysiological characteristics of invasive *Spartina alterniflora* and native species in salt marshes of Yangtze River estuary, China. Est Coast Shelf Sci 81:74–82

Jifon JL, Syvertsen JP (2003) Kaolin particle film applications can increase photosynthesis and water use efficiency of 'Ruby red' grapefruit leaves. J Am Soc Hortic Sci 128:107–111

Jones H (2004) What is water use efficiency? In: Bacon MA (ed) Water use efficiency in plant biology. Blackwell Publishing CRC Press, Oxford, pp 27–41

Jones HG (1992) Plants and microclimate: a quantitative approach to environmental plant physiology, 2nd edn. Cambridge University Press, Cambridge

Karabourniotis G, Bornman J (1999) Penetration of UV-A, UV-B and blue light through the leaf trichome layers of two xeromorphic plants, olive and oak, measured by optical fibre microprobes. Physiol Plantarum 105:655–661

Karl TR, Melillo JM, Peterson TC (eds) (2009) Global climate change impacts in the United States. Cambridge University Press, Cambridge

Kaya C, Higgs D, Kirnak H, Tas I (2003) Mycorrhizal colonisation improves fruit yield and water use efficiency in watermelon (*Citrullus lanatus* Thunb.) grown under well-watered and water-stressed conditions. Plant Soil 253:287–292

Keller M (2005) Deficit irrigation and vine mineral nutrition. Am J Enol Vitic 56(3):267–283

Kloeppel BD, Abrams MD (1995) Ecophysiological attributes of the native *Acer saccharum* and the exotic *Acer platanoides* in urban oak forests in Pennsylvania, U S A. Tree Physiol 15:739–746

Kozlowski TT, Kramer PJ, Pallardy SG (1991) The physiological ecology of woody plants. Academic, San Diego

Kozlowski TT, Pallardy SG (1997) Physiology of woody plants, 2nd edn. Academic, San Diego

Kriedemann PE, Goodwin I (2003) Regulated deficit irrigation and partial root-zone drying. Irrigation insights No.3. In: Currey A (ed) Land and water Australia, Canberra, p 102

Lake JC, Leishman MR (2004) Invasion success of exotic plants in natural ecosystems: the role of disturbance, plant attributes and freedom from herbivores. Biol Conserv 117:215–226

Larcher W (1995) Physiological plant ecology, 3rd edn. Springer, Berlin

Lee D, Graham R (1986) Leaf optical properties of rainforest sun and extreme shade plants. Am J Bot 73:1100–1108
Leishman MR, Haslehurst T, Ares A, Baruch Z (2007) Leaf trait relationships of native and invasive plants: community- and global-scale comparisons. New Phytol 176:635–643
Leon J, Bukovac M (1978) Cuticle development and surface morphology of olive leaves with reference to penetration of foliar-applied chemicals. J Am Soc Hortic Sci 103:465–472
Levitt J (1972) Responses of plants to environmental stresses. Academic, New York
Liakoura V, Stavrianakou S, Liakopoulos G, Karabourniotis G, Manetas Y (1999) Effects of UV-B radiation on *Olea europaea*: comparisons between a greenhouse and a field experiment. Tree Physiol 19:905–908
Lipecki J, Berbeć S (1997) Soil management in perennial crops: orchards and hop gardens. Soil Till Res 43:169–184
Liu A, Plenchette C, Hamel C (2007) Soil nutrient and water providers: how arbuscular mycorrhizal mycelia support plant performance in a resource limited world. In: Hamel C, Plenchette C (eds) Mycorrhizae in crop production, Haworth Press, Inc, Philadelphia, pp 37–66
Lovisolo C, Schubert A (1998) Effects of water stress on vessel size and xylem hydraulic conductivity in *Vitis vinifera* L. J Exp Bot 49:693–700
Ludlow MM (1980) Adaptive significance of stomatal responses to water stress. In: Turner NC, Kramer PJ (eds) Adaptations of plants to water and high temperature stress. Wiley Interscience, New York, pp 123–138
Ludlow MM (1989) Strategies of response to water stress. In: Kreeb KH, Richter H, Hinckley TM (eds) Structural and functional responses to environmental stresses. SPB Academic Publishing, The Hague, pp 269–281
Maroco JP, Pereira JS, Chaves MM (2000) Growth, photosynthesis and water use efficiency of two C_4 Sahelian grasses subjected to water deficits. J Arid Environ 45:119–137
Masle J, Gilmore SR, Farquhar GD (2005) The ERECTA gene regulates plant transpiration efficiency in *Arabidopsis*. Nature 436:866–870
McAlpine KG, Jesson LK, Kubien DS (2008) Photosynthesis and water-use efficiency: a comparison between invasive (exotic) and non-invasive (native) species. Austral Ecol 33:10–19
McDowell SCL (2002) Photosynthetic characteristics of invasive and noninvasive species of Rubus (Rosaceae). Am J Bot 89(9):1431–1438
McNeil SD, Nuccio ML, Hanson AD (1999) Betaines and related osmoprotectants. Targets for metabolic engineering of stress resistance. Plant Physiol 120:945–949
Mediavilla S, Escudero A, Heilmeier H (2001) Internal leaf anatomy and photosynthetic resourse-use efficiency: interspecific and intraspecific comparisons. Tree Physiol 21:251–259
Mittler R (2006) Abiotic stress, the field environment and stress combination. Trends Plant Sci 11(1):15–19
Mohammadi Goltapeh E, Rezaee Danesh Y, Prasad R, Varma A (2008) Mycorrhizal fungi: what we know and what should we know? In: Varma A (ed) Mycorrhiza. Springer, Berlin, pp 3–27
Møller IM (2001) Plant mitochondria and oxidative stress: electron transport, NADPH turnover, and metabolism of reactive oxygen species. Annu Rev Plant Physiol Plant Mol Biol 52:561–591
Moran JF, Becana M, Iturbe-Ormaetxe I, Frechilla S, Klucas RV, Aparicio-Tejo P (1994) Drought induces oxidative stress in pea plants. Planta 194:346–352
Morison JI, Baker NR, Mullineaux PM, Davies WJ (2008) Improving water use in crop production. Philos T Roy Soc B 363:639–665
Moutinho-Pereira JM, Correia CM, Gonçalves B, Bacelar E, Torres-Pereira JM (2004) Leaf gas-exchange and water relations of grapevines grown in three different conditions. Photosynthetica 42(1):81–86
Moutinho-Pereira JM, Magalhães N, Gonçalves B, Bacelar E, Brito M, Correia C (2007) Gas exchange and water relations of three *Vitis vinifera* L. cultivars growing under Mediterranean climate. Photosynthetica 45(2):202–207

Moutinho-Pereira JM, Magalhães N, de Torres Castro LF, Chaves MM, Torres-Pereira JM (2001) Physiological responses of grapevine leaves to Bordeaux mixture under light stress conditions. Vitis 40(3):117–121

Murai-Hatano M, Kuwagata T, Sakurai J, Nonami H, Ahamed A, Nagasuga K, Matsunami T, Fukushi K, Maeshima M, Okada M (2008) Effect of low root temperature on hydraulic conductivity of rice plants and the possible role of aquaporins. Plant Cell Physiol 49:1294–1305

Nelson DE, Repetti PP, Adams TR, Creelman RA, Wu J, Warner DC, Anstrom DC, Bensen RJ, Castiglioni PP, Donnarummo MG, Hinchey BS, Kumimoto RW, Maszle DR, Canales RD, Kroliwoski KA, Dotson SB, Gutterson N, Ratcliffe OJ, Heard JE (2007) Plant nuclear factor Y (NF–Y) B subunits confer drought tolerance and lead to improved corn yields on water-limited acres. P Nat Acad Sci U S A 104:16450–16455

Noiraud N, Delrot S, Lemoine R (2000) The sucrose transporter of celery: identification and expression during salt stress. Plant Physiol 122:1447–1455

Northcote KH (1988) Soils and Australian viticulture. In: Coombe BG, Dry PR (eds) Viticulture, vol 1. Winetitles, Australia, pp 61–90

Pang J, Yang J, Ward P, Siddique Kadambot HM, Lambers H, Tibbett M, Ryan M (2011) Contrasting responses to drought stress in herbaceous perennial legumes. Plant Soil 348(1–2):299–314

Papageorgiou GC, Murata N (1995) The unusually strong stabilizing effects of glycine betaine on the structure and function of the oxygen evolving photosystem II complex. Photosynth Res 44:243–252

Passioura J (2002) Environmental biology and crop improvement. Funct Plant Biol 29:537–546

Passioura J (2007) The drought environment: physical, biological and agricultural perspectives. J Exp Bot 58(2):113–117

Passioura JB (1983) Roots and drought resistance. Agr Water Manage 7:265–280

Passioura JB (1982) Water in the soil-plant-atmosphere continuum. In: Lange OL, Nobel PS, Osmond CB, Ziegler H (eds) Physiological plant ecology II. Encyclopedia of plant physiology, vol 12 B. Springer, Berlin, pp 5–33

Patakas A, Noitsakis B (1999) Mechanisms involved in diurnal changes of osmotic potential in grapevines under drought conditions. J Plant Physiol 154:767–774

Peterson RL, Massicotte HB, Melville LH (2004) Mycorrhizas: anatomy and cell biology. NRC Research Press Ottawa/CABI Publishing, Wallingford

Possigham JV (1992) Factors affecting growth, flowering and fruit ripening in *Vitis vinifera*. In: Proceedings of the IV international symposium on grapevine physiology, Torino (Italia)

Rajendrakumar CSV, Reddy BDB, Reddy AR (1994) Proline-protein interactions: protection of structural and functional integrity of M4 lactate dehydrogenase. Biochem Biophy Res Commun 201:957–963

Rathinasabapathi B, Fouad WM, Sigua CA (2001) β-Alanine betaine synthesis in the Plumbaginaceae: purification and characterization of a trifunctional, S-adenosyl-L-methionine-dependent N-methyltransferase from Limonium latifolium leaves. Plant Physiol 126:1241–1249

Reynolds MP, Ortiz-Monasterio JI, McNab A (2001) Application of physiology in wheat breeding. CIMMYT, Mexico

Rhodes D, Hanson AD (1993) Quaternary ammonium and tertiary sulfonium compounds in higher plants. Annu Rev Plant Physiol Plant Mol Biol 44:357–384

Richards RA, Condon AG, Rebetzke GJ (2001) Traits to improve yield in dry environments. In: Reynolds MP, Ortiz-Monasterio JI, McNab A (eds) Application of physiology in wheat breeding. CIMMYT, Mexico, pp 88–100

Richardson A, Berlyn G (2002) Changes in foliar spectral reflectance and chlorophyll fluorescence of four temperate species following branch cutting. Tree Physiol 22:499–506

Rivero RM, Kojima M, Gepstein A, Sakakibara H, Mittler R, Gepstein S, Blumwald E (2007) Delayed leaf senescence induces extreme drought tolerance in a flowering plant. P Nat Acad Sci U S A 104:19631–19636

Roché BF, Roché CT, Chapman RC (1994) Impacts of grassland habitat on yellow starthistle (*Centaurea solstitialis* L.) invasion. Northwest Sci 68(2):86–96
Rodrigues MA, Arrobas M, Bonifácio N (2005) Análise de terras em olivais tradicionais de sequeiro. O efeito da aplicação localizada de fertilizantes. Revista de Ciências Agrárias 28:167–176
Rodrigues MA, Pavão F, Lopes JI, Gomes V, Arrobas M, Moutinho-Pereira JM, Ruivo S, Cabanas J, Correia CM (2011) Olive yields and tree nutritional status during a four-year period without nitrogen and boron fertilization. Communin Soil Sci Plant Anal 42:803–814
Rodríguez-Pérez JR, Riaño D, Carlisle E, Ustin S, Smart DR (2007) Evaluation of hyperspectral reflectance indexes to detect grapevine water status in vineyards. Am J Enol Viticult 58:302–317
Samuel D, Kumar TKS, Ganesh G, Jayaraman G, Yang P-W, Chang M-M, Trivedi VD, Wang S-L, Hwang K-C, Chang D-K, Yu C (2000) Proline inhibits aggregation during protein refolding. Protein Sci 9:344–352
Santos TP, Lopes CM, Rodrigues ML, Souza CR, Ricardo-da-Silva JM, Marococ JP, Pereira JS, Chaves MM (2007) Effects of deficit irrigation strategies on cluster microclimate for improving fruit composition of Moscatel field-grown grapevines. Sci Hortic 112(3):321–330
Schulte PJ (1993) Tissue hydraulic properties and the water relations of desert shrubs. In: Smith JAC, Griffith H (eds) Water deficits, plant responses from cell to community. BIOS, Oxford, pp 177–192
Schultz HR, Matthews MA (1988) Resistance to water transport in shoots of *Vitis vinifera* L. Plant Physiol 88:718–724
Schwabe W, Lionakis S (1996) Leaf attitude in olive in relation to drought resistance. J Hortic Sci 71:157–166
Seibt U, Rajabi A, Griffiths H, Berry JA (2008) Carbon isotopes and water use efficiency: sense and sensitivity. Oecologia 155:441–454
Shinozaki K, Yamaguchi-Shinozaki K (2007) Gene networks involved in drought stress response and tolerance. J Exp Bot 58:221–227
Shorter R, Lawn RJ, Hammer GL (1991) Improving genotypic adaptation in crops—a role for breeders, physiologists and modellers. Exp Agr 27:155–175
Smart RE (1974) Photosynthesis by grapevine canopies. J Appl Ecol 11:997–1006
Smirnoff N (1998) Plant resistance to environmental stress. Curr Opin Biotech 9:214–219
Sperry JS (2003) Evolution of water transport and xylem structure. Int J Plant Sci (Suppl.) 164:S115–S127
Sun W, Van Montagu M, Verbruggen N (2002) Small heat shock proteins and stress tolerance in plants. Biochim Biophys Acta 1577:1–9
Taiz L, Zieger E (1998) Plant Physiology, 2nd edn. Sinauer Associates Inc., Sunderland
Tinker NA (2002) Why quantitative geneticists should care about bioinformatics. In: Kang MS (ed) Quantitative genetics, genomics and plant breeding. CABI, Wallingford, pp 33–44
Tognetti R, d'Andria R, Morelli G, Alvino A (2005) The effect of deficit irrigation on seasonal variations of plant water use in *Olea europaea* L. Plant Soil 273:139–155
Tognetti R, Raschi A, Béres C, Fenyvesi A, Ridder HW (1996) Comparison of sap flow, cavitation and water status of *Quercus petraea* and *Quercus cerris* trees with special reference to computer tomography. Plant, Cell Environ 19:928–938
Tuberosa R, Salvi S (2006) Genomics approaches to improve drought tolerance in crops. Trends Plant Sci 11(8):405–412
Tubiello FN, Lin G, Druitt JW, Marino BEV (1990) Ecosystem-level evapotranspiration and water-use efficiency in the desert biome of Biosphere 2. Ecol Eng 13:263–271
Turner NC (1986) Crop water deficits: a decade of progress. Adv Agron 39:1–51
Tyerman SD, Niemietz CM, Bramley H (2002) Plant aquaporins: multifunctional water and solute channels with expanding roles. Plant, Cell Environ 25:173–194
Tyree MT, Ewers FW (1991) The hydraulic architecture of trees and other woody plants. New Phytol 119:345–360

Tyree MT, Sperry S (1989) Vulnerability of xylem to cavitation and embolism. Annu Rev Plant Physiol Mol Biol 40:19–38

Van Huyssteen E, Van Zyl JL, Koen AP (1984) The effect of cover crop management on soil conditions and weed control in a Colombard vineyard in Oudtshoorn S Afr. J Enol Vitic 5:7–17

Vijn I, Smeekens S (1999) Fructan: more than a reserve carbohydrate? Plant Physiol 120:351–360

Vilá M, Weiner J (2004) Are invasive plant species better competitors than native plant species?—evidence from pair-wise experiments. Oikos 105:229–238

Wan XC, Zwiazek JJ (2001) Root water flow and leaf stomatal conductance in aspen (*Populus tremuloides*) seedlings treated with abscisic acid. Planta 213:741–747

Wehmeyer N, Hernandez LD, Finkelstein RR, Vierling E (1996) Synthesis of small heat-shock proteins is part of the development program of late seed maturation. Plant Physiol 112:747–757

Yuste J (2007) Efectos de la forma de conducción del viñedo en el régime hídrico y en la respuesta agronómica. Fundamentos, aplicación y consecuencias del riego en la vid, Editorial Agrícola Española S.A., Madrid

Chapter 7
Effects of Drought on Nutrient Uptake and Assimilation in Vegetable Crops

Youssef Rouphael, Mariateresa Cardarelli, Dietmar Schwarz, Philipp Franken and Giuseppe Colla

Abstract Scarcity of water is a severe environmental limit to plant productivity. Drought-induced loss in crop yield probably exceeds losses from all other causes, since both the severity and duration of the stress are critical. Nutritional imbalance under drought conditions depresses plant growth and therefore productivity by affecting nutrient uptake, transport, and distribution. Despite contradictory reports on the effects of nutrient supply on plant growth under drought conditions, it is generally accepted that an increased nutrient supply will not improve plant growth when the nutrient is already present in sufficient amounts in the soil and the drought is severe. A better understanding of the role of mineral nutrients in plant resistance to drought will contribute to improve fertilizer management in arid and semiarid areas and in regions suffering from temporary drought. Considering that vegetables are concentrated in semiarid zones where water stress is frequent, it is important to ascertain how this type of stress affects the nutrient uptake and assimilation of these crops. This chapter starts with an overview of the recent literature on plant nutrition of vegetables under drought conditions, stressing mainly the effects of drought on nutrient availability, uptake, transport, and accumulation in plants, and also the interactions between nutrient supply and drought response; it then proceeds to identify the means to increase nutrient availability under drought conditions through breeding, grafting, and fertilization.

Y. Rouphael
Department of Crop Production, Faculty of Agricultural Engineering and Veterinary Medicine, Lebanese University, Beirut, Dekwaneh-El Maten, Lebanon

M. Cardarelli · G. Colla (✉)
Department of Agriculture, Forestry, Nature and Energy, University of Tuscia, 01100 Viterbo, Italy
e-mail: giucolla@unitus.it

D. Schwarz · P. Franken
Leibniz-Institute of Vegetable and Ornamental Crops, 14979 Großbeeren, Germany

R. Aroca (ed.), *Plant Responses to Drought Stress*,
DOI: 10.1007/978-3-642-32653-0_7,

7.1 Introduction

It is very likely that the climate change will have a major impact on the hydrological cycle and consequently on available water resources, flood and drought potentials, and agricultural productivity (Evans 1967). Faced with a scarcity of water resources, particularly in the warm, arid, and semiarid parts of the world, drought is the single most critical threat to world food security (Farooq et al. 2009). It has been the catalyst of the great famines of the past. Because the world's water supply is limited, the predictable pressure of food demand for a rapidly increasing population is likely to further aggravate the effects of drought (Somerville and Briscoe 2001). The severity of drought depends on many factors such as occurrence and distribution of rainfall, evaporative demands, and moisture storing capacity of soils (Wery et al. 1994). Under soil water deficiency, crop yield is reduced as a result of (1) reduced canopy absorption of photosynthetically active radiation, (2) decreased efficiency in the use of radiation, and (3) reduced harvest index (Earl and Davis 2003). Water deficit has also an important effect on the acquisition of nutrients through the roots and their transport to the shoots. Decreasing soil water availability generally results in a limited total nutrient uptake and diminished tissue concentrations of nutrients in plants. A poorer absorption of the inorganic nutrients can result from interference in nutrient uptake and in the unloading mechanism, and reduced transpiration flow (Garg 2003; McWilliams 2003). Considering that vegetables are concentrated in semiarid zones where water stress is frequent, it is important to ascertain how water stress affects the nutrient uptake and assimilation of these crops, and to propose strategies for improving crop nutrient status under drought stress. Efforts toward solving the problems arising from drought by improving nutrient availability, uptake, transport, and accumulation in plants are based primarily on the selection of tolerant genotypes (Waraich et al. 2011). The conventional selection method by crossing exotic germplasm and adapted elite material has been predominant until quite recent times. Over the past several years, however, molecular markers have been used for the identification of tolerance carrying genes, in addition to an adequate selection technology (Miletic et al. 2010). Apart from developing tolerant genotypes, rational agricultural practices, such as grafting, use of beneficial microorganisms, and application of organic matter, nutrients, and chemicals, such as proline, silicon, and other osmoprotectants (Folkert et al. 2001), have been recognized worldwide as additional strategies for improving nutrient uptake and assimilation under drought conditions.

The objective of this chapter is to review the current situation of research on plant nutrition under drought. We will examine in particular (1) the effects of drought on nutrient availability, uptake, transport, and accumulation in plants, (2) interactions between nutrient supply and drought stress response, and (3) means to increase nutrient availability under drought conditions by breeding, molecular approaches, grafting, and application of mineral nutrients.

7.2 Soil Nutrient Availability

A large proportion of the nitrogen, phosphorus, and sulfur present in soil occurs in organic form. Even though it is now generally acknowledged that organic forms of nitrogen in soil and the rhizosphere play a role in supplying plants with nitrogen (Kreuzwieser and Gessler 2010), the availability of inorganic forms of nitrogen, i.e., nitrate and ammonium, is still assumed to be crucial (Lucash et al. 2007). Phosphorus is also absorbed by plants in inorganic forms, such as orthophosphate ions ($H_2PO_4^-$ and HPO_4^{2-}) which are present in the soil solution. Inorganic sulfate is the most important form of sulfur available to plant roots in the soil (Herschbach et al. 2005), and most of the other elements are available to plants mainly in inorganic mineral forms (Marschner 2012). Thus, any negative effect of drought on the microbial mineralization of organic matter, which in turn affects the amount of inorganic nutrients available for plant uptake, can potentially impair the mineral nutrition of plants. Even when we might assume that the roots can absorb organic nutrients, mainly N in the form of amino acids, microbial depolymerization of proteins/peptides is still essential (Rennenberg et al. 2009). Reduced soil water availability limits microbial activity in soil and, according to the intensity and duration of the drought event, may lead to a total inhibition of microbial metabolism (Borken and Matzner 2009). Furthermore, the diffusion of the organic substrates for microbe-driven mineralization or depolymerization becomes restricted (Schimel et al. 2007). Moreover, the mobility of microorganisms in the soil as well as that of the excreted exoenzymes, such as proteases, decreases with increasing drought (van Meeteren et al. 2008). Also, the intracellular processes of microbes are generally inhibited when countermeasures (e.g., increasing the osmotic potential) are no longer sufficient to avoid loss of water from the cells. Thus, drought events cause at first a reduction in bacterial activity accompanied by dehydration and, in a prolonged drought, a dieback of soil microorganisms (Schimel et al. 2007). Clearly, the decrease in microbial activity is related to the length and intensity of the drought period as well as to the adaptation potential of the microorganisms (Jensen et al. 2003). In general, the specific effects of drought on the processes involved in mineralization, and consequently in the resupply of inorganic nutrients, are very uncertain since the results in the existing literature are inconsistent. Gessler et al. (2005) even found increased gross ammonification rates in summer on a drought-exposed beech site in comparison with a cool-moist site. Beier et al. (2008) reported that carbon and nitrogen mineralization were affected differently by drought; while the decomposition of organic carbon was mainly sensitive to temperature, ammonification depended only slightly on temperature but was strongly inhibited by reduced soil water availability. It has also been observed that the amount of organic nitrogen dissolved in the soil increases during drought events, presumably due to dieback of the microbial biomass (Borken and Matzner 2009; Dannenmann et al. 2009). No research has been conducted on whether plants might benefit from this potential organic N source under drought conditions or whether diffusion limitations or other restrictions might prevent these compounds from being absorbed. In conclusion, it has not yet been established to what extent drought events affect the resupply of soil nutrients for plants by microorganisms. Therefore, it is necessary to implement researches so as to systematically target gross and net nutrient mineralization fluxes as affected by drought intensity and duration. It is also important in this context to assess the effects of drought on the spatial distribution of nutrients in the soil and relate this pattern to the effect of reduced water availability on the ability of roots to actively access areas of soil having higher nutrient concentrations. It is also important to differentiate between short- and long-term effects of drought on nutrient availability in soils. While occasional short summer drought periods and subsequent recovery times with sufficient rainfall might only lead to transient changes in soil nutrient conditions, severe long-term water restriction over years will most probably have stronger and more permanent effects.

7.3 Nutrient Uptake of Roots

The morphological, physiological, and genetic characteristics of roots influence strongly their capacity to absorb water and nutrients. Even if the nutrient is available in the soil solution, the root needs to absorb it. This depends mainly on the water moving into the roots. Under conditions of no or low drought, e.g., high transpiration and open stomata, water moves passively to and across the roots at a high rate (Steudle 2000). So far, the movement of low-molecular-weight solutes (e.g., ions, organic acids, amino acids, sugars) concurs with the movement of water as far as to the walls of the root cells and is a nonmetabolic, passive process, driven by mass flow and diffusion (Marschner 2012). Under conditions of reduced soil moisture, the stomata may close, transpiration decreases, and thus the water and nutrient flow diminish due to a reduced diffusion rate of nutrients in the soil to the absorbing root surface (Pinkerton and Simpson 1986). If the transpiration rate drops to zero, the ascending flow of water is sustained only by root pressure (Steudle 2000). The more severe the drought the lower the water and nutrient flow and the more limited is the availability of nutrients for absorption by the root system. This important limitation for plant growth and development has been extensively studied also by molecular methods. Numerous genes and regulating pathways were identified revealing the processes by which plants react to low amounts of, e.g., phosphate, nitrogen , potassium and trace elements. However, surprisingly little is known about the behavior of these molecular processes under drought conditions, particularly with reference to vegetables. Although large-scale transcriptomic approaches have identified numerous genes being differentially expressed in several herbaceous and tree crops, only very few were found encoding mineral nutrient transporters normally induced in roots by nutrient deficiency, and some results were in contrast with what one would expect. Out of approximately 2,200 differentially expressed genes of sorghum, no nutrient transporter genes were reported to respond to dehydration (Buchanan et al. 2005).

More than 2,600 genes were analyzed in maize and NRAMP encoding a metal ion transporter and a zinc transporter gene were downregulated in roots with a decreasing water supply (Li et al. 2009). Furthermore, the expression of a silicon transporter was repressed in rice roots exposed to drought (Yamaji and Ma 2007). In grapevine, however, three genes could be identified encoding a sulphate, a nitrate, and a nitrite transporter which were induced by water shortage (Cramer et al. 2007).

The roots themselves are considered the key organ for plant adaptation to drought. The more extensive and prolific a root system is, the more capable it is of absorbing relatively more water from deeper soils and/or absorbing water relatively rapidly (Subbarao et al. 1995; Turner et al. 2001). High rates of water uptake enhance the ability of a plant to acquire nutrients from the soil. Therefore, root growth, size, proliferation, density, and distribution are key factors for the response of plants to drought stress with respect to water and nutrient acquisition (Turner et al. 2001; Kavar et al. 2007). These root characteristics are genetically fixed and vary widely between plant species and even between genotypes of a species. Greenwood et al. (1982) derived equations describing root distribution of seven vegetable crops on a sandy loam. They confirmed that root density often declines exponentially with soil depth. Moreover, they showed that root length strongly depends on the weight of the plant. This relationship did not differ much between species within the legume (pea, broad bean) and the nonlegume group (lettuce, cauliflower, parsnip, and turnip). With respect to the rooting patterns, nonlegumes develop more extensive and denser root systems than legumes. Cauliflower roots, after 50 days under drought conditions, reached 90 cm depth, and under irrigated conditions 70 cm (Kage et al. 2004). Hence the growth into deeper soil layers was much faster under drought conditions. It appears from these characteristics that nonlegumes are better suited to face drought periods. *Allium* spp. differs from other vegetables; the increase in root length and density of onion, for example, was only from 4 to 18 cm over a period of 80 days and the roots did not reach the deep soil layers. This is because the roots sprout primarily from the base of the bulb and branch much less than the other crops. Root growth of fruit vegetables is described under many different environmental conditions and cultivation treatments, including different types of irrigation, particularly from experiments in protected and soilless cultivation. Unfortunately, rooting data in drought conditions are rare (Kage et al. 2004). Pepper and tomato are described as having a shallow root system, which extracts 70–80 % of the water requirement in the top 30 cm soil layer (Dimitrov and Ovtcharrova 1995). Therefore, they are more sensitive to drought. It has been demonstrated that up to a certain threshold of reduced water content in the soil the root weight, length, and surface area increase. This indicates that drought tolerance of vegetables is increased by improved root growth and root functioning. Therefore, selection for a deep and extensive root system has been suggested as a means to increase crop productivity (Subbarao et al. 1995).

Once ions arrive into the roots' free space, cations can accumulate in the cell wall continuum and other plant tissues since the carboxylic group acts as a cation exchanger, whereas anions are repelled (Marschner 2012). The apoplasmic free space can serve as a transient storage pool for essential cations which can be mobilized, for example, by specific root exudates such as phytosiderophores, and subsequently translocated to the shoots (Liu et al. 2010). Plant species and even genotypes differ considerably in the cation exchange capacity (CEC) of the cell wall continuum in roots. As the external pH decreases, the effective CEC is reduced. Since root CEC can affect the rate and selectivity of ion influx into root cells, and indirectly the apoplasmic ion movements, drought conditions result in a decreased cation supply (White and Broadley 2003). The uptake of a nutrient requires that it passes the cell membranes. The rate of transport of a solute and its concentration is described by the Michaelis–Menten equation. Important variables are the solute concentration (S), the maximal rate of solute transport (V_{max}), and the Michaelis constant (K_m), which is the solute concentration at which half the maximal transport rate is reached (Marschner 2012). The C_{min} concentration is an important factor in the ion uptake from dry soils, because it is the lowest concentration at which roots can extract an ion from the soil solution. It varies considerably between plant species. For example, for P a value of 0.12 μM has been found in tomato (Itoh and Barber 1983). For nitrate C_{min} can vary between more than 50 and 0.1 μM (Tinker and Nye 2000). Besides the differences between plant species, environmental conditions also can have an influence. For instance, C_{min} decreases significantly with the decrease of soil water content (Marschner et al. 1991). As a rule, the demand for nutrient uptake is driven primarily by shoot growth because the canopy forms the greatest N sink, especially during vegetative growth, as a result of its high nutrient content (Thorup-Kristensen and Sorensen 1999; Kage et al. 2004).

For many years, researchers have tried to assess whether the quantity or the quality of the root system was more important to fulfill shoot demands. De Willigen and Van Nordwijk (1987) calculated a minimal root surface area for tomato nutrient demands. Input variables are dry mass production over time, nutrient concentration in the plant, and plant density, while V_{max} of a nutrient is an equation parameter. Based on this calculation, the size of the root system is the main limiting factor for P and Ca uptake and less for NO_3 and K uptake. The variable that influences this calculation is mainly V_{max} indicating the importance of the root quality (i.e., the distribution and structure).

Furthermore, the root system functioning for nutrient uptake is directly affected by drought as follows:

- the hydraulic conductance of roots is reduced leading to a decrease of nutrient uptake and transport to the shoots (Aroca et al. 2006);
- cell elongation can be inhibited by the interruption of water flow (Nonami 1998) leading to changes in root morphology such as decrease in specific root length and surface area, as described for cauliflower (Kage et al. 2004), lettuce (Schwarz et al. 1995), and tomato (Agele et al. 2011);
- root hair length is reduced as a result of the abscisic acid (ABA) produced by dehydrating roots (Bibikova and Gilroy 2003);

- ABA signaling brings about stomata closure and consequently reduces transpiration, thereby restricting nutrient uptake by the roots and transport to the shoot (Turner et al. 2001);
- root respiration is increased and consequently, the energy demand is also increased (Balogh et al. 2008);
- oxidative stress is increased (Reddy et al. 2004) and consequently leakage of the cell membranes further impairs nutrient uptake.

Drought effects on the shoot and above ground plant parts feedback on the root system, and thus influence its capacity for nutrient uptake. Such indirect effects can be:

- photosynthesis is reduced (Reddy et al. 2004) leading to a reduction in the energy supply to the roots which contrasts with their increased energy consumption (e.g., for respiration);
- dry matter allocation toward the roots is enhanced (lower shoot/root ratio), which can increase water and nutrient uptake (Brouwer 1983; Kage et al. 2004);
- production of endogenous auxins is limited, usually when contents of ABA and ethylene increase (Nilsen and Orcutte 1996), leading to a reduction in the formation of new roots which is important for drought tolerance.

7.4 Crop Nutrient Status

7.4.1 Nitrogen

Nitrogen is one of the macronutrient that plants require in the large amount and it is a constituent of many plant cell components, including amino and nucleic acids. Nitrogen requirement for optimal plant growth is in the range of 2–5 % of the dry weight depending on the plant species and development stage (Marschner 2012). Drought conditions may reduce soil-N mineralization, thus lowering N availability (Bloem et al. 1992). The absorption of N by roots requires the presence of water in the soil, since it is the agent that transports solutes to the soil–root interface (Garwood and Williams 1967). Under drought conditions, reduced crop transpiration rate decreases N transport from roots to shoots, thereby limiting N uptake (Tanguilig et al. 1987). According to Dalla Costa et al. (1997) and Dalla Costa and Gianquinto (2002), the water regime is crucial in determining the ability of potato and pepper to absorb the nitrogen available, since a well-watered crop is more capable of benefiting from the applied fertilizer. Kirnak et al. (2002) showed that water stress treatments (treatments receiving 80 and 90 % replenishment of the Class A pan evaporation) significantly reduced watermelon leaf macronutrient concentrations, particularly N, compared with the control treatment (100 % replenishment of Class A pan evaporation). The authors also showed that leaf nutrient deficiency was directly related to decreasing soil moisture content in the root zone, since a close relationship between chlorophyll content, water deficit, and leaf nutrient concentration was reported. This is in agreement with a previous work on bell pepper (Simonne et al. 1998), indicating that water stress has a crucial role in reducing leaf macronutrient concentrations, particularly N, in vegetable crops. Moreover, Sanchez-Rodriguez et al. (2010) analyzed the variations in foliar concentrations of macro- and micronutrients as well as the transport of these nutrients in five cherry tomato cultivars under well watered and moderately water-stressed conditions with the aim of establishing whether the ionome of the plants is related to the degree of sensitivity or tolerance to this type of stress. The results show a general reduction in growth together with a lower concentration and uptake of macronutrients, especially N in all the cultivars studied, except for cv. Zarina, which showed better growth and increased concentration and uptake of nitrogen, compared to other genotypes. Substantial decrease in nitrate reductase activity has also been reported in leaves of plants exposed to drought (Ruiz-Lozano and Azcón 1996), whereas total amino acid levels may increase in more advanced stages of drought because of proteolysis (Fukutoku and Yamada 1984) and perturbations in the translocation of amino acids from shoots to roots (Larsson 1992). Under drought conditions, reduced N uptake, assimilation and transport, depresses plant growth. Typical N-deficiency symptoms are enhanced senescence of older leaves (due to N remobilized from old leaves to young leaves) causing leaf chlorosis, and decrease of leaf area. Moreover, N deficiency leads to a reduction of the shoot/root ratio which is more favorable for the acquisition of nutrients and water from soil. Nitrogen influences plant composition much more than any other mineral nutrient as a consequence of competition for photosynthates among the various metabolic pathways (Marschner 2012). Benar et al. (2009) reported that a low nitrogen supply decreased the acid content and increased the soluble sugar content in tomato fruits. Moreover, the content of some phenolic compounds (rutin, a caffeic acid glycoside, and a caffeic acid derivate) and total ascorbic acid tended to be higher in the tomato fruits with the lowest nitrogen supply. Drought conditions induce an increase in nitrate concentration in several crop species (Jones et al. 1980; Kameli and Lösel 1995) as a result of a rapid decline in nitrate reductase activity even under mild water deficit (Sivaramakrishnan et al. 1988). Moreover, nitrates act as an important osmoticum and contribute to osmotic adjustment in the droughted plants. Nitrate content is an important quality characteristic in vegetables. Nitrate itself is relatively nontoxic but its metabolites may produce a number of negative health effects. Therefore, increased nitrate content in droughted leafy vegetable crops represent a negative aspect from a nutritional point of view.

7.4.2 Phosphorus

Phosphorus is a constituent of nucleic acids, phospholipids, phosphoproteins, dinucleotides, and adenosine triphosphate. Hence, P is required for processes including the storage and transfer of energy, photosynthesis, the regulation of some enzymes, and the transport of carbohydrates. Phosphorus requirement for optimal plant growth is in the range from 0.3 to 0.5 % of the dry weight during the vegetative stage of growth (Marschner 2012). Soils in arid areas are often calcareous and have high pHs (e.g., the soils in Mediterranean regions). This type of soils shows a strong tendency to P fixation. It is generally accepted that the uptake of P by crop plants is reduced in dry-soil conditions (Pinkerton and Simpson 1986) and the translocation of P to the shoots is severely restricted even under relatively mild drought stress (Rasnick 1970). Kirnak et al. (2002) and Sanchez-Rodriguez et al. (2010) observed a lower concentration and uptake of P in watermelon and cherry tomato under water deficit conditions. However, Liebersbach et al. (2004) reported that the large amount of molecular exudates (e.g., mainly mucilage) from plants in dry-soil counteract the reduced mobility of P under such conditions. Turner (1985) pointed out that P deficiency appears to be one of the earliest effects of mild to moderate drought stress in soil-grown plants. In plants suffering from phosphorus deficiency, reduction in leaf expansion, leaf area, and number of leaves are the most common effects. In comparison to shoot growth, root growth is much less inhibited by P deficiency, and thus the shoot/root ratio decreases (Marschner 2012). Moreover, under P deficiency flower initiation is delayed, the number of flowers decreases and seed formation is particularly restricted. Premature senescence of leaves is another factor limiting seed yield in P-deficient plants (Marschner 2012). Furthermore, reduced P content in vegetables decreases the nutritional quality, because it diminishes the contribution of vegetables to the total P dietary intake. This decline in the nutritional quality is particularly significant, because fruits and vegetables usually contribute about 11 % of the total P dietary intake (Levander 1990).

7.4.3 Potassium

Potassium is the cationic mineral nutrient that plants require in the largest amounts (Pujos and Morard 1997). Potassium requirement for optimal plant growth is in the range of 2–5 % of the dry weight of vegetative parts, fleshy fruits, and tubers (Marschner 2012). This element is essential for many physiological processes, such as photosynthesis, enzyme activation, protein synthesis, osmoregulation (cell expansion), cell turgor and ion homeostasis in plant cells, energy status, and a competition of Na under saline conditions (Fournier et al. 2005; Kanai et al. 2007, 2011). It is essential that cells maintain a constant level of K^+ in the cytosol which should range between 80 and 150 mM, while vacuolar K concentrations vary dramatically from being extremely high (up to 600 mM) to almost zero (Shabala et al. 2003; Marschner 2012). The availability of K to the plant decreases with decreasing soil water content, due to the decreasing mobility of K under these conditions. Kuchenbuch et al. (1986) showed that low levels of soil moisture reduced root growth and the rate of potassium inflow in onion plants in terms of both per unit of root weight and per unit of root length. On the other hand, reducing the irrigation in soybeans by increasing the interval from 7 to 11 days increased K concentration to 22 % but decreased the total content since the dry weight produced was much less (Aliasgharzad et al. 2009). This indicates that the ratio between water and K uptake and transfer into the shoot is important for K concentration in plant tissue. If the duration and intensity of the drought is short and low, the K concentration may even rise. However, when the duration and impact of drought increase, K concentration decreases (McWilliams 2003). The level of K deficiency determines the growth and quality characteristics of the product. A weak decrease in K is less restrictive, and depending on its duration may not result in yield loss since vegetable plants are able to accumulate this element in the vacuole and remobilize it from older organs, and thus provide enough K to the sink organs. In the case of fruit vegetables, K is remobilized from the leaves and the stem and transferred largely to the fruits (Pujos and Morard 1997). Fruit vegetables have a higher K demand when they start the generative phase and grow fruits; this results in an enhanced sensitivity to drought and other factors reducing the supply. Deficient K may induce wilting in plants (Beringer and Trolldenier 1978). Genes encoding K transporters were repressed by drought (Cramer et al. 2007; Li et al. 2009) and an inward K channel in grapevine was drought-repressed in roots, but induced in leaves (Cuellar et al. 2010). This channel is regulated by the protein kinase CIPK23 which in turn interacts with calcineurin B-like calcium sensors. Interestingly, a mutation in CIPK23 enhanced drought tolerance in plants, but at the same time caused severe growth defects at low potassium concentrations in the soil (Cheong et al. 2007). Impact of K deficiency on vegetable growth is described in several papers and reviews (Pujos and Morard 1997; Kanai et al. 2007, 2011). The main effects are stomata closure, lowered transpiration rates, and consequently higher dry matter percentage in plant organs (Hsiao and Läuchli 1986). However, Fournier et al. (2005) explain that K deficiency does not necessarily result in a decreased water uptake but can have also the opposite effect if the apoplastic water flow is not affected. Diminished assimilate partitioning from the source into the fruits results in decreased crop growth and inhibits fruit production (Marschner 2012). Potassium is the cation that influences most strongly the quality attributes that determine vegetable marketability, consumer preference, and the concentration of critically important human–health associated phytonutrients (Lester et al. 2010). A lowered supply of K as caused by drought may result in the reduction of several quality attributes as summarized by Lester et al. (2010). A deficiency level may reduce fruit shape, red color for tomato and pepper, fruit acidity, lycopene content, and shelf life (Serio et al. 2007), while augmenting ripening disorders (Peet 2009), and increasing sugars or total soluble solid contents, ascorbic acid, and β-carotene content (Schwarz and Krumbein 2012). Since K deficiency alters the water relations of the plant and the more so by a reduced water uptake at drought conditions, dry matter percentage of fruits may increase and stimulate these effects. On the other hand, quality changes could be originated from the enhanced activity of acid invertase, since fruits were identified as being the strongest sinks among plant organs (Kanai et al. 2011). Low K supply can interfere with and increase the uptake and mobility of Ca and Mg because of an antagonistic effect between cations as observed in many plants including tomatoes (Pujos and Morard 1997; Kanai et al., 2007). Consequently, the occurrence of blossom-end rot in tomato or pepper is reduced (Peet 2009) and rigidity of the tissue structure increased, resulting in higher measurable firmness and puncture force (Buescher and Hobson 1982). Similar improvements were noticed in lettuces, cabbages, and even carrots (Olle and Bender 2009). The impact on quality depends on the level of K deficiency, the available source of potassium salt, the plant species and even cultivar, and additional effects of climate conditions, such as temperature and radiation. Finally, taking into consideration that fruits and vegetables usually contribute 35 % of the total K to the dietary intake of humans (Levander 1990), the decrease of K content in vegetable products as a result of water stress can be considered a negative aspect from a nutritional point of view.

7.4.4 Calcium

Calcium is an essential plant nutrient and is required for structural roles in the cell wall and membranes, as a counteraction for inorganic and organic anions in the vacuole and as an intracellular messenger in the cytosol (White and Broadley 2003; Plieth 2003; Marschner 2012). Moreover, it has an important role in response to stresses included drought (Bush 1995). For these processes, the availability of Ca to the cells seems to be fundamental. The Ca content of plants varies between 0.1 and 5 % of dry weight depending on the growing conditions, plant species, and plant organ (Marschner 2012). At the cellular level, the apoplastic Ca concentration must be greater than 0.1 mM Ca to maintain the integrity and selectivity of the plasma membrane, cytosolic Ca concentration must be maintained in the range of 0.1–0.2 mM to avoid toxicity, and Ca storage organelles (vacuole, endoplasmic reticulum) must contain 1–10 mM Ca, which is required for signaling responses and charge balance (Plieth 2001; White and Broadley 2003). The general lack of appreciation of the limiting role of Ca is due in part to the fact that some important plant functions are controlled by changes in very small physiologically active pools of Ca within the cytoplasm. Such responses were detected in vivo in intact whole Arabidopsis seedlings at an elevation level of around 1.5 pM (Knight et al. 1997). As such, whole-leaf Ca levels might not reflect any potential limitations (McLaughlin and Wimmer 1999). The major site of entry for Ca is the root tip area, in which the cell walls of the epidermis are not yet suberized (Kirkby and Pilbeam 1984). The zone of uptake for Ca is likely to be reduced under drought (Scaife and Clarkson 1978). Additional stress impairs the ability of roots to absorb Ca even more. Although Ca uptake decreases under drought condition, overall Ca accumulation is only slightly depressed in comparison with P and K (Jenne et al. 1958). However, the low mobility of Ca renders its uptake and distribution rates limiting processes for many key plant functions. Nevertheless, Ca flux in the xylem may show a significant increase or remain unaffected for a period of several hours after Ca starvation. This was related, as shown for excised tomato roots, to a reuse of the Ca from the apoplastic root stores (Morard et al. 1996). Since Ca is reported to be phloem immobile (Marschner 2012) it does not redistribute in plants. The only path for upward Ca transport is through the transpiration stream, thus the transpiration rate is a significant determinant of Ca distribution within the plant (De Freitas et al. 2011). Furthermore, during drought and other conditions suppressing transpiration, root-pressure flow may be conducive in Ca translocation to the heart leaves, as has been established for lettuce and cabbages (Cox and Dearman 1981). The role of Ca in the life cycle of plants is well established and has been extensively reviewed (Bush 1995; Plieth 2001; White and Broadley 2003; Hirschi 2004; Marschner 2012). Calcium ion binding pectins in the middle lamella are known to be essential for strengthening the cell wall (Marschner 2012). Calcium bound to the outer surface of the plasma membrane maintains membrane stability and cell integrity (Hirschi 2004). When Ca was withheld, deficiency symptoms became visible after a short period; e.g., tomato seedlings after only 1 day showed a reduction in leaf area, photosynthesis, and transpiration (Del Amor and Marcelis 2003). Root growth was impaired because of reduced water and Ca uptake and since the demand on Ca for cell wall stabilization and membrane integrity was higher than in shoots. Unlike other nutrients, dry matter partitioning changes in favor of the shoots (Olle and Bender 2009). On the other hand, any factor that inhibits root growth also reduces Ca absorption (Kirkby and Pilbeam 1984). Low-transpiring organs such as the fruits of vegetables or potato tubers are known to suffer most from calcium deficiency (Busse and Palta 2006). Calcium is responding also as an intracellular messenger on drought-induced proline accumulation for osmotic adjustment (Xiong et al. 2006). Drought stress initiates a signal transduction pathway, in which increased cytosolic Ca activates Ca/calmodulin-independent glutamate decarboxylase activity, leading to γ-aminobutyric acid synthesis (Shelp et al. 1999). Experimental evidence supports the involvement of γ-aminobutyric acid in pH regulation, nitrogen storage, plant development and defense, as well as a compatible osmolyte and an alternative pathway for glutamate utilization (Shelp et al. 1999; Wahid et al. 2007). Aquaporin functionality may be also controlled by Ca. Some results have suggested how submicromolar changes of Ca are involved, via protein kinase, in the opening and closing of aquaporins, for example in vitro in spinach (Johansson et al. 1998). On the other hand, Plieths (2003) supports the hypothesis that it is the cell that determines the appropriate kinetics of $[Ca^{2+}]$cyt for the final response, and not the $[Ca^{2+}]$cyt kinetics that dictate the final response of the cell. It is probable that only in very few cases, a $[Ca^{2+}]$cyt transient is sufficient and necessary to form a cellular response. Ca deficiency is usually related to the inability of the plant to translocate adequate Ca to the affected part. Many vegetables develop unique symptoms: for example black-heart in celery (Bible and Stiehl 1986), tipburn in lettuce (Collier and Tibbits 1984), tipburn in chervil (Kleemann 1999), tipburn in Chinese cabbage (Aloni 1986), and blossom-end rot in tomato (Peet 2009). The unpredictability of the occurrence of Ca deficiency and the absence of any effective control procedures make this a serious problem (Olle and Bender 2009). Drought increases symptoms of Ca deficiency because water deficiency slows transpiration. A reduction in water movement within the plant reduces the amount of Ca carrying water that reaches the developing leaves and fruits (Bradfield and Guttridge 1984). Drought might also be a direct cause of disorders by causing the deterioration of cell membranes with subsequent loss of turgor and leakage of cell liquids related with increase in physiologically active gibberellins. In turn, they trigger vegetable fruits to enhance, initiate specific changes in $[Ca2+]_{cyt}$, which effect appropriate developmental responses, and induce production of reactive oxygen species (White and Broadley 2003). Limited Ca availability and transport have also been shown to diminish simultaneously firmness, strengthening of cell walls and cellular integrity (Tabatabaie et al. 2004; Ho and White 2005; Lee and Kim 2010), and decrease the total soluble solids and sugars (Rubio et al. 2009). For several diseases, such as *Phytium* spp., *Botrytis* spp., *Ralstonia solanacearum*, infection rate and plant susceptibility may be higher under Ca-deficiency conditions, because of cell wall impairment or debilitated plant defense (Yamazaki 2001; Hiltunen and White 2002). The impact of Ca deficiency increases as a consequence of drought because of several potential interactions with other factors, such as imbalances with other ions, especially with cations (Voogt 1993), increased radiation (Pujos and Morard 1997), deficiency in micronutrients, particularly manganese (Aktas et al. 2005), low humidity, high temperature, and CO_2-concentration (Bar Tal et al. 2001), low soil pH, strong fluctuations in soil water potential (Morard et al. 1996), and the available form of Ca (Alarcon et al. 1999).

7.4.5 Other Elements

Little information is available on the effect of drought on Mg and S nutrition in vegetable crops. However, a reduction both in Mg and S uptake in vegetable crops grown under drought conditions is expected. The magnesium requirement for optimal plant growth ranges from 0.15 to 0.35 % of the dry weight in the vegetative plant parts (Marschner 2012). When Mg is deficient, the chlorophyll content is reduced and export of carbohydrates from source to sink sites is impaired, causing a decrease of starch content in storage tissue as in potato tubers (Marschner 2012). Moreover, Mg deficiency reduces the nutritional quality of vegetables; this has attracted considerable attention, because fruits and vegetables usually contribute about 24 % of the total Mg to the dietary intake of humans (Levander 1990). However, high Mg content in leaves might be critical under drought conditions because as the leaf water potential falls, the Mg concentration increases in chloroplasts inhibiting photophosphorylation and photosynthesis (Rao et al. 1987). The sulphur requirement for optimal growth varies between 0.1 and 0.5 % of the dry weight of plants (Marschner 2012). Sulfur deficiency leads to a decrease in shoot/root ratio, root hydraulic conductivity, chlorophyll content of leaves, stomatal aperture, and net photosynthesis (Karmoket et al. 1991). Moreover, sulphur deficiency inhibits protein synthesis thereby leading to chlorosis. Changes in the protein composition with a decrease of sulfur-rich proteins are a typical feature of sulfur-deficient plants. The lower sulphur content of proteins considerably reduces their nutritional quality. Methionine is an essential sulfur-containing amino acid in human nutrition and often a limiting factor in diet. Moreover, it has been demonstrated that sulfur has a strong influence on the glucosinolate content and its volatile metabolites in the *Brassica* species; a decrease of S availability causes a reduction of glucosinolates which can be considered a negative aspect from a nutritional point of view due to the potential anticancer activity of these phytochemical compounds (Omirou et al. 2009). Since the transport of micronutrients to the plant roots occurs via diffusion, low soil moisture content will reduce micronutrient uptake. However, since plants require much smaller quantities of micronutrients, the effects of drought stress on micronutrient uptake are not as great as that on P and K uptake. Temporary B deficiency is common under drought and is related both to low mineralization, because much of the B is in the organic matter, and also to the lower presence of B in subsoil in some areas (compared to that present in surface soil), because water uptake occurs predominantly from the subsoil under dry conditions. By contrast, excessive rainfall can leach some of the available soil B in sandy soils. Low soil moisture can also induce deficiencies in Mn, Mo, Cu, Fe, and Zn (Sanchez-Rodriguez et al. 2010). On the other hand, Mn and Fe become increasingly available under moist conditions because of their conversion to reduced and more soluble forms (Havlin et al. 1999).

7.5 Strategies to Improve Crop Nutrient Status

7.5.1 Breeding and Biotechnologies

There are genotypic differences in drought tolerance between cultivars, although specific studies on mineral element nutrition are rare (Gunes et al. 2006; Sanchez-Rodriguez et al. 2010) and efficiency in nutrient use and drought tolerance do not always occur together (Gorny 2001). Genotypic variability concerning N use under moderate water stress was observed in a cherry tomato trial, where cultivar Zarina showed the greatest tolerance to water deficit by presenting the greatest biomass, relative growth rate, leaf relative water content, total N content, and N-uptake efficiency values under water deficit (Sanchez-Rodrıguez et al. 2010). One important morphological trait associated with drought tolerance is root architecture, aimed at creating plants which are able to reach areas of soil with favorable conditions for the uptake of nutrients (Qu et al. 2008; Kell 2011). When such genotypes are found they can be used both for breeding programs (Fleury et al. 2010) and as rootstocks for grafting in those plant species where this is feasible (Isakaaidis et al. 2004). Alternatively, genetic engineering can be applied if the right targets are defined. Reducing specifically the cytokinin content in roots leads to increased root growth and branching and so to a higher content of mineral elements in leaves which makes the plants more resistant to drought

(Werner et al. 2010). Another strategy was followed by Bao et al. (2009). They overexpressed an H+− pyrophosphatase in alfalfa which leads to an increased rhizosphere acidification enabling a higher uptake of cations, thus making plants more tolerant to drought. Recently, it has been demonstrated that the NH_4^+ transporter AMT1 functions in a high-affinity transport system and that the NO_3^- transporters ANRT1 and ANRT2 function in low- and high-affinity transport systems, respectively (Dunlop and Phung 2002). Thus, the beneficial effects of overexpressing an NO_3^- or NH_4^+ transporter would be greatest under conditions where the external NO_3^- or NH_4^+ levels are low, which is true for drought conditions. Since drought induces a rapid decrease in nitrate reductase activity through the inhibition of nitrate reductase gene transcription and a decrease of the stability of nitrate reductase mRNAs, it has been reported that replacing the native nitrate reductase gene with the constitutively expressed gene 35S-NR (Vincentz and Caboche 1991) delayed drought-induced losses in nitrate reductase activity of tobacco, allowing a more rapid recovery of N assimilation following short-term water deficit (Ferrario-Méry et al. 1998). Moreover, because of drought-induced expression of Ca-dependent protein kinases (Urao et al. 1994) and hyperosmotic shock-induction of putative Ca-binding proteins (Ko and Lee 1996), overexpression of CBL5 encoding one of the Ca sensors also leads to less sensitivity to water deficiency (Cheong et al. 2010). Smith (2002) suggested that strategies for increasing nutrient uptake by overexpressing genes encoding for high-affinity P transporters might also be important to increase drought tolerance, especially in the light of the increasing problems caused by P-deficient soils in the semiarid areas. Apart from concentrating on the plant site, the application of plant-interacting microorganisms is also a possibility to improve drought tolerance. This is discussed in detail in another chapter of this book. However, the aspect of mineral nutrition must be mentioned also here. Since the addition of phosphate can attenuate the consequences of drought (Shangguan et al. 2007), it can be assumed that the effect of arbuscular mycorrhiza on drought tolerance is also due to the improved supply of this mineral element. However, it is difficult to distinguish this from other mechanisms discussed in the other chapter. In a compartment experimental system where nutrient deficiency was separated from water deficiency, it appeared clearly that mycorrhizal fungi enhance drought tolerance by absorbing phosphate, especially from dry soils, and transporting it to the plant (Neumann and George 2004).

7.5.2 Grafting

Grafting is the union of two or more pieces of living plant tissue so that they grow as a single plant. Grafting vegetables, including cucurbits (watermelon, melon, and cucumber) and *Solanaceae* (tomato, eggplant, and pepper), is a common practice in Japan, Korea, China, in the Mediterranean basin, and several European countries. As with other vegetables, the main purpose of employing this technology is to control soilborne diseases (Lee et al. 2010). Grafting helps reduce the input of agrochemicals against soilborne pathogens, and is therefore considered an environment friendly cultivation technique, strongly recommended for integrated crop management systems (Rivard and Louws 2008). However, the impact of grafting in *Solanaceae* and *Cucurbitaceae* families is not confined to a stronger resistance against pathogens but also concerns a higher tolerance to abiotic stress conditions such as salinity, heavy metal, nutrient stress, water stress, thermal stress, organic pollutants, and alkalinity (Colla et al. 2010; Savvas et al. 2010; Schwarz et al. 2010), and could improve fruit quality (Rouphael et al. 2010). Vegetable grafting onto some rootstocks can enhance the uptake and/or utilization efficiency of macro and micronutrients. This is mainly due to the root characteristics of these rootstocks, which are more vigorous than those of the highly productive cultivated varieties. However, other mechanisms implicated in the efficiency of roots in active nutrient absorption, as well as signals arising from the scion, which are mainly governed by sink demand, may also enhance nutrient uptake and utilization. The higher efficiency of some graft combinations of fruit vegetables in absorbing and utilizing nutrients can mitigate yield losses due to shortage of these nutrients in the root environment of plants (Savvas et al. 2010). In a recent study, Rouphael et al. (2008) found that mini-watermelons grafted onto a commercial rootstock 'PS 1313' (*Cucurbita maxima* Duchesne × *Cucurbita moschata* Duchesne), when grown under conditions of deficit irrigation (50 % of evapotranspiration), resulted in a marketable yield more than 60 % higher than ungrafted mini-watermelons. The higher marketable yield recorded with grafting was mainly due to an improvement in water and nutrient uptake, indicated by higher N, K, and Mg concentrations in the leaves, and higher CO_2 assimilation rates. However, to date there are no additional reports in the international literature on the response of grafted vegetable plants to drought. Thus, further researches in this field are required to expand the possible applications of grafting in fruit–vegetable production under drought conditions.

7.5.3 Fertilization

Optimization between water and nutrient supply, uptake, and transfer is important and necessary for both growth and quality, since irrigation and fertilization are linked intrinsically and interact (Garg 2003; McWilliams 2003). Soil application of nutrients under low water availability conditions is not always effective in enhancing uptake and translocation of nutrients to the shoot. In some cases, application of fertilizers under drought conditions can even be harmful for plants due to an increase in the salinity of the soil solution. Under these conditions, foliar application of nutrients is much more effective than soil application to improve the crop nutritional status (Marschner 2012). Foliar application provides the required nutrient directly to the location of demand in the leaves and results in rapid absorption, being independent of root activity and soil water availability (Römheld and El-Fouly 1999). Foliar application of nutrients has been recommended as a means of improving plant growth in the early stages of development because foliar absorption rates in younger leaves are more favorable than in older leaves. Beneficial effects of foliar application of mineral nutrients on vegetable crops have been studied under water stress induced by salinity, while there is a lack of information on vegetable crop response to foliar fertilization under drought conditions. Del Amor and Cuadra-Crespo (2011) reported that foliar application of urea in broccoli maintained growth and photosynthesis in plants exposed to moderate salt stress. Moreover, foliar application of KH_2PO_4 increased the dry matter and chlorophyll concentration in tomato (Kaya et al. 2001). Similarly, weekly foliar application of $CaCl_2$ to tomato plants reduced fruit blossom-end rot symptoms by about 50 % under low Ca availability in the nutrient solution (Schmitz-Eiberger et al. 2002). Similarly, foliar application of $Ca(NO_3)_2$ appeared to protect lettuce against blackheart and tip-burn caused by salinity-induced Ca deficiency (Tzortzakis 2009). Drought tolerance can also be increased by enhancing the nutritional status of the crop prior to water stress. Accumulation of K in plant tissue produced by potassium fertilization mitigated the adverse effects of drought on plant growth (Andersen et al. 1992; Sangakkara et al. 2001). Potassium increases the plant's drought resistance through its functions as stated above (Beringer and Trolldenier 1978; Aliasgharzad et al. 2009). It also maintains turgor pressure (Mengel and Arneke 1982) and reduces transpiration under drought conditions (Andersen et al. 1992). In plants exposed to drought stress, the accumulation of K may be more important than the production of organic solutes during the initial adjustment phase, because osmotic adjustment through ion uptake such as K is more energy efficient (Hsiao 1973; Ma et al. 2004). Moreover, application of P fertilizer can also improve plant growth considerably under drought conditions (Garg et al. 2004). The positive effects of P on plant growth under drought have been attributed to an increase in the efficiency of water use, stomatal conductance, and photosynthesis (Ackerson 1985), to higher cell-membrane stability, and to effects on water relations (Sawwan et al. 2000). Seed priming with solutions containing the nutrients limited under drought conditions (e.g., P, K, Zn) might be another interesting strategy for conferring drought tolerance in directly seeded vegetable crops, as observed in other herbaceous crops like barley (Ajouri et al. 2004). Beneficial elements can also promote plant growth under drought conditions (Pilon-Smits et al. 2009). For instance, several recent studies have shown that an increased silicon accumulation in plants through root or foliar Si supply can improve growth under water stress (Epstein 1999; Ma et al. 2004). This beneficial effect may result from better and more efficient osmoregulation, improved plant water status, reduction in water loss by transpiration, maintenance of an adequate supply of essential nutrients, restriction in the uptake of toxic ions, and efficient functioning of antioxidative mechanisms. From the current knowledge, it can be concluded that the role of Si in plants is not restricted to the formation of a physical or mechanical barrier (such as precipitated amorphous silica) in cell walls, lumens, and intercellular voids. Silicon can also modulate the plant's metabolism and alter physiological activities, particularly in plants subjected to stress conditions, such as salinity (Savvas et al. 2009). However, in some plants, an increased silicon supply does not improve plant growth. Therefore, a better understanding of the interactions between silicon application and plant responses would contribute to more efficient fertilizer practices, especially under water stress conditions.

7.6 Conclusions and Future Prospects

In conclusion, there is no straightforward picture of the effects of drought on the nutritional status of vegetable crops. There appears to be an impairment of the nutrient contents and concentrations under drought conditions, but that might not hold true for all nutrients. Therefore, more information is needed on the combined effects of drought on the availability of nutrients and the plant uptake capacity for various elements. In addition, experiments are necessary to assess how changes in these parameters affect the whole vegetable nutritional status and growth. Experiments should also be made to target the effects of dynamically changing drought conditions and assess the nutrient balance during repeated drought and recovery phases. Further researches are also essential in the fields of plant genetics and breeding for development of cultivars and rootstocks tolerant to drought. Finally, it is advisable to integrate the different approaches for improving drought tolerance in vegetable crops in order to explore their possible synergic effects on sustaining growth and yield under water stress (Sacala 2009).

References

Ackerson RC (1985) Osmoregulation in cotton in response towater-stress. 3. Effects of phosphorus fertility. Plant Physiol 77:309–312.

Agele SO, Iremiren GO, Ojeniyi SO (2011) Evapotranspiration, water use efficiency and yield of rainfed and irrigated tomato. Intern J Agric Biol 13:469–476

Ajouri A, Haben A, Becker M (2004) Seed priming enhances germination and seedling growth of barley under conditions of P and Zn deficiency. J Plant Nutr Soil Sci 167:630–636

Aktas H, Karni L, Chang D-C, Turhan E, Bar-Tal A, Aloni B (2005) The suppression of salinity-associated oxygen radical production, in pepper (*Capsicum annuum*) fruit, by manganese, zinc and calcium, in relation to its sensitivity to blossom end rot. Physiol Plant 123:67–74

Alarcon AL, Madrid R, Egea C, Guillen I (1999) Calcium deficiency provoked by the application of different forms and concentrations of Ca^{2+} to soil-less cultivated muskmelons. Sci Hortic 81:89–102

Aliasgharzad N, Bolandnazar SA, Neyshabouri MR, Chaparzadeh N (2009) Impact of soil sterilization and irrigation intervals on P and K acquisition by mycorrhizal onion (*Allium cepa*). Biologia 64:512–515

Aloni B (1986) Enhancement of leaf tipburn by restricting root growth in Chinese cabbage plants. J Hortic Sci 61:509–513

Andersen MN, Asch F, Wu Y, Jensen CR, Næsted H, Mogensen VO, Koch KE (1992) Soluble invertase expression is an early target of drought stress during the critical, abortion–sensitive phase of young ovary development in maize. Plant Physiol 130:591–604

Aroca R, Ferrante A, Vernieri P, Chrispeels MJ (2006) Drought, abscisic acid and transpiration rate effects on the regulation of PIP aquaporin gene expression and abundance in *Phaseolus vulgaris* plants. Ann Bot 98:1301–1310

Balogh J, Biró M, Pintér K (2008) Root respiration in dry grassland. Cereal Res Comm 36: 355–358

Bao AK, Wang SM, Wu GQ, Xi JJ, Zhang JL, Wang CM (2009) Overexpression of the Arabidopsis H(+)-PPase enhanced resistance to salt and drought stress in transgenic alfalfa (*Medicago sativa* L.). Plant Sci 176:232–240

Bar-Tal A, Keinan M, Oserovitz Y, Gantz S, Avidan A, Posalski I, Aloni B, Karni L, Hazan A, Itach M, Shaham IP, Tartakovski N (2001) Relationships between blossom-end rot and water availability and Ca fertilization in bell pepper fruit production. Acta Hortic 554:97–105

Beier C, Emmett BA, Penuelas J (2008) Carbon and nitrogen cycles in European ecosystems respond differently to global warming. Sci Total Environ 407:692–697

Bénard C, Gautier H, Bourgaud F, Grasselly D, Navez B, Caris-Veyrat C, Weiss M, Génard M (2009) Effects of low nitrogen supply on tomato (*Solanum lycopersicum*) fruit yield and quality with special emphasis on sugars, acids, ascorbate, carotenoids, and phenolic compounds. J Agric Food Chem 57:4112–4123

Beringer H, Trolldenier G (1978) Influence of K nutrition on the response to environmental stress. In: Potassium research-review and trends. International Potash Institute, Berne, pp 189–222

Bibikova T, Gilroy S (2003) Root hair development. J Plant Growth Regul 21:383–415

Bible BB, Stiehl B (1986) Effect of atmospheric modification on the incidence of blackheart and the cation content of celery. Sci Hortic 28:19–28

Bloem J, Deruiter PC, Koopman GJ, Lebbink G, Brussaard L (1992) Microbial numbers and activity in dried and rewetted arable soil under integrated and conventional management. Soil Biol Biochem 24:655–665

Borken W, Matzner E (2009) Reappraisal of drying and wetting effects on C and N mineralization and fluxes in soils. Glob Chang Biol 15:808–824

Bradfield and Guttridge (1984) The results support the hypothesis that a positive root pressure at night promotes transport of calcium into tissues and organs that have restricted transpiration. Sci Hortic 22:207–217

Brouwer R (1983) Functional equilibrium: sense or nonsense? Neth J Agric Sci 31:335–348

Buchanan CD, Lim SY, Salzman RA, Kagiampakis L, Morishige DT, Weers BD, Klein RR, Pratt LH, Cordonnier-Pratt MM, Klein PE, Mullet JE (2005) Sorghum bicolor's transcriptome response to dehydration, high salinity and ABA. Plant Molec Biol 58:699–720

Buescher RW, Hobson GE (1982) Role of calcium and chelating agents in regulating the degradation of tomato fruit tissue by polygalacturonase. J Food Biochem 6:147–160

Bush DS (1995) Calcium regulation in plant cells and its role in signaling. Annu Rev Plant Physiol Plant Mol Biol 46:95–122

Busse JS, Palta JP (2006) Investigating the in vivo calcium transport path to developing potato tuber using 45Ca: a new concept in potato tuber calcium nutrition. Physiol Plant 128:313–323

Cheong YH, Pandey GK, Grant JJ, Batistic O, Li L, Kim BG, Lee SC, Kudla J, Luan S (2007) Two calcineurin B-like calcium sensors, interacting with protein kinase CIPK23, regulate leaf transpiration and root potassium uptake in Arabidopsis. Plant J 52:223–239

Cheong YH, Sung SJ, Kim BG, Pandey GK, Cho JS, Kim KN, Luan S (2010) Constitutive overexpression of the calcium sensor CBL5 confers osmotic or drought stress tolerance in Arabidopsis. Mol Cells 29:159–165

Colla G, Rouphael Y, Leonardi C, Bie Z (2010) Role of grafting in vegetable crops grown under saline conditions. Sci Hortic 127:147–155

Collier GF, Tibbits TW (1984) Effects of relative humidity and root temperature on calcium concentration and tipburn development in lettuce. J Am Soc Hortic Sci 109:128–131

Cox EF, Dearman AS (1981) The effect of trickle irrigation, misting and row position on the incidence of tipburn of field lettuce. Sci Hortic 15:101–106

Cramer GR, Ergul A, Grimplet J, Tillett RL, Tattersall EAR, Bohlman MC, Vincent D, Sonderegger J, Evans J, Osborne C, Quilici D, Schlauch KA, Schooley DA, Cushman JC

(2007) Water and salinity stress in grapevines: early and late changes in transcript and metabolite profiles. Func Integr Gen 7:111–134
Cuellar T, Pascaud F, Verdeil JL, Torregrosa L, Adam-Blondon AF, Thibaud JB, Sentenac H, Gaillard IA (2010) grapevine Shaker inward K(+) channel activated by the calcineurin B-like calcium sensor 1-protein kinase CIPK23 network is expressed in grape berries under drought stress conditions. Plant J 61:58–69
Dalla Costa L, Gianquinto G (2002) Water stress and watertable depth influence yield, water use efficiency, and nitrogen recovery in bell pepper: Lysimeter studies. Aust J Agr Res 53:201–210
Dalla Costa L, Delle Vedove G, Gianquinto G, Giovanardi R, Peressotti A (1997) Yield, water use efficiency and nitrogen uptake in potato: influence of drought stress. Potato Res 40:19–34
Dannenmann M, Simon J, Gasche R (2009) Tree girdling provides insight on the role of labile carbon in nitrogen partitioning between soil microorganisms and adult European beech. Soil Biol Biochem 41:1622–1631
De Freitas ST, Shackel KA, Mitcham EJ (2011) Abscisic acid triggers whole-plant and fruit-specific mechanisms to increase fruit calcium uptake and prevent blossom end rot development in tomato fruit. J Exp Bot 62:2645–2656
De Willigen P, Van Noordwijk M (1987) Roots plant production and nutrient use effciency. Thesis University, Wageningen, p 282
Del Amor FM, Cuadra-Crespo P (2011) Alleviation of salinity stress in broccoli using foliar urea or methyl-jasmonate: analysis of growth, gas exchange, and isotope composition. Plant Growth Reg 63:55–62
Del Amor FM, Marcelis LFM (2003) Regulation of nutrient uptakenutrient uptake, water uptake and growth under calcium starvation and recovery. J Hortic Sci Biotechnol 78:343–349
Dimitrov Z, Ovtcharrova A (1995) The productivity of peppers and tomatoes in case of insufficient water supply. Proceedings of ICID special technical session on the role of advanced technologies in irrigation and drainage system, vol 1. pp ft9.1–ft9.5
Dunlop J, Phung T (2002) Transporter genes to enhance nutrient uptakenutrient uptake: opportunities and challenges. Plant Soil 245:115–122
Earl H, Davis RF (2003) Effect of drought stress on leaf and whole canopy radiation use efficiency and yield of maizemaize. Agron J 95:688–696
Epstein E (1999) Silicon. Annu Rev Plant Physiol 50:641–664
Evans TE (1967) The effects of changes in the World Hydrological Cycle on the availability of water resources. In: Bazzaz F, Sombroek W (eds) Global climate change and agricultural production, Wiley, Chichester, 1996, pp 15–48
Farooq M, Wahid A, Kobayashi N, Fujita D, Basra SMA (2009) Plant drought stress: effects, mechanisms and management. Agron Sustain Dev 29:185–212
Ferrario-Mery S, Valadier M-H, Foyer CH (1998) Overexpression of nitrate reductase in tobacco delays drought-induced decreases in nitrate reductase activity and mRNA. Plant Physiol 117:293–302
Fleury D, Jefferies S, Kuchel H, Langridge P (2010) Genetic and genomic tools to improve drought tolerance in wheat. J Exp Bot 61:3211–3222
Folkert AH, Elena AG, Buitink J (2001) Mechanisms of plant desiccation tolerance. Trends Plant Sci 6:431–438
Fournier JM, Roldán AM, Sánchez C, Alexandre G, Benlloch M (2005) K^+ starvation increases water uptake in whole sunflower plants. Plant Sci 168:823–829
Fukutoku Y, Yamada Y (1984) Sources of proline-nitrogennitrogen in water-stressed soybean (Glycine max). II. Fate of 15 N-labelled protein. Physiol Plant 61:622–628
Garg BK (2003) Nutrient uptake and management under drought: nutrient-moisture interaction. Curr Agric 27:1–8
Garg BK, Burman U, Kathju S (2004) The influence of phosphorus nutrition on the physiological response of moth bean genotypes to drought. J Plant Nutr Soil Sci 167:503–508
Garwood EA, Williams TE (1967) Growth, water use and nutrient uptakenutrient uptake from the subsoil by grass swards. J Agr Sci 69:125–130

Gessler A, Jung K, Gasche R, Papen H, Heidenfelder A, Borner E, Metzler B, Augustin S, Hildebrand E, Rennenberg H (2005) Climate and forest management influence nitrogen balance of European beech forests: microbial N transformations and inorganic N net uptake capacity of mycorrhizal roots. Eur J For Res 124:95–111

Gorni AG (2001) Variation in utilization efficiency and tolerance to reduced water and nitrogennitrogen supply among wild and cultivated barleys. Euphyt 117:59–66

Greenwood DJ, Gerwitz A, Stone DA, Barnes A (1982) Root development of vegetable crops. Plant Soil 68:75–96

Gunes A, Cicek N, Inal A, Alpaslan M, Eraslan F, Guner E, Guzelordu T (2006) Genotypic response of chickpea (*Cicer arietinum* L.) cultivars to drought stress implemented at pre- and post-anthesis stages and its relations with nutrient uptake and efficiency. Plant Soil Envir 52:368–376

Havlin JL, Beaton JD, Tisdale SL, Nelson WL (1999) Soil fertility and fertilizers: an introduction to nutrient management, 6th edn. Macmillan Inc, London, p 409

Herschbach C, Mult S, Kreuzwieser J, Kopriva S (2005) Influence of anoxia on whole plant sulphur nutrition of flooding tolerant poplar (*Populus tremula* x P. alba). Plant Cell Environ 28:167–175

Hiltunen LH, White JG (2002) Cavity spot of carrot (Daucus carota). Ann Appl Bot 141:201–223

Hirschi KD (2004) The calcium conundrum. Both versatile nutrient and specific signal. Plant Physiol 136:2438–2442

Ho LC, White PJ (2005) A cellular hypothesis for the induction of blossom-end rot in tomato fruit. Ann Botany 95:571–581

Hsiao TC (1973) Plant responses to water stress. Annu Rev Plant Physiol 24:519–570

Hsiao TC, Lauchli A (1986) Role of potassium in plant–water relations. In: Tinker B, Lauchli A (eds) Advances in plant nutrition, vol 2. Praeger Scientific Publication, New York, pp 281–312

Isaakidis A, Sotiropoulos T, Almaliotis D, Therios I, Stylianidis D (2004) Response to severe water stress of the almond (*Prunus amygdalus*) 'Ferragnes' grafted on eight rootstocks. New Zealand J Crop Hortic Sci 32:355–362

Itoh S, Barber SA (1983) A numerical solution of whole plant nutrient uptakenutrient uptake for soil–root systems with root hairs. Plant Soil 70:403–413

Jenne E, Rhoades H, Yien C, Howe O (1958) Change in nutrient element accumulation by corn with depletion of soil moisture. Agron J 50:71–80

Jensen K, Beier C, Michelsen A, Emmett B (2003) Effects of experimental drought on microbial processes in two temperate heathlands at contrasting water conditions. Appl Soil Ecol 24:165–176

Johansson I, Karlsson M, Shukla VK, Chrispeels MJ, Larsson C, Kjellbom P (1998) Water transport activity of the plasma membrane aquaporin PM28A is regulated by phosphorylation. Plant Cell 10:451–459

Jones MM, Osmond CB, Turner NC (1980) Accumulation of solutes in leaves of sorghum and sunflower in response to water deficits. Aust J Plant Physiol 7:193–205

Kage H, Kochler M, Stützel H (2004) Root growth and dry matter partitioning of cauliflower under drought stress conditions: measurements and simulations. Europ J Agron 20:379–394

Kameli A, Lösel DM (1995) Contribution of carbohydrates and other solutes to osmotic adjustment in wheat leaves under water stress. J Plant Physiol 145:363–366

Kanai S, Ohkura K, Adu-Gyamfi JJ, Mohapatra PK, Nguyen NT, Saneoka H, Fujita K (2007) Depression of sink activity precedes the inhibition of biomass production in tomato plants subjected to potassium deficiency stress. J Exp Bot 58:2917–2928

Kanai S, Moghaieb RE, El-Shemy HA, Panigrahi R, Mohapatra PK, Ito J, Nguyen NT, Saneoka H, Fujita K (2011) Potassium deficiency affects water status and photosynthetic rate of the vegetative sink in green house tomato prior to its effects on source activity. Plant Sci 180:368–374

Karmoket JL, Carkson DL, Saker LR, Rooney JM, Purves JV (1991) Sulphate deprivation depresses the transport of nitrogennitrogen to the xylem and the hydraulic conductivity of barley (*Hordeum vulgare* L.) roots. Planta 185:269–278

Kavar T, Maras M, Kidric M, Sustar-Vozlic J, Meglic V (2007) Identification of genes involved in the response of leaves of *Phaseolus vulgaris* to drought stress. Mol Breed 21:159–172

Kaya C, Kirnak H, Higgs D (2001) Enhancement of growth and normal growth parameters by foliar application of potassium and phosphorus in tomato cultivars grown at high (NaCl) salinity. J Plant Nutr 24:357–367

Kell DB (2011) Breeding crop plants with deep roots: their role in sustainable carbon, nutrient and water sequestration. Ann Bot 108:407–418

Kirkby EA, Pilbeam DJ (1984) Calcium as a plant nutrient. Plant Cell Environ 7:397–405

Kirnak H, Tas I, Kaya C, Higgs D (2002) Effects of deficit irrigation on growth, yield, and fruit quality of eggplant under semi-arid conditions. Aust J Agric Res 53:1367–1373

Kleemann M (1999) Development of calcium deficiencydeficiency symptoms in chervil (Anthriscus cerefolium (L.) Hoffm.) and curled parsley (Petroselinum crispum (Mill.) Nym. Convar. Crispum). In: Natural and economic importance of alternative plants, Warsaw, vol 468. pp 335–348

Knight H (1997) Calcium signaling during abiotic stress in plants. Int Rev Cytol 195:269–325

Ko JH, Lee SH (1996) Biochemical studies of purified 23 kD calcium-binding protein in *Dunaliella salina* and its cDNA cloning. Plant Physiol 111:714–720

Kreuzwieser J, Gessler A (2010) Global climate change and tree nutrition: influence of water availability. Tree Physiol 30:1221–1234

Kuchenbuch R, Claassen N, Jungk A (1986) Potassium availability in relation to soil-moisture. 1. Effect of soil-moisture on potassium diffusion, root-growth and potassium uptake of onion plants. Plant Soil 95:221–231

Larsson M (1992) Translocation of nitrogennitrogen in osmotically stressed wheat seedlings. Plant Cell Environ 15:447–453

Lee JG, Kim TJ (2010) Effect of the use of additional calcium sources on growth and fruit characteristics in tomato hydroponics. Hort Environ Biotechnol 51:360–366

Lee JM, Kubota C, Tsao SJ, Bie Z, Echevarria PH, Morra L, Oda M (2010) Current status of vegetable grafting: diffusion, grafting techniques, automation. Sci Hort 127:93–105

Lester GE, Jifon JL, Makus DJ (2010) Impact of potassium nutrition on postharvest fruit quality: Melon (*Cucumis melo* L) case study. Plant Soil 335:117–131

Levander OA (1990) Fruit and vegetable contributions to dietary mineral intake in human health and disease. HortSci 25:1486–1488

Li YZ, Sun CB, Huang ZB, Pan JL, Wang L, Fan XW (2009) Mechanisms of progressive water deficit tolerance and growth recovery of Chinese maize foundation genotypes Huangzao 4 and Chang 7–2, which are proposed on the basis of comparison of physiological and transcriptomic responses. Plant Cell Phys 50:2092–2111

Liebersbach H, Steingrobe B, Claassen N (2004) Roots regulate ion transport in the rhizosphere to counteract reduced mobility in dry soil. Plant Soil 260:79–88

Liu TY, Chang CY, Chiou TJ (2010) The long-distance signaling of mineral macronutrients. Curr Opin Plant Biol 12:312–319

Lucash MS, Eissenstat DM, Joslin JD, McFarlane KJ, Yanai RD (2007) Estimating nutrient uptakenutrient uptake by mature tree roots under field conditions: challenges and opportunities. Trees 21:593–603

Ma QF, Turner DW, Levy D, Cowling WA (2004) Solute accumulation and osmotic adjustment in leaves of *Brassica* oilseeds in response to soil water deficit. Aust J Agric Res 55:939–945

Marschner P (2012) Marschner's mineral nutritionmineral nutrition of higher plants. Academic, London, p 651

Marschner H, Haussling M, George E (1991) Ammonium and nitrate uptake rates and rhizosphere-pH in non-mycorrhizal roots of Norway spruce (*Picea abies* (L.) Karst.). Trees 5:14–21

McLaughlin SB, Wimmer R (1999) Transley review No. 104—Calcium physiology terrestrial ecosystem processes. New Phytol 142:373–417

McWilliams D (2003) Drought strategies for cotton, cooperative extension service circular 582, College of Agriculture and Home Economics. New Mexico State University, USA

Mengel K, Ameke WW (1982) Effect of potassium on the water potential, the pressure potential, the osmotic potential and cell elongation in leaves of *Phaseolus vulgaris*. Physiol Plant 54:402–408

Miletic RPR, Dodig D, Milutinovic S, Mihajlovic I, Nikodijevic SM (2010) Strategies for solving the problem of drought in Eastern Serbia. http://www.wgcrop. icidonline.org/40doc.pdf

Morard P, Pujos A, Bernadac A, Bertoni G (1996) Effect of temporary calcium deficiency on tomato growth and mineral nutrition. J Plant Nutr 19:115–127

Neumann E, George E (2004) Colonisation with the arbuscular mycorrhizal fungus Glomus mosseae (Nicol. & Gerd.) enhanced phosphorus uptake from dry soil in *Sorghum bicolor* (L.). Plant Soil 261:245–255

Nilsen ET, Orcutte DM (1996) Phytohormones and plant responses to stress. In: Nilsen ET, Orcutte DM (eds) Physiology of plant under stress: abiotic factors. Wiley, New York, pp 183–198

Nonami H (1998) Plant water relations and control of cell elongation at low water potentials. J Plant Res 111:373–382

Olle M, Bender I (2009) Causes and control of calcium deficiency disorders in vegetables: a review. J Hortic Sci Biotechnol 84:577–584

Omirou MD, Papadopoulou KK, Papastylianou I, Constantinou M, Karpouzas DG, Asimakopoulos I, Ehaliotis C (2009) Impact of nitrogennitrogen and sulfur fertilization on the composition of glucosinolates in relation to sulfur assimilation in different plant organs of broccoli. J Agric Food Chem 57:9408–9417

Peet MM (2009) Physiological disorders in tomato fruit development. Acta Hortic 821:151–160

Pilon-Smits EAH, Quinn CF, Tapken W, Malagoli M, Schiavon M (2009) Physiological functions of beneficial elements. Curr Opin Plant Biol 12:267–274

Pinkerton A, Simpson JR (1986) Interactions of surface drying and subsurface nutrients affecting plant-growth on acidic soil profiles from an old pasture. Aust J Exp Agric 26:681–689

Plieth C (2001) Plant calcium signaling and monitoring: pros and cons and recent experimental approaches. Protoplasma 218:1–23

Plieth C (2003) Calcium: just another regulator in the machinery of life? Ann Bot 96:1–8

Pujos A, Morard P (1997) Effects of potassium deficiencydeficiency on tomato growth and mineral nutrition at the early production stage. Plant Soil 189:189–196

Qu YY, Mu P, Zhang HL, Chen CY, Gao YM, Tian YX, Wen F, Li ZC (2008) Mapping QTLs of root morphological traits at different growth stages in rice. Genetica 133:187–200

Rao IM, Sharp RE, Boyer JS (1987) Leaf magnesium alters photosynthetic response to low water potentials in sunflower. Plant Physiol 84:1214–1219

Rasnick M (1970) Effect of mannitol and polyethylene glycol on phosphorus uptake by maizemaize plants. Ann Bot 34:497–502

Reddy AR, Chaitanya KV, Vivekanandan M (2004) Drought-induced responses of photosynthesis and antioxidant metabolism in higher plants. J Plant Physiol 161:1189–1202

Rennenberg H, Dannenmann M, Gessler A, Kreuzwieser J, Simon J, Papen H (2009) Nitrogen balance in forest soils: nutritional limitation of plants under climate change stresses. Plant Biol 11:4–23

Rivard CL, Louws FJ (2008) Grafting to manage soilborne diseases in heirloom tomato production. HortSci 43:204–211

Römheld V, El-Fouly MM (1999) Foliar nutrient application: challenge and limits in crop production. In: Proceedings of the 2nd international workshop on foliar fertilization, Bangkok, pp. 1–34, 4–10 April 1999

Rouphael Y, Cardarelli M, Rea E, Colla G (2008) Yield, mineral composition, water relations, and water use efficiency of grafted mini-watermelon plants under deficit irrigation. HortSci 43:730–736

Rouphael Y, Schwarz D, Krumbein A, Colla G (2010) Impact of grafting on product quality of fruit vegetables. Sci Hortic 127:172–179

Rubio JS, Garcia-Sanchez F, Rubio F, Martinez V (2009) Yield, blossom-end rot incidence, and fruit quality in pepper plants under moderate salinity are affected by K^+ and and Ca^{2+} fertilization. Sci Hortic 119:79–87

Ruiz-Lozano JM, Azcon R (1996) Mycorrhizal colonization and drought stress as factors affecting nitrate reductase activity in lettuce plants. Agric Ecosyst Environ 60:175–181

Sacała E (2009) Role of silicon in plant resistance to water stress. J Elementology 14:619–630

Sanchez-Rodriguez E, Rubio-Wilhelmi MD, Cervilla LM, Blasco B, Rios JJ, Leyva R, Romero L, Ruiz JM (2010) Study of the ionome and uptake fluxes in cherry tomato plants under moderate water stress conditions. Plant Soil 335:339–347

Sangakkara UR, Frehner M, Nosberger J (2001) Influence of soil moisture and fertilizer potassium on the vegetative growth of mungbean (*Vigna radiata* L.) and cowpea (*Vigna unguiculata* L.). J Agron Crop Sci 186:73–81

Savvas D, Giotis D, Chatzieustratiou E, Bakea M, Patakioutas G (2009) Silicon supply in soilless cultivations of zucchini alleviates stress induced by salinity and powdery mildew infections. Env Exp Bot 65:11–17

Savvas D, Colla G, Rouphael Y, Schwarz D (2010) Amelioration of heavy metal and nutrient stress in fruit vegetables by grafting. Sci Hortic 127:156–161

Sawwan J, Shibli RA, Swaidat I, Tahat M (2000) Phosphorusregulates osmotic potencial and growth of African violet under in vitro-induced water deficit. J Plant Nutr 23:759–771

Scaife MA, Clarkson DT (1978) Short communication. Calcium related disorders in plants—A possible explanation for the effect of weather. Plant Soil 50:723–725

Schimel J, Balser TC, Wallenstein M (2007) Microbial stress response physiology and its implications for ecosystem function. Ecol 88:1386–1394

Schmitz-Eiberger M, Haefs R, Noga G (2002) Reduction of calcium deficiency symptoms by exogenous application of calcium chloride solutions. Acta Hort 594:535–540

Schwarz D, Krumbein A (2012) Grafting can enhance tomato growth and quality characteristics at low potassium supply. Sci Hortic (accepted)

Schwarz D, Heinen M, Van Noordwijk M (1995) Rooting characteristics of lettuce grown in irrigated sand beds. Plant Soil 176:205–217

Schwarz D, Rouphael Y, Colla G, Venema JH (2010) Grafting as a tool to improbe tolerance of vegetables to abiotic stresses: Thermal stress, water stress and organic pollutants. Sci Hortic 127:162–171

Serio F, Leo L, Parente A, Santamaria P (2007) Potassium nutrition increases the lycopene content of tomato fruit. J Hort Sci Biotechnol 82:941–944

Shabala S (2003) Regulation of potassium transport in leaves: from molecular to tissue level. Ann Bot 92:627–634

Shangguan ZP, Lei TW, Shao MA, Xue QW (2007) Effects of phosphorus nutrient on the hydraulic conductivity of sorghum (*Sorghum vulgare* Pers.) seedling roots under water deficiency. J Integr Plant Biol 49:127

Shelp BL, Bown AW, McLean MD (1999) Metabolism and functions of gammaaminobutyric acid. Trends Plant Sci 11:446–452

Simonne EH, Joseph DE, Harris CE (1998) Effects of irrigation and nitrogen rates on foliar mineral composition of bell pepper. J Plant Nutr 21:2545–2555

Sivaramakrishnan S, Patell VZ, Flower DJ, Peacock JM (1988) Proline accumulation and nitrate reductase activity in contrasting sorghum lines during mild-season drought stress. Physiol Plant 74:418–426

Smith FW (2002) The phosphate uptake mechanism. Plant Soil 245:105–114

Somerville C, Briscoe J (2001) Genetic engineering and water. Science 292:2217

Steudle E (2000) Water uptake by plant roots: an integration of views. Plant Soil 226:45–56

Subbarao GV, Johansen C, Slinkard AE, Rao RCN, Saxena NP, Chauhan YS (1995) Strategies and scope for improving drought resistance in grain legumes. Crit Rev Plant Sci 14:469–523

Tabatabaie SJ, Gregory PJ, Hadley P (2004) Uneven distribution of nutrients in the root zone affects the incidence of blossom end rot and concentration of calcium and potassium in fruits of tomato. Plant Soil 258:169–178

Tanguilig VC, Yambao EB, O'Toole JC, Dedatta SK (1987) Water-stress effects on leaf elongation, leaf water potential, transpiration, and nutrient-uptake of rice, maize, and soybean. Plant Soil 103:155–168

Thorup-Kristensen K, Sorensen JN (1999) Soil nitrogen depletion by vegetable crops with variable root growth. Acta Agric Scand Sec B 49:92–97

Tinker BT, Nye PH (2000) Solute movement in the rhizosphere. Oxford University press, Oxford

Turner LB (1985) Changes in the phosphorus content of Capsicum annuum leaves during water-stress. J Plant Physiol 121:429–439
Turner NC, Wright GC, Siddique KHM (2001) Adaptation of grain legumes (pulses) to water-limited environments. Adv Agron 71:123–231
Tzortzakis NG (2009) Influence of NaCl and calcium nitrate on lettuce and endive growth using nutrient film technique. Int J Veg Sci 15(1):44–56
Urao T, Katagiri T, Mizoguchi T, Yamaguchishinozaki K, Hayashida N, Shinozaki K (1994) Two genes that encode Ca^{2+-}dependent protein-kinases are induced by drought and high-salt stresses in Arabidopsis thaliana. Mol Gen 244:331–340
van Meeteren MM, Tietema A, van Loon EE, Verstraten JM (2008) Microbial dynamics and litter decomposition under a changed climate in a Dutch heathland. Appl Soil Ecol 38:119–127
Vincentz M, Caboche M (1991) Constitutive expression of nitrate reductase allows normal growth. EMBO J 10:1027–1035
Voogt W (1993) Nutrient uptake of year round tomato crops. Acta Hortic 339:99–112
Wahid A, Gelani S, Ashraf M, Foolad MR (2007) Heat tolerance in plants: an overview. Environ Exp Bot 61:199–223
Waraich EA, Ahmad R, Saifullah, Ashraf MY, Ehsanullah (2011) Role of mineral nutrition in alleviation of drought stress in plants. Aust J Crop Sci 5:764–777
Werner T, Nehnevajova E, Kollmer I, Novak O, Strnad M, Kramer U, Schmülling T (2010) Root-specific reduction of cytokinin causes enhanced root growth, drought tolerance, and leaf mineral enrichment in arabidopsis and tobacco. Plant Cell 22:3905–3920
Wery J, Silim SN, Knights EJ, Malhotra RS, Cousin R (1994) Screening techniques and sources and tolerance to extremes of moisture and air temperature in cool season food legumes. Euphytica 73:73–83
White PJ, Broadley MR (2003) Calcium in plants. Ann Bot 92:487–511
Xiong L, Wang R, Mao G, Koczan JM (2006) Identification of drought tolerance determinants by genetic analysis of root response to drought stress and abscisic acid. Plant Physiol 142:1065–1074
Yamaji N, Ma JF (2007) Spatial distribution and temporal variation of the rice silicon transporter Lsi1. Plant Physiol 143:1306–1313
Yamazaki H (2001) Relation between resistance to bacterial wilt and calcium nutrition in tomato seedlings. Jpn Agric Res Quart 35:163–169

Part III
Biochemical and Molecular Responses

Chapter 8
Osmotic Adjustment Under Drought Conditions

Gregor J. Sanders and Stefan K. Arndt

Abstract In broad terms, plants adapt to drought either by decreasing water loss (reduced stomatal conductance) or by maintaining water uptake. The latter process is facilitated within plant cells by osmotic adjustment (OA), a biochemical mechanism that helps plants to acclimatize to dry and saline conditions. OA results in a net increase of the number of osmotically active substances in the cell. This increase in solutes leads to a more negative osmotic potential, which in turn can improve the degree of cell hydration, maintaining turgor in leaf tissue and in other metabolically active cells. In other words, plants can survive longer and maintain metabolic processes in drying soil if OA occurs. In particular crop cultivars, OA has positively affected growth and yield under drought stress. A wide range of substances can contribute to OA, including inorganic cations and anions, organic acids, carbohydrates, and amino acids. OA is often associated with an accumulation of specific solutes with protective functions. These compatible solutes—rich in hydroxyl (−OH) groups—such as sugars, cyclitols, proline and glycine betaine, can accumulate in the cytoplasm and help to protect cellular proteins, enzymes, and cellular membranes against dehydration. Still, it is important to recognize that individual solutes do not contribute greatly to OA in many species and that OA is mainly achieved by the accumulation of a multitude of solutes. As OA requires the metabolism or uptake of solutes it is generally a slow process, and is sensitive to the timing and intensity of stress. Adding to this inherent variation in expression of OA, there is some evidence that studies have underestimated leaf relative water content (RWC) when quantifying OA. The correct measurement of plant water

G. J. Sanders · S. K. Arndt (✉)
Department of Forest Ecosystem Science, University of Melbourne, 500 Yarra Boulevard, Richmond, VIC 3121, Australia
e-mail: sarndt@unimelb.edu.au

G. J. Sanders
e-mail: gregorjs@gmail.com

R. Aroca (ed.), *Plant Responses to Drought Stress*,
DOI: 10.1007/978-3-642-32653-0_8,

status, such as RWC, is vital to ensuring an accurate assessment of the relative capacity for OA in different plants.

8.1 Introduction

This chapter will discuss the significance of osmotic adjustment (OA) as a fundamental mechanism of drought adaptation in higher plants. There has been extensive research on OA in many plant species and it was specifically investigated in crop species. The first in depth review dates back to Hsiao (1973), highlighting some of the basic functions of OA. Turner and Jones (1980) and Morgan (1984) reviewed within-species variation in OA, drawing attention to the potential for improvement of OA capacity in crop cultivars and focusing on linkages between OA and growth and yield. These reviews described the complexity of interactions involving, among other factors, the stage of ontogeny when the drought stress is imposed. Munns (1988) and Serraj and Sinclair (2002) provided critical reviews of OA in connection with growth and yield, and Blum (1996, 2011b) emphasized the evidence for consequential effects on productivity. OA in woody plants was discussed by Abrams (1988), Gebre and Tschaplinski (2000), and Kozlowski and Pallardy (2002) with a focus on European and North American trees.

To complement these comprehensive reviews, this chapter will provide the context for OA in the drought response of plants and describe some important considerations involved in its measurement and interpretation. It will also discuss the significance of particular organic solutes generally associated with OA.

8.2 Definition of Osmotic Adjustment

OA involves the accumulation or de novo synthesis of solutes within the cell that causes a change in the cell osmotic potential (Boyer et al. 2008). It is important to emphasize the de novo synthesis or the net increase of solutes within the cell—this is also called “active” OA. This net increase in osmotically active solutes (hereinafter referred to as “solutes”) leads to a decrease in the osmotic potential of the cell and, therefore, also to a decrease of the cell’s total water potential. Hence, OA is best viewed as a drought adaptation, not simply a drought response.

Some researchers refer to a second form of OA, also called “passive OA”, which is associated with water loss and therefore a reduction in overall cell volume (Fig. 8.1a, case 3). As water is lost from the cell there is a passive concentration of solutes which, in turn, leads to a lower osmotic potential (Ψs). However, this passive change in Ψs occurs to some extent in all nonsucculents as they lose water under high evaporative demand, and is a normal diurnal and drought response, but not a plant adaptation. As the water status of plant tissue changes, the contribution

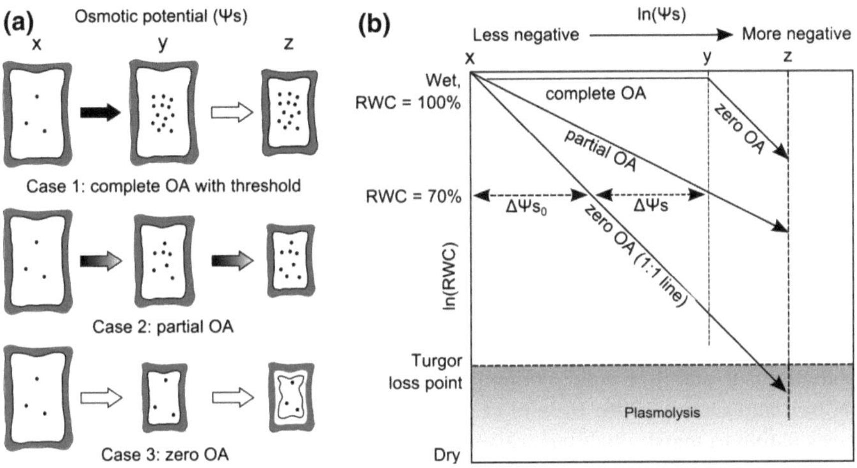

Fig. 8.1 Three potential relationships between a decrease in osmotic potential and relative water content in plants: **a** shows a simplified cartoon of how cell volume can change with decreasing osmotic potential in plants that express complete OA (case 1), partial *OA* (case 2), or no OA (case 3); **b** shows the relationship between the ln(Ψs) and ln(RWC) for these three cases. With complete OA (case 1), the increase of solutes leads to a lower (more negative) osmotic potential without any changes in cell volume (x–y). At a certain threshold (y) the cell volume is reduced and Ψs falls to z due to water loss from the cell rather than by solute accumulation. Partial OA (case 2) shows that OA can be successful in slowing the loss of water (reduction in cell volume), but not preventing it. If there is no OA (case 3), the decrease of Ψs is solely driven by cell water loss until cell plasmolysis occurs. The magnitude of OA in developing stress studies is calculated at an arbitrary RWC (shown here as 70 %) from the sum of $\Delta\Psi s_0$ and ΔΨs. Alternatively, an index has been proposed by Turner (2006) based on the slope of a given relationship, and is independent of the arbitrary standard for RWC

of active solute accumulation (OA) can be calculated as the difference between the measured Ψs and that expected from the concentrating effect of water loss from the tissue (Wright et al. 1983):

$$\mathrm{OA} = \Psi s - (\Psi s_{\mathrm{ref}} \mathrm{RWC}_{\mathrm{ref}})/\mathrm{RWC} \qquad (8.1)$$

Where Ψs_{ref} and RWC_{ref} are the osmotic potential and RWC at an arbitrary reference point.

OA can be described as an adaptation in both drought and salt affected plants (Boyer et al. 2008). Both stresses cause lower water potential affecting plant tissue cells, which can be countered by the accumulation of solutes within the symplast in the form of inorganic ions or organic solutes. Such accumulation effectively increases the osmotic force or tension that cells can exert on their surroundings to increase water uptake (Kramer and Boyer 1995). The expression of OA is stress induced and therefore an adaptive trait, and is distinguished from plants that accumulate high concentrations of solutes independent of water status (Kozlowski and Pallardy 2002). This constitutive solute accumulation occurs in many species—including summer ephemerals, xeromorphic shrubs, and halophytes (Jones et al. 1981).

8.2.1 Elastic Adjustment as a Related Adaptation

Elastic adjustment of the cell wall alters the relationship between cell volume and turgor (Smith et al. 1997). Elastic adjustment is often associated with a "softening" of cell walls: in response to drought stress the cell wall can become more elastic. In comparison with OA, which directly promotes turgor, this increased cell wall elasticity (decreased bulk elastic modulus) delays the loss of turgor with a reduction in the water potential (Wenkert et al. 1978; Saliendra and Meinzer 1991). In other words, cells can lose more water before they lose turgor.

A more rigid cell wall (or higher bulk elastic modulus) has also been considered as a possible drought response. Its theoretical effect with a reduction in water potential is to stabilize water content and symplast volume at the expense of turgor (Nilsen and Orcutt 1996; Hessini et al. 2009). However, in a perennial deciduous grass native to marshy habitats, severe water stress was linked to reduced elasticity along with reduced photosynthesis and transpiration, and loss of leaf turgor (Hessini et al. 2009). In contrast, Saliendra and Meinzer (1991) showed that water content and symplast volume in sugar cane was maintained by an increase in elasticity in combination with OA. Kozlowski and Pallardy (2002) compare the ecological significance of OA and elastic adjustment in different environments.

8.3 Osmotic Adjustment and Water Potential

The water potential (Ψ) in plant tissue cells must ultimately become more negative along with the water potential in the surrounding environment during a drought episode. OA allows this to occur while also avoiding or reducing loss of cellular water. Ψ can be zero or negative and is properly expressed as the sum of component potentials (Jones 2007):

$$\Psi = \Psi s + \Psi p + \Psi g, \tag{8.2}$$

where Ψs is the osmotic potential, Ψp is turgor pressure (positive), and Ψg is the gravitational potential (which can be ignored at the cellular level). Ψs is normally uniform within the main symplast compartments: the vacuole and the metabolically active cytosol (Kramer and Boyer 1995). The apoplast can also contain solutes, although the concentration is normally quite low in leaves (Wardlaw 2005).

In response to drought, Ψ in the cell will equilibrate with Ψ in the surrounding xylem as it falls. The two active processes by which a cell can react to this energy balance are OA and redistribution of water from the symplast to the apoplast (Joly and Zaerr 1987). In effect, both processes can maintain Ψp by reducing Ψs without loss of water from the cell, and may occur in combination (Nilsen and Orcutt 1996). However, as the buffering capacity of the apoplast is relatively limited, this is reasonably viewed as a secondary process to OA.

A reduction in the volume of the symplast as water is lost from the cell will also increase the concentration of solutes. As described in Sect. 8.2, this is not OA, but a passive change in Ψs associated with a reduction or total loss of turgor pressure, leading to plasmolysis (Fig. 8.1a, case 3). In contrast, by actively increasing the concentration of solutes in the symplast without a loss of water, OA can maintain cell volume and turgor as Ψs (and Ψ) become more negative (Fig. 8.1a, case 1 and 2).

8.4 Determining Osmotic Adjustment

The debate surrounding the effectiveness of OA has been complicated by the various approaches for assessing the magnitude of OA in several hundred published studies. This section describes the different experimental approaches and then introduces the specific techniques for measuring Ψs, from which OA is determined.

OA can be determined by either comparing control and stressed plants at a certain stage of stress or by taking consecutive measurements over the course of a developing stress. Babu et al. (1999) describe variations in each of these approaches and consider the latter approach the most thorough despite the intensity of time and plant material required. However, both approaches are valid and usually researchers select the method that best suits the individual research questions.

8.4.1 Comparing Control and Stressed Plants

This is the method most commonly used for practical reasons and is otherwise known as *Ludlow's method* (Blum 2011b). At a certain point in the stress cycle the Ψs of plants that are not stressed (control) is compared against the Ψs of plants that are subjected to stress. In glasshouse or potted trials, comparisons are typically made using well-watered and drought stressed plants. In field trials without irrigation and/or rainout shelters, a seasonal comparison requires a significant soil moisture difference between measurements (e.g., spring vs summer). The main advantage of this method is that it is very fast and many samples can be analyzed. The main difficulty is to pick the right time or stress level for the analysis. The extent of OA can differ depending on when it is measured in the stress cycle.

Assessing OA at a single stress point requires the difference between Ψs of stressed and control samples typically at **full hydration**, also called **full turgor** (Turner 1981). It is important to compare OA between different plants at the same water status, because water status itself will influence Ψs. It is best to compare the plants at full hydration because this condition is repeatable. Full hydration occurs when the cells cannot absorb more water without intercellular spaces being filled. In this condition, the relative water content (RWC) of the cell is 100 % at the point when Ψ as measured approaches zero. To determine Ψs at full hydration, sample tissue must be rehydrated before measurement or measured in conjunction with RWC at partial hydration. Both approaches require care in rehydrating plant tissue to be free of significant error, and this is discussed in Sect. 8.4.3. The measurement of Ψs itself follows in Sect. 8.4.4.

It is possible that the change in Ψs at full hydration (Ψs_{100}) is due (at least partially) to a change in cell volume rather than solute accumulation. Solute accumulation can be confirmed if the osmotically active dry weight of the tissue is correlated with the change in Ψs (Kassam and Elston 1976). In practical terms, if the turgid weight to dry weight ratio (TW/DW) of tissue is significantly reduced following drought, a lower Ψs_{100} may be the result of a smaller cell size (Myers and Neales 1986). Redistribution of water from the symplast to the apoplast can also lower Ψs_{100} (Nilsen and Orcutt 1996) without a change in solutes. Two methods are available to assess the apoplastic water fraction (AWF), as discussed in Sect. 8.4.4. Both TW/DW and AWF may change in response to seasonal water availability, indicating changes in symplast volume rather than accumulation of osmolytes (Callister et al. 2008).

8.4.2 Developing Stress Measurements

Although highly demanding in time and resources, Morgan (1992, 1995) proposed measuring Ψs and water status (RWC) repeatedly as a drought stress develops to determine OA. In this case, the degree of OA is likely to vary with the stress until measurements stop at the wilted or sub-lethal stage. However, the relationship between Ψs and RWC as the stress increases is usually sufficient to detect OA. A control may be necessary to account for any effect of plant developmental stage on OA over the course of a study. The main advantage of this method is that Ψs is continuously measured throughout the stress cycle, and hence OA will be detected if it is expressed. The main disadvantage is that measurements consume a lot of time and plant material during the course of the experiment.

The method was further refined by Turner (2006) with the development of a unitless index between 0 and 1 to quantify OA. This avoids the need to choose a reference stress level and standardize Ψs to full hydration (Fig. 8.1b). The index is based on the slope of the line represented by $\ln(\Psi s)/\ln(RWC)$, and accounts for the change in OA with stress level. An index allows different studies to be compared on the same terms provided that the plants have been stressed sufficiently that the linear regression can be determined. However, the index also remains subject to errors in RWC.

8.4.3 Determining Relative Water Content

The basis for RWC is described by the following equation (Barrs 1968):

$$\text{RWC}\ (\%) = [(\text{FW} - \text{DW})/(\text{TW} - \text{DW})] \times 100, \qquad (8.3)$$

where FW is sample fresh weight, TW is sample turgid weight, and DW is sample dry weight. Each sample must be fully rehydrated from distilled water before measuring TW. This can be achieved by different methods, which in some cases depend on the type of sample. Whole leaves or small branches can be rehydrated by placing the petiole of the leaf into distilled water. Partial samples, such as leaf discs, can be floated on water or submerged. In each case, there is a chance of oversaturation, leading to an artificially high TW. This is known as the plateau effect when observed from a linear regression of FW against leaf water potential, where Ψ is stable as FW begins to decrease (Parker and Pallardy 1987). Unless identified and corrected, this in turn results in artificially low RWC. A correction factor can be calculated and applied to all RWC measurements (Dreyer et al. 1990), which effectively extrapolates the true saturated weight (Ladiges 1975).

Eckardt (1965) first reported a study comparing the effect of rehydrating whole leaves with floating leaf discs. Standing leaves with the cut end of the petiole placed in water produced consistently higher (more reliable) RWC values compared with floating leaf discs. Dreyer et al. (1990) found that the correction required for RWC in four European oaks was highest when the TW had been determined from a submerged sample. Lafitte (2002) compared rehydrating rice leaves by standing, floating, or submerging under two different temperature regimes. Time had the most impact, with increases in TW still occurring after 24 h of rehydration in all cases except for standing petioles in water at 20°C. This suggests that oversaturation to some degree may be relatively common with most rehydration methods.

Table 8.1 shows the effect of TW rehydration time on RWC in some published studies. Many plants that were rehydrated using the submerging technique had RWC values which were lower than 80 % and in some instances even as low as 50–60 %. A review of typical turgor loss points in many higher plants (Table 8.2) shows that most plants will lose turgor at RWC values between 80 and 90 %. It is unlikely that the plants listed in Table 8.1 could have shown positive values for turgor and photosynthetic rate or recovered from rewatering if such low RWCs were indeed accurate. Consequently, in many cases oversaturation would have occurred during rehydration leading to un-naturally low RWC. This highlights that care must be taken when RWC is determined, especially if OA is determined during developing stress measurements, because a constant underestimation of RWC while measuring Ψs correctly will lead to erroneous regressions.

A comparison of eucalypts from different environments indicated that some genotypes are more susceptible to oversaturation, even when rehydrated through the petiole (Fig. 8.2). *Eucalyptus cladocalyx*, a species adapted to semiarid conditions in Australia, showed the greatest tendency for oversaturation to affect

Table 8.1 A comparison of the RWC derived from different methods of tissue rehydration under nonlethal stress (with physiological evidence of measurement at or above turgor loss point)

Plant species	RWC (%)	Evidence of physiological activity	Rehydration method	Rehydration time (h)	Reference
Wheat (*Triticum aestivum*)	84[a]	ABA production	Standing	6	(Wright 1977)
Cotton (*Gossypium hirsutum*)	83	Recovery after rewatering	Floating	4	(Olsen et al. 1983)
Sorghum cultivars (*Sorghum bicolor*)	80	Positive turgor measured	Floating	4	(Wright et al. 1983)
Perennial ryegrass (*Lolium perenne*)	82	Partial wilting	Submerged	12	(Eerens et al. 1998)
Soybean cultivars and related *Glycine spp.*	76[b]	50 % of plants with positive turgor	Floating	4	(James et al. 2008)
Mango cultivars (*Mangifera indica*)	71	Photosynthetic rate positive	Submerged	12	(Elsheery and Cao 2008)
Rice cultivars (*Oryza sativa*)	70[b]	Positive turgor measured	Floating	4	(Babu et al. 1999)
Fava bean (*Vicia faba*)	66	At turgor loss point	Standing	22–24	(Kassam and Elston 1974)
Eucalyptusobliqua	65	Photosynthetic rate positive	Submerged	12	(Merchant et al. 2007)
Olive (*Olea europaea*)	60[b]	Photosynthetic rate positive	Submerged?	24	(Ahmed et al. 2009)
Arabidopsis	59	>70 % survival	Submerged	24	(Ding et al. 2011)
Triticale	55[b]	Shoot growth measured	Floating	24	(Kaydan and Yagmur 2008)
Barrel medic (*Medicago truncatula*)	50	Recovery after rewatering	Submerged	24	(Trindade et al. 2010)

[a] Values derived from interpretation of figure
[b] Average value from multiple plant varieties

subsequent RWC measurements. This species is known to have a higher constitutive osmotic concentration compared with eucalypts from mesic environments (Merchant et al. 2007). Boyer et al. (2008) reported that OA itself can lead to oversaturation, at least when leaf discs are floated.

Hence, the optimal RWC method for each plant species needs to be determined. Rehydration via the petiole has the advantage that Ψ can be repeatedly measured as RWC increases due to rehydration. The time needed to reach full hydration ($\Psi = 0$) can be accurately determined and oversaturation avoided. In the case of the four species studied in Fig. 8.2, full rehydration was achieved after only one hour. This method will not always be appropriate but floating or

Table 8.2 Relative water content and osmotic potential at turgor loss point (RWC_{TLP}, Ψs_{TLP}) as derived from P–V curves of drought stressed plants

Plant species	RWC_{TLP} (%)	Ψs_{TLP} (MPa)	Reference
Galax urceolata	95	−1.70	(Hughes et al. 2010)
Vinca minor	90	−2.00	(Hughes et al. 2010)
Hedera helix	90	−2.10	(Hughes et al. 2010)
Rhododendron spp	90	−1.00	(Hughes et al. 2010)
Eucalyptusobliqua	89	−1.80	(Merchant et al. 2010)
Eucalyptusrubida	89	−1.87	(Merchant et al. 2010)
Eucalyptusspp	88	−2.37	(Callister et al. 2008)
Vismia guianensis	89	−1.45	(Dias-Filho and Dawson 1995)
Wheat (*Triticum aestivum*)	89	−1.80	(Quarrie and Lane 1983)
Grape(*Vitis vinifera*)	88	−2.25	(Patakas and Noitsakis 1999)
Lonicera japonica	88	−1.50	(Hughes et al. 2010)
Eucalyptus calophylla	87	−1.90	(Szota et al. 2011)
Glycine max (soybean)	87	?	(Zur and Boote 1981)
Rosa hybrida	86	−1.86	(Augé et al. 1986)
Quercus ilex	86	−2.52	(Castro-Díez and Navarro 2007)
Quercus coccifera	86	−2.27	(Castro-Díez and Navarro 2007)
Sorghum (*Sorghum bicolor*)	86	−1.43	(Girma and Krieg 1992)
Eucalyptus melliodora	86	−2.78	(Merchant et al. 2010)
Quercus faginea	85	−2.13	(Castro-Díez and Navarro 2007)
Eucalyptus marginata	85	−1.80	(Szota et al. 2011)
Eucalyptus saligna	85	−2.13	(White et al. 2000)
Pistachio (*Pistacia vera*)	85	−2.7	(Gijón et al. 2011)
Eucalyptus camaldulensis	84	−2.35	(White et al. 2000)
Eucalyptus leucoxylon	82	−3.18	(White et al. 2000)
Eucalyptus platypus	75	−2.78	(White et al. 2000)
Olive (*Olea europaea*)	75	−3.9	(Dichio et al. 1997)

submersion time will need to be calibrated for each plant species. Minimizing the rehydration period is also sensible because respiration and carbohydrate metabolism during this time can change the concentration of soluble sugars (Moreton and Munns 2010), possibly affecting the magnitude of OA detected.

Despite potential difficulties, RWC of sample tissue is a suitable indicator of plant water status because it reflects a physical change in cell volume. It is also useful to know how this varies with Ψ, since it is possible by plotting RWC against $1/\Psi$ to determine Ψs of stressed tissue at full hydration without a direct measurement. This relationship produces a pressure–volume (P–V) curve which, although time consuming to complete, provides information to correct RWC for oversaturation. Figure 8.3a illustrates the use of a P–V curve, and the calculation of OA is described in Fig. 8.3b.

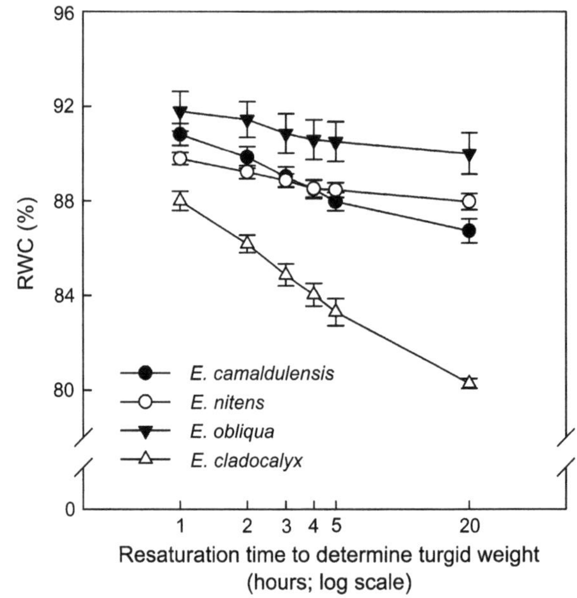

Fig. 8.2 The affect of the time taken to fully rehydrate leaves on RWC measurements in four eucalyptus species (SK Arndt, unpublished results). Leaves were taken from well-watered plants of each species at midday. Plants were fully rehydrated after 1 h. Means and standard errors shown (n = 5)

8.4.4 Measuring Osmotic Potential

Measurement of Ψs in either experimental approach (control-stress or developing stress) can be achieved by direct methods as an alternative to derived Ψs from a P–V curve. Choosing P–V curves over direct measurement to determine Ψs and OA will depend on the experimental design and available time and resources. P–V curves have the advantage of providing other useful water relation parameters in addition to OA, such as elastic adjustment, turgor loss point, and AWF (Fig. 8.3a). Hence, P–V curves allow for a more comprehensive analysis of the drought response of plants and tools for the automated analysis of P–V curves are also available (http://landflux.org/). See Cheung et al. (1975) for a description of the preparation, analysis, and application of P–V curves and Richter (1997) for a detailed critique of the water relations parameters obtained from them.

The main disadvantage of P–V curves is the time cost: only a limited number of samples can be analyzed concurrently (10–15 leaves/samples per person per day). For screening purposes commonly associated with crop research and sometimes in ecophysiological studies, direct measurement methods can be more efficient for larger sample sizes.

Direct measurement methods determine the concentration of solutes within the cell sap (osmolality), and can be classified according to:

1. expressed cell sap (ES) or hot water extracts (HWE) measured by vapor pressure or freezing point osmometer;
2. freeze-thaw leaf discs measured with thermocouple psychrometer.

Table 8.3 Advantages and disadvantages of the different methods for measuring Ψs, summarized from Callister et al. (2006)

Method	Leaf Form	Suitability for experimental approach		Pros	Cons
		Control/ stress	Developing stress		
Expressed sap	All	Yes	Yes	Fast; Can correct for AWF	AWF determination poorly researched and compared with P–V alternative
Hot water extract	All	Yes	Yes	Fast	No AWF determination. May reduce Ψs by introducing solutes through physical disturbance during sample preparation
Psychrometer of leaves	All	Yes	Yes		Slow; no AWF determination; very sensitive to external conditions at the time of measurement
P–V curve	Most	Yes	Impractical	Determines AWF (imprecisely); Other water relation parameters obtained	Slowest; practically not suitable for large sample sizes in most studies or regularly repeated measurements with moderate sample sizes

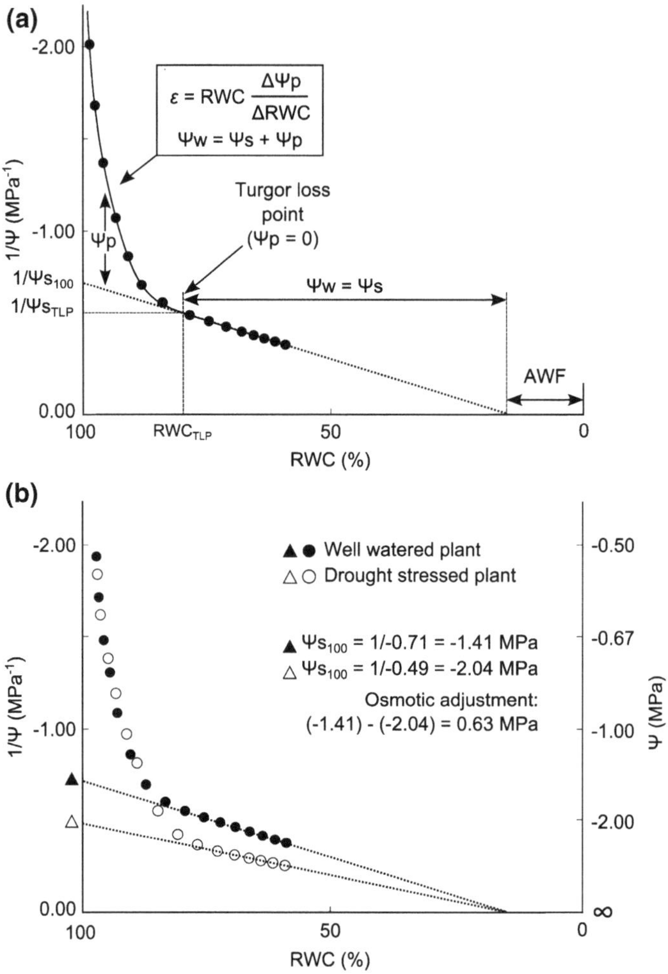

Fig. 8.3 An example P–V curve **a** indicating the relationship between the inverse of the plant water potential (1/Ψ) and RWC. Ψ is a function of both the turgor potential—also called turgor pressure—(Ψp) and the osmotic potential (Ψs) at high RWC. The steep initial decline in 1/Ψ at high RWC is driven by a rapid drop of Ψp until at a certain RWC the turgor pressure is lost (turgor loss point, TLP). The linear decline of 1/Ψ with decreasing RWC past TLP is now only driven by the passive concentration of solutes with water loss and consequently the water potential will equal the osmotic potential ($\Psi = \Psi s$). The linear extension of this "osmotic line" will result in the osmotic potential at full turgor (Ψs_{100}) at a RWC of 100 % and the AWF at a RWC of zero. The osmotic potential at zero turgor (Ψs_{TLP}) can be determined by extending a horizontal line from RWC_{TLP} to the y-axis. The slope of the initial turgor loss line allows the bulk modulus of elasticity (ε) the elasticity of the cell walls to be calculated. A steep slope results in a high ε or rigid cell walls, whereas a shallow slope will indicate a low ε and elastic cell walls. **b** Shows a theoretical example how OA can be calculated from P–V curves of a well-watered plant (*closed symbols*) and a drought-stressed plant (*open symbols*). The osmotic potential at full turgor (Ψs_{100}) for the stressed plant (*open triangle*, −2.04 MPa) is subtracted from the Ψs_{100} for the well-watered plant (*closed triangle*, −1.41 MPa) and the obtained difference of 0.63 MPa is the magnitude of OA

Table 8.4 Contribution of glycine betaine and proline to osmotic adjustment (OA) in higher plants

Species	Provenance/ cultivar	Compound	Contribution to OA (%)	OA (MPa)[a]	Reference
Spear grass (*Heteropogon contortus*)		Glycine betaine	2	−1.01	(Ford and Wilson 1981)
Atriplex halimus	Sbika	Glycine betaine	2	−1.41	(Ben Hassine et al. 2008)
Atriplex halimus	Monastir	Glycine betaine	5	−0.87	(Ben Hassine et al. 2008)
Jatropha curcas (root)		Glycine betaine	7	−0.21	(Silva et al. 2010)
Laurus nobilis	Tunisia	Glycine betaine	8	−0.45	(Maatallah 2010)
Jatropha curcas (leaf)		Glycine betaine	9	−0.65	(Silva et al. 2010)
Jatropha curcas (root)		Glycine betaine	9	−0.14	(Silva et al. 2010)
Laurus nobilis	Algeria	Glycine betaine	9	−0.48	(Maatallah 2010)
Jatropha curcas (leaf)		Glycine betaine	10	−0.38	(Silva et al. 2010)
Buffel grass (*Cenchrus ciliaris*)		Glycine betaine	14	−0.83	(Ford and Wilson 1981)
Green Panic(*Panicum maximum* var. *trichoglume*)		Glycine betaine	15	−0.65	(Ford and Wilson 1981)
Pea	Morón	Proline	2	−0.47	(Sánchez et al. 1998)
Laurus nobilis	Tunisia	Proline	3	−0.45	(Maatallah 2010)
Pea	Progress 9	Proline	3	−0.80	(Sánchez et al. 1998)
Coconut (*Cocus nucifera*)	UGD	Proline	4	−0.22	(Gomes et al. 2010)
Sorghum (*Sorghum bicolor*)	RS610	Proline	5	−0.49	(Jones et al. 1980)

(continued)

Table 8.4 (continued)

Species	Provenance/ cultivar	Compound	Contribution to OA (%)	OA (MPa)[a]	Reference
Spear grass (*Heteropogon contortus*)		Proline	5	−1.04	(Ford and Wilson 1981)
Atriplex halimus	Monastir	Proline	6	−0.87	(Ben Hassine et al. 2008)
Spartina alternifolia		Proline	6	−0.39	(Hessini et al. 2009)
Atriplex halimus	Sbika	Proline	7	−1.41	(Ben Hassine et al. 2008)
Buffel grass (*Cenchuus ciliaris*)		Proline	7	−0.76	(Ford and Wilson 1981)
Pearl Millet (*Pennisetum glaucum*)	IP8210	Proline	7	−0.71	(Kusaka et al. 2005)
Green Panic (*Panicum maximum* var. *trichoglume*)		Proline	17	−0.66	(Ford and Wilson 1981)
Laurus nobilis	Algeria	Proline	20	−0.48	(Maatallah 2010)
Maize (*Zea mays*)	CvWF9xMo17	Proline	44	−0.65	(Voetberg and Sharp 1991)

[a] OA represents the concentration of solutes in the symplast (vacuole and cytosol)

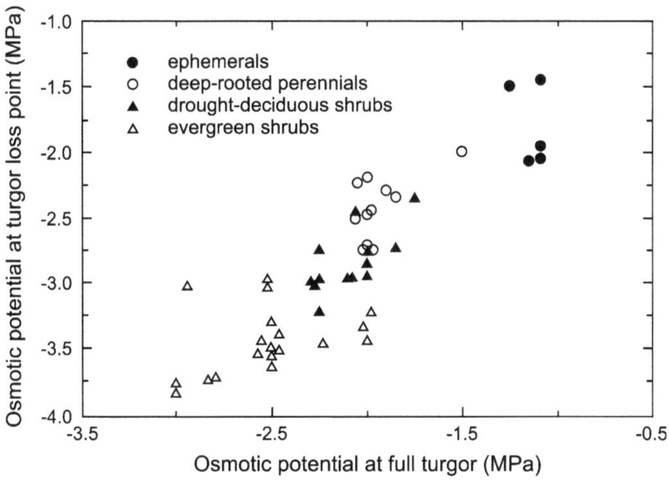

Fig. 8.4 The relationship between osmotic potential at the turgor loss point (incipient plasmolysis) and osmotic potential at full turgor in plants from the Sonoran Desert. Data pooled from the annual growth cycle of seven species with different life forms. (Redrawn from Smith et al. 1997 based on data in Monson and Smith 1982)

Callister et al. (2006) reviewed these methods in detail and recommended osmometry by ES for measuring Ψs over other direct measurement methods, because it was found to be comparable in reliability with a P–V curve. The main difference in these methods relates to the effect of apoplastic water on Ψs (Table 8.3). A completed P–V curve accounts for the symplastic (osmotic) water fraction separately, whereas direct measurement methods are subject to dilution of the symplastic water by apoplastic water. Apoplastic water usually contains a negligible amount of solute in leaves; hence, uncorrected Ψs by direct measurement will be less negative compared with Ψs derived from a P–V curve of the same tissue. The AWF can be determined from P–V curves by extending the osmotic line. However, this can be subject to error when few points are obtained past turgor loss point during the sample drying phase (Richter 1997). Caution (and sufficient replication) is, therefore, needed before ascribing a change in AWF to drought stress.

An alternate method to determine AWF and therefore apoplastic dilution involves comparing Ψ with Ψs from ES in tissues that are dried past the turgor loss point. If there is no dilution by apoplastic water, then $\Psi = \Psi s$ as the turgor pressure will be zero. However, in most cases Ψ will be more negative than Ψs because of dilution of the extracted cell sap with apoplastic water. Hence, the difference between Ψ and Ψs can be used to calculate AWF (Andersen et al. 1991; Urban et al. 1993):

$$\mathrm{AWF} = \mathrm{RWC}(1 - \Psi s/\Psi) \tag{8.4}$$

This can then be used to correctly measure Ψs specific to the osmotically active symplast (Callister et al. 2006):

$$\Psi s_{corr} = \Psi s \,/\, (1 - \mathrm{AWF/RWC}) \tag{8.5}$$

Many studies ignore this correction, which is acceptable to calculate OA only if the AWF remains constant over the course of stress. This is not always the case.

8.5 Where and When Does Osmotic Adjustment Occur?

OA does occur in many plant species but has predominantly been studied in agricultural plants. Important field crops that have been identified with OA in response to drought include wheat (Morgan 2000), sorghum (Tangpremsri et al. 1991), maize (Chimenti et al. 1996), rice (Babu et al. 2001), and sunflower (Chimenti et al. 2002). A comprehensive account of OA as reported in crops is given in Serraj and Sinclair (2002) and Blum (2011b). Kozlowski and Pallardy (2002) provide an account of some different tree species that exhibit OA, including many angiosperms and gymnosperms. Table 8.4 also lists a range of woody plants and crop species where accumulation of specific organic solutes associated with OA has been identified. OA occurs in a range of desert plant life forms, although the degree of OA varies among them (Smith et al. 1997).

Rapid onset of drought stress can exceed the capacity for plants to adapt with OA (López et al. 2008; Hessini et al. 2009), and so it is not surprising that greater OA has been observed with gradual stress in some plants (Turner and Jones 1980). This makes drought stress experiments with potted plants in glasshouses particularly difficult. Care needs to be taken not to induce a rapid onset of drought stress in glasshouse experiments. Repeated drought episodes may also increase the level of OA observed, at least in trees (Myers and Neales 1986; Abrams 1988). The intensity of stress and the physiological stage of the plant are other important influences, as was suspected with 11 genotypes of chickpea that demonstrated significant variation in OA in two separate experiments (Basu et al. 2007).

8.6 Function of Osmotic Adjustment

OA maintains water absorption and cell turgor in drought conditions, and is considered a key mechanism enabling plants to sustain higher photosynthetic rate and expansion growth under drought (Cattivelli et al. 2008). Photosynthetic capacity is maintained by lowering Ψ at which stomatal closure occurs (Chaves 1991). By maintaining leaf hydration, OA can prolong plant survival at otherwise lethal levels of drought stress, or maintain growth by turgor maintenance at more moderate stress levels.

Hsiao et al. (1976) attributed continued photosynthesis under drought stress to the affect of OA on turgor maintenance at low Ψ. This view is still held, although the related maintenance of symplast and chloroplast volume by OA may be as important (Gupta and Berkowitz 1987; Santakumari and Berkowitz 1991; Nilsen and Orcutt 1996). Reduced chloroplast volume can affect chloroplast pH and ion concentrations, which in turn can impact on the activity of photosynthetic enzymes.

It is also important to consider alternate responses and hydraulic limitations when a plant is subjected to low Ψ. The potential effectiveness of OA for photosynthesis may be redundant, where structural limits to hydraulic conductivity in leaves or other conducting tissue are exceeded by very high tension—very low Ψ (Ennajeh et al. 2008). Plants susceptible to the associated cavitation avoid irreversible drought damage by leaf shedding, whereas OA can maintain leaf area and so allow a fast recovery in photosynthesis and growth when conditions improve. Yazaki et al. (2010) demonstrated that a trade-off between these alternate drought responses may have been responsible for the competitive dominance of an exotic tree in a habitat prone to periodic drought. The progression toward lower osmotic potentials and greater OA in desert species lacking drought avoidance mechanisms (Smith et al. 1997) indicate that OA could have greater significance in drought-susceptible life forms (Fig. 8.4).

8.6.1 Turgor Maintenance

The increase of the number of solutes in the symplast of the cell as a consequence of OA results in turgor pressure (Ψp) being maintained at least partially as Ψ becomes more negative (Morgan 1984; Turner and Jones 1980; Kramer and Boyer 1995). Despite imposed drought stress, turgor is maintained if OA can maintain RWC with a reduction in Ψs (e.g., Fig. 8.1b, *complete OA* case).

Although features of the cell wall may delay water loss and loss of turgor, it will ultimately occur without sufficient osmotic tension within the cell to counter the external water tension. This external tension results from drying soil and so the tension created by OA must exceed the water removing strength of the environment to be effective at maintaining turgor. Cell sap can contain enough solutes to reduce Ψs to as low as −5.0 MPa, or twice the salinity of sea water (Meinzer et al. 1986; Nilsen and Orcutt 1996).

The maintenance of turgor at low Ψ can maintain a range of physiological processes as outlined below. As a competitive adaptation, OA and turgor maintenance may ultimately depend on whether growth rate in adverse conditions confers an ecological advantage, at least in woody plants (Kozlowski and Pallardy 2002). In field crops, a reduction of approximately 0.3 MPa is considered a minimum level to provide a growth advantage (Blum 1993), presumably through turgor maintenance.

Turgor may be the primary variable that controls OA in higher plants in response to osmotic stress (such as that imposed by drought), through pressure sensing mechanisms in the cell membrane (Zimmermann 1978).

8.6.1.1 Transpiration/Assimilation

Although stomatal conductance and maintenance of leaf turgor by OA are considered to be coupled, the connection is not universal (Munns 1988; Saliendra and Meinzer 1991). For example, OA associated with decreases in Ψ did not maintain stomatal conductance in a range of heath species during summer drought (Mitchell et al. 2008). In addition to leaf water status and regular photosynthetic factors (light and CO_2), guard cell turgor pressure can be affected by chemical signaling from the roots (Nilsen and Orcutt 1996; Chaves et al. 2003), which can occur in the early stages of drought before loss of leaf turgor.

Turgor loss in guard cells relative to epidermal cells causes stomatal closure (Chaves et al. 2003; Brodribb and Holbrook 2003), which in turn maintains hydration and turgor in remaining leaf tissue (Serraj and Sinclair 2002). In this regard, early stomatal closure is seen as an alternative stress adaptation to OA (Ludlow 1989). However, it is yet unclear to what extent OA may be associated with variation in stomatal responses ranging in between the two extremes of isohydric and anisohydric behavior, as described by Tardieu and Simonneau (1998).

McCree and Richardson (1987) compared stomatal closure with OA to see firstly if there was an interaction and secondly whether the interaction resulted in any difference in carbon gain. The species with greater OA (sugar beet) had lower Ψ than the species with reduced OA (cowpea), which closed stomata at higher Ψ. And yet, there was no significant difference in carbon gains between species because Ψ fell much faster in association with OA. This was a controlled environment study, and the effect of OA on whole plant growth may differ when a greater soil volume is available under field conditions (Blum 2011b).

8.6.1.2 Shoot Growth

High turgor pressure in meristematic cells of shoots at low Ψ is a strong indication of stress tolerance. Cell volume expansion in these cells is a physical requirement of plant growth, and the rate of permanent cell volume growth is a product of effective turgor pressure (Ψp minus cell wall yield threshold) and the irreversible cell wall plasticity (Lockhart 1965; Johnsen and Major 1999). Reversible cell wall hardening in response to particular solutes restricted leaf growth in maize in response to drought stress despite OA (Chazen and Neumann 1994). Therefore, although growth is highly sensitive to turgor pressure (Joly and Zaerr 1987), the evidence for leaf turgor maintaining growth under drought stress is equivocal (Munns 1988; Davies 1991). The connection between leaf turgor and leaf expansion (Turner 1986), biomass production (Hessini et al. 2009) or tree volume (Callister et al. 2008) may be lost, although long term growth responses have shown a strong correlation with leaf turgor (Johnsen and Major 1999).

Feng et al. (1994) showed that OA in sudangrass maintained nearly constant turgor while the relative rate of leaf area growth was significantly affected by Ψ. And so it is possible that turgor coregulates the rate of cell growth along with other variables responsive to a change in Ψ and temperature (Feng et al. 1994). Evidence is weak that plants respond or adapt to changes in Ψ *per se*, without associated physiological changes including the loss of turgor (Jones 2007).

8.6.1.3 Root Expansion

The effect of OA on turgor in root tips is a potential drought avoidance mechanism for maintaining water supply. Improved root development can raise Ψ by accessing supplementary soil water (Lambers et al. 2008). OA in roots, as in leaves, causes sustained turgor and allows the cell walls to loosen for the expansion phase (Hsiao and Xu 2000). Maintenance of growth in root tips allows the plant to occupy a greater volume of soil, and may also increase the density of roots within a given volume (Turner and Jones 1980).

There is evidence that root growth under drought stress is related to OA in a range of crops, including maize (Westgate and Boyer 1985; Voetberg and Sharp 1991; Chimenti et al. 2006; Ogawa and Yamauchi 2006), sorghum (Wright et al. 1983; Tangpremsri et al. 1991), wheat (Morgan 1995, 2000), and sunflower (Rauf and Ahmad Sadaqat 2008). In rainfed rice, however, OA genotypes maintained a lower allocation of biomass to roots in comparison with non-OA genotypes suggesting drought tolerance over drought avoidance (Wang et al. 2009). The same tendency is seen more generally in irrigated versus rainfed rice varieties (Babu et al. 1999, 2001). With such evidence it is important to consider that OA may be less dominant than other drought adaptations, and its expression curtailed (Blum 2011b).

Maintenance of turgor in roots by OA can contribute to the shift toward root growth generally observed in water-limited conditions (Westgate and Boyer 1985), although the internal balance of nutrients may also prove significant in this regard (Ericsson 1995). It is clear that increased cell elasticity and OA can sustain root growth with lower Ψ, and the reduction of leaf growth with even mild stress should make more photosynthates available for root growth (Hsiao and Xu 2000).

8.6.1.4 Metabolic Function

Positive cell turgor is required for membrane integrity and solute exchange between cells, including photosynthate transport (Thorpe and Minchin 1996). In turn, this function is essential for maintaining metabolic processes within the cell (Feng et al. 1994; Johnsen and Major 1999; Blum 2005), with implications for plant survival during drought stress after growth has ceased (Mitchell et al. 2008).

8.6.2 Whole Plant Productivity: Growth and Yield

In crops, OA is considered successful if it results in continued productivity under stress rather than from improved survival at extreme levels of dehydration (Munns and Richards 2007; Kramer and Boyer 1995). Consequently, crop resistance to drought stress effectively depends on maintaining photosynthesis and transpiration (Morgan 1984). Although the relationship between grain yield (or biomass production) and crop transpiration does vary within species, there are limits due to inflexibility in the basic biochemistry of photosynthesis (Parry et al. 2006). Serraj and Sinclair (2002) make this point in relation to OA: there is little benefit in OA for yield maintenance under drought unless plants can continue to transpire water, even though that water may be harder to obtain. In this regard, OA can be most successful in sustained droughts if it leads to increased access to available soil moisture (Blum 2009).

And so although an economic yield under drought stress is the ultimate measure of drought avoidance in crops, a link with OA depends on scaling up a particular biochemical adaptation to affect productivity (Blum 2011a). In some species, increased fruit yield may be improved by carefully timed drought conditions (scheduled deficit irrigation), and this response may not be linked with OA and the maintenance of turgor (Kozlowski and Pallardy 2002). In other cases, ABA can trigger abscission of fruit or flower buds in response to drought stress well before OA or stomatal closure has affected leaf water status (Blum 2011a).

Assessing the effectiveness of OA in crops may be complicated by the economic irrelevance of marginal differences noted under more intense stress levels when OA may be most effective (Serraj and Sinclair 2002). There also remains in the recent literature a long-held suspicion that yield improvement may actually run counter to the effect of OA when driven by accumulation of organic solutes. In some cases, a reduction in growth early in the onset of drought stress is a symptom/consequence of OA (Marigo and Peltier 1996; Osório et al. 1998). The lack of correlation between osmotic stress resistance (due to salinity or drought) and the accumulation of organic solutes may be the result of the high energy cost in synthesizing some solutes (Shabala and Shabala 2011).

Another plausible explanation for the dissociation between actual yield and known OA capacity (as established through genetic trials) is the impact of root chemical signals on shoot growth (Morgan 2000). This may prevent OA from being expressed if reduced leaf growth contributes to higher Ψ under conditions of low evaporative demand. Nevertheless, successful breeding for higher OA resulting in more growth under stress suggests that OA cannot be merely a consequence of unutilized organic solutes following growth reduction (Morgan 1984). Blum (2011b) lists crop studies where OA has been shown to positively affect yield under drought.

8.6.3 Genetic Enhancement of OA for Yield Improvement

Studies in a range of crops have identified varieties with a high capacity for OA, and in some cases with positive correlations with yield (Blum 2011b). A recently released dryland wheat variety has a high capacity for OA (Munns and Richards 2007; Blum 2011b). Transgenic approaches have yielded higher OA capacity and a reduction in growth deterioration during drought stress (Babu et al. 2004). However, in general terms, genetic modification for specific solute accumulation does not necessarily lead to OA in response to stress: see Valliyodan and Nguyen (2006) for a discussion.

8.7 Solutes

Solute accumulation or synthesis under drought stress (imposed by drought or salinity) can be achieved by different processes (Chaves 1991). This includes metabolites unused after a slowing of growth at the onset of stress (Munns 1988), the movements of existing solutes between cellular compartments (Morgan 1984), or the breakdown of storage molecules such as starch. Particular attention to the range of solutes involved in OA is given in reviews by Yancey et al. (1982), Hare et al. (1998), Chaves et al. (2003), Bartels and Sunkar (2005), and Chen and Jiang (2010). An introduction to the major classes of solutes is provided here, guided by their importance in contributing to a change in Ψs.

There are four classes of solutes that can lower Ψs of tissues: inorganic cations and anions, sugars and sugar alcohols, nonprotein amino acids, and organic acids (Nilsen and Orcutt 1996; Huang 2000). Inorganic ions, if available for translocation, come at a lower energy cost to the plant, and are often confined within the vacuole to protect the cytoplasm from possible toxic effects (Nilsen and Orcutt 1996). In contrast, organic solutes contributing to OA have not been associated with harmful metabolic effects, and some are known to protect cellular function during drought stress.

8.7.1 Compatible Solutes

Aside from maintaining cell hydration and turgor, organic solutes can also protect the cell membrane, stabilize enzymes or proteins, and protect against oxidation (Ashraf and Foolad 2007). In this capacity they are referred to as compatible solutes, because their accumulation does not interfere with normal metabolism by inhibiting cytosolic enzymes (Arndt et al. 2008). These compatible solutes are highly soluble, carry no net charge, and are nontoxic at high concentrations (McNeil and Nuccio 1999). A distinction between compatible solutes that are metabolically active and those that are relatively stable indicates a possible trade-off between stability and cost of production. It seems plausible that perennial species which are subjected to chronically low soil water potentials would maintain stable substances as solutes, compared with similar species adapted to more ideal growing conditions (Merchant et al. 2006).

Compatible solutes are represented by sugars (fructose, sucrose, glucose), sugar alcohols (polyols and cyclitols including glycerol, mannitol, sorbitol, quercitol, pinitol, etc.), amino acids (notably proline), and quaternary ammonium compounds (notably glycine betaine). All these solutes possess a multitude of hydroxyl (−OH) groups that help facilitate hydrogen bonds with water molecules in the cytoplasm. That way the solutes assist in keeping functional macromolecules in solution. As *osmoprotectants*, they can therefore stabilize proteins, enzymes and cell structures (Yancey et al. 1982; Morgan 1984; Travert et al. 1997; Martinez et al. 2004) and also function as antioxidants (Rhodes and Hanson 1993; Ashraf and Foolad 2007; Ben Hassine et al. 2008). Compatible solutes also lower Ψs in response to drought stress, and so contribute to OA.

It is generally assumed that organic solutes are preferentially accumulated in response to drought stress because they do not interfere with metabolism. This view is held despite the relative energy cost of compatible solute synthesis being more expensive than import and translocation of available inorganic ions (Raven 1985; Patakas et al. 2002; Shabala and Shabala 2011). However, amino acids and sugars contributed toward most of the OA in maize roots, depending on the distance from the apex (Voetberg and Sharp 1991). In eucalypts and acacias, all the solutes found to increase significantly in response to drought stress were closely associated with primary C metabolism (Warren et al. 2011). In that particular study, the energy cost of OA was weighed by the amount of nonreversible metabolites, which were estimated to require only 12–16 h of photosynthesis to produce.

Further discussion about the osmoregulatory and osmoprotective roles of various compatible solutes can be found in reviews by Heuer (2010), Chen and Murata (2002), Chaves et al. (2003), Shao et al. (2009), and Chen and Jiang (2010).

8.7.2 Relative Contribution of Solutes

A small increase in the measured symplastic concentration of a particular solute may have significance for OA if, for example, a correlation can be found using a scatter plot of the results from different cultivars (see Sánchez et al. 1998). It is possible that a small increase has been confined to the cytosol and therefore partly responsible for OA (Munns 2011), but only in conjunction with an accumulation of other solutes in the vacuole (Jones et al. 1980). The cytosolic volume is often about 10 % of the total volume of the symplast (from which cell sap is derived and measured) (Merchant and Adams 2005). For example, if there is a 5 mM increase in the cell sap concentration due to a particular solute, and if the solute is actually confined to the cytosol in situ, this represents a 50 mM increase in concentration in the cytosolic water (Munns 2002).

Several studies have quantified the relative contribution of a range of organic and inorganic solutes toward OA and/or constitutive Ψs in higher plants (see Jones et al. 1980; Ford and Wilson 1981; Kusaka et al. 2005; Raza et al. 2007; Warren et al. 2007; Silva et al. 2010; Maatallah 2010; Warren et al. 2011). In a complete metabolite profile of selected *Eucalyptus* and *Acacia* spp., inorganic ions did not increase significantly in response to drought stress, despite OA of up to 1.2 MPa in eucalypts and 1.5 MPa in acacias (Warren et al. 2011). Therefore, there are some cases in which an increase in organic solutes must occur in relatively large concentrations to account fully for OA, although this may involve the total contribution of many specific compounds. In many drought exposed eucalypts the cyclitol quercitol is observed in very high concentrations, yet the compound does not always increase in concentration in response to drought (Merchant et al. 2006) and a net increase is often observed only under severe stress (Arndt et al. 2008).

Table 8.4 shows that in a range of situations, the compatible solutes proline and glycine betaine do not make more than a 50 % contribution toward OA in the symplast as a whole. In fact, in most instances either solute contributed to less than 10 % of the OA of the measured tissue. Consequently, other solutes (individually or in combination) contribute more to OA in higher plants than proline or glycine betaine. Although there are suggestions to the contrary (Shabala and Shabala 2011), these other solutes may not always be inorganic ions when OA is associated with drought stress.

It will, therefore, be important to assess the real contribution of individual or specific solutes to OA, and a complete metabolic profile in response to drought stress will provide this. As the above mentioned evidence suggests, it is very unlikely for OA to be attributed to a single substance. In most cases, OA is the consequence of an increase of a multitude of substances and this will make the "genetic engineering" of OA more difficult. There have been many attempts to over-express specific compatible solutes (such as proline or glycine betaine) in genetically modified crops, but this rarely resulted in an improved drought response (Ashraf and Foolad 2007). In many instances, specific compatible solutes will have predominantly osmoprotective functions and these can be effective at lower concentrations if they are located in the smaller cytosol. I.e., the increases of proline and glycine betaine in the plants in Table 8.4 will be sufficient to provide increased osmoprotection for enzymes in the cytosol, but they play only a limited role in OA.

8.8 Conclusion

The primary effect of OA in maintaining cell hydration and turgor is a widely observed drought adaptation in both crops and woody plants. Depending on the timing and intensity of stress and a plant's inherent ability to use water effectively

(Levitt 1980; Blum 2009), OA can reduce the negative impact on crop yield and/or growth. Determining a plant's capacity for OA requires careful attention to measurement of plant water status and the imposition of stress. Accurately measuring water status of the plant in response to drought stress is critical for comparisons within or between species, and especially where interactions with growth or yield are of particular interest. Further, a focus on particular solutes with known drought performance in crops is discouraged in favor of broader profiling of solutes, including compatible solutes.

OA does play an important role in the drought response of plants and can be an important mechanism in improving performance in some crop species and cultivars, but is ineffective in others. However, the ecological role of OA is less well understood. It is not very clear under what circumstances OA is expressed in native or perennial species and how OA is related to different ecological and hydraulic strategies that plants use in response to water deficit, e.g., isohydric and anisohydric. The interplay of OA with other water potential adjustments like elastic adjustment has also not been investigated in detail. There are also open questions around the function of OA in the survival of predominantly perennial species that are subjected to drought stress. There have been extensive discussions around the observed increase of drought-induced tree death (Allen et al. 2010), with proposed mechanisms of tree death related to either carbon starvation(McDowell et al. 2008; McDowell and Sevanto 2010), or hydraulic failure (Sala et al. 2010; Anderegg et al. 2011; McDowell 2011). To what degree OA can contribute to an increased tolerance to drought and prolonged survival is unclear, and will be an important research question for future studies.

References

Abrams MD (1988) Sources of variation in osmotic potentials with special reference to North American tree species. For Sci 34:1030–1046

Ahmed CB, Rouina BB, Sensoy S, Boukhris M, Abdallah FB (2009) Changes in gas exchange, proline accumulation and antioxidative enzyme activities in three olive cultivars under contrasting water availability regimes. Environ Exp Bot 67:345–352

Allen CD, Macalady AK, Chenchouni H, Bachelet D, McDowell NG, Vennetier M, Kitzberger T, Rigling A, Breshears DD, Hogg EH et al (2010) A global overview of drought and heat-induced tree mortality reveals emerging climate change risks for forests. For Ecol Manage 259:660–684

Anderegg WRL, Berry JA, Smith DD, Sperry JS, Anderegg LDL, Field CB (2011) The roles of hydraulic and carbon stress in a widespread climate-induced forest die-off. Proc Nat Acad Sci 109:233–237

Andersen M, Jensen C, Losch R (1991) Derivation of pressure-volume curves by a non-linear regression procedure and determination of apoplastic water. J Environ Bot 42:159

Arndt SK, Livesley SJ, Merchant A, Bleby T, Grierson PF (2008) Quercitol and osmotic adaptation of field-grown Eucalyptus under seasonal drought stress. Plant Cell Environ 31:915–924

Ashraf M, Foolad MR (2007) Roles of glycine betaine and proline in improving plant abiotic stress resistance. Environ Exp Bot 59:206–216

Augé RM, Schekel KA, Wample RL (1986) Osmotic adjustment in leaves of VA mycorrhizal and nonmycorrhizal Rose plants in response to drought stress. Plant Physiol 82:765–770
Babu RC, Pathan MS, Blum A (1999) Comparison of measurement methods of osmotic adjustment in rice cultivars. Crop Sci 39:150–158
Babu RC, Shashidhar HE, Lilley JM, Thanh ND, Ray JD, Sadasivam S, Sarkarung S, O'Toole JC, Nguyen HT (2001) Variation in root penetration ability, osmotic adjustment and dehydration tolerance among accessions of rice adapted to rainfed lowland and upland ecosystems. Plant Breed 120:233–238
Babu RC, Zhang J, Blum A, David Ho T-H, Wu R, Nguyen HT (2004) HVA1, a LEA gene from barley confers dehydration tolerance in transgenic rice (*Oryza sativa* L.) via cell membrane protection. Plant Sci 166:855–862
Barrs HD (1968) Determination of water deficits in plant tissue. In: Kozlowski TT (ed) Water deficits and plant growth, vol 1. Academic Press, New York, pp 235–368
Bartels D, Sunkar R (2005) Drought and salt tolerance in plants. Crit Rev Plant Sci 24:23–58
Basu PS, Berger JD, Turner NC, Chaturvedi SK, Ali M, Siddique KHM (2007) Osmotic adjustment of chickpea (Cicerarietinum) is not associated with changes in carbohydrate composition or leaf gas exchange under drought. Ann Appl Biol 150:217–225
Ben Hassine A, Ghanem ME, Bouzid S, Lutts S (2008) An inland and a coastal population of the Mediterranean xero-halophyte species *Atriplex halimus* L. differ in their ability to accumulate proline and glycinebetaine in response to salinity and water stress. J Environ Bot 59:1315–1326
Blum A (1993) Selection for sustained production in water-deficit environments. In: Buxton DR, Shibles R, Forsberg RA (eds) International crop sci I. Crop Sci Soc Am, Madison, pp 343–347
Blum A (1996) Crop responses to drought and the interpretation of adaptation. J Plant Growth Regul 20:135–148
Blum A (2005) Drought resistance, water-use efficiency, and yield potential—are they compatible, dissonant, or mutually exclusive? Aust J Agric Res 56:1159–1168
Blum A (2009) Effective use of water (EUW) and not water-use efficiency (WUE) is the target of crop yield improvement under drought stress. Field Crop Res 112:119–123
Blum A (2011a) Drought resistance—is it really a complex trait? Funct Plant Biol 38:753–757
Blum A (2011b) Plant breeding for water limited environments. Springer, New York
Boyer JS, James RA, Munns R, Condon TA, Passioura JB (2008) Osmotic adjustment leads to anomalously low estimates of relative water content in wheat and barley. Funct Plant Biol 35:1172–1182
Brodribb TJ, Holbrook N (2003) Relations between stomatal closure, leaf turgor and xylem vulnerability in eight tropical dry forest trees. Plant Cell Environ 26:443–450
Callister AN, Arndt SK, Adams MA (2006) Comparison of four methods for measuring osmotic potential of tree leaves. Physiol Plant 127:383–392
Callister AN, Arndt SK, Ades PK, Merchant A, Rowell D, Adams MA (2008) Leaf osmotic potential of Eucalyptus hybrids responds differently to freezing and drought, with little clonal variation. Tree Physiol 28:1297–1304
Castro-Díez P, Navarro J (2007) Water relations of seedlings of three Quercus species: variations across and within species grown in contrasting light and water regimes. Tree Physiol 27:1011–1018
Cattivelli L, Rizza F, Badeck FW, Mazzucotelli E, Mastrangelo AM, Francia E, Marč C, Tondelli A, Stanca AM (2008) Drought tolerance improvement in crop plants: an integrated view from breeding to genomics. Field Crop Res 105:1–14
Chaves MM (1991) Effects of water deficits on carbon assimilation. J Environ Bot 42:1
Chaves MM, Maroco JP, Pereira JS (2003) Understanding plant responses to drought—from genes to the whole plant. Funct Plant Biol 30:239
Chazen O, Neumann PM (1994) Hydraulic signals from the roots and rapid cell-wall hardening in growing maize (*Zea mays* L.) leaves are primary responses to polyethylene glycol-induced water deficits. Plant Physiol 104:1385–1392

Chen H, Jiang J-G (2010) Osmotic adjustment and plant adaptation to environmental changes related to drought and salinity. Environ Rev 18:309–319

Chen THH, Murata N (2002) Enhancement of tolerance of abiotic stress by metabolic engineering of betaines and other compatible solutes. Curr Opin Plant Biol 5:250–257

Cheung YNS, Tyree MT, Dainty J (1975) Water relations parameters on single leaves obtained in a pressure bomb and some ecological interpretations. Can J of Bot 53:1342–1346

Chimenti CA, Cantagallo J, Guevara E (1996) Osmotic adjustment in maize: genetic variation and association with water uptake. In: Edmeades GO, Bänzinger M, Mickelson HR, Peña-Valdivia CB (eds) Developing drought and low N-tolerant maize: Proceedings of a symposium 25–29 March 1996, CIMMYT, El Batán, Mexico, pp 200–203. Int Maize and Wheat Improvement Center

Chimenti CA, Pearson J, Hall AJ (2002) Osmotic adjustment and yield maintenance under drought in sunflower. Field Crop Res 75:235–246

Chimenti CA, Marcantonio M, Hall AJ (2006) Divergent selection for osmotic adjustment results in improved drought tolerance in maize (*Zea mays* L.) in both early growth and flowering phases. Field Crop Res 95:305–315

Davies W (1991) Root signals and the regulation of growth and development of plants in drying soil. Annu Rev Plant Biol 42:55–76

Dias-Filho MB, Dawson TE (1995) Physiological responses to soil moisture stress in two Amazonian gap-invader species. Funct Ecol 9:213

Dichio B, Nuzzo V, Xiloyannis C, Celano G, Angelopoulos K (1997) Drought stress-induced variation of pressure-volume relationships in *Olea Europaea* L. CV. "Coratina." In: Proceedings of 2nd international symposium on irrigation of hort crops, pp 401–409. ISHS (Acta Horticulturae 449)

Ding Y, Avramova Z, Fromm M (2011) The Arabidopsis trithorax-like factor ATX1 functions in dehydration stress responses via ABA-dependent and ABA-independent pathways. Plant J Cell Mol Biol 66:735–744

Dreyer E, Bousquet F, Ducrey M (1990) Use of pressure volume curves in water relation analysis on woody shoots: influence of rehydration and comparison of four European oak species. Ann For Sci 47:285–297

Eckardt FE (1965) Methodology of plant eco-physiology. In: Proceedings of Montpellier symposium. UNESCO, Paris, p 549

Eerens J, Lucas R, Easton S, White J (1998) Influence of the endophyte (Neotyphodiumlolii) on morphology, physiology, and alkaloid synthesis of perennial ryegrass during high temperature and water stress. N Z J Agric Res 41:219–226

Elsheery NI, Cao K-F (2008) Gas exchange, chlorophyll fluorescence, and osmotic adjustment in two mango cultivars under drought stress. Acta Physiologiae Plant 30:769–777

Ennajeh M, Tounekti T, Vadel AM, Khemira H, Cochard H (2008) Water relations and drought-induced embolism in olive (*Olea europaea*) varieties "Meski" and "Chemlali" during severe drought. Tree Physiol 28:971–976

Ericsson T (1995) Growth and shoot: root ratio of seedlings in relation to nutrient availability. Plant Soil 168–169:205–214

Feng Y, Li X, Boersma L (1994) Roles of osmotic adjustment and turgor potential in leaf area expansion. Physiol Plant 90:1–8

Ford CW, Wilson JR (1981) Changes in levels of solutes during osmotic adjustment to water stress in leaves of four tropical pasture species. Aust J Plant Physiol 8:77–91

Gebre G, Tschaplinski T (2000) Role of osmotic adjustment in plant productivity: a summary report and review of current literature. Oak Ridge National Laboratory, Oak Ridge

Gijón MDC, Gimenez C, Perez-López D, Guerrero J, Couceiro JF, Moriana A (2011) Water relations of pistachio (*Pistaciavera* L.) as affected by phenological stages and water regimes. Sci Hortic 128:415–422

Girma FS, Krieg DR (1992) Osmotic adjustment in sorghum: I. Mechanisms of diurnal osmotic potential changes. Plant Physiol 99:577–582

Gomes FP, Oliva M a, Mielke MS, Almeida A-AF, Aquino L a (2010) Osmotic adjustment, proline accumulation and cell membrane stability in leaves of *Cocos nucifera* submitted to drought stress. Sci Hortic 126:379–384

Gupta AS, Berkowitz GA (1987) Osmotic adjustment, symplast volume, and nonstomatally mediated water stress inhibition of photosynthesis in wheat. Plant Physiol 85:1040–1047

Hare PD, Cress W a, Van Staden J (1998) Dissecting the roles of osmolyte accumulation during stress. Plant Cell Environ 21:535–553

Hessini K, Maroco JP, Gandour M, Albouchi A, Soltani A, Abdelly C (2009) Effect of water stress on growth, osmotic adjustment, cell wall elasticity and water-use efficiency in Spartina alterniflora. Environ Exp Bot 67:312–319

Heuer B (2010) Role of proline in plant response to drought and salinity. In: Pessarakli M (ed) Handbook of plant and crop stress. CRC Press, Boca Raton, pp 213–238

Hsiao TC (1973) Plant responses to water stress. Annu Rev Plant Physiol 24:519–570

Hsiao TC, Xu LK (2000) Sensitivity of growth of roots versus leaves to water stress: biophysical analysis and relation to water transport. J Environ Bot 51:1595–1616

Hsiao TC, Acevedo E, Fereres E, Henderson DW (1976) Water stress, growth, and osmotic adjustment. Philos Trans R Soc Lond B Biol Sci 273:479–500

Huang X (2000) The growth potential generated in citrus fruit under water stress and its relevant mechanisms. Sci Hortic 83:227–240

Hughes NM, Reinhardt K, Feild TS, Gerardi AR, Smith WK (2010) Association between winter anthocyanin production and drought stress in angiosperm evergreen species. J Environ Bot 61:1699

James AT, Lawn RJ, Cooper M (2008) Genotypic variation for drought stress response traits in soybean. I. Variation in soybean and wild Glycine spp. for epidermal conductance, osmotic potential, and relative water content. Aust J Agric Res 59:656

Johnsen KH, Major JE (1999) Shoot water relations of mature black spruce families displaying a genotype x environment interaction in growth rate. I. Family and site effects over three growing seasons. Tree Physiol 19:367–374

Joly RJ, Zaerr JB (1987) Alteration of cell-wall water content and elasticity in douglas-fir during periods of water deficit. Plant Physiol 83:418–422

Jones HG (2007) Monitoring plant and soil water status: established and novel methods revisited and their relevance to studies of drought tolerance. J Exp Bot 58:119–130

Jones MM, Osmond C, Turner NC (1980) Accumulation of solutes in leaves of sorghum and sunflower in response to water deficits. Funct Plant Biol 7:193–205

Jones MM, Osmond C, Turner NC (1981) Mechanisms of drought resistance. In: Paleg LG, Aspinall D (eds) The physiology and biochemistry of drought resistance in plants. Academic Press, Sydney, pp 15–38

Kassam A, Elston J (1974) Seasonal changes in the status of water and tissue characteristics of leaves of *Vicia faba L.* Ann Bot 38:419

Kassam A, Elston J (1976) Changes with age in the status of water and tissue characteristics in individual leaves of *Vicia faba L.* Ann Bot 40:669–679

Kaydan D, Yagmur M (2008) Germination, seedling growth and relative water content of shoot in different seed sizes of triticale under osmotic stress of water and NaCl. J Biotechnol 7:2862–2868

Kozlowski TT, Pallardy SG (2002) Acclimation and adaptive responses of woody plants to environmental stresses. Bot Rev 68:270–334

Kramer PJ, Boyer JS (1995) Water relations of plants and soil, Academic Press, San Diego

Kusaka M, Ohta M, Fujimura T (2005) Contribution of inorganic components to osmotic adjustment and leaf folding for drought tolerance in pearl millet. Physiol Plant 125:474–489

Ladiges PY (1975) Some aspects of tissue water relations in three populations of *Eucalyptus Viminalis Labill.* New Phytol 75:53–62

Lafitte R (2002) Relationship between leaf relative water content during reproductive stage water deficit and grain formation in rice. Field Crop Res 76:165–174

Lambers H, Chapin FS, Pons TL (2008) Plant physiological ecology, Springer, New York

Levitt J (1980) Responses of plants to environmental stresses. Volume II: water, radiation, salt and other stresses, Academic Press, New York

Lockhart JA (1965) An analysis of irreversible plant cell elongation. J Theor Biol 8:264–275

López R, Rodríguez-Calcerrada J, Gil L (2008) Physiological and morphological response to water deficit in seedlings of five provenances of *Pinus canariensis*: potential to detect variation in drought-tolerance. Trees 23:509–519

Ludlow MM (1989) Strategies in response to water stress. In: Kreeb KH, Richter H, Hinckley TM (eds) Structural and functional responses to environmental stresses: water shortage: XIV international botanical congress, Berlin (West), Germany, 24 July–1 August 1987, SPB Academic Publishing, The Hague, pp 269–281

Maatallah S (2010) A greenhouse investigation of responses to different water stress regimes of Laurusnobilis trees from two climatic regions. J of Arid Environ 74:327–337

Marigo G, Peltier JP (1996) Analysis of the diurnal change in osmotic potential in leaves of *Fraxinus excelsior* L. J Environ Bot 47:763–769

Martinez J, Lutts S, Schanck A, Bajji M, Kinet J-M (2004) Is osmotic adjustment required for water stress resistance in the Mediterranean shrub *Atriplex halimus* L? J Plant Physiol 161:1041–1051

McCree KJ, Richardson SG (1987) Stomatal closure vs. osmotic adjustment: a comparison of stress response. Crop Sci 27:539–543

McDowell NG (2011) Mechanisms linking drought, hydraulics, carbon metabolism, and vegetation mortality. Plant Physiol 155:1051–1059

McDowell NG, Sevanto S (2010) The mechanisms of carbon starvation: how, when, or does it even occur at all? New Phytol 186:264–266

McDowell NG, Pockman WT, Allen CD, Breshears DD, Cobb N, Kolb TE, Plaut J, Sperry J, West A, Williams DG, Yepez EA (2008) Mechanisms of plant survival and mortality during drought: why do some plants survive while others succumb to drought? New Phytol 178:719–739

McNeil S, Nuccio M (1999) Betaines and related osmoprotectants. Targets for metabolic engineering of stress resistance. Plant Physiol 120:945–949

Meinzer FC, Rundel PW, Sharifi MR, Nilsen ET (1986) Turgor and osmotic relations of the desert shrub *Larrea tridentata*. Plant Cell Environ 9:467–475

Merchant A, Adams MA (2005) Stable osmotica in *Eucalyptus spathulata*—responses to salt and water deficit stress. Funct Plant Biol 32:797–805

Merchant A, Tausz M, Arndt SK, Adams MA (2006) Cyclitols and carbohydrates in leaves and roots of 13 Eucalyptus species suggest contrasting physiological responses to water deficit. Plant Cell Environ 29:2017–2029

Merchant A, Callister AN, Arndt SK, Tausz M, Adams MA (2007) Contrasting physiological responses of six Eucalyptus species to water deficit. Ann Bot 100:1507–1515

Merchant A, Arndt SK, Rowell DM, Posch S, Callister A, Tausz M, Adams MA (2010) Seasonal changes in carbohydrates, cyclitols, and water relations of 3 field grown Eucalyptus species from contrasting taxonomy on a common site. Ann For Sci 67:104

Mitchell PJ, Veneklaas EJ, Lambers H, Burgess SSO (2008) Leaf water relations during summer water deficit: differential responses in turgor maintenance and variation in leaf structure among different plant communities in south-western Australia. Plant Cell Environ 31:1791–1802

Monson RK, Smith SD (1982) Seasonal water potential components of Sonoran Desert plants. Ecology 63:113–123

Moreton N, Munns R (2010) Prometheus wiki: plant water content and relative water content. CSIRO Publishing. http://prometheuswiki.publish.csiro.au/tiki-index.php?page=Plant+water+content+and+relative+water+content. Cited 2 Dec 2011

Morgan JM (1984) Osmoregulation and water stress in higher plants. Annu Rev Plant Physiol 35:299–319

Morgan JM (1992) Osmotic components and properties associated with genotypic differences in osmoregulation in wheat. Funct Plant Biol 19:67–76

Morgan JM (1995) Growth and yield of wheat lines with differing osmoregulative capacity at high soil water deficit in seasons of varying evaporative demand. Field Crop Res 40:143–152
Morgan JM (2000) Increases in grain yield of wheat by breeding for an osmoregulation gene: relationship to water supply and evaporative demand. Aust J Agric Res 51:971–978
Munns R (1988) Why measure osmotic adjustment? Aust J Plant Physiol 15:717
Munns R (2002) Comparative physiology of salt and water stress. Plant Cell Environ 25:239–250
Munns R (2011) Plant adaptations to salt and water stress: differences and commonalities. Adv Bot Res 57:1–32
Munns R, Richards RA (2007) Recent advances in breeding wheat for drought and salt stresses. In: Jenks M, Hasegawa P, Jain S (eds) Advances in molecular breeding toward drought and salt tolerant crops. Springer, Dordrecht, pp 565–586
Myers B, Neales T (1986) Osmotic adjustment, induced by drought, in seedlings of three Eucalyptus species. Funct Plant Biol 13:597–603
Nilsen ET, Orcutt DM (1996) Physiology of plants under stress, Wiley, New York
Ogawa A, Yamauchi A (2006) Root osmotic adjustment under osmotic stress in maize seedlings: 1. Transient change of growth and water relations in roots in response to osmotic stress. Plant Prod Sci 9:27–38
Olsen M, Misaghi I, Goldstein D, Hine R (1983) Water relations in cotton plants infected with Phymatotrichum. Phytopathol 73:213–216
Osório J, Osório ML, Chaves MM, Pereira JS (1998) Water deficits are more important in delaying growth than in changing patterns of carbon allocation in *Eucalyptus globulus*. Tree Physiol 18:363–373
Parker WC, Pallardy SG (1987) The influence of resaturation method and tissue type on pressure-volume analysis of Quercus alba L seedlings. J Environ Bot 38:535–549
Parry M a J, Madgwick PJ, Carvalho JFC, Andralojc PJ (2006) Prospects for increasing photosynthesis by overcoming the limitations of Rubisco. J Agric Sci 145:31
Patakas A, Noitsakis B (1999) Osmotic adjustment and partitioning of turgor responses to drought in grapevines leaves. Am J Enol Vitic 50:76
Patakas A, Nikolaou N, Zioziou E, Radoglou K, Noitsakis B (2002) The role of organic solute and ion accumulation in osmotic adjustment in drought-stressed grapevines. Plant Sci 163:361–367
Quarrie SA, Lane M (1983) Characterization of spring wheat genotypes differing in drought-induced abscisic acid accumulation. Plant Breed 34:1528–1540
Rauf S, Ahmad Sadaqat H (2008) Effect of osmotic adjustment on root length and dry matter partitioning in sunflower (Helianthus annuus L.) under drought stress. Acta Agric Scand, Sect B 58:252–260
Raven JA (1985) Tansley review no. 2. Regulation of pH and generation of osmolarity in vascular plants: a cost-benefit analysis in relation to efficiency of use of energy nitrogen and water. New Phytol 101:25–77
Raza SH, Athar HR, Ashraf M, Hameed A (2007) Glycinebetaine-induced modulation of antioxidant enzymes activities and ion accumulation in two wheat cultivars differing in salt tolerance. Environ Exp Bot 60:368–376
Rhodes D, Hanson AD (1993) Quaternary ammonium and tertiary sulfonium compounds in higher plants. Annu Rev Plant Biol 44:357–384
Richter H (1997) Water relations of plants in the field: some comments on the measurement of selected parameters. J Exp Bot 48:1–7
Sala A, Piper F, Hoch G (2010) Physiological mechanisms of drought-induced tree mortality are far from being resolved. New Phytol 186:274–281
Saliendra NZ, Meinzer FC (1991) Symplast volume, turgor, stomatal conductance and growth in relation to osmotic and elastic adjustment in droughted sugarcane. J Environ Bot 42:1251
Sánchez FJ, Manzanares M, de Andres EF, Tenorio JL, Ayerbe L (1998) Turgor maintenance, osmotic adjustment and soluble sugar and proline accumulation in 49 pea cultivars in response to water stress. Field Crop Res 59:225–235

Santakumari M, Berkowitz GA (1991) Chloroplast volume: cell water potential relationships and acclimation of photosynthesis to leaf water deficits. Photosynthesis Res 28:9–20

Serraj R, Sinclair TR (2002) Osmolyte accumulation: can it really help increase crop yield under drought conditions? Plant Cell Environ 25:333–341

Shabala S, Shabala L (2011) Ion transport and osmotic adjustment in plants and bacteria. Biomol Concepts 2:407–419

Shao H-B, Chu L-Y, Jaleel CA, Manivannan P, Panneerselvam R, Shao M-A (2009) Understanding water deficit stress-induced changes in the basic metabolism of higher plants—biotechnologically and sustainably improving agriculture and the ecoenvironment in arid regions of the globe. Crit Rev Biotechnol 29:131–151

Silva EN, Ferreira-Silva SL, Viégas RA, Silveira JAG (2010) The role of organic and inorganic solutes in the osmotic adjustment of drought-stressed Jatrophacurcas plants. Environ Exp Bot 69:279–285

Smith SD, Monson RK, Anderson JE (1997) Physiological ecology of North American desert plants, Springer, Berlin

Szota C, Farrell C, Koch JM, Lambers H, Veneklaas EJ (2011) Contrasting physiological responses of two co-occurring eucalypts to seasonal drought at restored bauxite mine sites. Tree Physiol 31:1052–1066

Tangpremsri T, Fukai S, Fischer K, Henzell R (1991) Genotypic variation in osmotic adjustment in grain sorghum. II. Relation with some growth attributes. Aust J Agric Res 42:759–767

Tardieu F, Simonneau T (1998) Variability among species of stomatal control under fluctuating soil water status and evaporative demand: modelling isohydric and anisohydric behaviours. J Exp Bot 49:419–432

Thorpe MR, Minchin PEH (1996) Mechanisms of long- and short-distance transport from sources to sinks. In: Zamski E, Schaffer AA (eds) Photoassimilate distribution in plants and crops, New York, pp 261–282

Travert S, Valerio L, Fouraste I, Boudet AM, Teulieres C (1997) Enrichment in specific soluble sugars of two Eucalyptus cell-suspension cultures by various treatments enhances their frost tolerance via a noncolligative mechanism. Plant Physiol 114:1433–1442

Trindade I, Capitão C, Dalmay T, Fevereiro MPS, Santos DMD (2010) miR398 and miR408 are up-regulated in response to water deficit in *Medicago truncatula*. Planta 231:705–716

Turner NC (1981) Techniques and experimental approaches for the measurement of plant water status. Plant Soil 58:339–366

Turner NC (1986) Adaptation to water deficits: a changing perspective. Aust J Plant Physiol 13:175

Turner DW (2006) An index of osmotic adjustment that allows comparison of its magnitude across species and experiments. Physiol Plant 127:478–482

Turner NC, Jones MM (1980) Turgor maintenance by osmotic adjustment: a review and evaluation. In: Turner NC, Kramer PJ (eds) Adaptation of plants to water and high temperature stress. Wiley, New York, pp 155–172

Urban L, Jaffrin A, Chraibi A (1993) Analysis of pressure-volume curves of leaves of *Rosa hybrida* cv. Sonia. J Environ Bot 44:605

Valliyodan B, Nguyen HT (2006) Understanding regulatory networks and engineering for enhanced drought tolerance in plants. Curr Opin Plant Biol 9:189–195

Voetberg GS, Sharp RE (1991) Growth of the maize primary root at low water potentials: III. Role of increased proline deposition in osmotic adjustment. Plant Physiol 96:1125–1130

Wang H, Siopongco J, Wade LJ, Yamauchi A (2009) Fractal analysis on root systems of rice plants in response to drought stress. Environ Exp Bot 65:338–344

Wardlaw IF (2005) Viewpoint: consideration of apoplastic water in plant organs: a reminder. Funct Plant Biol 32:561

Warren CR, Bleby T, Adams MA (2007) Changes in gas exchange versus leaf solutes as a means to cope with summer drought in Eucalyptus marginata. Oecologia 154:1–10

Warren CR, Aranda I, Cano FJ (2011) Responses to water stress of gas exchange and metabolites in Eucalyptus and Acacia spp. Plant Cell Environ 34:1609–1629

Wenkert W, Lemon E, Sinclair T (1978) Water content-potential relationship in soya bean: changes in component potentials for mature and immature leaves under field conditions. Ann Bot 42:295

Westgate M, Boyer JS (1985) Osmotic adjustment and the inhibition of leaf, root, stem and silk growth at low water potentials in maize. Planta 164:540–549

White DA, Turner NC, Galbraith JH (2000) Leaf water relations and stomatal behavior of four allopatric Eucalyptus species planted in Mediterranean southwestern Australia. Tree Physiol 20:1157–1165

Wright S (1977) The relationship between leaf water potential (Ψ leaf) and the levels of abscisic acid and ethylene in excised wheat leaves. Planta 134:183–189

Wright G, Smith R, Morgan JM (1983) Differences between two grain sorghum genotypes in adaptation to drought stress. III. physiological responses. Aust J Agric Res 34:637

Yancey PH, Clark ME, Hand SC, Bowlus RD, Somero GN (1982) Living with water stress: evolution of osmolyte systems. Science 217:1214–1222

Yazaki K, Sano Y, Fujikawa S, Nakano T, Ishida A (2010) Response to dehydration and irrigation in invasive and native saplings: osmotic adjustment versus leaf shedding. Tree Physiol 30:597–607

Zimmermann U (1978) Physics of turgor-and osmoregulation. Annu Rev Plant Physiol 29:121–148

Zur B, Boote K (1981) Changes in internal water relations and osmotic properties of leaves in maturing soybean plants. J Environ Bot 32:1181–1191

Chapter 9
Antioxidant Defenses Against Drought Stress

Iker Hernández, Jana Cela, Leonor Alegre and Sergi Munné-Bosch

Abstract Plants possess a battery of structural, physiological, biochemical, and molecular mechanisms to withstand drought periods. During drought, stomatal limitation of photosynthesis, overreduction of the photosynthetic electron transport chain, enhanced photorespiration, and many other processes may result on enhanced formation of reactive oxygen species (ROS) and other oxidizing agents. One of the most important defense mechanisms against drought is the antioxidant system, which detoxifies prooxidants such as ROS and lipid peroxyl radicals, and keeps an adequate cellular redox balance. Antioxidants may be classified in enzymatic (e.g., ascorbate peroxidases, catalases, and superoxide dismutases) or nonenzymatic (syn. low molecular weight) antioxidants (e.g., ascorbate, glutathione, carotenoids, and tocopherols). Antioxidants may scavenge ROS directly or in co-operation with other antioxidants. This co-operation between antioxidants also allows re-cycling of oxidized antioxidants. Moreover, antioxidants are key sensors of the cellular redox status, so they trigger a number of signaling events intended to keep an adequate cellular redox balance. In this chapter, the function of the most important antioxidants in plants and the role of antioxidants in cellular redox homeostasis during drought stress will be reviewed.

Abbreviations

APx	Ascorbate peroxidase
CAT	Catalase
DHA	Dehydroascorbate
DHAR	DHA reductase
Fd_{red}	Reduced ferredoxin

I. Hernández · J. Cela · L. Alegre · S. Munné-Bosch (✉)
Departament de Biologia Vegetal, Facultat de Biologia,
Universitat de Barcelona, Avinguda Diagonal 643,
08028 Barcelona Spain
e-mail: smunne@ub.edu

R. Aroca (ed.), *Plant Responses to Drought Stress*,
DOI: 10.1007/978-3-642-32653-0_9, © Springer-Verlag Berlin Heidelberg 2012

FTR	Ferredoxin-thioredoxin reductase
GRx	Glutaredoxin
GSH	Reduced glutathione
GSSG	GSH disulfide (oxidized GSH)
GST	GSH-*S*-transferase
MDHA	Monodehydroascorbate
MDHAR	MDHA reductase
NPQ	Non-photochemical quenching
NTR	NADPH-thioredoxin reductase
PRx	Peroxiredoxin
PSI/II	Photosystem I or II
ROS	Reactive oxygen species
qE	ΔpH-dependent NPQ
SOD	Superoxide dismutase
TRx	Thioredoxin

9.1 Introduction

Oxygen in aerobic organisms shows redox states between molecular oxygen (O_2) and water (H_2O). Reactive oxygen species (ROS), including $O_2^{-\cdot}$ (superoxide anion), H_2O_2 (hydrogen peroxide) and OH^- (hydroxyl radical), are partially reduced forms of oxygen that are extremely reactive and tend to completely reduce to H_2O very quickly (between milliseconds and picoseconds), thereby oxidizing lipids, proteins, sugars, nucleic acids, and other neighboring molecules. Another very reactive ROS is singlet oxygen (or $^1O_2$), the common name used foa an electronically excited state of molecular oxygen (O_2), that is usually used from light-harvesting chlorophyll molecules.

Among membrane components, poly-unsaturated fatty acids (PUFA) and proteins are the most common ROS targets. Lipid peroxides, which are products of lipid peroxidation, can in turn oxidize neighboring PUFAs, establishing a chain reaction that may lead to the dysfunction of biological membranes. Sulphur-containing amino acids, such as cysteine and methione, are also particularly prone to oxidation. The oxidation of these amino acids is sequential, from sulfhydryl to disulfide, sulfenic acid, sulfinic acid, and sulfonic acid, the first three oxidations (to disulfide, sulfenic, and sulfinic acids) being reversible. The oxidation to sulfinic acid is also reversible in some particular cases (Møller et al. 2007). After oxidation of thiol groups, protein carbonylation that occurs in lysine, arginine, proline, and threonine is the second most common protein oxidation reaction. Moreover, other aminoacids such as tryptophan and tyrosine are also common targets of ROS (Rinalducci et al. 2008; Spoel and Loake 2011). In plants, ROS additionally cause DNA base deletions, pyrimidin dimers, strand breaks, and base modifications such

Table 9.1 Main sources of ROS in mesophyll cells of drought-stressed plants

Cell compartment	Main sources of ROS	Brief description of the reaction	ROS formed
Chloroplast	PSII/PSI	Energy transfer from triplet state chlorophyll	$^1O_2$ $O_2^{-\bullet}$
	PSI (Mehler reaction)	Electron transfer to O_2 as alternative electron acceptor	
Peroxisome	Glycolate oxidase	Oxidation of glycolate from photorespiration	H_2O_2
Mitochondria	Respiratory electron transport chain	Electron transference from different complexes to O_2	$O_2^{-\bullet}$
Apoplast	NADPH oxidase	Oxidation of symplastic NADPH to generate O_2^-in the apoplast	$O_2^{-\bullet}$
	Polyamine oxidases		H_2O_2
	Class III peroxidases	Catabolism of polyamines	
		The catalytic cycle of class III peroxidases	$HOO^\bullet$, $O_2^{-\bullet}$
Ubiquitous	Fenton reaction	Generation of HO by the oxidation of transition metals	$HO^\bullet$
	Haber–Weiss cycle	Catalytic activity of transition metals in presence of O_2^- and H_2O_2	$HO^\bullet$
	SOD	Dismutation of O_2^-	H_2O_2

as alkylation and oxidation (Gill and Tuteja 2010). Moreover, products of the ROS-dependent PUFA peroxidation, such as malondialdehyde, can form adducts with DNA bases (preferentially guanine).

In addition to their deleterious effects leading to oxidative damage and destruction of several cellular components when found at high concentrations, ROS can also play an important role in cellular signaling in plant responses to environmental stresses, including drought (Møller et al. 2007). It is therefore essential that plants possess mechanisms that finely control ROS levels in various cellular compartments. The formation of ROS occurs in plants under optimum growth conditions in almost all subcellular compartments, but its production can be boosted during drought (Table 9.1). Plants contain a powerful antioxidant repertoire that is finely regulated in time and space in order to keep ROS levels under tight control.

An antioxidant is defined as a molecule that donates electrons or hydrogen atoms (i.e., has low reduction potential) to yield a radical that is either harmless or efficiently quenched by other electron donors, and the properties of which are displayed in a spatial and temporal correlation with oxidative stress (Hernández et al. 2009). Plants possess both nonenzymatic and enzymatic antioxidants. The formers donate electrons or hydrogen atoms to the oxidizing agent. The oxidized antioxidants may be converted to harmless products or they may be recycled back to the reduced antioxidant either spontaneously or through enzyme-catalyzed reactions. Enzymatic antioxidants are proteins that catalyze the scavenging of prooxidants using electrons provided by nonenzymatic antioxidants or other electron donors such as water, NAD(P)H or ferredoxin (Fd). Also, one can distinguish between primary antioxidants, which are those that scavenge prooxidants; and secondary antioxidants, those that recycle primary antioxidants. We will discuss here the origin of oxidative stress, but also at the same time the antioxidant mechanisms operating in the different cellular compartments to avoid oxidative damage during drought stress.

9.2 Oxidative Stress and Antioxidants in Chloroplasts

Chloroplasts are quantitatively and qualitatively one of the most important sources of ROS in illuminated plant cells (Foyer and Noctor 2003). During drought stress, the stomatal closure prevents the diffusion of CO_2 to the carboxylation sites, which avoids its utilization by the enzyme RuBisCO. Under this condition, NADPH and ATP are not consumed in the Calvin cycle and can over-accumulate. If this occurs, drought results in the saturation of the photosynthetic electron transport, especially when it is combined with high light or other conditions that result on excess excitation energy in chloroplasts. At the level of the photosystem II (PSII) and some recent evidence suggests also this may also occur at the level of photosystem I (PSI, cazzaniga et al. 2012), energy can be transferred from triplet state chlorophyll (excited chlorophyll; $^3Chl^*$) directly to O_2 in its basal state (triplet; $^3O_2$) to yield $^1O_2$ (Table 9.1). At the reducing side of the photosystem I (PSI), in the so-called Mehler reaction, membrane-bound photosynthetic electron transporters such as reduced ferredoxin (Fd_{red}) can transfer one electron to O_2, generating O_2^-. This is quickly converted to H_2O_2, either spontaneously or in a reaction catalyzed by superoxide dismutases (SODs; Table 9.1). In order to cope with the enhanced formation of ROS that occurs during drought in chloroplasts plants has evolved a broad spectrum of antioxidants.

9.2.1 Carotenoids: Carotenes and Xanthophylls

Carotenoids are tetraterpenes unsaturated to a different extent, synthesized, and accumulated in plastids (Fig. 9.1). In chloroplasts, carotenoids are associated to both PSI and PSII, particularly to their respective light harvesting complexes, forming pigment-protein complexes. Carotenoids are the main factors responsible for the qE component of the nonphotochemical quenching (NPQ): the quenching of chlorophyll *a* fluorescence by processes other than photochemistry, which involves mainly energy dissipation as heat. During drought the factors limiting photosynthesis provoke the accumulation of singlet state excited chlorophylls ($^1Chl^*$), increasing the probability of $^3Chl^*$ formation by intersystem crossing, and subsequently that of $^1O_2$ by energy transfer to O_2. Lutein (and other carotenoids with nine or more conjugated double bounds; Fig. 9.1) is the main compound

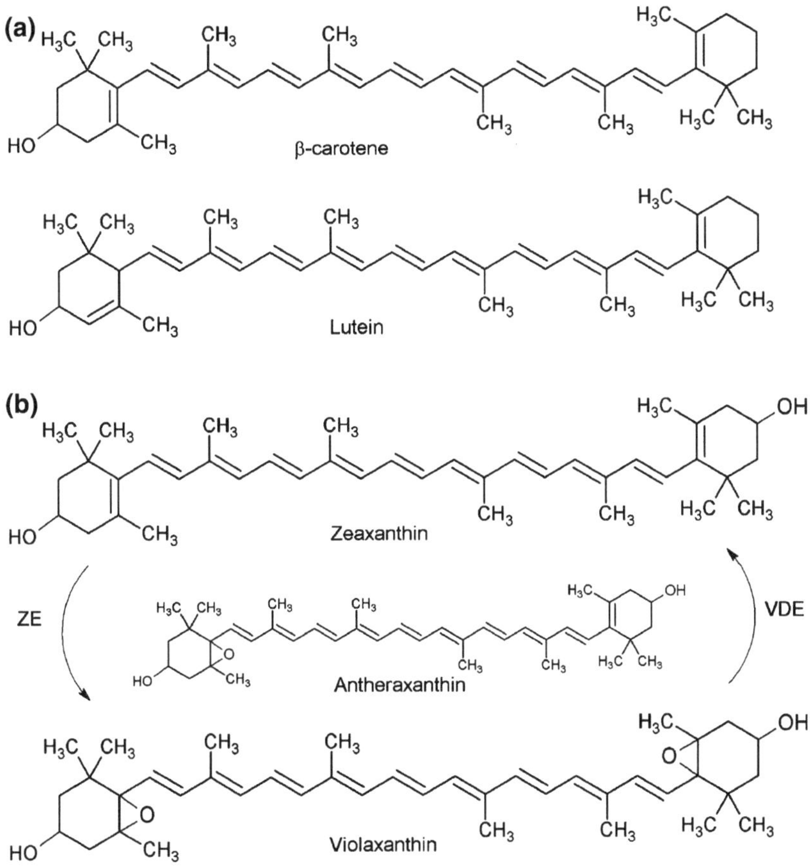

Fig. 9.1 Molecular formula of some representative nonenzymatic antioxidants in plant cells. First, some relevant carotenoids are shown (**a**). In **b**, the xanthophyll cycle is outlined: zeaxanthin is transformed into violaxanthin by the action of zeaxanthin epoxidase (*ZE*) through the intermediary antheraxanthin; the reverse reactions are catalyzed by violaxanthin de-epoxidase (*VDE*). Next, the molecular formula of tocopherols is depicted (**c**). In **d**, ascorbic acid and its oxidation products, monodehydroascorbic and dehydroascorbic acids, are shown. **e** shows the chemical formula of glutathione (γ-L-glutamyl-L-cysteinylglycine). Gallic acid is one of the most common hydrolysable tannins, while quercetin is also a representative flavonoid; both phenolics with high antioxidant capacity (**f**). Finally the most important polyamines, putrescine, spermine and spermidine, are shown in **g**

responsible for the quenching of ^{3}Chl* by harvesting its energy and releasing it as heat by thermal relaxation (Jahns and Holzwarth 2012), thereby preventing the formation of $^1O_2$. In addition, other carotenoids (especially the xanthophyll zeaxanthin) are able to quench ^{1}Chl*, yielding an excited singlet state carotenoid (for instance, ^{1}Zeaxanthin*). These excited singlet state carotenoid molecules return to their basal state by thermal relaxation as well. Similarly, most carotenoids, and particularly *β*-carotene (Fig. 9.1), quench $^1O_2$ via energy transfer and subsequent thermal relaxation. Plants have evolved a sophisticated method for the control of

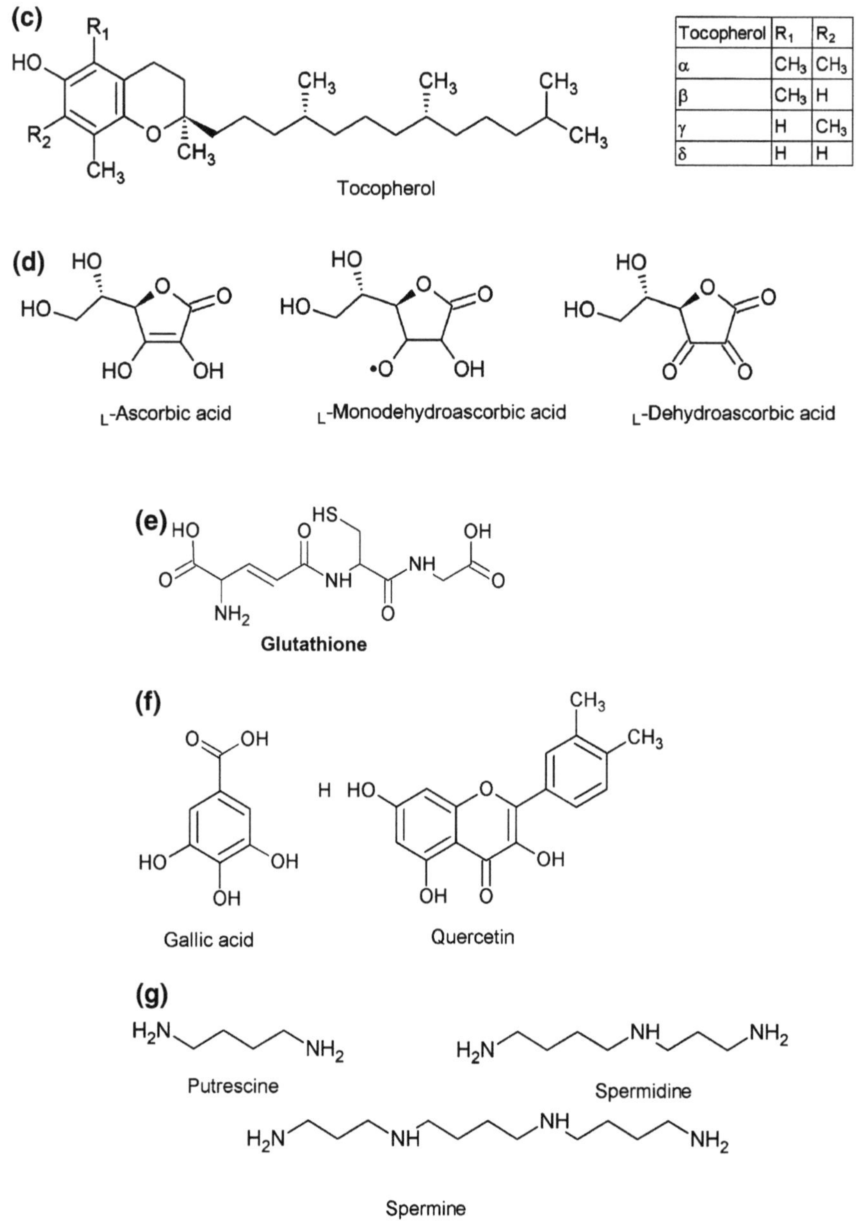

Tocopherol	R_1	R_2
α	CH_3	CH_3
β	CH_3	H
γ	H	CH_3
δ	H	H

Fig. 9.1 (continued)

the qE consisting on a series of de-epoxidation and epoxidation reactions involving different xanthophylls, commonly called as the “xanthophyll cycle” (Fig. 9.1). The de-epoxidation reactions consist on the de-epoxidation of violaxanthin to zeaxanthin through an intermediary, antheraxanthin, by the action of violaxanthin de-epoxidase (VDE) which uses electrons donated by ascorbate. The reverse reaction, the epoxidation reaction, is catalyzed by zeaxanthin epoxidase (ZE). Similar sets of reactions have been identified for lutein epoxide and lutein and for diadinoxanthin and datoxanthin as substrates for ZE and VDE, respectively, but these xanthophyll cycles are restricted to certain taxa (reviewed by Esteban et al. 2009). VDE shows highest activity when the pH in thylakoid lumen is lowest (i.e., when light is high), whereas ZE requires neutral pH in the thylakoid lumen (i.e. darkness). Violaxanthin is bound to PSII and PSI antenna complexes, and upon the activation of VDE it is replaced by zeaxanthin, which quenches $^1Chl^*$ by energy transfer and subsequent thermal relaxation, preventing the formation of $^3Chl^*$ and $^1O_2$ (reviewed by Demmig-Adams et al. 1996). Moreover, it has been also shown that carotenoids quench chemically, i.e. scavenge, free radicals such as $^1O_2$, O_2^- and lipid peroxyl radicals (Burton and Ingold 1984; Conn et al. 1992; Ramel et al. 2012). It has been generally shown that in drought-resistant plants, carotenoid levels increase under drought stress when expressed per chlorophyll unit, thus indicating an increased photoprotection per amount of light absorbed (Munné-Bosch and Alegre 2000). In addition, a number of *Arabidopsis* mutants, such as *npq1*, *npq4*, *lut2,* and *szl1*, have shown that carotenoids such as lutein and zeaxanthin are essential for the qE component of NPQ, so carotenoid-deficient plants are hypersensitive to photooxidative stress (Havaux and Kloppstech 2001). To our knowledge, the improvement of plant performance through engineering carotenoid levels or composition has not been reported so far. But enhanced tolerance to photooxidative stress, a condition tightly associated with drought stress, by such means have been achieved by many researchers. Plants with enhanced content in xanthophylls show enhanced resistance to photooxidative stress (Johnson et al. 2007). In addition, plants overexpressing β-carotene hydroxylase, which catalyzes the hydroxylation of β-type carotenes thus leading to the accumulation of zeaxanthin, are more tolerant to that stress factor (Davidson et al. 2002) and concomitantly, plants lacking simultaneously all β-type xanthophylls (such as zeaxanthin) are more sensitive to photooxidative stress than plants lacking only zeaxanthin or lutein (Pogson and Risler 2000). *Arabidopsis* mutants in which all leaf xanthophylls have been substituted by zeaxanthin show reduced levels of lipid peroxidation and enhanced tolerance to high light stress (Giuliano et al. 2008).

9.2.2 Tocopherols

Tocopherols (including α-, β-, γ-, and δ-tocopherol) are chloroplast synthesized and -located amphipathic molecules consisting on a hydrophobic prenyl chain and a hydrophilic chromanol ring (Fig. 9.1). The chromanol head is able to donate single electrons to various acceptors, yielding a resonance-stabilized tocopheroxyl radical. Some of the best electron acceptors of tocopherols in such reactions are lipid peroxyl radicals, which are converted into hydroperoxides, preventing the lipid peroxidation propagation in thylakoids. Tocopherols can also physically quench (by resonance energy transfer) and chemically scavenge $^1O_2$ in PSII reaction centers, the latter resulting in the formation of tocopherol quinone or quinone epoxides. Moreover, O_2^-, HOO,$^-$ and OH^- can be scavenged by tocopherols in vitro, although whether these reactions occur in vivo is still unknown (Nishikimi et al. 1980; Fukuzawa and Gebicki 1983). Therefore, tocopherols can play a role in drought-stressed plants by dissipating excess excitation energy during photooxidative stress. In addition, since tocopherols modulate the levels of ROS and therefore the extent of lipid peroxidation, they also modulate the accumulation of PUFA oxidation products some of which —named oxylipins—play key signaling functions during drought and other stress factors (Sattler et al. 2006; Munné-Bosch et al. 2007; Cela et al. 2011). The most clear example of a PUFA peroxidation product involved in stress signaling, including drought, is the phytohormone jasmonic acid, which is formed by the peroxidation of linolenic acid and other tri-unsaturated fatty acids (Fonseca et al. 2009; Reinbothe et al. 2009; Munemasa et al. 2011). The levels of tocopherols, in agreement with their antioxidant function, increase in plants adapted to drought (e.g., Hernández et al. 2004; Munné-Bosch and Alegre 2003). Successful efforts to improve plant performance against drought through engineering tocopherol levels and composition have been reported in the literature. For instance, a recent study showed that tobacco plants overexpressing the *Arabidopsis VTE1* gene (tocopherol cyclase) accumulated 10–30 fold wild-type α-tocopherol levels, which led to reduced lipid peroxidation, electrolyte leakage, and H_2O_2 levels, and increased chlorophyll content, under water deficit (Liu et al. 2008).

9.2.3 Superoxide Dismutases

SODs (EC 1.15.1.1) are considered to be the first enzymatic antioxidant barrier of aerobic organisms; they are the fastest enzymes known and are present in all aerobic organisms as well as in some anaerobes (McCord and Fridovich 1969). The activity of SODs in chloroplasts (and in all other organelles in which they are present), aside of detoxifying $O_2^{-\cdot}$, is of capital importance to avoid the formation of $HO^\cdot$ by the Fenton reaction and the Haber–Weiss cycle, wich consist on the formation of OH from H_2O_2 in the presence of transition metals such as Fe or Cu (Fenton reaction), that may become catalytic in the presence of $O_2^{-\cdot}$ (Haber–Weiss cycle):

$$\begin{aligned}
&\text{Fenton reaction (1):} && 38; H_2O_2 + Fe^{2+} \rightarrow OH^\bullet + OH^- + Fe^{3+}\\
&\text{Recycling the transition metal (2):} && 38; O_2^{-\bullet} + Fe^{3+} \rightarrow O_2 + Fe^{2+}\\
&\text{NetHaber - Weiss cycle (1 + 2):} && 38; H_2O_2 + O_2^{-\bullet} \rightarrow O_2 + OH^\bullet + OH^-
\end{aligned}$$

Chloroplasts show Cu/Zn- and Fe-SODs (Alscher et al. 2002). The increment in endogenous levels of chloroplastic SODs during drought has been extensively reported (e.g., Salekdeh et al. 2002; Fulda et al. 2011), although the opposite trend has been also reported (Cruz de Carvalho 2008). Furthermore, the overexpression of chloroplastic SODs has been proven a successful way to improve plant responses to different sources of oxidative stress (O_3, methyl viologen and chilling temperatures) (Perl et al. 1993, Van Camp et al. 1994, 1996). However, the overexpression of cytosolic SOD does not confer tolerance to drought, while the simultaneous overexpression of chloroplastic SODs and APX enhances drought stress tolerance (Lee et al. 2007; Faize et al. 2011). These reports highlight the complexity of the antioxidant machinery and the fact that the interaction between different antioxidants are still far from being fully understood.

9.2.4 Ascorbate, APxs, and Related Enzymes

L-Ascorbate (ascorbate herein, vitamin C) is a small carbohydrate that is present in all plant species, tissues and organs, except in dormant seeds, showing the maximum concentrations, between 20 and 300 mM, in illuminated chloroplasts (Noctor and Foyer 1998). Ascorbate can be oxidized to monodehydroascorbate radical (MDHA; Fig. 9.1) by monovalent electron transfer, for instance, to H_2O_2. This reaction can be either spontaneous or catalyzed by ascorbate peroxidases (APxs; EC.1.11.1.11). APxs are class I heme-containing peroxidases present in most subcellular compartments. Chloroplasts bear, at least, three APx isoforms: thylakoidal APx is bound to the thylakoidal membrane, while stromal APx and lumen APx are soluble proteins. In some cases (for instance *Arabidopsis*) these isoforms are encoded by different genes, but in other cases (e.g., pumpkin and spinach) a single gene encodes the different APx isoforms by alternative splicing (Mano et al. 1997; Ishikawa et al. 1997). APxs have high affinity for H_2O_2 and ascorbate, which suggests that APxs not only detoxify H_2O_2 but also control H_2O_2 levels for signaling purposes (Mittler and Poulos 2005). Overexpression of APX has been shown to increase drought tolerance, while ascorbate deficient plants such as *Arabidopsis vtc1* mutants are known to be hypersensitive to drought stress (see for instance, Pastori et al. 2003; López-Carbonell et al. 2006; Faize et al. 2011). However, the overexpression of *Escherichia coli* catalase in tobacco chloroplasts led to the suppression of chloroplast APX gene expression while increased the tolerance of the transgenic plants to drought, thus indicating that the role of APxs is more complex than simply scavenging ROS (Shikanai et al. 1998).

Beside the total ascorbate amount, the redox state of the ascorbate pool (ascorbate/total ascorbate, where total ascorbate is the sum of reduced plus oxidized forms), has been shown to change differentially in resistant and sensitive plants in response to drought, so that drought-sensitive species usually have an ascorbate pool more shifted toward its oxidized forms, MDHA and particularly DHA (Jubany-Marí et al. 2010). MDHA, the primary product of ascorbate oxidation, is a relatively stable radical; but two MDHA molecules can yield ascorbate and DHA spontaneously in aqueous solutions or in a reaction catalyzed by monodehydroascorbate reductases (MDHARs; EC 1.6.5.4). MDHARs are FAD enzymes that obtain the reduction equivalents from NADH or NADPH (2 MDHA + NAD(P)H + $H^+ \rightarrow$ 2 Ascorbate + $NADP^+$). The reduction of MDHA to ascorbate may occur also nonenzymatically with electrons provided by Fd_{red} or MDHA itself (2 MDHA $\rightarrow$ ascorbate + DHA). Computational analyses though suggest that the majority of MDHA in chloroplasts are reduced to ascorbate by MDHARs (Polle 2001). If DHA is formed it can also be enzymatically converted back to ascorbate by DHA reductases (DHARs), which use glutathione (GSH) as an electron donor.

9.2.5 Glutathione and Glutathione-Related Enzymes

Glutathione is a tripeptide (γ-glutamylcysteinyl glycine or γ-ECG; Fig. 9.1) of enzymatic biosynthesis. Although ubiquitous, GSH shows its highest concentration, between 1 and 4.5 mM, in chloroplasts (Meyer 2008). GSH is well known to act as an antioxidant and redox buffer. When the sulfhydryl groups of the cysteine residues of two GSH molecules are oxidized they form a disulfide bond between each other to yield GSSG (GSH disulfide; *syn.* oxidized GSH). GSH can directly scavenge ROS, particularly peroxides, and NO (nitric oxide), but the best known electron acceptor for such reaction is DHA.

A number of studies have reported on the dynamics of GSH levels and GSH-utilizing enzymes under drought stress in different species (e.g., Pastori and Trippi 1992; Galle et al. 2009). The set of reactions in which ascorbate is used to detoxify H_2O_2 and recycled by GSH receives the name of ascorbate-GSH cycle (Dalton et al. 1993; Jiménez et al. 1997; Fig. 9.2), which has also been shown to be involved in drought stress tolerance in several species (reviewed by Jubany-Marí et al. 2010). Furthermore, great efforts have been devoted to understand the water–water cycle, which couples part of the linear photosynthetic electron transport with the ascorbate-GSH cycle to dissipate excess energy (reviewed by Asada 1999). This cycle starts with the photolysis of water in PSII reaction centers and ends with the water resulting from the ascorbate-GSH cycle. The key point of the cycle is the transfer of electrons in PSI from Fd_{red} to O_2 to yield O_2^-, so that NADPH resulting from complete linear electron transport is not formed. Instead, O_2^- is formed, which by the action of SODs quickly disproportionates to H_2O_2, which is scavenged by ascorbate with the involvement of APxs to yield MDHA and H_2O. The oxidized forms of ascorbate are then recycled to reduced ascorbate by the means of GSH through the action of DHARs, MDHARs (see previous section), and GRs. GRs recycle GSSG to GSH with electrons provided by NADPH, which is reduced by water after its split at the level of PSII. Thus, GRs act as the link between the primary electron donor, water (NADPH), and antioxidant recycling. The ability of GRs to keep the GSH/GSSG redox balance under control has been shown to be determinant for the tolerance of the plant to drought stress (Torres-Franklin et al. 2007; Cruz de Carvalho and Contaour-Ansel 2008).

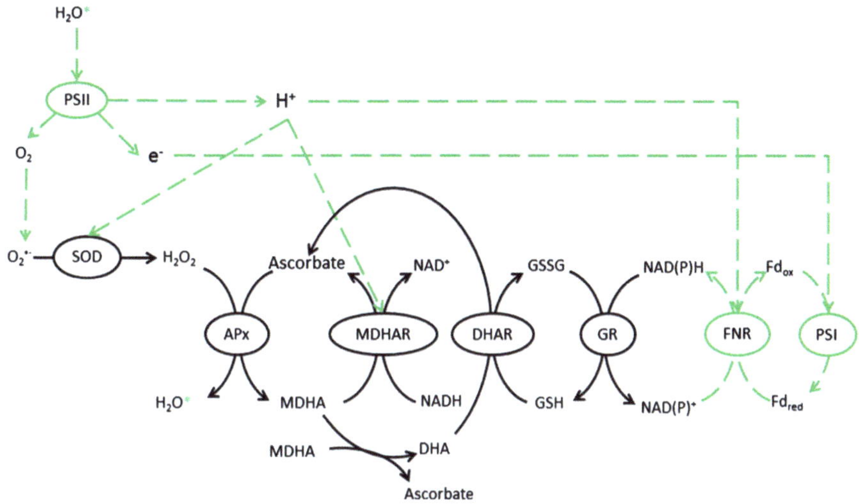

Fig. 9.2 The ascorbate–glutathione pathway and its variation, the water–water cycle (green dashed lines and circles). Water is split in the oxygen evolving complex of the PSII, yielding molecular oxygen (O_2), electrons (e^-), and protons (H^+). When the photosynthetic electron transport chain is over-reduced, the electrons may be transferred to O_2 yielding superoxide anion (O_2^-). O_2^- is quickly disproportionated to H_2O_2 (hydrogen peroxide) by the action of superoxide dismutases (*SOD*) in many different subcellular compartments. H_2O_2 is detoxified by ascorbate peroxidases (*APxs*) with electrons from ascorbate, yielding monodehydroascorbate (*MDHA*) and water. MDHA can be re-cycled to ascorbate by the action of MDHA reductases (MDHAR) or spontaneously yield dehydroascorbate (*DHA*), which is re-cycled to ascorbate by DHA reductases (*DHARs*), and ascorbate. DHARs obtain the reducing equivalents from glutathione (GSH; GSSG stands for oxidized GSH). GSSG is reduced back to GSH with reducing equivalents from NAD(P)H by the action of GSH reductases (GR), which obtain the reducing equivalents from NAD(P)H. In chloroplasts, $NAD(P)^+$ is reduced to NAD(P)H with electron from reduced ferredoxin (Fd_{red}), which obtains the electrons from the photosynthetic electron transport chain, i.e. from the electrons generated by splitting water. The stoichiometry of the reactions is not adjusted in this chart, but it is shown in Asada (1999) that the substrate of the whole set of reactions is water, and the only product of the net set of reactions is water (asterisks)

In addition, GSH participates in the posttranscriptional modification of several proteins. Glutathionylation is a type of posttranscriptional modification that consists on the formation of a mixed disulfide bond between a cysteine residue of a protein and GSH. Glutathionylation can occur spontaneously in the presence of GSSG, or in the presence of GSH and ROS. Thus, the enhanced formation of ROS, as it occurs in chloroplasts during drought, promotes glutathionylation (Fig. 9.3). The glutathionylation of proteins is a well-known regulatory mechanism, and a number of glutathionylation targets have been described in chloroplasts so far. These include the Calvin cycle enzymes triose phosphate isomerase, FBP aldolase, phosphoglycerate kinase, and ribose-5-phosphate isomerase (Dixon et al. 2002; Ito et al. 2003; Mohr et al. 1999; Michelet et al. 2008), and some thioredoxins (TRxs), which in turn are known to regulate the activity of several enzymes of the same cycle (reviewed by Rouhier et al. 2008).

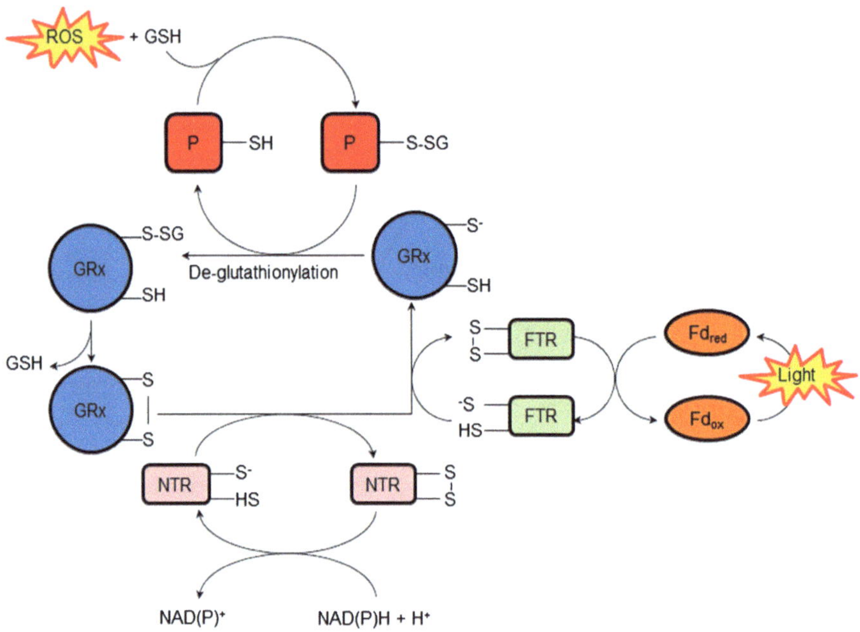

Fig. 9.3 De/glutathionylation of proteins and the role of the ferredoxin-thioredoxin and NADPH-thioredoxin systems in the reduction of oxidized glutaredoxins. Fd stands for ferredoxin (Fd_{red}, reduced ferredoxin; Fd_{ox}, oxidized ferredoxin); FTR, for Fd-thioredoxin reductase; GSH, for reduced glutathione; NTR, for NADPH-thioredoxin reductase; P indicates a protein susceptible of de/glutathionylation; ROS stands for reactive oxygen species

Moreover, GSH can be oxidized in reactions catalyzed by a number of enzymes including glutaredoxins (GRxs), GSH-*S*-transferases (GSTs), peroxiredoxins (PRxs), and TRxs. GSH peroxidase (GPx) is a general term for the enzymes that catalyze the reduction of H_2O_2 and organic hydroperoxides to water or their respective alcohols using GSH as electron donor ($H_2O_2 + 2$ GSH $\rightarrow 2$ H_2O + GSSG). However, plant GPxs show weak affinity for GSH, with K_m values over the physiological GSH concentration (Herbette et al. 2002). Thus, GPxs are nowadays included in the thiol peroxidase family of proteins (Sztajer et al. 2001, Herbette et al. 2002; Maiorino et al. 2007; Navrot et al. 2006). Still, GSH-dependent peroxidase activity exists in plant cells and is carried out by other enzymes such as GRxs, GSTs, or some PRxs (Rouhier et al. 2008). GRxs are small proteins that catalyze the reduction of other proteins or mixed disulfides and that are reduced nonenzymatically by GSH itself. Among other functions, GRxs can act as antioxidants by reducing directly peroxides, DHA, or TRxs, which in turn reduce H_2O_2 and alkyl hydroperoxides (Lee et al. 2002; Rouhier et al. 2001). Moreover, some GRxs are able to reduce methionine sulfoxide, the product of methionine oxidation by some ROS, back to methionine thus providing an antioxidant mechanism (Stadtman 2006). Oxidized GRxs can be also reduced back to their thiol form with electrons donated by Fd_{red} through the Fd-TRx (FTR;

Table 9.2 Overview of the antioxidant mechanisms in different subcellular compartments of plant cells under drought stress. Dots indicate major antioxidant mechanisms; dots in brackets indicate presence but not of major relevance, or the importance of which are unknown; crosses indicate the absence of the mechanism and zeros indicate data unknown; and asterisks indicate enzymatic activities rather than particular enzymes. CAT; catalase; SOD, superoxide dismutase; APx, ascorbate peroxidase; MDHAR, monodehydroascorbate reductase; DHAR, dehydroascorbate reductase; ER, endoplasmic reticulum, and Golgi apparatus; GPx, glutathione peroxidase GR, glutathione reductase; GRx, glutaredoxin; GSH, glutathione; GST, glutathione-*S*-transferase; NADPH ox., NADPH oxidase; POx, peroxidase; PRx, peroxiredoxin; TRx, thioredoxin

	Chloroplast	Peroxisome	Mitochondria	Apoplast	Nucleus	Cytosol	ER
Tocopherols	•	X	X	X	X	X	X
Carotenoids	•	X	X	X	X	X	X
CAT	X	•	X	X	X	X	X
SOD	•	(•)	•	•	(•)	•	O
Ascorbate	•	(•)	•	•	(•)	•	(•)
APx	•	(•)	•	(•)	X	•	(•)
MDHAR	•	(•)	•	X	X	•	X
DHAR*	•	(•)	•	X	X	•	X
GSH	•	(•)	•	(•)	(•)	•	(•)
GR	•	(•)	•	X	O	•	O
GST	(•)	(•)	(•)	(•)	(•)	•	(•)
GPx*	(•)	(•)	(•)	(•)	X	(•)	(•)
GRx	(•)	X	(•)	O	(•)	(•)	(•)
TRx	(•)	X	(•)	(•)	(•)	(•)	(•)
PRx	(•)	X	(•)	X	(•)	(•)	X
Class III POx	X	X	X	•	X	X	(•)
NADPH ox.	X	X	X	•	X	X	X
Phenolics	(•)	(•)	(•)	•	(•)	(•)	O
Polyamines	(•)	O	(•)	•	(•)	(•)	O

Fig. 9.3, Dai et al. 2000). In addition, GRxs are involved in protein deglutathionylation. In support of a role of GRxs in plant resistance to drought, *Arabidopsis* plants devoid of a chloroplastic GRx, GRXS14, are hypersensitive to oxidative stress (Feng et al. 2006), but no study has reported up to date on enhanced tolerance to drought itself through engineering GRx levels. Finally, GSTs are a large family of proteins that catalyze the conjugation of GSH to electrophilic substrates. Although they are most common in the cytosol (see Sect. 9.6), they are also present in chloroplasts (Table 9.2). GSTs can transfer GSH to organic hydroperoxides such as lipid peroxides; thus exerting GSH-dependent peroxidase activity (GPx). Moreover, some GSTs have been shown to have GSH-dependent DHAR activity, thus catalyzing the recycling DHA back to its reduced form ascorbate. Two recent studies have shown that overexpressing GSTs of unknown function from soybean and *Prosopis juliflora* resulted on enhanced drought tolerance in tobacco plants (Suja et al. 2010; Ji et al. 2010). In agreement, it has been recently shown that the overexpression of a chloroplast-localized GST confers tolerance to osmotic stress to the transgenic plants (George et al. 2010).

9.2.6 Peroxiredoxins: A Diverse Subfamily of Thiol Peroxidases

Thiol peroxidases are a large family of nonheme peroxidases that catalyze the reduction of peroxides using catalytic cysteine residues and thiol-containing proteins as reductants (in contrast to heme-peroxidases, which use Fe in their catalytic mechanism). Thiol peroxidases include five types of peroxiredoxins (PRxs; EC 1.11.1.15) that vary in their sequence and mechanism of action; and nonselenium GPxs. PRxs are ubiquitous thiol-dependent peroxidases that reduce a wide range of peroxides from H_2O_2 to organic hydroperoxides. PRx lack a prosthetic group, so the oxidation of PRx thiols to disulfides, sulfenic, or even to sulfinic acids yield inactive forms of PRxs that need to be recycled back to their active, reduced form by additional electron donors such as thioredoxins (TRxs) or GRx. It is extremely difficult to determine unequivocally the reductant that recycles oxidized PRxs in vivo, since PRxs co-localize with a high number of possible electron donors (e.g., 26 TRxs and 31 GRxs). The mutation of an *Arabidopsis* 2-cysteine PRxs, a class of chloroplastic PRxs, results in impaired photosynthesis and accumulation of damaged proteins in chloroplasts. This suggests that these enzymes protect chloroplast proteins against photooxidative damage (Baier and Dietz 1999; Dietz et al. 2002; Baier et al. 2000). Potato mutants devoid of CDSP35, a chloroplastic TRx, show strong overoxidation of the 2-cysteine PRx pool. These plants are hypersensitive to drought stress or methyl viologen treatment (a promoter of O_2^- formation in chloroplasts), and show enhanced lipid peroxidation under those conditions which suggests that CDSP35 reduces oxidized 2-cysteine PRxs that are oxidized to terminate the lipid peroxidation reaction (Broin and Rey 2003). In addition, it has been suggested that 2-cysteine PRxs, with the involvement of TRxs, may take over ascorbate as primary antioxidant in the water–water cycle (Rey et al. 2005; Dietz et al. 2006, Vieira Dos Santos and Rey 2005).

9.2.7 TRxs: A Redox Regulatory Hub

TRxs are ubiquitous small proteins with a redox-active dithiol/disulfide group in their structure that reduce disulfide groups of other proteins to dithiols (TRx-$(SH)_2$ + Protein-S_2 → TRx-S_2 + Protein-$(SH)_2$). TRxs show a CxxC motif (being x any amino acid and C a cysteine), that forms an intermolecular disulfide bond when oxidized, but they do not show further sequence or structural homology between each other (Meyer and Hell 2005). Plants have a great amount of different TRxs (42 Trx genes in the *Arabidopsis thaliana* genome) that are found in plastids, cytosol, nucleus, mitochondria, and the apoplast (Meyer and Hell 2005). Chloroplasts bear four TRx types: TRx *f*, *m*, x and *y*, (Meyer and Hell 2005). In chloroplasts, oxidized TRxs are recycled by TRx reductases, which ultimately use the reduction equivalents from Fd_{red} (Fd-dependent TRx reductases, FTR; Fig. 9.3). Although TRxs do not play an antioxidant function sensu stricto, they display important redox properties that enable them to fulfill a variety of functions. There have been identified over 500 TRx targets in oxygenic photosynthetic organisms, including enzymes capital for plant responses to drought such as Calvin cycle enzymes (e.g., RuBisCO, fructose-1,6-bisphosphatase, and glyceraldehyde-3-phosphate dehydrogenase), antioxidant enzymes (e.g. CATs, SODs, and MDHAR) and proteins involved in photosynthetic electron transport and light harvesting (e.g., LHCIIb, Fd and psaK) (Montrichard et al. 2009). TRxs, together with GSH and GRxs, are the major thiol-based regulatory systems in plants. The expression of some plastidic TRxs has been shown to be upregulated during drought (Rey et al. 1998). In agreement, transgenic plants lacking a plastidic TRx show enhanced sensitivity to photooxidative stress and in some cases this protective effect is exerted by recycling oxidized PRxs (Broin et al. 2002).

9.3 Photorespiratory H_2O_2 Production in Peroxisomes

Peroxisomes are quantitatively the most important source of ROS, particularly H_2O_2, in illuminated plant cells (Foyer and Noctor 2003). Under drought stress, the stomatal limitation of photosynthesis reduces the availability of CO_2 thus promoting the oxygenase activity of RuBisCO (in chloroplasts), which yields 2-phosphoglycolate. In the photorespiratory cycle, glycolate is oxidized to glyoxylate by the action of glycolate oxidase, producing H_2O_2 (Foyer and Noctor 2003; Table 9.1). In addition, there are other sources of ROS in peroxisomes (e.g., xanthine oxidase and fatty acid β-oxidation), but their relevance under drought stress, if any, is unknown. As in other subcellular compartments, the presence of O_2^- (for instance, generated by xanthine oxidase) together with transition metals in peroxisomes can lead to the formation of HO by the Fenton reaction and the Haber–Weiss cycle. To avoid the formation of $HO^\cdot$ or $O_2^{-\cdot}$ is quickly disproportionated to H_2O_2 by peroxisomal Mn-SODs (Sandalio et al. 1987). However, O_2^- is not the main ROS formed in peroxisomes during drought. Most of the photorespiratory H_2O_2 is scavenged by catalases (CATs). Still, it has been shown that this H_2O_2 can diffuse to other subcellular locations such as the nucleus and cytosol, and play a key role in intracellular signaling during acclimation to high light stress, a process that is tightly associated to drought (Vanderauwera et al. 2005).

9.3.1 *Catalases: The Main Mechanism for Photorespiratory H_2O_2 removal*

Catalases (CATs, *syn.* hydroperoxydases; EC 1.11.1.6) are enzymes that are ubiquitous among aerobic organisms (Feierabend 2005). CAT activity consists on reducing two H_2O_2 molecules to two molecules of H_2O and O_2 ($2\ H_2O_2 \rightarrow 2\ H_2O + O_2$). There are three main CAT types (nonheme Mn CATs, bifunctional

CATs (catalase-peroxidase), and monofunctional CATs), but only monofunctional CATs (they are not monofunctional strictly though) can be found in land plants (Klotz and Loewen 2003; Nicholls et al. 2001, Carpena et al. 2003). The mechanism of CAT activity lays on the heme group they bear, and since the heme group alone can do it, many heme-containing proteins such as methemoglobin, metmyoglobin, cytochrome *c* oxidase and chloroperoxidases exert CAT activity at very low rates (Keilin and Hartreef 1950, 1955; Bickar et al. 1982; Sun et al. 1994; Paco et al. 2009). Plant CATs show high turnover but low affinity toward H_2O_2, which makes them optimum for gross removal of H_2O_2 (Nicholls et al. 2001). It has been shown that downregulation of CAT gene expression leads to hypersensitivity to drought and other stress factors (reviewed by Smirnoff 2005).

9.3.2 Other Antioxidants in Peroxisomes

It has been also shown that the ascorbate-GSH cycle (Fig. 9.2), including ascorbate and GSH themselves as well as APxs, MDHARs, DHARs, and GRs, is fully active in peroxisomes, thereby providing an additional antioxidative protection to this organelle under drought stress (Jiménez et al. 1997). Furthermore, recent studies have shown that the expression of drought-responsive genes in plants with impaired root peroxisomal polyamine oxidase is altered, suggesting that polyamines (Fig. 9.1) may play a role in keeping ROS levels under control (Kamada-Nobusada et al. 2008).

9.4 Mitochondrial Respiration, Oxidative Stress, and Antioxidants

Aerobic metabolism leads to the production of ROS also in mitochondria, which may be the main ROS sources in plant cells in the dark. Complexes I and III of the mitochondrial electron transport chain are the main sources of ROS in these organelles (Table 9.1). The ubisemiquinone intermediary formed in these complexes can transfer a single electron to O_2 to yield O_2^- when the electrical and pH gradients are too steep and oxidized electron acceptors are not available. The main ROS formed in mitochondria is O_2^-, but it is quickly disproportionated to H_2O_2 by the action of mitochondrial Mn-SODs. Mitochondrial production of ROS under normal conditions is about 2–6 % of the consumed O_2, and in many cases increases during drought (Bartoli et al. 2004). It has been also shown that in some cases mitochondrial respiration decreases during drought, but, nevertheless, the ratio photosynthesis/respiration decreases almost invariably during drought (reviewed by Atkin and Macherel 2009). It is noteworthy that on the other hand, the respiration rates in roots of drought-stressed plants decrease (or transiently increase and then decrease) in most studies performed to date, a trend that has been ascribed to a substrate limitation (reviewed by Atkin and Macherel 2009).

Plant mitochondria possess several energy-dissipating mechanisms such as the ATP-sensitive plant mitochondrial potassium channel, the plant uncoupling proteins, the rotenone-insensitive type II NAD(P)H dehydrogenases, and the alternative oxidase (for review, see Atkin and Macherel 2009, and Millar et al. 2011). It is out of the scope of this chapter to review ROS formation avoidance mechanisms, but in mitochondria these mechanisms have been proven of capital importance in avoiding drought-induced oxidative stress. For instance, the overexpression of an *Arabidopsis* uncoupling protein (AtUCP) results in enhanced tolerance to drought (Begcy et al. 2011) and *Arabidopsis* plants devoid of alternative oxidase 1a (AOX1a) are hypersensitive to the combination of water deficit and excess light, and show altered expression of genes involved in the chloroplastic and mitochondrial antioxidant machineries (Giraud et al. 2008). Still, despite the ROS formation-avoidance mechanisms existing in mitochondria, this organelle is one of the most important sources of ROS, and it can be quantitatively the most important one in the dark.

9.4.1 Antioxidant Mechanisms Operating in Mitochondria Under Drought

As mentioned before, Mn-SODs disproportionate O_2^- to H_2O_2 thus preventing the Fenton reaction and the Haber–Weiss cycle, and mitochondria bear a fully operative ascorbate-GSH cycle, including ascorbate, GSH, APxs, MDHARs, DHARs, and GRs, that quenches H_2O_2 produced by the dismutation of O_2^-. In addition, as in chloroplasts, it has been suggested that PRxs, with the involvement of TRxs for the re-cycling of oxidized PRxs, may take over ascorbate in such cycle.

GRxs are also present in mitochondria. In other organelles, it is known that GRxs are able to reduce methionine sulfoxides and to catalyze protein de-glutathionylation, but it is so far unknown whether or not mitochondrial GRxs may fulfill these or any other function in plant responses to drought.

Polyamines (Fig. 9.1) have been also shown to be present in mitochondria, particularly bound to the membrane fraction (Votyakova et al. 1999). Liu et al. (2004) showed that during PEG-induced osmotic stress polyamine conjugation increases in wheat seedlings, and suggested that this phenomenon may be associated with an improved performance of mitochondrial membrane ATPase activity. Aside of this study there is no report on the role of mitochondrial polyamines in plant responses to drought.

9.5 Oxidative Stress and Antioxidant Defenses in the Apoplast

The production of ROS in the apoplast is boosted during plant responses to drought. The primary ROS formed in the apoplast under drought stress is O_2^-, formed by the action of plant NADPH oxidases (Table 9.1). H_2O_2 in the apoplast induces the opening of plasma membrane Ca^{2+} channels in guard cells, which results in increased cytosolic Ca^{2+} levels and, ultimately, in stomatal closure. Thus, the production of H_2O_2 in the apoplast is key for ABA-mediated stomatal closure (Pei et al. 2000), and therefore for plant responses to drought, since stomatal function is the main mechanism for plants to regulate transpiration.

The catabolism of polyamines occurs mainly in the apoplast by the means of apoplastic polyamine oxidases (PAOs; Table 9.1), and the action of PAOs generates H_2O_2 (Moschou et al. 2008). It has been recently found that drought, by the means of ABA signaling, induces polyamines export to the apoplast, where these compounds are degraded producing H_2O_2 that may act as a cell signal for drought acclimation (Toumi et al. 2010). Plants possess also apoplastic SOD isoenzymes that dismutate O_2^- into H_2O_2.

9.5.1 Plant NADPH Oxidases

The mammalian respiratory burst oxidase multiproteic complex is a nonheme peroxidase composed of a membrane-bound NADPH binding flavocytochrome b_{558} and a number of cytosolic accessory proteins. The membrane-bound flavocytochrome b_{558} is comprised of two peptides: one of them, gp91^{phox}, contains all necessary elements to bind NADPH and to transfer one electron from it to O_2 thereby yielding O_2^-. Plant NADPH oxidases (also named Rboh after *respiratory burst oxidase homologue*) encode proteins homolog to gp91^{phox} that are sufficient to transport electrons from NADPH to O_2 and generate O_2^- (1998; Torres et al. 1998; Foreman et al. 2003). Thus, plant NADPH oxidases are membrane proteins with the catalytic domain in the apoplast and a cytosolic N terminus end that contains two Ca^{2+} binding domains (EF-hand motifs) (Keller et al. 1998, Torres et al. 1998). It is well documented that ROS generation by plant NADPH oxidases in the apoplast increases during drought (see for example Duan et al. 2009). Plant NADPH oxidases are involved in plant responses to drought by generating the H_2O_2 in the apoplast necessary for the stomatal function and monolignol cross-linking (see Marjamaa et al. 2009 for review).

9.5.2 Lignification: A Structural Modification Depending on Free Radicals

Lignin is a highly branched polymer of phenylpropanoid units (monolignols; e.g., *p*-coumaryl, coniferyl, and sinapyl alcohols) cross-linked by oxidative coupling, which is deposited in secondary cell walls (for review, see Vanholme et al. 2010). The high evaporative demand that occurs during drought results in highly negative pressures in xylem cell walls that may result in cavitation or in the collapse (implosion) of the xylem vessels (Hacke et al. 2001). It is widely accepted that lignification reduces xylem vulnerability by strengthening the secondary cell walls of xylem vessels (Raven 1987; Cochard et al. 2004).

The cross-linking of monolignols requires a monolignol radical that can be formed by the action of class III peroxidases (Vanholme et al. 2010). These enzymes are encoded by large multigene families (for instance, there are 73 class III peroxidases encoded in the *Arabidopsis* genome), that show variable sequence homologies that are highest in their active centers. For their regular peroxidase activity ("catalytic cycle"), class III peroxidases can use monolignols as substrates to reduce H_2O_2, yielding monolignol radicals that subsequently polymerize ($H_2O_2 + 2AH \rightarrow 2H_2O + 2A^{\bullet}$). Class III peroxidases, aside of their catalytic cycle, can catalyze a set of reactions so-called "hydroxylic cycle". In this cycle, the interconversion between several redox states of the heme group leads to the generation of $HO^{\bullet}$ and $HOO^{\bullet}$ (perhydroxyl radical): two of the most reactive ROS (Liszkay et al. 2003; Marjamaa et al. 2009). In addition, the downstream modification of peroxidase-oxidized products can generate O_2^- and subsequently H_2O_2. Thus, depending on the substrates and reaction conditions, class III peroxidases can scavenge H_2O_2 and $O_2^{-\cdot}$, or generate H_2O_2, $O_2^{-\cdot}$, $HO^{\cdot}$ or $HOO^{\cdot}$, aside of generating a substrate radical such as monolignol radicals (Møller and McPherson 1998; Caliskan and Cuming 1998; Barceló et al. 2002). Llorente et al. (2002) reported that AtPrx03, and *Arabidopsis* class III peroxidase, was induced by cold stress and that plants overexpressing this peroxidase showed increased tolerance to dehydration and salt stress, most likely due to improved lignification.

9.5.3 Other Antioxidants Operating in the Apoplast

Ascorbate is present in the apoplast, while GSH is either absent or at very low concentrations. The enzymes required for re-cycling oxidized ascorbate are also absent in the apoplast (Hernández et al. 2001; Table 9.2). Still, specific transporters exchange ascorbate from the symplast by DHA from the apoplast, which once in the symplast can undergo re-cycling by the ascorbate-GSH cycle, this way providing a mechanism for regenerating ascorbate oxidized in the apoplast (Horemans et al. 2000).

9.6 Other Organelles

It is obvious that the nucleus has an outstanding importance in plant responses to drought since it hosts DNA and gene expression machinery. ROS can induce a number of alterations in nucleic acids such as deletions, pyrimidin dimers, strand breaks, nucleic acid-protein crosslinks, nucleic-acid MDA adducts, sister chromatid exchange, and base modifications such as alkylation and oxidation (Sohal and Weindruch 1996; Markesbery and Lovell 2006, Roldán-Arjona and Ariza 2009; Gill and Tuteja 2010). Several proteins (e.g., maturases, late embryogenesis abundant (LEA) proteins, and helicases) are involved in maintaining nuclear functions during desiccation by other means different from an antioxidative protection. However, the overexpression of a citrus LEA protein in tobacco led to the inhibition of lipid peroxidation of the transgenic plants under cold stress (Hara et al. 2003), so an antioxidant role of these proteins cannot be discarded. In nuclei one can find a broad repertoire of antioxidants (Table 9.2), including the complete set for the ascorbate-GHS cycle, GPx activity, GSTs, GRx, TRxs, PRxs, phenolic compounds (including flavonoids), and polyamines. Still, little is known about their possible antioxidant roles in nuclei of drought-stressed plants. The expression of a nuclear-located PRx (1-C PRx) from *Xerophyta viscosa*, a resurrection plant that keeps its viability at relative leaf water content as low as 5 %, was shown to be upregulated by dehydration, heat, high light stress, salt stress, and ABA (Mowla et al. 2002). Furthermore, the promoter of the *Arabidopsis* nuclear-located 1-C PRx shows ABA-responsive elements (Haslekås et al. 2003), so PRxs may pose an antioxidant mechanism in nuclei of drought-stressed plants.

It is worth to highlight that protein folding during drought stress is severely affected, and a number of chloroplast chaperones, including sulfhydryl oxidases, have been proven of capital importance in maintaining the correct functionality of many proteins under decreasing water potential (Wang et al. 2004; Stengel et al. 2010). The endoplasmic reticulum and Golgi apparatus host many antioxidant systems as well (see Table 9.1), but their role in plant responses to drought is still to be determined.

Finally, several sources of ROS, such as cytochromes P450 are well described in the cytosol of plant cells (Lewis 2002). However, it is largely unknown whether the production of ROS by these sources is enhanced during drought. Nevertheless, some of the ROS formed in many subcellular compartments (mainly H_2O_2 formed in peroxisomes, chloroplast, and mitochondria) can diffuse to the cytosol. Since ROS formation in most of these organelles is boosted during drought, ROS levels in the cytosol may increase as well during drought if ROS-scavenging mechanisms in the source organelles fail. Plant cell cytosol accounts for a broad array of antioxidants, including the complete set of elements of the ascorbate-GSH cycle (ascorbate, glutathione, APxs, MDHARs, DHARs, and GRs), GPx activity, GRxs, TRxs, PRxs, phenolic compounds, and polyamines (Table 9.2). Miao et al. (2006) showed that *Arabidopsis* cytosolic nonselenium GPx, AtGPX3, is essential for H_2O_2 scavenging and ABA-dependent and H_2O_2-mediated stomatal closure, and concomitantly plants overexpressing AtGPX3 are more tolerant to water deficit. But as mentioned in Sect. 9.2, plant GPxs show little affinity for GSH, and a higher affinity for other thiol-containing compounds such as TRxs, which might serve as an electron donor. Indeed, a number of TRxs have been reported in the cytosol of plant cells. They might reduce oxidized PRxs by using electrons from NADPH. Cytosolic PRxs are a target of GRxs in poplar (Rouhier et al. 2001). PRx in the cytosol of plant cells exert peroxidase activity by using reducing equivalents from GRxs or TRxs, although their importance in plant responses to drought is largely unknown. Furthermore, most GSTs are localized in the cytosol (Dixon et al. 2009). The overexpression of GSTs has been successfully applied to improve plant performance during drought but, although predicted to be cytosolic in some cases, the subcellular localization of the ectopically overexpressed GST is not detailed in these reports (e.g., Ji et al. 2010; Jha et al. 2011). Similarly, other putative antioxidants such as polyamines and phenolic compounds such as flavonoids are known to be present in the cytosol of plant cells. However, their importance in plant responses to drought stress is still unknown.

9.7 Concluding Remarks

We have discussed here the origin of oxidative stress and the antioxidant mechanisms operating in different cellular compartments to avoid oxidative damage during drought stress. Most of the research performed thus far has been focused on better understanding oxidative stress and antioxidant protection mechanisms operating in chloroplasts, organelles with an extraordinary battery of antioxidants. It should be noted that several antioxidants, such as carotenoids and tocopherols, have been exclusively found in these organelles, which suggests that due to their photosynthetic function makes them require higher antioxidant protection compared to other organelles. Peroxisomes and mitochondria due to photorespiration and respiration, respectively, also appear to be organelles producing high amounts of ROS that contain at the same time an important set of antioxidant defenses. It should be noted, however, that the set of antioxidants required in these organelles will be more modest. This does not mean, however, that these organelles or others producing ROS at even lowest rates do not need antioxidants. The fact that ROS are used by aerobic organisms for cellular signaling forces these organisms, including plants, to finely modulate ROS levels in time and space with a complex array of antioxidants in all cellular compartments, as it has been summarized in this chapter.

Transgenic approaches to improve plant performance to drought through engineering antioxidant mechanisms have been proven useful. However, in many cases the output is far from what was expected, indicating our understanding of the antioxidant machinery as it works during drought stress is still far from being completely understood. A paradigmatic example is the case of the *Arabidopsis apx1/cat2* (CAT and cytosolic APx) double mutants. *apx1* and *cat2* single mutants are hypersensitive to oxidative stress, while *apx1/cat2* double mutants are more tolerant (Rizhsky et al. 2002). The double mutants show the DNA damage response constitutively activated, which contributes to the stress tolerance (Vanderauwera et al. 2011). Future research will undoubtedly be directed to better understand this fine tuning of ROS levels in different subcellular compartments and the relative contribution of different ROS in signaling and oxidative damage in drought-stressed plants.

References

Alscher RG, Erturk N, Heath LS (2002) Role of superoxide dismutases (SODs) in controlling oxidative stress in plants. J Exp Bot 53:1331–1341

Asada K (1999) The water–water cycle in chloroplasts: scavenging of active oxygens and dissipation of excess photons. Annu Rev Plant Physiol Plant Mol Biol 50:601–639

Atkin OK, Macherel D (2009) The crucial role of plant mitochondria in orchestrating drought tolerance. Ann Bot 103:581–597

Baier M, Dietz KJ (1999) Protective function of chloroplast 2- cysteine peroxiredoxin in photosynthesis. Evidence from transgenic Arabidopsis. Plant Physiol 119:1407–1414

Baier M, Noctor G, Foyer CH, Dietz KJ (2000) Antisense suppression of 2-cysteine peroxiredoxin in Arabidopsis specifically enhances the activities and expression of enzymes associated with ascorbate metabolism but not glutathione metabolism. Plant Physiol 124:823–832

Barceló AR, Pomar F, Ferrer MA, Martínez P, Ballesta MC, Pedreño MA (2002) In situ characterization of a NO-sensitive peroxidase in the lignifying xylem of *Zinnia elegans*. Physiol Plant 114:33–40

Bartoli CG, Gómez F, Martínez DE, Guiamet JJ (2004) Mitochondria are the main target for oxidative damage in leaves of wheat (*Triticum aestivum* L.). J Exp Bot 55:1663–1669

Begcy K, Mariano ED, Mattiello L, Nunes AV, Mazzafera P, Maia IG, Menossi M (2011) An Arabidopsis mitochondrial uncoupling protein confers tolerance to drought and salt stress in transgenic tobacco plants. PLoS ONE 6:e23776

Bickar D, Bonaventura J, Bonaventura C (1982) Cytochrome c oxidase binding of hydrogen peroxide. Biochemistry 21:2661–2666

Broin M, Cuiné S, Eymery F, Rey P (2002) The plastidic 2-cysteine peroxiredoxin is a target for a thioredoxin involved in the protection of the photosynthetic apparatus against oxidative damage. Plant Cell 14:1417–1432

Broin M, Rey P (2003) Potato plants lacking the CDSP32 plastidic thioredoxin exhibit overoxidation of the BAS1 2-cysteine peroxiredoxin and increased lipid peroxidation in thylakoids under photooxidative stress. Plant Physiol 132:1335–1343

Burton GW, Ingold KU (1984) β-Carotene: an unusual type of lipid antioxidant. Science 224:569–573

Caliskan M, Cuming AC (1998) Spatial specificity of H_2O_2-generating oxalate oxidase gene expression during wheat embryo germination. Plant J 15:165–171

Carpena X, Soriano M, Klotz MG, Duckworth HW, Donald LJ, Melik-Adamyan W, Fita I, Loewen PC (2003) Structure of the Clade 1 catalase, CatF of *Pseudomonas syringae*, at 1.8 Å resolution. Proteins 50:423–436

Cazzaniga S, Li Z, Niyogi KK, Bassi R, Dall'Osto L (2012) The Arabidopsis sz/1 mutant reveals a critical role of β-carotene in photosystem I photoprotection. Plant Physiol 159:1745–1758

Cela J, Chang C, Munné-Bosch S (2011) Accumulation of γ- rather than α-tocopherol alters ethylene signaling gene expression in the *vte4* mutant of *Arabidopsis thaliana*. Plant Cell Physiol 52:1389–1400

Cochard H, Froux F, Mayr S, Coutand C (2004) Xylem wall collapse in water-stressed pine needles. Plant Physiol 134:401–408

Conn PF, Lambert C, Land EJ, Schalch W, Truscott TG (1992) Carotene–oxygen radical interactions. Free Radic Res Commun 16:401–408

Cruz de Carvalho MH (2008) Drought stress and reactive oxygen species. Production, scavenging and signaling. Plant Signal Behav 3:156–165

Cruz de Carvalho MH, Contour-Ansel D (2008) (h)GR, beans and drought stress. Plant Signal Behav 3:834–835

Dai S, Schwendtmayer C, Schürmann P, Ramaswamy S, Eklund H (2000) Redox signaling in chloroplasts: cleavage of disulfides by an iron-sulfur cluster. Science 287:655–658

Dalton DA, Baird LM, Langeberg L, Taugher CY, Anyan WR, Vance CP, Sarath G (1993) Subcellular localization of oxygen defense enzymes in soybean (*Glycine max* [L.] Merr.) root nodules. Plant Physiol 102:481–489

Davison PA, Hunter CN, Horton P (2002) Overexpression of β-carotene hydroxylase enhances stress tolerance in Arabidopsis. Nature 418:203–206
Demmig-Adams B, Gilmore AM, Adams WW III (1996) In vivo functions of carotenoids in higher plants. FASEB J 10:403–412
Dietz KJ, Horling F, König J, Baier M (2002) The function of the chloroplast 2-cysteine peroxiredoxin in peroxide detoxification and its regulation. J Exp Bot 53:1321–1329
Dietz KJ, Jacob S, Oelze ML, Laxa M, Tognetti V, de Miranda SM, Baier M, Finkemeier I (2006) The function of peroxiredoxins in plant organelle redox metabolism. J Exp Bot 57:1697–1709
Dixon DP, Davis BG, Edwards R (2002) Functional divergence in the glutathione transferase superfamily in plants. Identification of two classes with putative functions in redox homeostasis in *Arabidopsis thaliana*. J Biol Chem 277:30859–30869
Dixon DP, Hawkins T, Hussey PJ, Edwards R (2009) Enzyme activities and subcellular localization of members of the Arabidopsis glutathione transferase superfamily. J Exp Bot 60:1207–1218
Duan ZQ, Bai L, Zhao ZG, Zhang GP, Cheng FM, Jiang LX, Chen KM (2009) Drought-stimulated activity of plasma membrane nicotinamide adenine dinucleotide phosphate oxidase and its catalytic properties in rice. J Integr Plant Biol 51:1104–1115
Esteban R, Becerril JM, García-Plazaola JI (2009) Lutein epoxide cycle, more than just a forest tale. Plant Signal Behav 4:342–344
Faize M, Burgos L, Faize L, Piqueras A, Nicolas E, Barba-Espin G, Clemente-Moreno MJ, Alcobendas R, Artlip T, Hernández JA (2011) Involvement of cytosolic ascorbate peroxidase and Cu/Zn-superoxide dismutase for improved tolerance against drought stress. J Exp Bot 62:2599–2613
Feierabend J (2005) Catalases in plants: molecular and functional properties and role in stress defence. In: Smirnoff N (ed) Antioxidants and reactive oxygen species in plants. Blackwell, Oxford, pp 101–140
Feng Y, Shong N, Rouhier N, Hase T, Kusunoki N, Jacquot JP, Jin C, Xia B (2006) Structural insight into poplar glutaredoxin C1 with a bridging iron-sulfur cluster at the active site. Biochemistry 45:7998–8008
Fonseca S, Chico JM, Solano R (2009) The jasmonate pathway: the ligand, the receptor and the core signalling module. Curr Opin Plant Biol 12:539–547
Foreman J, Demidchik V, Bothwell JH, Mylona P, Miedema H, Torres MA, Linstead P, Costa S, Brownlee C, Jones JD, Davies JM, Dolan L (2003) Reactive oxygen species produced by NADPH oxidase regulate plant cell growth. Nature 422:442–446
Foyer CH, Noctor G (2003) Redox sensing and signalling associated with reactive oxygen in chloroplasts, peroxisomes and mitochondria. Physiol Plant 119:355–364
Fukuzawa K, Gebicki JM (1983) Oxidation of α-tocopherol in micelles and liposomes by the hydroxyl, perhydroxyl and superoxide free radicals. Arch Biochem Biophys 226:242–251
Fulda S, Mikkat S, Stegmann H, Horn R (2011) Physiology and proteomics of drought stress acclimation in sunflower (*Helianthus annuus* L.). Plant Biol 13:632–642
Galle A, Csiszar J, Secenji M, Guoth A, Cseuz L, Tari I (2009) Glutathione transferase activity and expression patterns during grain filling in flag leaves of wheat genotypes differing in drought tolerance: Response to water deficit. J Plant Physiol 166:1878–1891
George S, Venkataraman G, Parida A (2010) A chloroplast-localized and auxin-induced glutathione S-transferase from phreatophyte *Prosopis juliflora* confer drought tolerance on tobacco. J Plant Physiol 167:311–318
Gill SS, Tuteja N (2010) Reactive oxygen species and antioxidant machinery in abiotic stress tolerance in crop plants. Plant Physiol Biochem 48:909–930
Giraud E, Ho LH, Clifton R, Carroll A, Estavillo G, Tan YF, Howell KA, Ivanova A, Pogson BJ, Millar AH, Whelan J (2008) The absence of ALTERNATIVE OXIDASE1a in Arabidopsis results in acute sensitivity to combined light and drought stress. Plant Physiol 147:595–610
Giuliano G, Tavazza R, Diretto G, Beyer P, Taylor MA (2008) Metabolic engineering of carotenoid biosynthesis in plants. Trends Biotechnol 26:139–145

Hacke UG, Sperry JS, Pockman WT, Davis SD, McCulloch KA (2001) Trends in wood density and structure are linked to prevention of xylem implosion by negative pressure. Oecologia 126:457–461

Hara M, Terashima S, Fukaya T, Kuboi T (2003) Enhancement of cold tolerance and inhibition of lipid peroxidation by citrus dehydrin in transgenic tobacco. Planta 217:290–298

Haslekås C, Grini PE, Nordgard SH, Thorstensen T, Viken MK, Nygaard V, Aalen RB (2003) ABI3 mediates expression of the peroxiredoxin antioxidant AtPER1 gene and induction by oxidative stress. Plant Mol Biol 53:313–326

Havaux M, Kloppstech K (2001) The protective functions of carotenoid and flavonoid pigments against excess visible radiation at chilling temperature investigated in Arabidopsis *npq* and *tt* mutants. Planta 213:953–966

Herbette S, Lenne C, Leblanc N, Julien JL, Drevet JR, Roeckel-Drevet P (2002) Two GPX-like proteins from *Lycopersicon esculentum* and *Helianthus annuus* are antioxidant enzymes with phospholipid hydroperoxide glutathione peroxidase and thioredoxin peroxidase activities. Eur J Biochem 269:2414–2420

Hernández JA, Ferrer MA, Jiménez A, Barceló AR, Sevilla F (2001) Antioxidant systems and O_2^{-} H_2O_2 production in the apoplast of pea leaves. Its relation with salt-induced necrotic lesions in minor veins. Plant Physiol 127:817–831

Hernández I, Alegre L, Munné-Bosch S (2004) Drought-induced changes in flavonoids and other low molecular weight antioxidants in *Cistus clusii* grown under Mediterranean field conditions. Tree Physiol 24:1303–1311

Hernández I, Alegre L, Van Breusegem F, Munné-Bosch S (2009) How relevant are flavonoids as antioxidants in plants? Trends Plant Sci 14:125–132

Horemans N, Foyer CH, Asard H (2000) Transport and action of ascorbate at the plant plasma membrante. Trends Plant Sci 5:263–267

Ishikawa T, Yoshimura K, Tamoi M, Tamoi M, Takeda T, Shigeoka S (1997) Alternative mRNA splicing of 3'-terminal exons generates ascorbate peroxidase isoenzymes in spinach (*Spinacia oleracea*) chloroplasts. Biochem J 328:795–800

Ito H, Iwabuchi M, Ogawa K (2003) The sugar-metabolic enzymes aldolase and triose-phosphate isomerase are targets of glutathionylation in *Arabidopsis thaliana*: detection using biotinylated glutathione. Plant Cell Physiol 44:655–660

Jahns P, Holzwarth AR (2012) The role of the xanthophyll cycle and of lutein in photoprotection of photosystem II. Biochim Biophys Acta 1817:182–193

Jha B, Sharma A, Mishra A (2011) Expression of SbGSTU (tau class glutathione S-transferase) gene isolated from *Salicornia brachiata* in tobacco for salt tolerance. Mol Biol Rep 38: 4823–4832

Ji W, Zhu Y, Li Y, Yang LA, Zhao XW, Cai H, Bai X (2010) Over-expression of a glutathione S-transferase gene, GsGST, from wild soybean (*Glycine soja*) enhances drought and salt tolerance in transgenic tobacco. Biotechnol Lett 32:1173–1179

Jiménez A, Hernández A, del Río LA, Sevilla F (1997) Evidence for the presence of the ascorbate-glutathione cycle in peroxisomes of pea leaves. Plant Physiol 114:275–284

Johnson MP, Havaux M, Triantaphylidés C (2007) Elevated zeaxanthin bound to oligomeric LHCII enhances the resistance of Arabidopsis to photooxidative stress by a lipid-protective, antioxidant mechanism. J Biol Chem 282:22605–22618

Jubany-Marí T, Munné-Bosch S, Alegre L (2010) Redox regulation of water stress responses in field-grown plants. Role of hydrogen peroxide and ascorbate. Plant Physiol Biochem 48: 351–458

Kamada-Nobusada T, Hayashi M, Fukazawa M, Sakakibara H, Nishimura M (2008) A putative peroxisomal polyamine oxidase, AtPAO4, is involved in polyamine catabolism in *Arabidopsis thaliana*. Plant Cell Physiol 49:1272–1282

Keilin D, Hartreef EF (1950) Reaction of methaemoglobin with hydrogen peroxide. Nature 166:513–514

Keilin D, Hartreef EF (1955) Catalase, peroxidase and metmyoglobin as catalysts of coupled peroxidatic reactions. Biochem J 60:310–325

Keller T, Damude HG, Werner D, Doerner P, Dixon RA, Lamb C (1998) A plant homolog of the neutrophil NADPH oxidase gp91phox subunit gene encodes a plasma membrane protein with Ca^{2+} binding motifs. Plant Cell 10:255–266
Klotz MG, Loewen PC (2003) The molecular evolution of catalatic hydroperoxidases: evidence for multiple lateral transfer of genes between prokaryota and from bacteria into eukaryota. Mol Biol Evol 20:1098–1112
Lee KO, Lee JR, Yoo JY, Jang HH, Moon JC, Jung BG, Chi YH, Park SK, Lee SS, Lim CO, Yun DJ, Cho MJ, Lee SY (2002) GSH-dependent peroxidase activity of the rice (*Oryza sativa*) glutaredoxin, a thioltransferase. Biochem Biophys Res Commun 296:1152–1156
Lee SH, Ahsan N, Lee KW, Kim DH, Lee DG, Kwak SS, Kwon SY, Kim TH, Lee BH (2007) Simultaneous overexpression of both CuZn superoxide dismutase and ascorbate peroxidase in transgenic tall fescue plants confers increased tolerance to a wide range of abiotic stresses. J Plant Physiol 164:1628–1638
Lewis DFV (2002) Oxidative stress: the role of cytochromes P450 in oxygen activation. J Chem Technol Biotechnol 77:1095–1100
Liszkay A, Kenk B, Schopfer P (2003) Evidence for the involvement of cell wall peroxidase in the generation of hydroxyl radicals mediating extension growth. Planta 217:658–667
Liu X, Hua X, Guo J, Qi D, Wang L, Liu Z, Jin Z, Chen S, Liu G (2008) Enhanced tolerance to drought stress in transgenic tobacco plants overexpressing VTE1 for increased tocopherol production from *Arabidopsis thaliana*. Biotechnol Lett 30:1275–1280
Liu HP, Liu J, Zhang YY, Liu YL (2004) Relationship between ATPase activity and conjugated polyamines in mitochondrial membrane from wheat seedling roots under osmotic stress. J Environ Sci 16:712–716
Llorente F, López-Cobollo RM, Catalá R, Martínez-Zapater JM, Salinas J (2002) A novel cold-inducible gene from Arabidopsis, *RCI3*, encodes a peroxidase that constitutes a component for stress tolerance. Plant J 32:13–24
López-Carbonell M, Munné-Bosch S, Alegre L (2006) The ascorbate-deficient *vtc-1* Arabidopsis mutant shows altered ABA accumulation in leaves and chloroplasts. J Plant Growth Regul 25:137–144
Maiorino M, Ursini F, Bosello V, Toppo S, Tosatto SC, Mauri P, Becker K, Roveri A, Bulato C, Benazzi L, De Palma A, Flohé L (2007) The thioredoxin specificity of Drosophila GPx: a paradigm for a peroxiredoxin-like mechanism of many glutathione peroxidases. J Mol Biol 365:1033–1046
Mano S, Yamaguchi K, Hayashi M, Nishimura M (1997) Stromal and thylakoid-bound ascorbate peroxidases are produced by alternative splicing in pumpkin. FEBS Lett 413:21–22
Marjamaa K, Kukkola EM, Fagerstedt KV (2009) The role of xylem class III peroxidases in lignification. J Exp Bot 60:367–376
Markesbery WR, Lovell MA (2006) DNA oxidation in Alzheimer's disease. Antiox Redox Signal 8:2039–2045
McCord JM, Fridovich I (1969) Superoxide dismutase. An enzymatic function for erythrocuperin (hemecuperin). J Biol Chem 244:6049–6055
Meyer AJ (2008) The integration of glutathione homeostasis and redox signaling. J Plant Physiol 165:1390–1403
Meyer AJ, Hell R (2005) Glutathione homeostasis and redox-regulation by sulfhydryl groups. Photosynth Res 86:435–457
Miao Y, Lv D, Wang P, Wang XC, Chen J, Miao C, Song CP (2006) An Arabidopsis glutathione peroxidase functions as both a redox transducer and a scavenger in abscisic acid and drought stress responses. Plant Cell 18:2749–2766
Michelet L, Zaffagnini M, Vanacker H, Le Maréchal P, Marchand C, Schroda M, Lemaire SD, Decottignies P (2008) In vivo targets of S-thiolation in *Chlamydomonas reinhardtii*. J Biol Chem 283:21571–21578
Millar AH, Whelan J, Soole KL, Day DA (2011) Organization and regulation of mitochondrial respiration in plants. Annu Rev Plant Biol 62:79–104

Mittler R, Poulos TL (2005) Ascorbate peroxidase. In: Smirnoff N (ed) Antioxidants and reactive oxygen species in plants. Blackwell, Oxford, pp 87–100

Mohr S, Hallak H, de Boitte A, Lapetina EG, Brüne B (1999) Nitric oxide-induced S-glutathionylation and inactivation of glyceraldehyde-3-phosphate dehydrogenase. J Biol Chem 274:9427–94230

Møller IM, Jensen PE, Hansson A (2007) Oxidative modifications to cellular components in plants. Annu Rev Plant Biol 58:459–481

Møller SG, McPherson MJ (1998) Developmental expression and biochemical analysis of the Arabidopsis *atao1* gene encoding an H_2O_2-generating diamine oxidase. Plant J 13:781–791

Montrichard F, Alkhalfioui F, Yano H, Vensel WH, Hurkman WJ, Buchanan BB (2009) Thioredoxin targets in plants: the first 30 years. J Proteomics 72:452–474

Moschou PN, Paschalidis KA, Delis ID, Andriopoulou AH, Lagiotis GD, Yakoumakis DI, Roubelakis-Angelakis KA (2008) Spermidine exodus and oxidation in the apoplast induced by abiotic stress is responsible for H_2O_2 signatures that direct tolerance responses in tobacco. Plant Cell 20:1708–1724

Mowla SB, Thomson JA, Farrant JM, Mundree SG (2002) A novel stress-inducible antioxidant enzyme identified from the resurrection plant *Xerophyta viscosa* Baker. Planta 215:716–726

Munemasa S, Mori IC, Murata Y (2011) Methyl jasmonate signaling and signal crosstalk between methyl jasmonate and abscisic acid in guard cells. Plant Signal Behav 6:939–941

Munné-Bosch S, Alegre L (2000) Changes in carotenoids, tocopherols and diterpenes during drought and recovery, and the biological significance of chlorophyll loss in *Rosmarinus officinalis* plants. Planta 210:925–931

Munné-Bosch S, Alegre L (2003) Drought-induced changes in the redox state of α-tocopherol, ascorbate, and the diterpene carnosic acid in chloroplasts of Labiatae species differing in carnosic acid contents. Plant Physiol 131:1816–1825

Munné-Bosch S, Weiler EW, Alegre L, Müller M, Düchting P, Falk J (2007) α-Tocopherol may influence cellular signaling by modulating jasmonic acid levels in plants. Planta 225:681–691

Navrot N, Collin V, Gualberto J, Gelhaye E, Hirasawa M, Rey P, Knaff DB, Issakidis E, Jacquot JP, Rouhier N (2006) Plant glutathione peroxidases are functional peroxiredoxins distributed in several subcellular compartments and regulated during biotic and abiotic stresses. Plant Physiol 142:1364–1379

Nicholls P, Fita I, Loewen P (2001) Enzymology and structure of catalases. Adv Inorg Chem 51:51–106

Nishikimi M, Yamada H, Yagi K (1980) Oxidation by superoxide of tocopherols dispersed in aqueous media with deoxycholate. Biochim Biophys Acta 627:101–108

Noctor G, Foyer CH (1998) Ascorbate and glutathione: keeping active oxygen under control. Annu Rev Plant Physiol Plant Mol Biol 49:249–279

Paco L, Galarneau A, Drone J, Fajula F, Bailly C, Pulvin S, Thomas D (2009) Catalase-like activity of bovine met-hemoglobin: interaction with the pseudo-catalytic peroxidation of anthracene traces in aqueous medium. Biotechnol J 4:1460–1470

Pastori GM, Kiddle G, Antoniw J, Bernard S, Veljovic-Jovanovic S, Verrier PJ, Noctor G, Foyer CH (2003) Leaf vitamin C contents modulate plant defense transcripts and regulate genes that control development through hormone signaling. Plant Cell 15:939–951

Pastori GM, Trippi VS (1992) Oxidative stress induces high-rate if glutathione reductase synthesis in a drought-resistant maize strain. Plant Cell Physiol 33:957–961

Pei ZM, Murata Y, Benning G, Thomine S, Klüsener B, Allen GJ, Grill E, Schroeder JI (2000) Calcium channels activated by hydrogen peroxide mediate abscisic acid signalling in guard cells. Nature 406:731–734

Perl A, Perl-Treves R, Galili S, Aviv D, Shalgi E, Malkin S, Galun E (1993) Enhanced oxidative-stress defense in transgenic potato expressing tomato Cu/Zn superoxide dismutases. Theor Appl Genet 85:568–576

Pogson BJ, Rissler HM (2000) Genetic manipulation of carotenoid biosynthesis and photoprotection. Philos Trans R Soc London 355:1395–1403

Polle A (2001) Dissecting the superoxide dismutase-ascorbate-glutathione-pathway in chloroplasts by metabolic modeling. Computer simulations as a step towards flux analysis. Plant Physiol 126:445–462

Ramel F, Birtic S, Cuiné S, Triantaphylidès C, Ravanat JL, Havaux M (2012) Chemical quenching of singlet oxygen by carotenoids in plants. Plant Physiol 158:1267–1278

Raven JA (1987) The evolution of vascular land plants in relation to supracellular transport processes. Adv Bot Res 87:1287–1299

Reinbothe C, Springer A, Samol I, Reinbothe S (2009) Plant oxylipins: role of jasmonic acid during programmed cell death, defence and leaf senescence. FEBS J 276:4666–4681

Rey P, Cuiné S, Eymery F, Garin J, Court M, Jacquot JP, Rouhier N, Broin M (2005) Analysis of the proteins targeted by CDSP32, a plastidic thioredoxin participating in oxidative stress responses. Plant J 41:31–42

Rey P, Pruvot G, Becuwe N, Eymery F, Rumeau D, Peltier G (1998) A novel thioredoxin-like protein located in the chloroplast is induced by water deficit in *Solanum tuberosum* L. plants. Plant J 13:97–107

Rinalducci S, Murgiano L, Zolla L (2008) Redox proteomics: basic principles and future perspectives for the detection of protein oxidation in plants. J Exp Bot 59:3781–3801

Rizhsky L, Hallak-Herr E, Van Breusegem F, Rachmilevitch S, Barr JE, Rodermel S, Inzé D, Mittler R (2002) Double antisense plants lacking ascorbate peroxidase and catalase are less sensitive to oxidative stress than single antisense plants lacking ascorbate peroxidase or catalase. Plant J 32:329–342

Roldán-Arjona T, Ariza RR (2009) Repair and tolerance of oxidative DNA damage in plants. Mutat Res 681:169–179

Rouhier N, Gelhaye E, Sautiere PE, Brun A, Laurent P, Tagu D, Gerard J, de Faÿ E, Meyer Y, Jacquot JP (2001) Isolation and characterization of a new peroxiredoxin from poplar sieve tubes that uses either glutaredoxin or thioredoxin as a proton donor. Plant Physiol 127:1299–1309

Rouhier N, Lemaire SD, Jacquot JP (2008) The role of glutathione in photosynthetic organisms: emerging functions for glutaredoxins anf glutathionylation. Annu Rev Plant Biol 59:143–166

Salekdeh GH, Siopongco J, Wade LJ, Ghareyazie B, Bennett J (2002) Proteomic analysis of rice leaves during drought stress and recovery. Proteomics 2:1131–1145

Sandalio LM, Palma JM, del Río LA (1987) Localization of manganese superoxide-dismutase in peroxisomes isolated from *Pisum sativum* (L). Plant Sci 51:1–8

Sattler SE, Mène-Saffrané L, Farmer EE, Krischke M, Mueller MJ, DellaPenna D (2006) Nonenzymatic lipid peroxidation reprograms gene expression and activates defense markers in Arabidopsis tocopherol-deficient mutants. Plant Cell 18:3706–3720

Shikanai T, Takeda T, Yamauchi H, Sano S, Tomizawa KI, Yokota A, Shigeoka S (1998) Inhibition of ascorbate peroxidase under oxidative stress in tobacco having bacterial catalase in chloroplasts. FEBS Lett 428:47–51

Smirnoff N (2005) Ascorbate, tocopherol and carotenoids: metabolism, pathway engineering and functions. In: Smirnoff N (ed) Antioxidants and reactive oxygen species in plants. Blackwell, Oxford, pp 53–86

Sohal RS, Weindruch R (1996) Oxidative stress, caloric restriction, and aging. Science 273: 59–63

Spoel SH, Loake GJ (2011) Redox-based protein modifications: the missing link in plant immune signaling. Curr Opin Plant Biol 14:358–364

Stadtman ER (2006) Protein oxidation and aging. Free Radic Res 40:1250–1258

Stengel A, Benz JP, Soll J, Bölter B (2010) Redox-regulation of protein import into chloroplasts and mitochondria: similarities and differences. Plant Signal Behav 5:105–109

Suja G, Gayatri V, Ajay P (2010) A chloroplast-localized and auxin-induced glutathione S-transferase from phreatophyte *Prosopis juliflora* confer drought tolerance on tobacco. J Plant Physiol 167:311–318

Sun W, Kadima TA, Pickard MA, Dunford HB (1994) Catalase activity of chloroperoxidase and its interaction with peroxidase activity. Biochem Cell Biol 72:321–331

Sztajer H, Gamain B, Aumann KD, Slomianny C, Becker K, Brigelius-Flohé R, Flohé L (2001) The putative glutathione peroxidase gene of *Plasmodium falciparum* codes for a thioredoxin peroxidase. J Biol Chem 276:7397–7403

Torres MA, Onouchi H, Hamada S, Machida C, Hammond-Kosack KE, Jones JD (1998) Six *Arabidopsis thaliana* homologues of the human respiratory burst oxidase (gp91phox). Plant J 14:365–370

Torres-Franklin ML, Contour-Ansel D, Zuily-Fodil Y, Pham-Thi AT (2007) Molecular cloning of glutathione reductase cDNAs and analysis of GR gene expression in cowpea and common bean leaves during recovery from moderate drought stress. J Plant Physiol 165:514–521

Toumi I, Moschou PN, Paschalidis KA, Bouamama B, Ben Salem-Fnayou A, Ghorbel AW, Mliki A, Roubelakis-Angelakis KA (2010) Abscisic acid signals reorientation of polyamine metabolism to orchestrate stress responses via the polyamine exodus pathway in grapevine. J Plant Physiol 167:519–525

Van Camp W, Capiau K, Van Montagu M, Inzé D, Slooten L (1996) Enhancement of oxidative stress tolerance in transgenic tobacco overproducing Fe-superoxide dismutase in chloroplasts. Plant Physiol 112:1703–1714

Van Camp W, Willekens H, Bowler C, Van Montagu M, Inze D (1994) Elevated levels of superoxide dismutase protect transgenic plants against ozone damage. Bio/Technology 12:165–168

Vanderauwera S, Zimmermann P, Rombauts S, Vandenabeele S, Langebartels C, Gruissem W, Inzé D, Van Breusegem F (2005) Genome-wide analysis of hydrogen peroxide-regulated gene expression in Arabidopsis reveals a high light-induced transcriptional cluster involved in anthocyanin biosynthesis. Plant Physiol 139:806–821

Vanderauwera S, Suzuki N, Miller G, van de Cotte B, Morsa S, Ravanat JL, Hegie A, Triantaphylidès C, Shulaev V, Van Montagu MC, Van Breusegem F, Mittler R (2011) Extranuclear protection of chromosomal DNA from oxidative stress. Proc Natl Acad Sci USA 108:1711–1716

Vanholme R, Demedts B, Morreel K, Ralph J, Boerjan W (2010) Lignin biosynthesis and structure. Plant Physiol 153:895–905

Vieira Dos Santos C, Cuiné S, Rouhier N, Rey P (2005) The Arabidopsis plastidic methionine sulfoxide reductase B proteins. Sequence and activity characteristics, comparison of the expression with plastidic methionine sulfoxide reductase A, and induction by photooxidative stress. Plant Physiol 138:909–922

Votyakova TV, Wallace HM, Dunbar B, Wilson SB (1999) The covalent attachment of polyamines to proteins in plant mitochondria. Eur J Biochem 260:250–257

Wang W, Vinocur B, Shoseyov O, Altman A (2004) Role of plant heatshock proteins and molecular chaperones in the abiotic stress response. Trends Plant Sci 9:244–252

Chapter 10
Understanding and Exploiting Plant Hormone Biology to Enhance Crop Production Under Water Scarcity

W. J. Davies and Sally Wilkinson

Abstract There is much interest in the concept of sustainable intensification of agriculture as one of the actions that can contribute to a reduction in food insecurity for the future. However, drought can significantly limit plant productivity through an impact on a variety of processes which contribute to carbon gain, to the development of the canopy and the establishment and filling of reproductive structures. If we are to minimize these restrictions to growth and yielding, it is important to understand the mechanistic basis of these responses. For more sustainable use of water in crop production, it is also important to understand the biology behind resource use efficiency by crops. We highlight here the importance of the influence of both hydraulic and chemical signaling on the regulation of functioning, growth, and development of plants under stress and suggest that by processing and responding to these signals, plants are able to regulate growth and development relative to the availability of water and other resources. We propose that via enhancement or suppression of different 'root signal' cascades we can intervene to sustain plant yielding under drought. We review the identity of signals limiting leaf conductance, leaf growth, and harvest index and consider prospects for manipulation of signaling, with special reference to the impact of changes in sap pH, modified fluxes of abscisic acid (ABA) and of the ethylene precursor ACC. While there are many genetic opportunities to modify stress signaling cascades to increase crop yield, crop improvement cannot immediately combat changes in the climate and increases in food demand that are happening now. It is therefore important that we use novel crop management techniques to impact some of the same biology. Such manipulations may be a cost-effective way to sustain yielding and increase water use efficiency in dryland agriculture.

W. J. Davies (✉) · S. Wilkinson
Lancaster Environment Centre,
Lancaster University, Lancaster LA1 4YQ, UK
e-mail: w.davies@lancaster.ac.uk

R. Aroca (ed.), *Plant Responses to Drought Stress*,
DOI: 10.1007/978-3-642-32653-0_10,

10.1 Introduction

Severe drought stress can greatly limit crop growth and yielding and while the mechanisms behind drought tolerance are multiple, they themselves can often reduce yield. Realistically, in many cropping systems, a drought tolerance phenotype is defined as one that can maintain or enhance crop productivity when water is in short supply (a high water productivity). Importantly, a water productive genotype should not show a restriction in yielding when water is freely available. There is a strong argument that selection for high yielding genotypes in favorable conditions (high yield potential) can also lead to high productivity in less favorable environments. Nevertheless, there is still much interest in identifying the mechanisms behind good performance under drought in the hope that such knowledge can be exploited in plant improvement programmes and in novel methods of crop management.

As the climate changes, water in many cropping regions of the world will be in increasingly restricted supply and this will often mean reduction in yield. We argue here that some knowledge of the plant's drought stress biology can help producers to minimize the effects of soil drying on yielding (Fereres and Soriano 2007). Many plant responses to drought are mediated via changes in plant water relations but mild drought can directly restrict both root growth and shoot growth and gas exchange, sometimes even before plant water deficits are detectable. One explanation for these effects can be chemical regulation of growth and functioning (see e.g. Wilkinson and Davies 2002).

10.2 Early Work on Chemical Regulation of Plant Growth, Development and Gas Exchange

Since the early 1980s, much research work has provided us with a novel view of how above-ground plant parts can "sense" changes in both their atmospheric and edaphic environments. The work has also provided insight into the ways in which different environmental stresses interact in their effects on plant growth and functioning through changes in both chemical and hydraulic regulation. In recent years, these discoveries have led to radical changes in irrigation practice and to the introduction of crop management techniques which contribute to sustainable intensification of agriculture in drought-prone environments.

Until the 1980s, the predominant view in the literature was that soil drying and other stresses modified plant growth, functioning, and development only *after* uptake of water was reduced and water deficits had developed in the leaves. There has since been a paradigm shift in our thinking about plant responses to drought. This has led not only to entirely new perspectives on what actually constitutes the commonly occurring phenomenon of 'plant water stress', but also has pointed to new possibilities for the modification of plant growth and food production via simple agronomical strategies as well as innovative biotechnological manipulations. Further, our understanding of the ways in which plants can detect and respond to very mild soil drying provides some explanation of how plant communities and landscape may be affected by modified rainfall patterns under a changing climate.

There is now unequivocal evidence of regulation of stomatal functioning and leaf growth via a chemical message moving from the roots to the shoots in the xylem stream. It is clear that leaf water balance of plants in drying soil can be controlled *as a result of* stomatal closure and reduced leaf expansion. The view originally virtually unchallenged in the literature was that a reduction in the water status of the plant was necessary *before* physiological controls such as stomatal closure were initiated. What then is restricting shoot growth and functioning? Several groups have now shown that the causal agents are chemicals delivered to the shoots in the xylem stream which constituted a 'signal' generated as a result of the interaction between some of the plant's roots and drying soil. This hypothesis was substantiated in experiments with plants with divided root systems. Those roots in contact with drying soil were excised from the plants (removing the source of the chemical inhibitor) and this simple manipulation resulted in an increase in leaf growth rate (Gowing et al. 1990) and caused a re-opening of stomata.

10.3 The Nature of Chemical Signaling in Droughted Plants

There is now strong evidence that the plant hormone abscisic acid (ABA) can play a central part in the long distance chemical signaling process in plants in drying soil. Work by Zhang and Davies (1989) showed that enhanced xylem ABA fluxes from roots in contact with drying soil, could provide shoots with a measure of the extent of soil drying. This ABA can be both root-synthesized and re-circulated from shoots via roots in quantities reflecting the plant's access to soil water. Importantly, this means that evidence that ABA may be synthesized in very limited quantities in roots of some plant species (e.g., Christmann et al. 2007) is not evidence against the root signaling hypothesis

ABA is clearly not the only chemical signal responsible for regulation of gas exchange and growth and some other ionic signals (Davies et al. 2002) moving from roots to shoots via the xylem can impact on leaf growth and functioning via a local redistribution of ABA from anion traps in leaves to sites of action for regulation. A hypothesis for stomatal and growth control based on ionic effects on pH gradients and ABA redistribution is now generally accepted in the literature (Wilkinson and Davies 1997).

10.4 Signaling in Crops and Natural Vegetation in the Field

Recent work shows how the plant's chemical signaling capabilities can combine to allow integration of the effects of a range of environmental signals through at least the operation of the stomata and the regulation of leaf growth. Control of reproductive processes can also be integrated into this system of control. A simple model (Tardieu and Davies 1993) describes the regulation of stomata by drought via an ABA response of stomata modulated by plant water status, while other models address the interaction between different hormones. Another model (Dodd et al. 2008) shows how heterogeneous soil drying can influence gas exchange via variation in ABA flux. These signal transduction systems provide excellent opportunities for manipulation of important drought responses of plants by genetic and agronomic means (see e.g. Wilkinson et al. 2012) and can be useful because the ultimate impacts of hormones on yielding of crops are not always open to intuition. This should not be surprising because the G x E interactions (genetics x environment) can often be very significant. This may be one of the reasons why drought-related transgenes have not yet yielded too many benefits in the commercial sector.

10.5 Critical Analysis of the Chemical Signaling Hypothesis

In the 30 years since the first elucidation of the chemical signaling hypotheses described above, there has been some critical analysis of both the general concept and the mechanistic basis of the proposed ideas.

10.5.1 Is There Sufficient Chemical Regulator in the Xylem of Droughted Plants to Account for the Response Observed?

A substantial body of work has now demonstrated that there is always enough root-sourced ABA in the xylem (both root-synthesized and re-circulated) to account for stomatal closure and leaf growth limitation in response to drought (Wilkinson and Davies 1997). The key issue in this controversy seems to be whether or not the ABA arriving in the leaf in the transpiration stream gets to the sites of action for regulation of leaf growth and functioning (e.g., the guard cell plasma membrane for regulation of stomatal behavior). In well-watered plants, ABA will partition to alkaline compartments (e.g. the symplast of the mesophyll cells) and much hormone will never reach the guard cells and other sites of action in the leaf. Soil drying can cause alkalinization of the xylem sap, and alkaline plant sap (with no extra ABA added above well-watered concentrations) is an effective stomatal closing agent in a whole leaf transpiration bioassay (Jia and Davies 2007).

10.5.2 *Are Isohydric Regulation of Shoot Water Status and the Removal of Roots in Drying Soil Critical Tests of the Signaling Hypothesis?*

Because of the subtle regulation of cellular water status and the difficulty of measuring accurately the water potential gradients in plants, it is difficult to rule out unequivocally the hydraulic disruption of plant growth and functioning in experiments where environments are perturbed. Indeed, it would be perverse to argue that chemical and hydraulic regulation does not work in concert to deliver subtle changes in growth and functioning of plants. This case is based on an accumulation of evidence of different types, e.g., (a) quantitative assessment of hormone fluxes through the stele (Dodd et al. 2008), (b) quantification of hormones on a subcellular basis and the impact of the environment on the redistribution of hormones between plant compartments (Jia and Davies 2007), (c) studies with mutants, transgenics, and focussed chemical treatments to allow quantification of plant responses without the involvement of hormones (Bacon et al. 1998), (d) consideration of the interactions between different chemicals and hydraulic influences to change the sensitivity of the response (Tardieu and Davies 1992), (e) application of irrigation, root pressure, and crop management techniques to generate functional and developmental responses in plants where even precision assessments of shoot water balance cannot account for observed changes in growth and gas exchange (Davies et al. 2002).

10.6 Exploitation of the Signaling Mechanism

Although the mechanistic bases of chemical regulation of plant responses to environmental stress are not fully elucidated, there are nevertheless substantial opportunities for manipulation of these signals for increasing crop yield in situations where water for agriculture is in short supply. The aim is to produce more 'crop per drop' using low technology exploitation of novel plant science to impact on drought limitations to food security. Water saving in agriculture has become unavoidable in most regions of the world but the use of controlled soil drying to enhance chemical signaling is now commonly used as a crop management tool to deliver extra value via enhanced harvest index (HI) and increases in the quality of the plant product (see e.g. Dodd 2009; Davies et al. 2011).

In addition to this, a knowledge of plant signaling biology can be helpful in a plant improvement programme which includes physiological traits in a phenotyping program (see e.g. Wilkinson et al. 2012). Richards (2004) has suggested that to be useful in breeding, a trait must be:

- Easy and inexpensive to measure;
- Highly heritable;
- Should not result in yield penalties in the absence of drought;
- Should not give rise to negative pleiotropic effects on other important crop characteristics.

Biomass accumulation is intrinsically linked to transpiration because stomatal aperture and leaf area determine the rate of photosynthesis (C gain) as well as the rate of transpiration. Thus, anything that reduces transpiration has the potential to reduce biomass accumulation; however, it may be possible to uncouple these responses for particular plant parts by changing Harvest Index or by identifying genotypes with high water use efficiency such that they can maintain more open stomata and/or continue to grow under drought, to continue to fix carbon for sustained productivity.

Impacts of chemical signaling such as regulation of stomatal conductance, shoot and canopy development, and modification of root architecture may all be important targets in programs aimed at producing more water productive plants (e.g. Mittler and Blumwald 2010; Travaglia et al. 2010). It seems likely that such chemical regulation will also regulate reproductive development in plants that are well supplied with water (e.g. Ashikari et al. 2005) and may thus be suitable targets to increase yield potential under favorable conditions (e.g. Foulkes et al. 2011). For example, there appears to be excess photosynthetic capacity during grain filling in modern wheat varieties (Reynolds et al. 2005) which suggests that wheat aborts more florets than necessary, a process that is likely regulated by spikelet hormone balance and which might be targeted to increase grain yield.

In a look forward at opportunities for exploitation of plant hormone biology in agriculture we focus attention on ABA, ethylene, and cytokinin signaling in response to environmental stress and developmental change. Our proposal is that these signals act antagonistically or synergistically in regulating the plant's response to environmental change or to developmental cues (e.g., Wilkinson and Davies 2010; Ghanem et al. 2011). We consider below the evidence for this proposal with a focus on the production of yield in cereal crops, the staple food for most of the world.

10.7 Ethylene/ABA Biology

Ethylene accumulation under environmental stress can reduce root and shoot growth and biomass accumulation via direct effects on growth processes (Sharp 2002) and via reductions in photosynthesis (Rajala and Peltonen-Sainio 2001). The reduction in root growth and biomass is particularly significant because water capture by roots can drive yield under many water limited scenarios (Blum 2009), and can also do so under hot conditions even when plants are irrigated to field capacity. In maturing wheat plants, increased ethylene production triggers plant senescence and hastens maturity, shortens the grain filling period and filling rate, increases embryo abortion, and decreases thousand-grain weight ((Hays et al. 2007). It also disrupts control of gas exchange and leaf growth by drought and ABA (Wilkinson and Davies 2010; Fig. 10.1), which is significant as the "drought stress" hormone ABA usually closes stomata and can limit leaf growth to reduce plant water loss via transpiration (and thus reduce water use in agriculture).

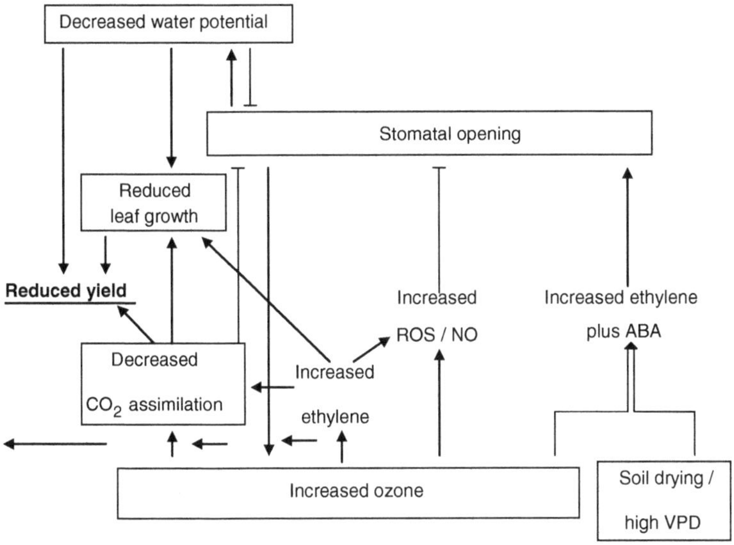

Fig. 10.1 Schematic summary of effects of drought and ozone stresses on plant growth and functioning. Effects are proposed as manifest through changes in ethylene/ABA ratios (modified from Wilkinson and Davies 2010)

Furthermore, significant genetic variability in stress ethylene generation exists within several species (including wheat, e.g. Balota et al. 2004) and this has been linked to stress sensitivity in terms of yield, leaf injury, senescence, and growth of plants exposed to heat, drought, ozone, and salinity stress (e.g. Hays et al. 2007; Balota et al. 2004). Many of the effects of drought, heat, and O_3 stress on plants mirror those known to be ethylene induced and there is now general acceptance that ethylene responses may unpin much of the limitation of seed crop yield which occurs under abiotic stress. Nevertheless, direct evidence for this is still fairly sparse (e.g. Naik and Mohapatra 2000; Yang et al. 2006), and this knowledge has not yet been fully exploited in agriculture. Evidence is accumulating that sensitivity of stress responses is dependent on a ratio between ethylene (and/or its precursor ACC—1-aminocyclopropane) and ABA concentration, rather than on a given concentration of either hormone alone. We have found that ACC interferes with the action of ABA on stomata, greatly impacting transpiration and in grains, this ratio influences grain size and number .

High levels of ethylene synthesis seem to promote kernel abortion (Hays et al. 2007) and ABA/ethylene ratios can also be important in determining (a) spikelet fertility in maize and rice (Wang et al. 2005; Yang et al. 2007) and (b) the rate of grain filling in wheat (Yang et al. 2006). Post-anthesis application (directly to developing grains) of ABA and ethylene biosynthesis inhibitors showed that an increased ratio of ABA to ethylene increased grain filling rate (Yang et al. 2006). Ethylene seems to reduce stomatal conductance and biomass growth even when plants are fully watered (Vysotskaya et al. 2011), and to antagonize stomatal

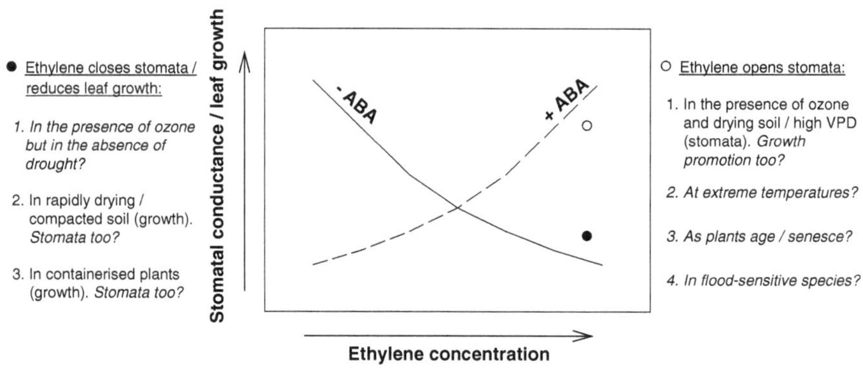

Fig. 10.2 Proposed interaction between ABA and ethylene in the regulation of stomatal behavior and leaf growth (from Wilkinson and Davies 2010) which may also apply to the regulation of reproductive development

closure under drought/ABA (Tanaka et al. 2005). It seems likely that this antagonism will also occur in well-watered plants, during, for example leaf aging where stomata loses the capacity to respond to ABA (e.g., Rodriguez and Davies 1982) (Fig. 10.2)

Sometimes high ABA levels can be associated with growth maintenance (Tardieu et al. 2010) and increased yield (Travaglia et al. 2010) and has been attributed variously to links between ABA accumulation and synthesis of compatible solutes, increased aquaporin activity in roots and shoots ,and deeper root penetration. However, it may also be the case that the positive effects of ABA can be attributed to a reduction in ethylene production (and the various negative effects that ethylene induces), because there is some evidence to suggest that ABA antagonizes ethylene synthesis both in roots and shoots (Sharp 2002). For these effects to develop, extra ABA must be present at moderate levels over a prolonged period (i.e., when positive effects on plant water status negate initial direct closing effect on stomata). This, alongside clear benefits of ABA on grain filling (by negating effects induced by ethylene), demonstrates that accurate timing and extent of up-regulation of ABA synthesis could be beneficial for yield through an increase in the ABA:ethylene ratio of the shoot, rather then through an increase in ABA per se.

As well as impacting negatively on grain filling and grain number under stress (see above), ethylene accumulation can be directly inhibitory of grain filling even under well-watered conditions (Mohapatra et al. 2011). This is because there exist naturally a range of different sized grains on one ear (wheat) or spikelet (rice) even under non-stressful conditions, where the smaller ones have been termed "inferior". Slow filling rates in these inferior grains, which are found lower down on the ear/spikelet correlate with low cytokinin and ABA concentrations and greater ethylene production, than the "superior" grains higher up the ear/spikelet (Yang et al. 2000; Zhao et al. 2007; Zhang et al. 2009). It is thought that, because these inferior grains are confined in the leaf sheath for a longer period, simply as a result of being lower down the ear (i.e. as a result of typical plant architecture), they are thus exposed to more ethylene, presumably as it cannot dissipate freely from the confined grains (note, ethylene is a gas).. The extent of this effect may vary with genotype as a result of variations in ear/panicle architecture. Ethylene inhibitors improved the growth and development of the inferior spikelets, while application of ethylene promoters has the opposite effect (Mohapatra et al. 2000). Ethylene inhibitors also improved starch biosynthesis and the activity of sucrose synthase and invertase enzymes in the kernels of inferior spikelets (Naik and Mohapatra 2000). Thus, yield improvement in grain crops may be accessible via screening genotypes for architectural traits of the ear/spikelet (shorter leaf sheaths, more rapid ear emergence) and/or for reduced flag leaf ethylene production post anthesis. Also, ethylene reduces ABA concentration in wheat shoots (Wilkinson et al. 2012), and high ABA has been associated with improved grain filling, attesting to the greater influence of the ratio between the two hormones rather than to changes in one hormone in isolation.

10.8 Cytokinins and Ck/ABA Ratios

The yielding of transgenic plants overexpressing CK can be improved as a result of an enhanced stay-green capacity under stress. When the *IPT* gene, which drives CK biosynthesis, is overexpressed using a stress-inducible promoter, the performance of the transgenic plants can be improved (reviewed by Peleg and Blumwald 2011), although results can be negative in roots.

Tomato plants grafted onto rootstocks constitutively expressing *IPT* (Ghanem et al. 2011) yielded 30 % more than the wild-type plants under salinity stress. Silencing of cytokinin oxidase genes increased CK levels in non-stressed barley and can lead to higher plant productivity (Zalewski et al. 2010). Qin et al. (2011) found that maturation-induced and stress-induced promoters resulted in enhanced drought tolerance and yield in peanut crops grown in the field, by delaying stress-induced senescence. The key feature of this plant is that the manipulation has been tested in the field. See also Peleg and Blumwald (2011) for effects in other crop plants under more controlled conditions. Other positive effects in the transgenic peanut plants were higher rates of photosynthesis and transpiration in drying soil.

High concentrations of cytokinins are known to inhibit root elongation and branching (e.g., Ghanem et al. 2011), which may decrease water and nitrogen capture by plants and unfavorably affect productivity, although reduced resource allocation to roots is one way to increase harvest index which will likely increase yield potential of plants under favorable environmental conditions. Transgenic plants overproducing cytokinin oxidase in roots, with reduced cytokinin concentrations, had increased root growth and greater plant survival under drought (Werner et al. 2010).

Cytokinin content usually declines in stressed plants co-incidentally with abscisic acid accumulation. This is believed to result from an ABA-induced increase in the expression of the gene coding for cytokinin oxidase (Brugiere et al. 2003) and/or an increase in the activity of enzymes catalyzing irreversible cytokinin degradation (Havlova et al. 2008; Vysotskaya et al. 2009). Treatment of plants with ABA synthesis inhibitors diminished both the activation of cytokinin oxidase in stressed plants, and the decrease in cytokinin accumulation (Vysotskaya et al. 2009). A reduction in cytokinin content in stressed plants will tend to sensitize ABA-induced stomatal closure, which may contribute to plant survival under severe drought, although it may be beneficial to prevent any cytokinin decline in plants (e.g., with aging) if it would be desirable to reduce stomatal closing sensitivity to drying soil where this is relatively mild or of short duration, to prevent reductions in carbon gain.

It still remains to be seen whether yield might be increased via a decrease in the ABA:cytokinin ratio, which may be the result if carbon gain and/or leaf cooling through more open stomata occurs, and it appears logical to assume that maintaining cytokinin levels would be directly beneficial for grain filling and number. This is because, in rice and wheat, yield can be negatively affected by the strong metabolic dominance of the apical spikelets during grain filling, to the detriment of that in the inferior grains in basal regions of ears or spikes (see above). Low filling rate in inferior spikelets is not due to a lower assimilate supply, and Mohapatra et al. (2011) suggest instead that a transient cytokinin spike after flowering could enhance the activities of cell cycle genes and increase the number of cells and, as such, the sink capacity of the developing seeds. Alternatively, these authors suggest that cytokinins could improve phloem unloading. Slow grain filling rates in inferior grains have been linked with a variety of changes in hormonal balance, namely, low zeatin, zeatin riboside, and IAA contents (Yang et al. 2000; Zhao et al. 2007; Zhang et al. 2009); and also with low ABA content and high rates of ethylene evolution (Zhao et al. 2007; Zhang et al. 2009).

Our general understanding is that the development of a plant organ is regulated by a balance between hormones that promote and those that inhibit development. A variety of changes have been proposed but success stories are relatively few and far between in the literature (but see Ashikari et al. (2005) and Zalewski et al. (2010) who have transgenically increased CK and linked this effect to positive changes in the influoresence (grain number) of rice and increased root biomass of barley). It would seem that increased cytokinin content should not occur at the expense of ABA to maintain high grain filling rate (Zhao et al. 2007; Zhang et al. 2009). Again, breeding for high ABA plus high CK seems preferable where stress-induced senescence is not necessarily an issue, and where stress is only mild or of short duration, as both improved inferior spikelet grain filling and reduced stomatal sensitivity to ABA can be coincidental. The high ABA concentration may also be beneficial in regard to root extension rate, aquaporin activity and synthesis of solutes involved in turgor maintenance as discussed above.

10.9 Conclusions

To understand the role of phytohormones in reproductive development of wheat, it is important to discriminate effects of altered chemical status and water status, as many studies indicate that cereal crops can maintain turgor in vegetative and reproductive structures in the field in response to a range of environmental challenges (e.g., Michelena and Boyer 1982). Environmental effects on yield development in cereals can be extremely subtle. For example, Boyer (1982) has shown that even well-irrigated, well-fed crops in the USA may yield only 20 % of potential yield values. We have shown that fully fed and irrigated greenhouse grown non-competing lettuce respond to ethylene perception inhibitors with a biomass increase of 20 % (Vysotskaya et al. 2011). We believe that hormonal and other chemical signals have an important role to play in regulating yield, even in plants that are to all appearances 'unstressed'.

10.9.1 Science for Improved Crop Management

Chemical manipulation of ethylene biology is an important target for at least one major chemical company seeking to develop products to increase drought tolerance of important food crops. Other researchers have proposed the application of plant growth-promoting rhizobacteria (PGPR) to soil to reduce ethylene accumulation under drought and increase yields of peas in the field (Belimov et al. 2009). Other PGPRs used over several seasons in the field have also repeatedly showed positive impacts on wheat yield while Bacillus spp which may be used to augment the plant's CK concentrations have been shown to have positive effects on yields of wheat under mild drought in the field in Russia (Wilkinson et al. 2012).

The production of new genotypes with improved yielding and water productivity is a major target in our fight to address the challenge of global food security. We suggest that progress might be made via exploitation of new understanding of the involvement of ethylene in yield regulation. Modified trait screening for drought productivity might include physiological variables such as maximal stress ethylene production ("eth-max"), ABA-ACC/ethylene concentration ratios (Yang and Zhang 2010, 2006), and sensitivity of plant development and physiology to ethylene and A:E). Maximal ethylene production can be wheat genotype dependent in response to stress (Balota et al. 2004).

This strategy would enable the development of a crop protection/improvement strategy that is applicable to a combination of stresses. Trait selection is the cornerstone of modern plant breeding and has made continual progress through incorporating simply inherited agronomic characteristics such as height and flowering time, resistance to a spectrum of prevalent diseases, quality parameters determined by end use, and yield based on multi-location trials. To accelerate genetic gains in yield in the future, it is now accepted that complex physiological traits (PTs) must also be incorporated as additional criteria in trait selection programmes. The main objective of the breeding approach suggested above will be to combine relevant PTs deterministically whereby progeny will encompass expression of low eth-max and/or appropriate A:E with additional stress adaptive traits in elite agronomic backgrounds. PT-based breeding approaches are new but have been implemented successfully by CIMMYT, leading to international distribution of a new generation of elite drought-adapted lines (Reynolds et al. 2009).

References

Ashikari M, Sakakibara H, Lin S, Yamamoto T, Takashi T, Nishimura A, Angeles ER, Qian Q, Kitano H, Matsuoka M (2005) Cytokinin oxidase regulates rice grain production. Science 309:741–745

Bacon MA, Wilkinson S, Davies WJ (1998) pH regulated leaf cell expansion in droughted plants is ABA-dependent. Plant Physiol 118:1507–1515

Balota M, Cristescu S, Payne WA, Lintel Hekkert S, Laarhoven LJJ, Harren FJM (2004) Ethylene production of two wheat cultivars exposed to desiccation, heat, and paraquat-induced oxidation. Crop Sci 44:812–818

Belimov AA, Dodd IC, Hontzeas N, Theobald JC, Safronova VI, Davies WJ (2009) Rhizosphere bacteria containing 1-aminocyclopropane-1-carboxylate deaminase increase yield of plants grown in drying soil via both local and systemic hormone signalling. New Phytol 181:413–423

Boyer JS (1982) Plant productivity and environment. Science 218:443–448

Brugiere N, Jiao SP, Hantke S, Zinselmeier C, Roessler JA, Niu XM, Jones RJ, Habben JE (2003) Cytokinin oxidase gene expression in maize is localized to the vasculature, and is induced by cytokinins, abscisic acid, and abiotic stress. Plant Physiol 132:1228–1240

Christmann A, Weiler EW, Steudle E, Grill E (2007) A hydraulic signal in root-to-shoot signalling of water shortage. Plant J 52:167–174

Davies WJ, Wilkinson S, Loveys B (2002) Stomatal control by chemical signalling and the exploitation of this mechanism to increase water use efficiency in agriculture. New Phytol 153:449–460

Davies WJ, Zhang J, Yang J, Dodd IC (2011) Novel crop science to improve yield and resource use efficiency in water-limited agriculture. J Agric Sci 149:123–131

Dodd IC (2009) Rhizosphere manipulations to maximize 'crop per drop' during deficit irrigation. J Exp Bot 60:2454–2459

Dodd IC, Egea G, Davies WJ (2008) ABA signalling when soil moisture is heterogeneous: decreased photoperiod sap flow from drying roots limits ABA export to the shoots. Plant, Cell Environ 31:1263–1274

Fereres E, Soriano MA (2007) Deficit irrigation for reducing agricultural water use. J Exp Bot 58:147–159

Foulkes MJ, Slafer GA, Davies WJ, Berry PM, Sylvester-Bradley R, Martre P, Calderini DF, Reynolds MP (2011) Raising yield potential in wheat: optimizing partitioning to grain yield while maintaining lodging resistance. J Exp Bot 62:469–486

Ghanem ME, Albacete A, Smigocki AC, Frebort I, Pospısilova H, Martınez-Andujar C, Acosta M, Sanchez-Bravo J, Lutts S, Dodd ID, Perez-Alfocea F (2011) Root-synthesized cytokinins improve shoot growth and fruit yield in salinized tomato (Solanum lycopersicum L.) plants. J Exp Bot 62:125–140

Gowing DJ, Davies WJ, Jones HG (1990) A positive root-sourced signal as an indicator of soil drying in apple, malus x domestica bork. J Exp Bot 41:1535–1540

Havlová M, Dobrev PI, Motyka V, Storchová H, Libus J, Dobrá J, Malbeck J, Gaudinová A, Vanková R (2008) The role of cytokinins in responses to water deficit in tobacco plants over-expressing trans-zeatin O-glucosyltransferase gene under 35S or SAG12 promoters. Plant, Cell Environ 31:341–353
Hays DB, Do JH, Mason RE, Morgan G, Finlayson SA (2007) Heat stress induced ethylene production in developing wheat grains induces kernel abortion and increased maturation in a susceptible cultivar. Plant Sci 172:1113–1123
Jia W, Davies WJ (2007) Modification of leaf apoplastic pH in relation to stomatal sensitivity to root sourced ABA signals. Plant Physiol 143:68–77
Michelena VA, Boyer JS (1982) Complete turgor maintenance at low water potentials in the elongating region of maize leaves. Plant Physiol 65:1145–1149
Mittler R, Blumwald E (2010) Genetic engineering for modern agriculture: challenges and perspectives. Annu Rev Plant Biol 61:443–462
Mohapatra PK, Naik PK, Patel R (2000) Ethylene inhibitors improve dry matter partitioning and development of late flowering spikelets on rice panicles. Aust J Plant Physiol 27:311–323
Mohapatra PK, Panigrahi R, Turner NC (2011) Physiology of spikelet development on the rice panicle: is manipulation of apical dominance crucial for grain yield improvement? Adv Agron 110:333–360
Naik PK, Mohapatra PK (2000) Ethylene inhibitors enhanced sucrose synthase activity and promoted grain filling of basal rice kernels. Aust J Plant Physiol 27:997–1008
Peleg Z, Blumwald E (2011) Hormone balance and abiotic stress tolerance in crop plants. Curr Opin Plant Biol 14:290–295
Qin H, Gu Q, Zhang J, Sun L, Kuppu S, Zhang Y, Burow M, Payton P, Blumwald E, Zhang H (2011) Regulated expression of an Isopentenyltransferase gene (IPT) in peanut significantly improves drought tolerance and increases yield under field conditions. Plant Cell Physiol 52:1904–1914
Rajala A, Peltonen-Sainio P (2001) Plant growth regulator effects on spring cereal root and shoot growth. Agron J 93:936–943
Reynolds MP, Mujeeb-Kazi A, Sawkins M (2005) Prospects for utilizing plant-adaptive mechanisms to improve wheat and other crops in drought and salinity-prone environments. Ann Appl Biol 146:239–259
Reynolds MP, Manes Y, Izanloo A, Langridge P (2009) Phenotyping for physiological breeding and gene discovery in wheat. Ann Appl Biol 155:309–320
Richards RA (2004) Physiological traits used in the development of new cultivars for water-use environments. In: New directions for a diverse planet. Proceedings of the 4th international crop science congress. Brisbane, Queensland, Australia. 26 Sept–1 Oct 2004
Rodriguez JL, Davies WJ (1982) The effects of temperature and ABA on stomata of Zea mays L. J Exp Bot 33:977–987
Sharp RE (2002) Interaction with ethylene: changing views on the role of abscisic acid in root and shoot growth responses to water stress. Plant, Cell Environ 25:211–222
Tanaka Y, Sano T, Tamaoki M, Nakajima N, Kondo N, Hasezawa S (2005) Ethylene inhibits abscisic acid-induced stomatal closure in Arabidopsis. Plant Physiol 138:2337–2343
Tardieu F, Davies WJ (1992) Stomatal response to ABA is a function of current plant water status. Plant Physiol 98:540–545
Tardieu F, Davies WJ (1993) Integration of hydraulic and chemical signalling in the control of stomatal conductance and water status of droughted plants. Plant, Cell Environ 16:341–350
Tardieu F, Parent B, Simmoneau T (2010) Control of leaf growth by abscisic acid: hydraulic or non-hydraulic processes? Plant Cell Environ 33:636–647
Travaglia C, Reinoso H, Cohen A, Luna C, Tommasino E, Castillo C, Bottini R (2010) Exogenous ABA increases yield in field-grown wheat with moderate water restriction. J Plant Growth Regul 29:366–374
Vysotskaya LB, Korobova AV, Veselov SY, Dodd IC, Kudoyarova GR (2009) ABA mediation of shoot cytokinin oxidase activity: assessing its impacts on cytokinin status and biomass allocation of nutrient deprived durum wheat. Funct Plant Biol 36:66–72

Vysotskaya L, Wilkinson S, Davies WJ, Arkhipova T, Kudoyarova G (2011) The effect of competition from neighbours on stomatal conductance in lettuce and tomato plants. Plant, Cell Environ 34:729–737

Wang Y, Ying J, Kuzma M, Chalifoux M, Sample A, McArthur C, Uchacz T, Sarvas C (2005) Molecular tailoring of farnesylation for plant drought tolerance and yield protection. Plant J 43:413–424

Werner T, Nehnevajova E, Köllmer I, Novak O, Strnad M, Krämer U, Schmülling T (2010) Root-specific reduction of cytokinin causes enhanced root growth, drought tolerance, and leaf mineral enrichment in Arabidopsis and tobacco. Plant Cell 22:3905–3920

Wilkinson S, Kudoyarova G, Veselov DS, Archipova TN, Davies WJ (2012) Plant hormone interactions—innovative targets for crop breeding and management. J Exp Bot 63:3499–3509

Wilkinson S, Davies WJ (1997) Xylem sap pH increase: a drought signal received at the apoplastic face of the guard cell which involves saturable ABA uptake by the epidermal symplast. Plant Physiol 113:559–573

Wilkinson S, Davies WJ (2002) ABA-based chemical signalling: the co-ordination of responses to stress in plants. Plant, Cell Environ 25:195–210

Wilkinson S, Davies WJ (2010) Drought, ozone, ABA and ethylene: new insights from cell to plant to community. Plant, Cell Environ 33:510–525

Yang J, Zhang J (2010) Grain-filling problems in 'super' rice. J Exp Bot 61:1–5

Yang J, Peng S, Visperas RM, Sanico AL, Zhu Q, Gu S (2000) Grain filling pattern and cytokinin content in the grains and roots of rice plants. Plant Growth Regul 30:261–270

Yang J, Zhang J, Wang Z, Liu K, Wang P (2006) Abscisic acid and ethylene interact in wheat grains in response to soil-drying during grain filling. New Phytol 171:293–303

Yang J, Zhang J, Liu K, Wang Z, Liu L (2007) Abscisic acid and ethylene interact in rice spikelets in response to water stress during meiosis. J Plant Growth Regul 26:318–328

Zalewski W, Galuszka P, Gasparis S, Orczyk W, Nadolska-Orczyk A (2010) Silencing of the HvCKX1 gene decreases the cytokinin oxidase/dehydrogenase level in barley and leads to higher plant productivity. J Exp Bot 61:1839–1851

Zhang J, Davies WJ (1989) Abscisic acid produced in dehydrating roots may enable the plant to measure the water status of the soil. Plant, Cell Environ 12:73–81

Zhang H, Tan G, Wang Z, Yang J, Zhang J (2009) Ethylene and ACC levels in developing grains are related to the poor appearance and milling quality of rice. Plant Growth Regul 58:85–96

Zhao B, Liu K, Zhang H, Zhu Q, Yang J (2007) Causes of poor grain plumpness of two line hybrids and their relationships to the contents of hormones in rice grain. Agricultural Sciences in China 6:930–940

Chapter 11
Genome-Wide Transcriptional Reprogramming Under Drought Stress

Hao Chen and Liming Xiong

Abstract Soil water deficit is one of the major factors limiting plant productivity. Plants cope with this adverse environmental condition by coordinating the up- or downregulation of an array of stress responsive genes. Reprogramming the expression of these genes leads to rebalanced development and growth that are in concert with the reduced water availability and that ultimately confer enhanced stress tolerance. Currently, several techniques have been employed to monitor genome-wide transcriptional reprogramming under drought stress. The results from these high throughput studies indicate that drought stress-induced transcriptional reprogramming is dynamic, has temporal and spatial specificity, and is coupled with the circadian clock and phytohormone signaling pathways.

11.1 Introduction

As global climates change, plants are more frequently exposed to adverse growth conditions, among which water deficit is the most devastating and restricting in terms of plant growth, development, and crop productivity (Boyer 1982; Chaves et al. 2003). Anchored to the soil through their root systems, plants must respond and adapt to the reduced water availability in order to survive. Many physiological and developmental changes could occur when plants encounter water deficit that mainly results from drought conditions. Short-term responses of plants to drought stress include the closure of stomata pores, leaf rolling, and withering. In a longer

H. Chen · L. Xiong (✉)
Plant Stress Genomics Laboratory and Division of Chemical and Life Sciences and Engineering, King Abdullah University of Science and Technology (KAUST), Thuwal 23955-6900, Saudi Arabia
e-mail: liming.xiong@kaust.edu.sa

R. Aroca (ed.), *Plant Responses to Drought Stress*,
DOI: 10.1007/978-3-642-32653-0_11, © Springer-Verlag Berlin Heidelberg 2012

term, changes in the root architecture, deposits of more leaf waxes, and promoting flowering can also occur. However, one of the major responses of plants to drought stress that occurs at the molecular level is the induction of drought-responsive genes. The products of these genes have been shown in many transgenic studies to be able to mitigate damages caused by drought stress through reconfiguring growth and development (Bohnert et al. 1995; Ingram and Bartels 1996; Hasegawa et al. 2000; Finkelstein et al. 2002; Xiong et al. 2002; Zhu 2002; Shinozaki et al. 2003; Yamaguchi-Shinozaki and Shinozaki 2006; Nakashima et al. 2009). But this protection and increased survivability come at a price, as ectopic expression of such drought-responsive genes often leads to undesired effects, including retarded growth and/or yield loss (Kasuga et al. 1999; Zhang et al. 2004; Umezawa et al. 2006). It is therefore imperative that we decipher the modes of action for such drought-responsive genes so that we can uncouple the undesired side effects such as growth retardation from the enhanced drought tolerance traits. To accomplish this, we need a much deeper understanding of plant drought stress response and adaptation.

One way to gain comprehensive knowledge of the genetic and molecular mechanisms underlying drought adaptation is to study the genome-wide transcriptional reprogramming that occurs due to drought stress. The transcript levels of tens of thousands of genes can be monitored simultaneously using one of several methods to compare either drought and well-watered conditions or drought-tolerant and -sensitive cultivars. The current methods can be grouped into two major categories: hybridization-based and direct sequencing. Hybridization-based assays, such as microarray analysis, are relatively cost-effective but are only suitable for species with more or less decoded genomes: DNA sequences must be known to make the microarray chips and only the nucleic acids present on the chips can be assessed. On the other hand, direct sequencing using next-generation technology can identify totally unknown genes from species without any previous knowledge of its genome. Theoretically, all transcripts including not only protein-coding mRNAs but also long non-coding RNAs and precursors of small RNAs can be detected. Next-generation sequencing is currently expensive, although it is becoming more affordable as the technology continues developing and its application becomes more widespread.

As a model organism and the first plant species with a sequenced genome, Arabidopsis has been the subject of large-scale gene expression analyses under drought stress conditions (Seki et al. 2001; Cheong et al. 2002; Hoth et al. 2002; Kreps et al. 2002; Seki et al. 2002; Liu et al. 2008; Matsui et al. 2008; Zeller et al. 2009; Wilkins et al. 2010). As more expressed sequence tag (EST) libraries or genomes of plant species especially crop plants become available, more and more high-throughput analyses under drought stress have been done with other plant species such as rice (Rabbani et al. 2003; Hazen et al. 2005; Zhou et al. 2007; Degenkolbe et al. 2009; Narsai et al. 2010; Lenka et al. 2011; Wang et al. 2011a), maize (Yu and Setter 2003; Hayano-Kanashiro et al. 2009; Li et al. 2009; Marino et al. 2009; Luo et al. 2010; Zheng et al. 2010), sorghum (Buchanan et al. 2005; Pratt et al. 2005), *Thellungiella* (Wong et al. 2005, 2006), *Medicago* (Buitink et al.

2006), wheat (Xue et al. 2006; Aprile et al. 2009; Ergen et al. 2009), poplar (Bogeat-Triboulot et al. 2007; Caruso et al. 2008; Cohen et al. 2010; Hamanishi et al. 2010), peanut (Drame et al. 2007), soybean (Irsigler et al. 2007), chickpea (Mantri et al. 2007; Molina et al. 2008; Jain and Chattopadhyay 2010), sugarcane (Rocha et al. 2007), sunflower (Roche et al. 2007), barley (Talame et al. 2007; Guo et al. 2009; Abebe et al. 2010), potato (Schafleitner et al. 2007; Watkinson et al. 2008; Evers et al. 2010), *Brassica rapa* (Lee et al. 2008), citrus (Gimeno et al. 2009), Bermudagrass (Kim et al. 2009), moss (Cuming et al. 2007), *Jatropha curcas* (Costa et al. 2010), tomato (Gong et al. 2010), millet (Lata et al. 2010), ryegrass (Liu and Jiang 2010), loblolly pine (Lorenz et al. 2011), and cotton (Payton et al. 2011). In this chapter, we provide an overview of genome-wide transcriptome studies in both model and non-model organisms, delineate how drought signaling interacts with other signaling pathways, and discuss how drought transcriptome reprogramming can be achieved through a central regulatory network and epigenetic modifications.

11.2 Transcriptional Reprogramming Landscapes Under Drought Stress

Although certain genes that are highly up- or downregulated by drought stress in one plant species are regulated in a similar way in other plant species, significant differences in the expression patterns of many genes have been observed from large-scale gene expression-profiling studies performed with different species or, sometimes, even the same species. Generally, relatively low correspondence has been observed among different species or between different experiments on the same species, which may partially be due to differences in experimental procedures, developmental stages, and/or tissue types. It is also possible that genome-wide transcriptional reprogramming induced by drought stress is very plastic. In yeast exposed to new challenging growth conditions, it has been shown that the expression patterns of a large fraction of genes (out of >1200 responding genes) could not be reproduced in repeated experiments (Stern et al. 2007). Likewise in plants, drought stress responses can be represented by varied combinations of the expression of drought-responsive genes. This suggests that multiple pathways are activated in parallel and that these pathways can confer drought tolerance independently or cooperatively. In line with this notion, Arabidopsis *NUCLEAR FACTOR Y, SUBUNIT B1 (NF-YB1)* overexpression lines with improved performance under drought conditions showed altered expression for less than 0.5 % of the 23,000 genes in the genome. There was no significant overlap found between the transcriptome of the *NF-YB1* overexpression line and that of either a *C-REPEAT BINDING FACTOR* 4 *(CBF4)* overexpression line or wild-type seedlings treated with abscisic acid (ABA) (Nelson et al. 2007). Moreover, significant divergence between the Arabidopsis and rice drought transcriptomes also indicates that unique combinations of genes are utilized by different plant species to respond to drought stress challenges (Narsai et al. 2010).

It is worth noting that many transcriptome analyses were performed with phenotypically divergent accessions, such as drought-sensitive and -tolerant cultivars, in parallel (Hazen et al. 2005; Wong et al. 2006; Xue et al. 2006; Roche et al. 2007; Watkinson et al. 2008; Aprile et al. 2009; Guo et al. 2009; Hayano-Kanashiro et al. 2009; Gong et al. 2010; Hamanishi et al. 2010; Jain and Chattopadhyay 2010; Secenji et al. 2010; Zheng et al. 2010; Lenka et al. 2011). These studies revealed two common scenarios in plants with contrasting stress resistance. One scenario is that drought-sensitive lines respond to drought stress more slowly and with a lower capacity, since fewer genes are activated and most often they are expressed at lower levels under water deficit conditions (Hazen et al. 2005; Hayano-Kanashiro et al. 2009; Hamanishi et al. 2010; Jain and Chattopadhyay 2010). The second scenario is that drought-tolerant and -sensitive genotypes display opposite expression profiles for the majority of differentially expressed genes (Roche et al. 2007). So far, only a few differentially expressed genes that also correlate with drought tolerance have been identified (Hazen et al. 2005; Xue et al. 2006; Roche et al. 2007; Guo et al. 2009; Gong et al. 2010; Jain and Chattopadhyay 2010). In the near future, comparative studies combining genome tiling array and RNAseq techniques with drought tolerant quantitative trait locus (QTL) association mapping may discover additional key factors responsible for drought tolerance.

Based on their biological functions, drought-stress regulated genes from these transcriptomic analyses can be grouped into the following functional categories: (a) Genes encoding components involved in the drought stress signal transduction; (b) Genes encoding stress damage mitigating agents; and (c) Genes encoding proteins related to growth and development.

Signal transduction components in drought stress response include transcription factors (such as AP2/EREBP, MYB, NAC, WRKY, zinc finger, bHLH, DREB/CBF, and bZIP/AREB/ABF proteins) that often act as master regulators to modulate multiple downstream genes, protein kinases, protein phosphatases, phospholipids metabolizing enzymes, and other components of calcium-coupled phosphoprotein cascades. Other genes encode proteins regulating RNA metabolism and stability (such as DEAD box RNA helicases), translation (ribosomal proteins), and protein degradation (proteases/protease inhibitors, ubiquitin ligase). These proteins could fine-tune the signaling output.

Genes encoding damaging mitigating agents may encode chaperone proteins, such as heat-shock proteins, KIN (cold-inducible) proteins, and late-embryogenesis-abundant (LEA) proteins, or enzymes involved in the synthesis of osmoprotectant compounds (amino acids such as proline, amines such as spermidine, sugars such as raffinose and trehalose, sugar alcohols such as galactinol, mannitol, and inositol). In addition, genes that are regulated by drought stress may encode enzymes involved in detoxification (such as aldehyde dehydrogenase, aminotransferases, and tocopherol enzymes), generation and scavenging of reactive oxygen species (ROS) (such as NADPH oxidase, catalase, peroxidase, and ascorbate peroxidase), or biosynthesis and catabolism of abscisic acid (ABA) (such as ZEP, NCED, ABA2, LOS5/ABA3, AAO, cytochrome p450, ABA hydroxylase). The additive effects of these proteins or their products can improve plant drought tolerance.

Genes encoding proteins related to growth and development can be regulated by drought stress. These include proteins that control cell division and cellular differentiation (such as cyclin-dependent protein kinases), components of the cytoskeleton (such as actin and tubulin), proteins in the light reaction of photosynthesis, enzymes for pigment biosynthesis or catabolism, enzymes involved in cell membrane biogenesis and cell wall modifications (such as proline- or glycine-rich proteins and expansins), and lignin and chalcone biosynthesis enzymes. The combined activation of these genes will collectively reduce the damaging effects of stress and increase the drought tolerance of the plants.

11.3 Temporal and Spatial Specificity of Drought Stress-Induced Transcriptional Reprogramming

In early studies, drought-induced transcriptomic changes were monitored only in whole seedlings or a single tissue type (Seki et al. 2001; Cheong et al. 2002; Hoth et al. 2002; Kreps et al. 2002; Seki et al. 2002). Over time, more and more studies were performed with samples from different tissue types and/or at different developmental stages (Kreps et al. 2002; Oztur et al. 2002; Yu and Setter 2003; Roche et al. 2007; Talame et al. 2007; Zhou et al. 2007; Ergen et al. 2009; Abebe et al. 2010; Cohen et al. 2010; Luo et al. 2010; Narsai et al. 2010; Payton et al. 2011; Wang et al. 2011a; Worch et al. 2011). The results from these detailed comparative studies indicate that drought stress does modulate a common set of genes but not all genes are regulated similarly in different tissue types and/or at different developmental stages.

Upon drought stress, genes in some functional categories may be turned on or off only in certain tissues or cell types or at particular developmental stages. There are genes in other functional categories that display opposite responses between two different tissues. For example, in barley, wheat, cotton, and poplar, the majority of genes induced in leaves by dehydration are only slightly induced or not regulated at all in roots, and vice versa for the genes induced in the roots (Kreps et al. 2002; Oztur et al. 2002; Ergen et al. 2009; Cohen et al. 2010; Payton et al. 2011). In the placenta and endosperm of maize, less than ten percent of drought stress-induced genes share similar regulation patterns (Yu and Setter 2003). Organ-specific expression profiles under drought conditions were also observed among rice shoot, flag leaf, and panicle (Zhou et al. 2007), rice leaf, young panicle, and root (Wang et al. 2011a), and barley lemma, palea, seed, and awn (Abebe et al. 2010; Worch et al. 2011). Moreover, converse gene expression patterns in response to drought stress were observed between sunflower embryos and leaves (Roche et al. 2007) and among rice leaves, young panicles and roots at the same developmental stage (Wang et al. 2011a). These tissue-specific expression patterns of stress-responsive genes may have resulted from tissue-specific signaling and transcription or from different susceptibility among tissues to the stress signal.

Drought stress-induced reprogramming of gene expression is affected by the developmental stage of plants as well. For instance, many of the 5284 genes differentially regulated by drought stress in rice display developmental stage-specific regulation under drought (Wang et al. 2011a). Since many of the transcriptome profiling studies were done with young seedlings, these data may not necessarily be applicable to plants at the reproductive stage when drought stress has a dramatic impact on yield. The temporal- and spatial- regulation of drought stress-modulated genes suggests that various developmental cues interact with drought response and thus determine the ultimate output of drought-induced transcriptional reprogramming.

11.4 Transcriptional Reprogramming Under Drought Stress Interacts With Phytohormone or Other Abiotic Stress Signaling Pathways

Transcriptomics studies also reveal that drought stress-induced transcriptional reprogramming overlaps with other abiotic stress and plant hormone signaling pathways. Since drought, cold, and salt stresses all lead to cellular dehydration (Bohnert et al. 1995; Ingram and Bartels 1996; Hasegawa et al. 2000; Xiong et al. 2002; Zhu 2002; Yamaguchi-Shinozaki and Shinozaki 2006), there is moderate overlap between drought stress-induced genes and cold-induced and higher overlap with salinity stress-induced (Cheong et al. 2002; Seki et al. 2002; Shinozaki et al. 2003; Shinozaki and Yamaguchi-Shinozaki 2007; Nakashima et al. 2009). Genes induced by these multiple abiotic stresses belong to all three above-mentioned drought stress regulated functional categories, suggesting that these abiotic stresses share some common components in all steps, from signal perception and transduction to signaling output. The concerted action of such gene products presumably protects cellular integrity from dehydration damage caused by these various stresses (Bohnert et al. 1995; Ingram and Bartels 1996; Hasegawa et al. 2000; Zhu 2002).

To a lesser extent, the drought stress-induced transcriptome overlaps with that of heat stress, oxidative stress, and biotic stress. Drought stress leads to reduced stomata aperture, carbon dioxide uptake, and photosynthesis efficiency, which in turn result in overaccumulation of ROS (Foyer and Shigeoka 2011). For example, heat-shock protein genes, including *HSP*2, *HSP*70, and *HSP*82, and defense-related genes, including lipid transfer proteins, pathogenesis-related proteins, proteases, protease inhibitors, and NADPH oxidase, are up-regulated by drought stress in barley lemma, palea, and awn (Abebe et al. 2010). The activation of defense genes may partly have to do with drought stress-induced accumulation of

ROS, which serves as a signal to activate MAPK cascades in biotic stress response as well (Foyer and Shigeoka 2011; Mittler et al. 2011).

The plant hormone ABA plays important roles in abiotic stress response and tolerance (Bohnert et al. 1995; Ingram and Bartels 1996; Xiong et al. 1999; Hasegawa et al. 2000; Finkelstein et al. 2002; Xiong et al. 2002; Zhu 2002; Shinozaki et al. 2003; Yamaguchi-Shinozaki and Shinozaki 2006). ABA biosynthesis is induced by drought and other abiotic stresses, especially high salinity (Xiong and Zhu 2003). This increased biosynthesis mainly results from increased expression of ABA biosynthetic genes. Indeed, genes encoding enzymes in ABA biosynthesis and metabolism including *ABA1*, *NCED*, *ABA2*, *LOS5/ABA3*, *AAO*, and ABA hydroxylase are induced by drought stress (Xiong et al. 2002; Zhu 2002; Xiong and Zhu 2003; Seki et al. 2007; Seiler et al. 2011). Furthermore, more than half of the genes regulated by drought stress are also responsive to ABA treatment (Seki et al. 2001; Oztur et al. 2002; Seki et al. 2002). Analysis of the *cis*-elements in promoters of genes regulated by drought stress indicates that two conserved motifs, ABA-responsive element (ABRE) and dehydration-responsive element (DRE)/C-repeat (CRT), are involved in ABA-dependent and ABA-independent activation of stress responsive genes (Shinozaki et al. 2003; Shinozaki and Yamaguchi-Shinozaki 2007). Transcription factors of the bZIP/AREB/ABF or DREB/CBF classes bind to and activate ABRE or DRE/CRT-containing promoters, respectively. Interestingly, the expression of some of these transcription factor genes is also induced by stress or ABA.

However, the ABA-dependent and ABA-independent pathways may not function independent of each other. For example, although genes such as *RD29A*, *KIN1*, and *COR47* harbor both ABRE and DRE/CRT elements in their promoters, their induction under osmotic stress is either severely impaired or virtually blocked by mutation in the ABA biosynthesis mutant *los5/aba3* (Xiong et al. 2001). Furthermore, ectopic expression of the transcription factor *DREB2A* alone is not able to activate downstream genes (Liu et al. 1998). Microarray analysis showed that overexpression of the transcription factor *ABF3* has no effect on the transcriptome in the absence of drought stress (Aprile et al. 2009). These results further support the notion that ABA-dependent and ABA-independent pathways, such as those involved in drought stress response, are interdependent on each other.

In addition to ABA metabolism and ABA-regulated genes, various genes involved in the metabolism, signaling, or responsiveness of other plant hormones are identified among drought stress-regulated genes. These include genes involved in jasmonic acid (JA), auxin, cytokinin, and ethylene metabolism and JA-responsive (such as AP2 transcription factors and coronatine-induced proteins), auxin-responsive (such as auxin response factors, SAUR), ethylene-responsive (such as ethylene response factor, AP2 transcription factors), cytokinin-responsive (such as type-A response regulators), GA-responsive (such as GA20ox1), salicylic acid-responsive (such as WAK2), and brassinosteroid-responsive (such as remorin) genes (Oztur et al. 2002; Seki et al. 2002; Buchanan et al. 2005; Buitink et al. 2006; Jain et al. 2006; Wong et al. 2006; Cramer et al. 2007; Huang et al. 2008; Aprile et al. 2009; Hayano-Kanashiro et al. 2009; Cohen et al. 2010; Gong et al.

2010). The phytohormones JA, ethylene, cytokinin, brassinosteroid, auxin, and GA have also been shown to modulate drought stress response to some extent, either negatively or positively (Creelman and Mullet 1997; Bianchi et al. 2002; Krishna 2003; Achard et al. 2006; Manavella et al. 2006; Kagale et al. 2007; Tran et al. 2007; Rivero et al. 2010; Zhang et al. 2010; Nishiyama et al. 2011).

11.5 Drought-Induced Transcriptional Reprogramming is Primed by the Biological Clock

Circadian rhythm, or the biological clock, acts as a multifunction timer to coordinate biochemical and physiological processes, modulate growth and development, regulate environmental responsiveness, and control hormone levels with an endogenously driven roughly 24 h cycle in all cellular organisms (Mas 2005; McClung 2008; Harmer 2009; Zhang and Kay 2010). Using GeneChip arrays representing about 8,200 different genes, the temporal patterns of gene expression in Arabidopsis were monitored at 4 h intervals over two days under normal growth conditions. The steady-state levels of several drought-induced genes, including those encoding LEA, ribosomal protein S6 kinase homolog (ATPK19), cysteine protease (RD19A) and COR6.6, were found to cycle with the circadian clock and peak at a certain time of the day (Harmer et al. 2000). Later, 306 abiotic stress (including cold, dehydration, and high salt)-induced genes were identified among 453 known circadian controlled genes (Kreps et al. 2002). Moreover, reanalysis of multiple circadian microarray experiments from publically available database suggests that from 30 to 40 % of expressed genes are clock regulated in seedlings and that clock-regulated genes are overrepresented among multiple hormone (including ABA) and abiotic stress (including drought) response pathways in Arabidopsis (Covington et al. 2008; Mizuno and Yamashino 2008). While studying TOC1 (timing of CAB expression 1), a key clock component, it was found that 39 or 37 % of genes regulated by drought stress are misregulated in overexpression lines or a TOC1 loss of function mutant, respectively (Legnaioli et al. 2009). By examining the drought-induced transcriptome at four time points—midday, late day, midnight, and pre-dawn—over a single period, multiple instances of time-of-day-specific, drought stress-regulated transcriptional reprogramming in both Arabidopsis and poplar were demonstrated and characterized (Wilkins et al. 2009; 2010). Taken together, these results indicate that circadian rhythm modifies the overall drought response by modulating drought-induced transcriptional reprogramming.

Studies with the ABA biosynthesis mutant *aba2* and with direct application of ABA showed that ABA also modulates circadian rhythm periodicity (Hanano et al. 2006). One interlocked feedback mechanism linking the circadian clock with plant responses to drought has been proposed. The transcription factor TOC1 binds to the promoter of one ABA-binding protein gene (*ABAR/CHLH/GUN5*) and controls its circadian expression. ABAR/CHLH/GUN5-mediated ABA signaling can induce TOC1 expression, and this induction increases TOC1 binding and again modulates *ABAR* circadian expression (Legnaioli et al. 2009).

11.6 Epigenetic Regulation of Drought-Induced Transcriptional Reprogramming

The mechanisms by which drought modulates transcriptional programming at the molecular level is largely unknown, although recently some novel mechanisms involved in this process have been identified. Studies on genome-wide DNA methylation found that about 12 % of site-specific methylation differences are induced by drought in the rice genome (Wang et al. 2011c). Interestingly, more than 70 % of the drought-induced methylation sites are reversed after recovery from the stress. Furthermore, drought-induced DNA methylation changes display significant levels of developmental and tissue specificity (Wang et al. 2011c). In drought-stressed poplar, change of the genome-wide DNA methylation pattern parallels with the change of the transcriptome level (Raj et al. 2011).

Drought stress also induces histone modifications, including changes in histone methylation and acetylation and histone displacement by a corresponding histone variant (Ascenzi and Gantt 1999; Kim et al. 2008). Such dynamic changes in histone status are associated with drought tolerance (Sokol et al. 2007; Chinnusamy and Zhu 2009; Granot et al. 2009).

In addition to DNA methylation and histone modifications, drought-induced transcriptional reprogramming may be modulated by non-coding RNAs as well. Drought stress-regulated small RNAs have been identified by large-scale gene expression studies (Kulcheski et al. 2011; Li et al. 2011; Qin et al. 2011; Wang et al. 2011b). Small RNAs including miRNAs and siRNAs may confer either transcriptional and/or post-transcriptional modulations of drought stress-reprogrammed genes (Yaish et al. 2011; Yan et al. 2011). On the other hand, long non-coding RNAs may regulate drought stress reprogrammed genes prominently at transcription level through chromatin remodeling (Umlauf et al. 2008; De Lucia and Dean 2011).

In summary, studies of genome-wide transcriptional reprogramming under drought stress have greatly advanced our understanding of the plant's response to drought stress. Figure 11.1 outlines some of these recent progresses as discussed above. Combining transcriptome studies with classical breeding techniques has identified a few genes differentially expressed between drought tolerant and drought sensitive genotypes that are associated with known drought tolerant QTLs. Multi-faced research such as this is both exciting and promising. Given that drought stress-induced transcriptomic changes are affected by many factors, it is important to consider such factors during the design and the implementation of the drought stress treatment to minimize the resulting variations among drought transcriptome data sets, which will also make it easier to integrate drought

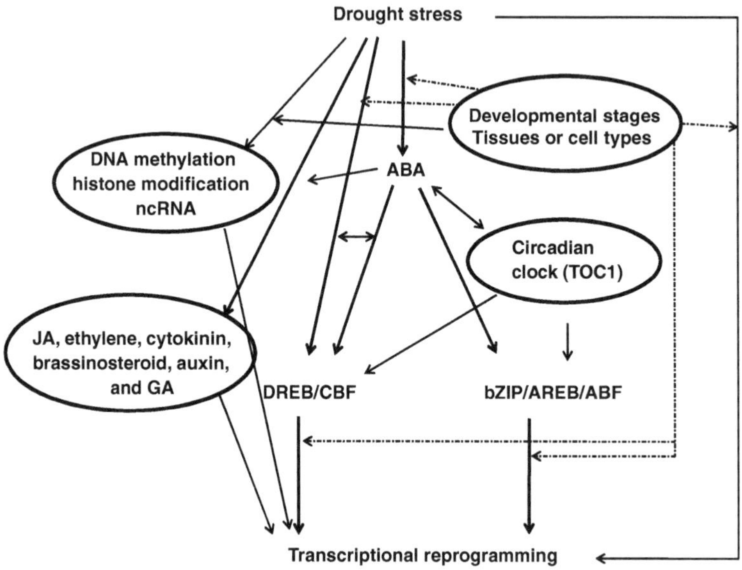

Fig. 11.1 Drought transcriptome reprogramming and its modulation by other signaling pathways. *Dashed lines* indicate that experimental evidence to demonstrate the indicated relations is still insufficient. *Double arrowed lines* indicate interactions of the pathways. Abbreviations: ABA, abscisic acid; DREB, dehydration responsive element-binding protein; CBF, C-repeat-binding factor; AREB, ABA-responsive element binding protein; bZIP, basic leucine zipper; ncRNA, non-coding RNA; TOC1, timing of CAB expression 1; JA, jasmonic acid; GA, gibberellic acid

transcriptomic data with other drought stress related "omics" data. Future research based on full-genome tiling array or next generation mRNA-Seq technology will result in a better appreciation of the landscape of drought stress-induced transcriptional reprogramming. In addition, systems biology that integrates transcriptomes, proteomes, metabolomes, and DNA methylomes may help to reveal ways to finally achieve drought tolerance without yield penalty or growth retardation through a coordinated fine-tuning of key drought-tolerant determinants.

References

Abebe T, Melmaiee K, Berg V, Wise RP (2010) Drought response in the spikes of barley: gene expression in the lemma, palea, awn, and seed. Funct Integr Genomics 10:191–205

Achard P, Cheng H, De Grauwe L, Decat J, Schoutteten H, Moritz T, Van Der Straeten D, Peng J, Harberd NP (2006) Integration of plant responses to environmentally activated phytohormonal signals. Science 311:91–94

Aprile A, Mastrangelo AM, De Leonardis AM, Galiba G, Roncaglia E, Ferrari F, De Bellis L, Turchi L, Giuliano G, Cattivelli L (2009) Transcriptional profiling in response to terminal drought stress reveals differential responses along the wheat genome. BMC Genomics 10:279

Ascenzi R, Gantt JS (1999) Molecular genetic analysis of the drought-inducible linker histone variant in *Arabidopsis thaliana*. Plant Mol Biol 41:159–169

Bianchi MW, Damerval C, Vartanian N (2002) Identification of proteins regulated by cross-talk between drought and hormone pathways in Arabidopsis wild-type and auxin-insensitive mutants, axr1 and axr2. Funct Plant Biol 29:55–61

Bogeat-Triboulot MB, Brosche M, Renaut J, Jouve L, Le Thiec D, Fayyaz P, Vinocur B, Witters E, Laukens K, Teichmann T, Altman A, Hausman JF, Polle A, Kangasjarvi J, Dreyer E (2007) Gradual soil water depletion results in reversible changes of gene expression, protein profiles, ecophysiology, and growth performance in *Populus euphratica*, a poplar growing in arid regions. Plant Physiol 143:876–892

Bohnert HJ, Nelson DE, Jensen RG (1995) Adaptations to environmental stresses. Plant Cell 7:1099–1111

Boyer JS (1982) Plant productivity and environment. Science 218:443–448

Buchanan CD, Lim S, Salzman RA, Kagiampakis I, Morishige DT, Weers BD, Klein RR, Pratt LH, Cordonnier-Pratt MM, Klein PE, Mullet JE (2005) Sorghum bicolor's transcriptome response to dehydration, high salinity and ABA. Plant Mol Biol 58:699–720

Buitink J, Leger JJ, Guisle I, Vu BL, Wuilleme S, Lamirault G, Le Bars A, Le Meur N, Becker A, Kuster H, Leprince O (2006) Transcriptome profiling uncovers metabolic and regulatory processes occurring during the transition from desiccation-sensitive to desiccation-tolerant stages in *Medicago truncatula* seeds. Plant J 47:735–750

Caruso A, Chefdor F, Carpin S, Depierreux C, Delmotte FM, Kahlem G, Morabito D (2008) Physiological characterization and identification of genes differentially expressed in response to drought induced by PEG 6000 in *Populus canadensis* leaves. J Plant Physiol 165:932–941

Chaves M, Maroco J, Pereira J (2003) Understanding plant responses to drought—from genes to the whole plant. Funct Plant Biol 30:239–264

Cheong YH, Chang HS, Gupta R, Wang X, Zhu T, Luan S (2002) Transcriptional profiling reveals novel interactions between wounding, pathogen, abiotic stress, and hormonal responses in Arabidopsis. Plant Physiol 129:661–677

Chinnusamy V, Zhu JK (2009) Epigenetic regulation of stress responses in plants. Curr Opin Plant Biol 12:133–139

Cohen D, Bogeat-Triboulot MB, Tisserant E, Balzergue S, Martin-Magniette ML, Lelandais G, Ningre N, Renou JP, Tamby JP, Le Thiec D, Hummel I (2010) Comparative transcriptomics of drought responses in *Populus*: a meta-analysis of genome-wide expression profiling in mature leaves and root apices across two genotypes. BMC Genomics 11:630

Costa GG, Cardoso KC, Del Bem LE, Lima AC, Cunha MA, de Campos-Leite L, Vicentini R, Papes F, Moreira RC, Yunes JA, Campos FA, Da Silva MJ (2010) Transcriptome analysis of the oil-rich seed of the bioenergy crop *Jatropha curcas* L. BMC Genomics 11:462

Covington MF, Maloof JN, Straume M, Kay SA, Harmer SL (2008) Global transcriptome analysis reveals circadian regulation of key pathways in plant growth and development. Genome Biol 9:R130

Cramer GR, Ergul A, Grimplet J, Tillett RL, Tattersall EA, Bohlman MC, Vincent D, Sonderegger J, Evans J, Osborne C, Quilici D, Schlauch KA, Schooley DA, Cushman JC (2007) Water and salinity stress in grapevines: early and late changes in transcript and metabolite profiles. Funct Integr Genomics 7:111–134

Creelman RA, Mullet JE (1997) Biosynthesis and action of jasmonates in plants. Ann Rev Plant Physiol Plant Mol Biol 48:355–381

Cuming AC, Cho SH, Kamisugi Y, Graham H, Quatrano RS (2007) Microarray analysis of transcriptional responses to abscisic acid and osmotic, salt, and drought stress in the moss, Physcomitrella patens. New Phytol 176:275–287

De Lucia F, Dean C (2011) Long non-coding RNAs and chromatin regulation. Curr Opin Plant Biol 14:168–173

Degenkolbe T, Do PT, Zuther E, Repsilber D, Walther D, Hincha DK, Kohl KI (2009) Expression profiling of rice cultivars differing in their tolerance to long-term drought stress. Plant Mol Biol 69:133–153

Drame KN, Clavel D, Repellin A, Passaquet C, Zuily-Fodil Y (2007) Water deficit induces variation in expression of stress-responsive genes in two peanut (*Arachis hypogaea* L.) cultivars with different tolerance to drought. Plant Physiol Biochem 45:236–243
Ergen NZ, Thimmapuram J, Bohnert HJ, Budak H (2009) Transcriptome pathways unique to dehydration tolerant relatives of modern wheat. Funct Integr Genomics 9:377–396
Evers D, Lefevre I, Legay S, Lamoureux D, Hausman JF, Rosales RO, Marca LR, Hoffmann L, Bonierbale M, Schafleitner R (2010) Identification of drought-responsive compounds in potato through a combined transcriptomic and targeted metabolite approach. J Exp Bot 61:2327–2343
Finkelstein RR, Gampala SS, Rock CD (2002) Abscisic acid signaling in seeds and seedlings. Plant Cell 14 Suppl:S15–S45
Foyer CH, Shigeoka S (2011) Understanding oxidative stress and antioxidant functions to enhance photosynthesis. Plant Physiol 155:93–100
Gimeno J, Gadea J, Forment J, Perez-Valle J, Santiago J, Martinez-Godoy MA, Yenush L, Belles JM, Brumos J, Colmenero-Flores JM, Talon M, Serrano R (2009) Shared and novel molecular responses of mandarin to drought. Plant Mol Biol 70:403–420
Gong P, Zhang J, Li H, Yang C, Zhang C, Zhang X, Khurram Z, Zhang Y, Wang T, Fei Z, Ye Z (2010) Transcriptional profiles of drought-responsive genes in modulating transcription signal transduction, and biochemical pathways in tomato. J Exp Bot 61:3563–3575
Granot G, Sikron-Persi N, Gaspan O, Florentin A, Talwara S, Paul LK, Morgenstern Y, Granot Y, Grafi G (2009) Histone modifications associated with drought tolerance in the desert plant *Zygophyllum dumosum* Boiss. Planta 231:27–34
Guo P, Baum M, Grando S, Ceccarelli S, Bai G, Li R, von Korff M, Varshney RK, Graner A, Valkoun J (2009) Differentially expressed genes between drought-tolerant and drought-sensitive barley genotypes in response to drought stress during the reproductive stage. J Exp Bot 60:3531–3544
Hamanishi ET, Raj S, Wilkins O, Thomas BR, Mansfield SD, Plant AL, Campbell MM (2010) Intraspecific variation in the Populus balsamifera drought transcriptome. Plant Cell Environ 33:1742–1755
Hanano S, Domagalska MA, Nagy F, Davis SJ (2006) Multiple phytohormones influence distinct parameters of the plant circadian clock. Genes Cells 11:1381–1392
Harmer SL (2009) The circadian system in higher plants. Annu Rev Plant Biol 60:357–377
Harmer SL, Hogenesch JB, Straume M, Chang HS, Han B, Zhu T, Wang X, Kreps JA, Kay SA (2000) Orchestrated transcription of key pathways in Arabidopsis by the circadian clock. Science 290:2110–2113
Hasegawa PM, Bressan RA, Zhu JK, Bohnert HJ (2000) Plant cellular and molecular responses to high salinity. Ann Rev Plant Physiol Plant Mol Biol 51:463–499
Hayano-Kanashiro C, Calderon-Vazquez C, Ibarra-Laclette E, Herrera-Estrella L, Simpson J (2009) Analysis of gene expression and physiological responses in three Mexican maize landraces under drought stress and recovery irrigation. PLoS ONE 4:e7531
Hazen SP, Pathan MS, Sanchez A, Baxter I, Dunn M, Estes B, Chang HS, Zhu T, Kreps JA, Nguyen HT (2005) Expression profiling of rice segregating for drought tolerance QTLs using a rice genome array. Funct Integr Genomics 5:104–116
Hoth S, Morgante M, Sanchez JP, Hanafey MK, Tingey SV, Chua NH (2002) Genome-wide gene expression profiling in Arabidopsis thaliana reveals new targets of abscisic acid and largely impaired gene regulation in the abi1-1 mutant. J Cell Sci 115:4891–4900
Huang D, Wu W, Abrams SR, Cutler AJ (2008) The relationship of drought-related gene expression in *Arabidopsis thaliana* to hormonal and environmental factors. J Exp Bot 59:2991–3007
Ingram J, Bartels D (1996) The molecular basis of dehydration tolerance in plants. Ann Rev Plant Physiol Plant Mol Biol 47:377–403
Irsigler AS, Costa MD, Zhang P, Reis PA, Dewey RE, Boston RS, Fontes EP (2007) Expression profiling on soybean leaves reveals integration of ER- and osmotic-stress pathways. BMC Genomics 8:431

Jain D, Chattopadhyay D (2010) Analysis of gene expression in response to water deficit of chickpea (*Cicer arietinum* L.) varieties differing in drought tolerance. BMC Plant Biol 10:24
Jain M, Tyagi AK, Khurana JP (2006) Molecular characterization and differential expression of cytokinin-responsive type-A response regulators in rice (*Oryza sativa*). BMC Plant Biol 6:1
Kagale S, Divi UK, Krochko JE, Keller WA, Krishna P (2007) Brassinosteroid confers tolerance in *Arabidopsis thaliana* and *Brassica napus* to a range of abiotic stresses. Planta 225:353–364
Kasuga M, Liu Q, Miura S, Yamaguchi-Shinozaki K, Shinozaki K (1999) Improving plant drought, salt, and freezing tolerance by gene transfer of a single stress-inducible transcription factor. Nat Biotechnol 17:287–291
Kim JM, To TK, Ishida J, Morosawa T, Kawashima M, Matsui A, Toyoda T, Kimura H, Shinozaki K, Seki M (2008) Alterations of lysine modifications on the histone H3N-tail under drought stress conditions in *Arabidopsis thaliana*. Plant Cell Physiol 49:1580–1588
Kim C, Lemke C, Paterson AH (2009) Functional dissection of drought-responsive gene expression patterns in *Cynodon dactylon* L. Plant Mol Biol 70:1–16
Kreps JA, Wu Y, Chang HS, Zhu T, Wang X, Harper JF (2002) Transcriptome changes for Arabidopsis in response to salt, osmotic, and cold stress. Plant Physiol 130:2129–2141
Krishna P (2003) Brassinosteroid-mediated stress responses. J Plant Growth Regul 22:289–297
Kulcheski FR, de Oliveira LF, Molina LG, Almerao MP, Rodrigues FA, Marcolino J, Barbosa JF, Stolf-Moreira R, Nepomuceno AL, Marcelino-Guimaraes FC, Abdelnoor RV, Nascimento LC, Carazzolle MF, Pereira GA, Margis R (2011) Identification of novel soybean microRNAs involved in abiotic and biotic stresses. BMC Genomics 12:307
Lata C, Sahu PP, Prasad M (2010) Comparative transcriptome analysis of differentially expressed genes in foxtail millet (*Setaria italica* L.) during dehydration stress. Biochem Biophys Res Commun 393:720–727
Lee SC, Lim MH, Kim JA, Lee SI, Kim JS, Jin M, Kwon SJ, Mun JH, Kim YK, Kim HU, Hur Y, Park BS (2008) Transcriptome analysis in *Brassica rapa* under the abiotic stresses using Brassica 24 K oligo microarray. Mol Cells 26:595–605
Legnaioli T, Cuevas J, Mas P (2009) TOC1 functions as a molecular switch connecting the circadian clock with plant responses to drought. EMBO J 28:3745–3757
Lenka SK, Katiyar A, Chinnusamy V, Bansal KC (2011) Comparative analysis of drought-responsive transcriptome in Indica rice genotypes with contrasting drought tolerance. Plant Biotech J 9:315–327
Li Y, Sun C, Huang Z, Pan J, Wang L, Fan X (2009) Mechanisms of progressive water deficit tolerance and growth recovery of Chinese maize foundation genotypes Huangzao 4 and Chang 7–2, which are proposed on the basis of comparison of physiological and transcriptomic responses. Plant Cell Physiol 50:2092–2111
Li B, Qin Y, Duan H, Yin W, Xia X (2011) Genome-wide characterization of new and drought stress responsive microRNAs in *Populus euphratica*. J Exp Bot 62:3765–3779
Liu S, Jiang Y (2010) Identification of differentially expressed genes under drought stress in perennial ryegrass. Physiol Plant 139:375–387
Liu Q, Kasuga M, Sakuma Y, Abe H, Miura S, Yamaguchi-Shinozaki K, Shinozaki K (1998) Two transcription factors, DREB1 and DREB2, with an EREBP/AP2 DNA binding domain separate two cellular signal transduction pathways in drought- and low-temperature-responsive gene expression, respectively, in Arabidopsis. Plant Cell 10:1391–1406
Liu HH, Tian X, Li YJ, Wu CA, Zheng CC (2008) Microarray-based analysis of stress-regulated microRNAs in *Arabidopsis thaliana*. RNA 14:836–843
Lorenz WW, Alba R, Yu YS, Bordeaux JM, Simoes M, Dean JF (2011) Microarray analysis and scale-free gene networks identify candidate regulators in drought-stressed roots of loblolly pine (*P. taeda* L.). BMC Genomics 12:264
Luo M, Liu J, Lee RD, Scully BT, Guo B (2010) Monitoring the expression of maize genes in developing kernels under drought stress using oligo-microarray. J Integ Plant Biol 52:1059–1074

Manavella PA, Arce AL, Dezar CA, Bitton F, Renou JP, Crespi M, Chan RL (2006) Cross-talk between ethylene and drought signalling pathways is mediated by the sunflower Hahb-4 transcription factor. Plant J 48:125–137

Mantri NL, Ford R, Coram TE, Pang EC (2007) Transcriptional profiling of chickpea genes differentially regulated in response to high-salinity, cold and drought. BMC Genomics 8:303

Marino R, Ponnaiah M, Krajewski P, Frova C, Gianfranceschi L, Pe ME, Sari-Gorla M (2009) Addressing drought tolerance in maize by transcriptional profiling and mapping. Mol Gen Genomics 281:163–179

Mas P (2005) Circadian clock signaling in *Arabidopsis thaliana*: from gene expression to physiology and development. Internl J Dev Biol 49:491–500

Matsui A, Ishida J, Morosawa T, Mochizuki Y, Kaminuma E, Endo TA, Okamoto M, Nambara E, Nakajima M, Kawashima M, Satou M, Kim JM, Kobayashi N, Toyoda T, Shinozaki K, Seki M (2008) Arabidopsis transcriptome analysis under drought, cold, high-salinity and ABA treatment conditions using a tiling array. Plant Cell Physiol 49:1135–1149

McClung CR (2008) Comes a time. Curr Opin Plant Biol 11:514–520

Mittler R, Vanderauwera S, Suzuki N, Miller G, Tognetti VB, Vandepoele K, Gollery M, Shulaev V, Van Breusegem F (2011) ROS signaling: the new wave? Trends Plant Sci 16:300–309

Mizuno T, Yamashino T (2008) Comparative transcriptome of diurnally oscillating genes and hormone-responsive genes in *Arabidopsis thaliana*: insight into circadian clock-controlled daily responses to common ambient stresses in plants. Plant Cell Physiol 49:481–487

Molina C, Rotter B, Horres R, Udupa SM, Besser B, Bellarmino L, Baum M, Matsumura H, Terauchi R, Kahl G, Winter P (2008) SuperSAGE: the drought stress-responsive transcriptome of chickpea roots. BMC Genomics 9:553

Nakashima K, Ito Y, Yamaguchi-Shinozaki K (2009) Transcriptional regulatory networks in response to abiotic stresses in Arabidopsis and grasses. Plant Physiol 149:88–95

Narsai R, Castleden I, Whelan J (2010) Common and distinct organ and stress responsive transcriptomic patterns in *Oryza sativa* and *Arabidopsis thaliana*. BMC Plant Biol 10:262

Nelson DE, Repetti PP, Adams TR, Creelman RA, Wu J, Warner DC, Anstrom DC, Bensen RJ, Castiglioni PP, Donnarummo MG, Hinchey BS, Kumimoto RW, Maszle DR, Canales RD, Krolikowski KA, Dotson SB, Gutterson N, Ratcliffe OJ, Heard JE (2007) Plant nuclear factor Y(NF-Y) B subunits confer drought tolerance and lead to improved corn yields on water-limited acres. Proc Natl Acad Sci 104(42):16450–16455

Nishiyama R, Watanabe Y, Fujita Y, Le DT, Kojima M, Werner T, Vankova R, Yamaguchi-Shinozaki K, Shinozaki K, Kakimoto T, Sakakibara H, Schmulling T, Tran LS (2011) Analysis of cytokinin mutants and regulation of cytokinin metabolic genes reveals important regulatory roles of cytokinins in drought, salt and abscisic acid responses, and abscisic acid biosynthesis. Plant Cell 23:2169–2183

Oztur ZN, Talame V, Deyholos M, Michalowski CB, Galbraith DW, Gozukirmizi N, Tuberosa R, Bohnert HJ (2002) Monitoring large-scale changes in transcript abundance in drought- and salt-stressed barley. Plant Mol Biol 48:551–573

Payton P, Kottapalli KR, Kebede H, Mahan JR, Wright RJ, Allen RD (2011) Examining the drought stress transcriptome in cotton leaf and root tissue. Biotechnol Lett 33:821–828

Pratt LH, Liang C, Shah M, Sun F, Wang H, Reid SP, Gingle AR, Paterson AH, Wing R, Dean R, Klein R, Nguyen HT, Ma HM, Zhao X, Morishige DT, Mullet JE, Cordonnier-Pratt MM (2005) Sorghum expressed sequence tags identify signature genes for drought, pathogenesis, and skotomorphogenesis from a milestone set of 16,801 unique transcripts. Plant Physiol 139:869–884

Qin Y, Duan Z, Xia X, Yin W (2011) Expression profiles of precursor and mature microRNAs under dehydration and high salinity shock in *Populus euphratica*. Plant Cell Rep 30:1893–1907

Rabbani MA, Maruyama K, Abe H, Khan MA, Katsura K, Ito Y, Yoshiwara K, Seki M, Shinozaki K, Yamaguchi-Shinozaki K (2003) Monitoring expression profiles of rice genes under cold, drought, and high-salinity stresses and abscisic acid application using cDNA microarray and RNA gel-blot analyses. Plant Physiol 133:1755–1767

Raj S, Brautigam K, Hamanishi ET, Wilkins O, Thomas BR, Schroeder W, Mansfield SD, Plant AL, Campbell MM (2011) Clone history shapes Populus drought responses. Proc Natl Acad Sci U S A 108:12521–12526

Rivero RM, Gimeno J, Van Deynze A, Walia H, Blumwald E (2010) Enhanced cytokinin synthesis in tobacco plants expressing PSARK:IPT prevents the degradation of photosynthetic protein complexes during drought. Plant Cell Physiol 51:1929–1941

Rocha FR, Papini-Terzi FS, Nishiyama MY Jr, Vencio RZ, Vicentini R, Duarte RD, de Rosa VE Jr, Vinagre F, Barsalobres C, Medeiros AH, Rodrigues FA, Ulian EC, Zingaretti SM, Galbiatti JA, Almeida RS, Figueira AV, Hemerly AS, Silva-Filho MC, Menossi M, Souza GM (2007) Signal transduction-related responses to phytohormones and environmental challenges in sugarcane. BMC Genomics 8:71

Roche J, Hewezi T, Bouniols A, Gentzbittel L (2007) Transcriptional profiles of primary metabolism and signal transduction-related genes in response to water stress in field-grown sunflower genotypes using a thematic cDNA microarray. Planta 226:601–617

Schafleitner R, Gutierrez Rosales RO, Gaudin A, Alvarado Aliaga CA, Martinez GN, Tincopa Marca LR, Bolivar LA, Delgado FM, Simon R, Bonierbale M (2007) Capturing candidate drought tolerance traits in two native Andean potato clones by transcription profiling of field grown plants under water stress. Plant Physiol Biochem 45:673–690

Secenji M, Lendvai A, Miskolczi P, Kocsy G, Galle A, Szucs A, Hoffmann B, Sarvari E, Schweizer P, Stein N, Dudits D, Gyorgyey J (2010) Differences in root functions during long-term drought adaptation: comparison of active gene sets of two wheat genotypes. Plant Biol 12:871–882

Seiler C, Harshavardhan VT, Rajesh K, Reddy PS, Strickert M, Rolletschek H, Scholz U, Wobus U, Sreenivasulu N (2011) ABA biosynthesis and degradation contributing to ABA homeostasis during barley seed development under control and terminal drought-stress conditions. J Exp Bot 62:2615–2632

Seki M, Narusaka M, Abe H, Kasuga M, Yamaguchi-Shinozaki K, Carninci P, Hayashizaki Y, Shinozaki K (2001) Monitoring the expression pattern of 1,300 Arabidopsis genes under drought and cold stresses by using a full-length cDNA microarray. Plant cell 13:61–72

Seki M, Narusaka M, Ishida J, Nanjo T, Fujita M, Oono Y, Kamiya A, Nakajima M, Enju A, Sakurai T, Satou M, Akiyama K, Taji T, Yamaguchi-Shinozaki K, Carninci P, Kawai J, Hayashizaki Y, Shinozaki K (2002) Monitoring the expression profiles of 7,000 Arabidopsis genes under drought, cold and high-salinity stresses using a full-length cDNA microarray. Plant J 31:279–292

Seki M, Umezawa T, Urano K, Shinozaki K (2007) Regulatory metabolic networks in drought stress responses. Curr Opin Plant Biol 10:296–302

Shinozaki K, Yamaguchi-Shinozaki K (2007) Gene networks involved in drought stress response and tolerance. J Exp Bot 58:221–227

Shinozaki K, Yamaguchi-Shinozaki K, Seki M (2003) Regulatory network of gene expression in the drought and cold stress responses. Curr Opin Plant Biol 6:410–417

Sokol A, Kwiatkowska A, Jerzmanowski A, Prymakowska-Bosak M (2007) Up-regulation of stress-inducible genes in tobacco and Arabidopsis cells in response to abiotic stresses and ABA treatment correlates with dynamic changes in histone H3 and H4 modifications. Planta 227:245–254

Stern S, Dror T, Stolovicki E, Brenner N, Braun E (2007) Genome-wide transcriptional plasticity underlies cellular adaptation to novel challenge. Mol Sys Biol 3:106

Talame V, Ozturk NZ, Bohnert HJ, Tuberosa R (2007) Barley transcript profiles under dehydration shock and drought stress treatments: a comparative analysis. J Exp Bot 58:229–240

Tran LS, Urao T, Qin F, Maruyama K, Kakimoto T, Shinozaki K, Yamaguchi-Shinozaki K (2007) Functional analysis of AHK1/ATHK1 and cytokinin receptor histidine kinases in response to abscisic acid, drought, and salt stress in Arabidopsis. Proc Natl Acad Sci U S A 104:20623–20628

Umezawa T, Fujita M, Fujita Y, Yamaguchi-Shinozaki K, Shinozaki K (2006) Engineering drought tolerance in plants: discovering and tailoring genes to unlock the future. Curr Opin Biotechnol 17:113–122
Umlauf D, Fraser P, Nagano T (2008) The role of long non-coding RNAs in chromatin structure and gene regulation: variations on a theme. J Biol Chem 389:323–331
Wang D, Pan Y, Zhao X, Zhu L, Fu B, Li Z (2011a) Genome-wide temporal-spatial gene expression profiling of drought responsiveness in rice. BMC Genomics 12:149
Wang T, Chen L, Zhao M, Tian Q, Zhang WH (2011b) Identification of drought-responsive microRNAs in *Medicago truncatula* by genome-wide high-throughput sequencing. BMC Genomics 12:367
Wang WS, Pan YJ, Zhao XQ, Dwivedi D, Zhu LH, Ali J, Fu BY, Li ZK (2011c) Drought-induced site-specific DNA methylation and its association with drought tolerance in rice (*Oryza sativa* L.). J Exp Bot 62:1951–1960
Watkinson JI, Hendricks L, Sioson AA, Heath LS, Bohnert HJ, Grene R (2008) Tuber development phenotypes in adapted and acclimated, drought-stressed Solanum tuberosum ssp. andigena have distinct expression profiles of genes associated with carbon metabolism. Plant Physiol Biochem 46:34–45
Wilkins O, Waldron L, Nahal H, Provart NJ, Campbell MM (2009) Genotype and time of day shape the Populus drought response. Plant J 60:703–715
Wilkins O, Brautigam K, Campbell MM (2010) Time of day shapes Arabidopsis drought transcriptomes. Plant J 63:715–727
Wong CE, Li Y, Whitty BR, Diaz-Camino C, Akhter SR, Brandle JE, Golding GB, Weretilnyk EA, Moffatt BA, Griffith M (2005) Expressed sequence tags from the Yukon ecotype of *Thellungiella* reveal that gene expression in response to cold, drought and salinity shows little overlap. Plant Mol Biol 58:561–574
Wong CE, Li Y, Labbe A, Guevara D, Nuin P, Whitty B, Diaz C, Golding GB, Gray GR, Weretilnyk EA, Griffith M, Moffatt BA (2006) Transcriptional profiling implicates novel interactions between abiotic stress and hormonal responses in *Thellungiella*, a close relative of Arabidopsis. Plant Physiol 140:1437–1450
Worch S, Rajesh K, Harshavardhan VT, Pietsch C, Korzun V, Kuntze L, Borner A, Wobus U, Roder MS, Sreenivasulu N (2011) Haplotyping, linkage mapping and expression analysis of barley genes regulated by terminal drought stress influencing seed quality. BMC Plant Biol 11:1
Xiong L, Zhu JK (2003) Regulation of abscisic acid biosynthesis. Plant Physiol 133:29–36
Xiong L, Ishitani M, Zhu JK (1999) Interaction of osmotic stress, temperature, and abscisic acid in the regulation of gene expression in Arabidopsis. Plant Physiol 119:205–212
Xiong L, Ishitani M, Lee H, Zhu JK (2001) The Arabidopsis *LOS5/ABA3* locus encodes a molybdenum cofactor sulfurase and modulates cold stress- and osmotic stress-responsive gene expression. Plant Cell 13:2063–2083
Xiong L, Schumaker KS, Zhu JK (2002) Cell signaling during cold, drought, and salt stress. Plant Cell 14(Suppl):S165–S183
Xue GP, McIntyre CL, Chapman S, Bower NI, Way H, Reverter A, Clarke B, Shorter R (2006) Differential gene expression of wheat progeny with contrasting levels of transpiration efficiency. Plant Mol Biol 61:863–881
Yaish MW, Colasanti J, Rothstein SJ (2011) The role of epigenetic processes in controlling flowering time in plants exposed to stress. J Exp Bot 62:3727–3735
Yamaguchi-Shinozaki K, Shinozaki K (2006) Transcriptional regulatory networks in cellular responses and tolerance to dehydration and cold stresses. Ann Rev Plant Biol 57:781–803
Yan Y, Zhang Y, Yang K, Sun Z, Fu Y, Chen X, Fang R (2011) Small RNAs from MITE-derived stem-loop precursors regulate abscisic acid signaling and abiotic stress responses in rice. Plant J 65:820–828
Yu LX, Setter TL (2003) Comparative transcriptional profiling of placenta and endosperm in developing maize kernels in response to water deficit. Plant Physiol 131:568–582

Zeller G, Henz SR, Widmer CK, Sachsenberg T, Ratsch G, Weigel D, Laubinger S (2009) Stress-induced changes in the *Arabidopsis thaliana* transcriptome analyzed using whole-genome tiling arrays. Plant J 58:1068–1082

Zhang EE, Kay SA (2010) Clocks not winding down: unravelling circadian networks. Nat Rev Mol Cell Biol 11:764–776

Zhang JZ, Creelman RA, Zhu JK (2004) From laboratory to field. Using information from Arabidopsis to engineer salt, cold, and drought tolerance in crops. Plant Physiol 135:615–621

Zhang Z, Li F, Li D, Zhang H, Huang R (2010) Expression of ethylene response factor JERF1 in rice improves tolerance to drought. Planta 232:765–774

Zheng J, Fu J, Gou M, Huai J, Liu Y, Jian M, Huang Q, Guo X, Dong Z, Wang H, Wang G (2010) Genome-wide transcriptome analysis of two maize inbred lines under drought stress. Plant Mol Biol 72:407–421

Zhou J, Wang X, Jiao Y, Qin Y, Liu X, He K, Chen C, Ma L, Wang J, Xiong L, Zhang Q, Fan L, Deng XW (2007) Global genome expression analysis of rice in response to drought and high-salinity stresses in shoot, flag leaf, and panicle. Plant Mol Biol 63:591–608

Zhu JK (2002) Salt and drought stress signal transduction in plants. Annu Rev Plant Biol 53:247–273

Part IV
Ecophysiological Responses

Chapter 12
Drought Response in Forest Trees: From the Species to the Gene

I. Aranda, E. Gil-Pelegrín, A. Gascó, M. A. Guevara, J. F. Cano, M. De Miguel, J. A. Ramírez-Valiente, J. J. Peguero-Pina, P. Perdiguero, A. Soto, M. T. Cervera and C. Collada

Abstract Forest tree species, considering their long lifespan, symbolize one of the best biological examples of adaptation to a frequently changing harsh terrestrial environment. The adaptation to environments with water scarcity was the first challenge in the evolution of terrestrial photosynthetic organisms, and prompted the development of strategies and mechanisms to cope with drought. In this respect, the particular evolution and life history of forest tree species have brought about a plethora of specific adaptations to dry environments. The presence of a hydraulic system for long distance water transport and the need of maintaining functional tissues and organs for long periods of time are two important characteristics making forest tree species singular organisms within the plant kingdom. Selective pressure has prompted a variety of strategies in the control of water losses to maintain the functionality of the hydraulic system without compromising the carbon balance of the plant. These and other physiological responses focussed

I. Aranda (✉) · A. Gascó · M. A. Guevara · J. F. Cano · M. De Miguel · J. A. Ramírez-Valiente · M. T. Cervera
Instituto Nacional de Investigaciones Agrarias (INIA), Centro de Investigaciones Forestales (CIFOR), Carretera de la Coruña km 7,5, 28040, Madrid, Spain
e-mail: aranda@inia.es

E. Gil-Pelegrín
Centro de Investigación y Tecnología Agroalimentaria de Aragón (CITA de Aragón), Av. Montañana 930, 50059, Zaragoza, Spain

M. A. Guevara · J. F. Cano · P. Perdiguero · A. Soto · M. T. Cervera · C. Collada
Unidad Mixta de Genómica y Ecofisiología Forestal, INIA/UPM, Madrid, Spain

J. F. Cano · P. Perdiguero · A. Soto · C. Collada
GENFOR Grupo de investigación en Genética y Fisiología Forestal, Universidad Politécnica de Madrid, E-28040, Madrid, Spain

J. J. Peguero-Pina
Department de Biologia, Universitat de les Illes Balears, Ctra. Valldemossa km 7.5, 07122, Palma de Mallorca Illes Balears, Spain

R. Aroca (ed.), *Plant Responses to Drought Stress*,
DOI: 10.1007/978-3-642-32653-0_12, © Springer-Verlag Berlin Heidelberg 2012

to increase the dehydration tolerance of tissues (e.g., osmotic adjustment) have played an important role in the development of specific adaptations under water limiting conditions. The adaptive changes are observable at different scales: from the population to the species, from the individual to the gene. The advance of high-throughput technologies will enable to unveil the complex interplay between phenotype and genotype. Genomic, proteomic, transcriptomic, and metabolomic approaches are beginning to bring light to the molecular basis of adaptation to drought in forest tree species. These new technologies, combined with more traditional approaches, will improve our current knowledge of the functional and molecular basis underlying adaptation and evolution of forest tree species living under dry environments. In this respect, this chapter covers some aspects of adaptation to drought at different integrative levels, from an ecophysiological perspective to a molecular-based point of view.

12.1 Introduction

Forest tree species are considered some of the most long-living organisms. During their long lifetime, and considering their substantial developmental plasticity, trees need to overcome the restrictions imposed by harsh dry land media. This is probably one of the most relevant properties that make them ideal models to understand the complexity of adaptive processes through the life cycle of plants. Within the plant kingdom, forest tree species cover a particular ample range of adaptive solutions to the challenges of terrestrial media. The mid Devonian marked the appearance of true trees with homoisohydric lifestyle. Since then, the maintenance of a proper water economy has been one of the main challenges for these plant organisms in order to thrive under environments highly limited by water availability. The importance of this factor shaping forest landscapes was already recognized by naturalists of the nineteenth century such as Alexander von Humboldt (von Humboldt and Bonpland 2009). Water availability was soon viewed as the most important factor eliciting segregation of forest tree species along gradients of soil moisture availability.

The control of the hydration of tissues within ranges that are compatible with their functionality is important for any tree species. Though, various degrees of tolerance to the dehydration of leaf tissues have been developed within different *phyla* relying on common processes, such as osmotic adjustment or the regulation of water losses by an efficient stomatal control. In addition, forest tree species have required the evolutionary development of an efficient hydraulic system for transporting water from soil to leaves in order to control the hydration of aerial tissues. This system must fulfill the compromise of conductance efficiency versus security against the loss of its functionality in response to stressful conditions (Tyree and Sperry 1989; Aranda et al. 2005; Brodribb and Cochard 2009). The maintenance of the hydraulic function versus an optimum carbon balance could be the basis to understand survival and growth in most forest tree species (Breshears et al. 2009; McDowell 2011); but especially under dry conditions, where the optimum performance of both functions may be highly threatened and conditioned by biophysical compromises (McDowell et al. 2008).

In the last decades, an increasing number of reports have pointed out that some forests might become more vulnerable to drought in the close future (Seager et al. 2007; Allen et al. 2010). An increased vulnerability may not be only a consequence of the direct effect of dryness intensification in some areas, but also the result of the interaction between dry periods and other climatic factors, such as higher temperatures (Adams et al. 2009). Drastic changes in forest species composition in response to intense drought events have been already reported in different forest ecosystems worldwide in the last decades (Breshears et al. 2005, 2009). In other cases, altitudinal displacements of drought sensitive forest tree species, from lower and drier altitudes to higher and wetter ones, have been observed at the retreatment area in Europe (Jump et al. 2006; Peñuelas et al. 2008; Lindner et al. 2010). Furthermore, the increase in the recurrence and intensity of droughts is being a motive of concern even in areas of the world where dry periods have been considered atypical, such as temperate or tropical rain forests (van Mantgem et al. 2009; Newbery and Lingenfelder 2009; Allen et al. 2010). Therefore, it is important to frame the study of water use by forest trees into a proper perspective, considering not only evolutionary and ecological consequences of drought in drier areas, but its importance for most forest ecosystems from the Tropics to Mediterranean environments.

The analysis of drought response in forest trees may be considered at different scales, ranging from a species level to individuals (Aranda et al. 2000, 2010; Sánchez-Gómez et al. 2011). In this sense, forest tree species covering broad ranges of distribution are comprised of many local populations covering a high degree of genetic variability and adaptive solutions to cope with multiple environmental gradients (Zhang and Marshall 1995; Aspelmeier and Leuschner 2006; Ducrey et al. 2008). In addition, it is expected that a high degree of phenotypic plasticity will allow long-living sessile organisms withstand different environmental conditions varying from a seasonal to an annual basis (Bradshshaw 1965; Kremer 1995). Furthermore, forest tree species have to endure very adverse conditions such as singular extreme droughts, and throughout their life cycles spanning several centuries in many cases. These characteristics highlight the idiosyncrasy of forest tree species within the plant kingdom, and make them singular examples of adaptation to land media by a complex inter-play between mechanisms of local adaptation with a genetic basis (Linhart and Grant 1996; Savolainen et al. 2007), and phenotypic plasticity that partially relies on epigenetic mechanisms (Bossdorf et al. 2008; Nicotra et al. 2010). Both genetic variation and phenotypic plasticity are considered as fundamental to understand the future of forest tree species in their challenge to face the Global Change (Hamrick 2004; Kremer et al. 2010; Benito et al. 2011; Chmura et al. 2011). As a natural selection factor, drought may have modulated the relationship between genetic and epigenetic adaptive changes in forest tree species under water limiting conditions

(Raj et al. 2011). This turns the analysis of drought effects and the underpinning physiological and molecular mechanisms into a task of prime importance (Neale and Kremer 2011).

In this chapter, some basic mechanisms related to the capacity of forest tree species to cope with drought are summarized. The response of trees under water limiting conditions is briefly treated at different scales, from the species-specific performance to the molecular response. The fast development of genomic, proteomic, transcriptomic and metabolomic approaches is providing new insights into the molecular basis of adaptation to drought. The combination of these lines of research with more traditional approaches will open new perspectives in the understanding of the functional and molecular basis of adaptation and evolution of forest tree species in dry environments. This chapter ends with a brief overview of these new technological approaches applied to the study of drought responses in forest tree species. The ability to understand and advance the adaptive potential of forest tree species in response to the expected Climate Change requires the integration of the information gathered at different scales of study in the very close future. This chapter itself is an exercise of analyzing some aspects of drought response at different scales in forest tree species, but, furthermore, attempts to push forward the limits imposed by specific fields of research and investigation to the whole understanding of this topic.

12.2 Mechanisms of Ecophysiological Response to Drought in Forest Tree Species

Water is one of the main environmental factors conditioning the segregation of species into different biomes across ranges of wetness. Composition of forest ecosystems reflects different strategies and sensitivities to water stress displayed by each species. Even in ecosystems such as tropical rainforest or temperate forests, which are characterized by a seasonally high rainfall over most of the growing season, water availability can modulate community structure and function in the long term by infrequent but intense dry periods (Ciais et al. 2005; Breda et al. 2006; Newbery and Lingenfelder 2009).

The long-living nature of trees implies that their relationship with the environment changes throughout their life span. Environmentally induced developmental changes constitute an intrinsic process of permanent adjustment of plant performance (Day et al. 2002). Nevertheless, living for a long period of time, and the multiple contingencies trees must cope with in terms of extreme events, may not be enough to adequately be buffered by its phenotypic plasticity (Grether 2005). Therefore, extrinsic modifications in the habitat, due to perturbations or degradation (Camarero et al. 2011), changes in the silvicultural practices (Corcuera et al. 2006), or climate fluctuations (Foster et al. 2006), might lead forest tree species to their survival limits (Linares et al. 2009). Their capability to withstand these changes would ultimately establish the frontier between the survival of trees and their death, and relies in the adjustment of different functional and morphological traits, especially under water stressful conditions. These changes sum up in modifications of biomass investment in different plant organs, development of efficient hydraulic systems from roots to leaves, an efficient management of water losses, and the production of leaf tissues with a high degree of tolerance to dehydration.

12.2.1 Changes in Biomass Investment

The architecture, stratification, and lateral and vertical extent of root systems are key factors for understanding water relations of plants. Trees and shrubs clearly have the potential for developing deeper root systems than grasses, although the depth to which soil water depletion occurs varies widely among species (Canadell et al. 1996; Eggemeyer et al. 2009) and sites (Meier and Leuschner 2008a). The understanding of root system structure and function in trees is based largely on highly controlled seedling studies (Pemán et al. 2006). However, changes in function and allocation to roots with ontogeny must be considered when scaling from seedlings to mature trees (Topa 2004; Poorter et al. 2012). Root depth influences capacity to extract water from different soil horizons, being lower at the juvenile phase. In this way, Esteso-Martínez et al. (2006) showed that the minimum seasonal water potential in a stand of *Quercus faginea* were much less negative in adult trees that in seedlings, results that could explain the high percentage of cavitation found in seedling stems.

The investment of large amounts of reserves in the development of a large and deep root system may be considered crucial for trees in water-limited habitats according to the optimal partitioning theory (Bloom et al. 1985). A higher partitioning of biomass to belowground organs can be adaptive in relation to water stress, as observed from the inter-population variation within some forest tree species (Aranda et al. 2010). However, the investment in roots may condition the amount of reserves that can be used to produce shoots for the capture of aerial resources (Valladares and Pearcy 2002), as those under a Mediterranean type climate (Corcuera et al. 2005). The adjustment of the root to shoot ratio has been early recognized as a tradeoff between growth and survival under water limitations (Monk 1966). However, the production of a large root system does not ensure survival of trees under an extreme water deficit. Effectively, soil drought induces a loss of root hydraulic conductivity which can be due to changes in root anatomy (Nobel and Lee 1991), root xylem cavitation (Sperry and Ikeda 1997; Martínez-Vilalta and Piñol 2002), or changes in the expression of aquaporins (Secchi et al. 2007). The recovery from this lack of root conductivity usually implies the investment of new resources for reconstructing the root system (Lo Gullo et al. 1998), except for some woody plants that can recover the hydraulic conductivity by refilling xylem conduits through the generation of positive water pressures in the root (Tyree and Ewers 1991; Melcher et al. 2001). Recently, other mechanisms have also been proposed for recovering the hydraulic conductivity (see Zwieniecki and Holbrook 2009).

In addition, the ability to survive during severe drought events by shedding expendable organs has been pointed as other mechanism to cope with water stress. This would be at last explained in terms of a segmented vulnerability to cavitation across plant organs. Petioles would be more vulnerable than stems, and stems more than roots, in forest tree species that develop a low whole-plant hydraulic resistance such as *Acer saccharum* Marsh. (Tyree et al. 1991), *Junglans regia* L. (Tyree et al. 1993) or *Acer saccharinum* L. (Tsuda and Tyree 1997). However, petioles of the strict riparian *Betula occidentalis* Hook showed lower vulnerability to cavitation than stems, and stems were less vulnerable than roots (Sperry and Saliendra 1994). On the other hand, no differences in vulnerability were reported for other forest tree species (e.g., three *Quercus* species in Cochard et al. 1992).

12.2.2 Hydraulic System and Response to Water Stress

The movement of water according to a gradient of water potential through the xylem starts as water reaches the root stele. At this point, water moves through a path of conduits (Fig. 12.1) overcoming the hydraulic resistance imposed to water flow by very narrow elements, as explained by the Hagen–Poiseuille law (Tyree and Zimmermann 2002). According to Sperry et al. (2003), the restriction to flow through the nonvascular and vascular pathway can be of similar magnitude if expressed as conductances (volume flow rate per pressure difference) in well-watered plants, whereas the radial barrier to water movement in the roots shows an intrinsically low conductivity (volume flow rate per pressure gradient per cross-sectional area).

As a consequence of tree height growth, the hydraulic path is lengthened (King 1990), so a decrease in leaf-specific hydraulic conductance results due to the increased path length (Barnard and Ryan 2003). Accordingly, it has been proposed that tree growth may be hydraulically limited due to the drop in water potential through the path (Ryan and Yoder 1997; Becker et al. 2000; Ryan et al. 2006; Sperry et al. 2008). The diameter of conduits has been reported to increase with tree height (Zach et al. 2010), branching order (Mayr et al. 2003), or tree age (Corcuera et al. 2004a) as a way to increase the efficiency of the xylem to transport water. Nevertheless, drought can affect negatively the average diameter of xylem conduits as a plastic response that induces an unbalance between demands and flow, which may result in damaging consequences for plant survival (Corcuera et al. 2004b).

Significant differences in the vulnerability to xylem cavitation have been reported for a wide range of species widespread from tropical rain forests to deserts; resulting that those species showing higher drought tolerance generally exhibit hydraulic systems more resistant to cavitation (Maherali et al. 2004;

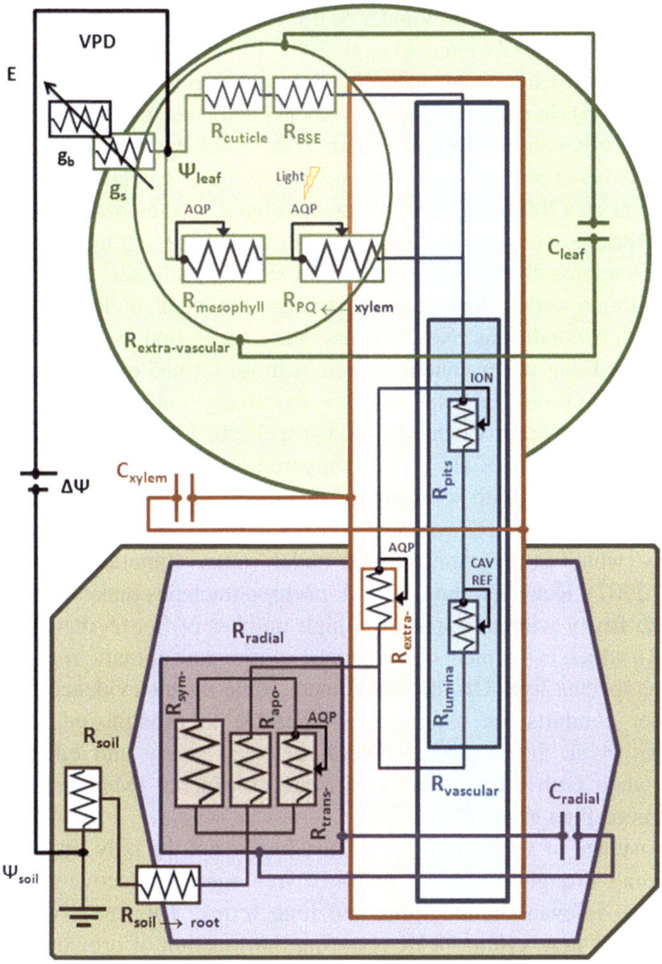

Fig. 12.1 Water transport pathways through trees following a classical Ohm's Law analogical representation. Resistances may be dynamically modified by cavitation (CAV) and refilling (REF) processes, ion-mediated changes of hydrogel configuration in inter-conduit pits (ION), or aquaporin gating (AQP) against environmental changes. In addition, some other components as the extra-vascular xylem resistance ($R_{extra\text{-}}$) and the resistance of the mesophyll ($R_{mesophyll}$) may be additionally decomposed into apoplastic ($R_{apo\text{-}}$), symplastic ($R_{symp\text{-}}$) and trans-membrane ($R_{trans\text{-}}$) components as it is represented for radial root resistance (R_{radial}). Water storage has been included as a series of capacitors (C) within different plant tissues (C_{leaf} and C_{radial}). Rest of abbreviators represent: g_s (stomatal conductance to water vapor), g_b (boundary layer conductance), E (transpiration), VPD (vapor pressure deficit between leaf and air), Ψ_{leaf} (leaf water potential), Ψ_{soil} (soil water potential). The relative size of each element is not proportional to its real average magnitude

Brodribb and Cochard 2009). In general, a functional tradeoff between hydraulic efficiency and safety, comprising support and water storage, results from considering the water transport through the xylem pipeline (Tyree and Zimmermann

2002), their size distribution (Vander Willigen and Pammenter 1998), configuration of pit membranes (Zwieniecki et al. 2001; Pittermann et al. 2006; Jansen et al. 2011), and the intrinsic mechanical reinforcement against lumen implosion (Hacke et al. 2001, 2005). In short, very small dimensional changes (e.g., vessel diameter, parenchyma cells volume, and pit structure) drive substantial changes in hydraulic conductivity, water storage capacity, and resistance to cavitation (Fig. 12.1).

Bhaskar et al. (2007) found that closely related species may strongly differ in terms of hydraulic conductivity, being higher in those living under drier climates. Hence, the increase in the efficiency of the xylem would serve as a way to cope with an extreme water flow demand through the plant under very high vapor pressure deficit conditions. Nevertheless, very early studies on plant hydraulics already showed that wider conduits, such as those formed during the early annual growth in ring-porous tree species, are extremely vulnerable to freeze–thaw-induced cavitation (e.g., Cochard and Tyree 1990; Lo Gullo and Salleo 1993). Recently, Peguero-Pina et al. (2011) compared tracheids dimension in two Mediterranean fir species with a strong phylogenetic link, resulting that the better adapted species to drier environments (*Abies pinsapo*) show wider tracheids than *Abies alba*, which is commonly found under wetter climatic regimes (Peguero-Pina et al. 2007). However, the size of *A. pinsapo* tracheids makes this species very sensitive to frosty winters imposing a high number of freeze–thaw cycles (Mayr et al. 2003), which is a typical characteristic of montane climatic regimes where its congeneric species live. On the other hand, some recent evidence indicates that these wider conduits are also more vulnerable to drought-induced cavitation (Pittermann et al. 2006); so, a tradeoff between safety and efficiency shapes tracheids when both drought and freeze are considered (Martínez-Vilalta et al. 2002; Peguero-Pina et al. 2011).

The resistance of xylem to functionality loss is not the only important trait for understanding drought tolerance of forest tree species. Recovering after water stress is also relevant in the short and long terms. This may be mediated by development of new xylem, or by restoring the function of previously embolized vessels (Resco et al. 2009; Brodribb et al. 2010). Although the refilling of embolized vessels is still far from being completely understood, this mechanism should be considered at least as important as the runoff of cavitation. Experimental evidence shows that plants are able to repair embolized xylem conduits by exerting enough root pressure during the night and/or along rainy seasons (Sperry et al. 1987; Hacke and Sauter 1996), or by pushing water from living conduit-associated parenchyma cells into gas-filled lumina when the bulk of water-transporting xylem is still under tension (novel refilling, Hacke and Sperry 2003; Bucci et al. 2003). The specific mechanisms of refilling are beginning to be elucidated. It seems to require the hydraulic isolation of embolized conduits, and some changes in the sugar metabolism of vessel-associated cells (evidenced through an observed consumption of starch) to provide the necessary driving force for water. In fact, novel refilling might be simplified to a particular case of phloem unloading (Nardini et al. 2011). A different refilling rate has been reported across several forest trees species (e.g., Hacke and Sperry 2003); and it has been suggested that refilling under tension might be operated somehow different in conifers, given its markedly different wood anatomy from angiosperms (Borghetti et al. 1991). In short, embolism repair may be more likely to occur in organs where there is a closer contact between phloem, xylem and other involved living cells, and/or if conduits are narrower and shorter (Clearwater and Goldstein 2005). Although there is much research to be performed in order to fully address this hypothesis, refilling might be more likely to occur in conifers, monocots, and protoxylem conduits, than in dicots, angiosperms, and metaxylem; as well as in smaller distal organs or leaf veins where cavitation vulnerability is also higher. Considering this hypothesis, xylem structures developed under drought conditions might be also tested for higher refilling occurrence.

Once water arrives to the leaves, it still flows through the xylem across their veins. As they are made up of xylem tissue, petioles (Bucci et al. 2003) and leaf veins (Nardini et al. 2001; Salleo et al. 2003) are also vulnerable to cavitation whether water flow to the atmosphere is high enough to generate critical water potential drops. Leaf hydraulic properties are receiving a special attention by plant physiologists that are unraveling the mechanism for preserving the integrity of other organs by limiting transpiration during water stress (Brodribb and Holbrook 2003; Zufferey et al. 2011). Considering the full hydraulic path, the hydraulics of the leaf lamina accounts for about the 25 % of the whole-plant resistance to water flow on average (Sack et al. 2003). This particular highlights the importance of any hydraulic dynamic change to whole-plant water balance, such as molecular mechanisms involved in aquaporin gating within the mesophyll of leaves (Shatil-Cohen et al. 2011). For instance, regulation of leaf hydraulic conductivity by light and mediated by aquaporin expression, would be a species-specific molecular mechanism allowing a fine tuning of water movement into the leaf lamina according to the light environment (Cochard et al. 2007; Baaziz et al. 2012). In addition, the resistance against water flowing out of the xylem to the nonvascular pathway of the leaf is also of major importance for understanding the overall contribution of leaves to whole-plant hydraulic resistance (Sack and Holbrook 2006; Johnson et al. 2009); especially, if the conclusions supported in Brodribb et al. (2005) relating leaf hydraulic conductivity and photosynthetic capacity are taken into consideration. In fact, leaf hydraulic conductance itself has been also addressed as highly coordinated with both stomatal conductance and net assimilation rate (e.g., Aasamaa et al. 2001 or Brodribb and Holbrook 2006). This coordination would be of major importance under water stress conditions (Shatil-Cohen et al. 2011).

12.2.3 Stomatal Control of Water Loss

Plants lose at least 100 times more water than they are able to gain carbon by gas exchange through stomata. Thus, net carbon assimilation is a very expensive process in terms of water consumption. While most assimilated carbon dioxide is integrated within tree biomass, water has a short lifetime inside plant tissue and requires continuous replenishment. Though nonstomatal mechanisms such as the leaf cuticle resistance plays an important role in limiting leaf water losses, probably stomatal regulation is the main point control in water use. Leaf stomatal closure is a common plant mechanism for water saving in drought stress periods at the expense of reducing net CO_2 assimilation (Chaves et al. 2003; Peguero-Pina et al. 2009). Moreover, stomata respond to very different environmental stimuli, and even within the same forest ecosystem it is possible to find a full range of strategies regarding water economy under drought conditions, varying from isohydric or "water saving", to anisohydric or "water spender" performance (Breda et al. 2006). However, in the last years, the idea that leaf stomatal control in forest tree species is constrained by the need to maintain the hydraulic function *versus* an optimum carbon balance has gained force (Campanello et al. 2008). In species with secondary growth, but especially in forest trees, maintenance of the hydraulic system within safety margins from water-stress-induced embolism is of prime importance (Tyree and Sperry 1989). Woody plants have shown their ability to avoid the partial cavitation of their xylem through a fine stomatal control of xylem pressures (Jones and Sutherland 1991). This regulation would be especially relevant under dry conditions where optimum in both functions is more threatened and conditioned by biophysical compromises (McDowell 2011). This has conducted to propose either hydraulic failure or carbon starvation (although see Sala et al. 2010), as the mechanistic basis explaining the phenomena of mortality after intense periods of drought (McDowell et al. 2008). In both processes, the particular strategy of forest tree species regarding stomatal regulation of water losses would be the keystone.

A consequence of stomatal closure under water stress to prevent a catastrophic loss of hydraulic conductance, and to minimize dehydration of leaf tissues, is an overheating of leaves. In addition, an excess of excitation energy cannot be directed to the photosynthetic electron transport chain (Demmig-Adams and Adams 1996). Electrons not consumed in CO_2 fixation may react with O_2 generating reactive oxygen species and increasing the possibility of oxidative damage. Under these conditions, both the pH and the de-epoxidation state of the xanthophyll cycle pigments increase, protecting the photosynthetic apparatus through a mechanism that dissipates excess of light as heat (Demmig-Adams and Adams 1996; Li et al. 2000a; Morales et al. 2006). However, there is no evidence for major sustained photodamage in water-stressed plants, as judged by the lack of effects of drought on the maximum potential PSII efficiency (F_V/F_M) even for very stressed leaves (Morales et al. 2006 and references therein). On the other hand, recent studies show evidence of drought-mediated down regulation of F_V/F_M in some Mediterranean forest tree species (Peguero-Pina et al. 2009), which seems to be related to an additional photoprotective mechanism that may play an important role for survival of species living in sites with long and intense summer drought periods.

12.2.4 Increase of Tolerance to Dehydration of Leaf Tissues

Besides mechanisms to optimize water capture, transport and control of losses under drought, forest tree species, as other plants, have developed means to increase the tolerance to dehydration of their tissues. Differences in leaf osmotic adjustment capacity, a well-known mechanism to increase drought tolerance in plants, frequently reflect also the dryness of the species habitat (Abrams 1988; Corcuera et al. 2002). Even within the same stand, it is possible to find forest tree species with a marked differentiation in tolerance to dehydration of their leaf tissues, reflecting different functional strategies regarding water economy. In this regard, Lenz et al. (2006) found in a meta analysis a good correlation between the osmotic potential at full and zero turgor and the degree of tolerance to drought for the pool of species analyzed. They found a continuous range of responses, with species originating from xeric sites showing lower values of leaf osmotic potential than those from wet sites (Abrams 1988). Relevance of the osmotic properties in leaf tissues is present even for ecosystems where drought is not a permanent handicap for growth and survival, as temperate forests, but where species from the same functional group maintain a different degree of leaf tolerance to dehydration through the same range of water stress and according to the species drought resistance (Fig. 12.2).

Differences in leaf osmotic potential at full turgor are underpinned by a complex molecular expression of different metabolites with osmotic activity. Important variations have been observed in the kind and amount of the metabolomic profiling according to the species and degree of water stress endured (Merchant et al. 2006; Warren et al. 2011, 2012), and with important consequences from the point of view of species strategy regarding drought adaptation (see Sect. 12.4.3).

12.3 Genetic and Phenotypic Variation in Response to Drought

In the last decades, study of forest tree response to drought has focused mostly in the functional analysis at the specific level (i.e. Breda et al. 2006). Though, in the case of forest tree species characterised by a high intra-specific variability, analysis of the response of different populations deserves especial consideration. The wide geographical range of many forest tree species suggests maintenance of a large adaptive genetic variability (Aitken et al. 2008; Savolainen et al. 2007). Species-specific response to drought would be modulated by the differential capacity of local populations to cope with water stressful environments. Consistently, differential response to drought has been reported for different geographical origins (Burczyk and Giertych 1991; Zhang et al. 1993; Lauteri et al. 1997; Ares et al. 2000; Ramírez-Valiente et al. 2009), and even within the same local population offspring from different

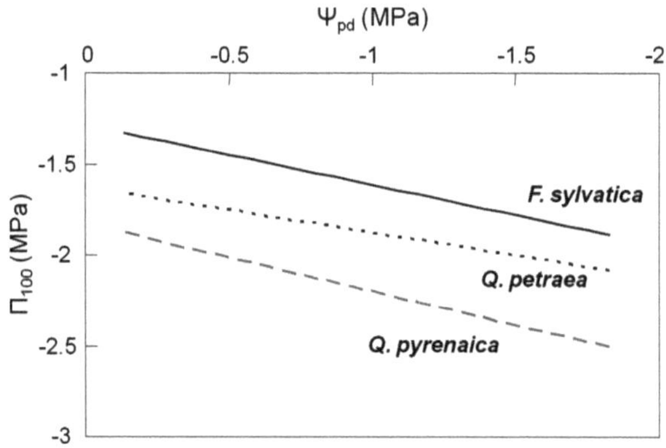

Fig. 12.2 The leaf osmotic potential at full turgor (π_{100}- MPa) is a good trait indicative of the leaf dehydration tolerance, even for seedlings of temperate forest tree species not adapted to dry environments. The parameter can be used as a proxy of tolerance to drought by different forest tree species (Lenz et al. 2006). Each line depicts a decreasing trend of π_{100} with leaf predawn water potential (Ψ_{pd}) as surrogate of the effective water stress endured by seedlings. *Fagus sylvatica* (continuous line) maintained a higher value of π_{100} whichever the Ψ_{pd} compared to the two oak species: *Quercus petraea* (dotted line) and *Quercus petraea* (dashed line). Redrawn from data in Aranda et al. (2001, 2002, 2004), Robson et al. (2009), Rodríguez-Calcerrada et al. (2010)

mother trees shows dissimilar performances (Major and Johnsenn 1996; Major and Johnsenn 1999; Aranda et al. 2010).

Changes reported recently for some forestlands seem to have been abrupt as a consequence of sporadic, but very intense droughts that have modified the competitive relation among species in a few years (Ciais et al. 2005; Breshears et al. 2005; Allen et al. 2010). Reports of drastic changes in the composition of forest vegetation seem already to confirm current standing genetic variability would not be enough to suit to the new environments (Hamrick 2004; Kremer 2010). Widespread phenomena in the last decade of forest die-back, increase in mortality rates, and altitudinal displacement are examples that seem to confirm local populations could not keep enough adaptive potential to overcome the awkward situation brought about by an increase in the dryness at the local scale. The issue is especially relevant for populations close to the trailing edge of the species distribution, and where larger changes in thermal and moisture regimes are expected in the next decades (IPCC 2007). This is the case, for instance, in some countries surrounding the Mediterranean basin where some changes in forest systems are already beginning to be observed (Peñuelas et al. 2001; Martinez-Villalta and Piñol 2002b; Jump et al. 2006; Peñuelas et al. 2008; Linares et al. 2010). Therefore, knowledge of the intra-specific variation in drought response emerges as a need for a better understanding of the microevolutive changes that could affect the sustainability of forest tree species in new climate contexts.

12.3.1 Within-Species Variability in Functional Traits Related to Drought

Forest tree species maintain high degrees of intra-specific adaptive variability for traits related to growth or phenology (Jensen 1993; Meier and Leuschner 2008b; Chambel et al. 2007; Vitasse et al. 2010), and with a wealth of examples dealing with specific local adaptation and variability in quantitative traits responding to different stresses, particularly drought (Aranda et al. 2005; Savolainen et al. 2007; Kremer et al. 2010). Well known is the high within-species genetic variability in phenotypic traits responding to water stress, such as water-use efficiency in *Populus sp* (Ceulemans and Impens 1980; Bassman and Zwier 1991; Monclus et al. 2006), *Eucalyptus sp* (Li et al. 2000b), *Pinus sp.* (Cregg and Zhang 2001; Guy and Holowachuk 2001; Voltas et al. 2008; Aranda et al. 2010;), *Pseudostuga sp.* (Zhang et al. 1993) or *Quercus sp.* (Arend et al. 2011). This variability has been the point of attention for tree breeding programs from the middle of the twentieth century (Langlet 1971; König 2005).

Although information related to the within-species functional response to water stress is scarcer than for growth or phenology, it is enough to conclude that it might be also under genetic control. Besides recognizing the expression of most functional traits is mostly conditioned by the environment, the genetic variation also underlies expression of different functional traits such as water use efficiency, stomatal control of water losses or net photosynthesis under a common environment (Table 12.1). Variation is observed at different genetic levels from population (Zhang and Marshall 1995; Benowicz et al. 2000; Ducrey et al. 2008) and open pollinated families within the same population (Prasolova et al. 2001), to clones (Aspelmeier and Leuschner 2006; De Miguel et al. 2012).

Genetic variance is high, for instance, when analyzing the expression of traits such as carbon isotope discrimination ($\Delta^{13}C$) related with intrinsic water-use efficiency (Zhang et al. 1993; Flanagan and Johnsen 1995; Prasolova et al. 2001; Voltas et al. 2008). No matter the high phenotypic variance reported for drought response, the proportion of this variance attributable to genetic factors is usually just moderate. In general, moderate values are found for narrow sense heritability in $\Delta^{13}C$ (the proportion of total variance attributable to additive genetic variance), as well as for other drought stress-related traits (Brendel et al. 2002). Analogously, relatively high Q_{ST} values (genetic parameter used as surrogate of distance among populations in quantitative traits) have been also reported for morpho-functional traits. This indicates a moderate among-population genetic differentiation for drought tolerance in some forest tree species (Ramírez-Valiente et al. 2009).

The relationship between the expression of morpho functional and growth traits and fitness is inferred from genetic correlations on most cases, but in general it is recognized functional traits could be submitted to strong selective pressure (Lamy et al. 2011). Traits such as water use efficiency, leaf size, or osmotic adjustment capacity, would maintain a putative adaptive value according to the environment, but especially in dry areas with a direct impact on plant fitness. The ecological and

Table 12.1 Variation within forest tree species of physiological and morphological traits related to drought response has been frequently reported at different genetic levels

Species	Genetic level	Characters	Level of variation	Reference
Pseudotsuga menziesii (Douglas-fir)	Population	Gas exchange, carbon isotope discrimination	Moderate	Zhang et al. (1993)
Picea abies (Norway spruce)	Populations	Growth parameters	Low	Burczyk and Giertych (1991)
Castanea sativa (Chesnut)	Population	Gas exchange, water use efficiency	Low	Lauteri et al. (1997)
Pinus sylvestris (Scotch pine)	Population	Gas exchange	Low	Cregg and Zhang (2001)
Alnus sinuata (Sitka alder)	Population	Gas exchange, growth	Moderate	Benowicz et al. (2000)
Betula papyrifera (Paper birch)				
Fagus sylvatica (Beech)	Population	Gas exchange, metabolites, and ABA	Moderate	Peuke et al. (2002) and Robson et al. (2012)
Thuja plicata (Western red cedar)	Population	Gas exchange, water use efficiency	Low	Grossnickle et al. (2005)
Cedrus brevifolia, C. libani (Cedrus)	Population	Gas exchange, carbon isotope discrimination, and hydraulic traits	Low	Ducrey et al. (2008)
Pinus pinaster (Maritime pine)	Population	Hydraulic system and xylem embolism	Low	Corcuera et al. (2011)
Picea glauca (white spruce)	Families	Gas exchange, chlorophyll fluorescence	Low	Bigras (2005)
Pseudotsuga menziesii (Douglas-fir)	Families	Hydraulic system and xylem cavitation, growth parameters	Moderate	Anekonda et al. (2002)
Picea mariana (Black spruce)	Family	Carbon isotope discrimination	Moderate	Flanagan and Johnsen (1995)
Larix occidentalis (Western Larch)	Family	Gas exchange	Low	Zhang et al. (1994)
Fagus sylvatica (Beech)	Family	Gas exchange, water use efficiency	Low	Leonardi et al. (2006)
Pinus pinaster (Maritime pine)	Population/family	Hydraulic system and xylem embolism	Moderate	Lamy et al. (2011)
Populus deltoides (Poplar)	Clones	Osmotic adjustment	Moderate	Gebre et al. (1994)

(continued)

Table 12.1 (continued)

Species	Genetic level	Characters	Level of variation	Reference
Hevea brasiliensis (Rubber tree)	Clones	Gas exchange, xylem embolism	Moderate	Sangsing et al. (2004)
Betula pendula (Silver birch)	Clones	Growth parameters	Moderate	Aspelmeier and Leuschner (2006)
Populus deltoides × nigra (Poplar)	Clones	Carbon Isotope discrimination	High	Monclus et al. (2006)
Populus tremula (Common Aspen)	Clones	Growth parameters, Gas exchange	High/ moderate	Possen et al. (2011)
Betula pendula (Silver birch)				
Pinus pinaster (Maritime pine)	Clones	Gas exchange, intrinsic water use efficiency	Moderate	De Miguel et al. (2012)

Although functional response is under high environmental influence, its control through experiments in common garden tests, or greenhouse and climatic chambers where environment is standardized, allows ascertaining different functional strategies by different genetic backgrounds. Level of variation for population was graded only as low or moderate evolutionary importance of morpho-functional traits, spanning from changes in biomass partitioning to gas exchange or water parameters, and in response to water limiting environments, has been tested for a wide range of annual model species or crops. Results are expressed in terms of the expression of a specific phenotype jointly a better performance under drought, and increase in fitness by using growth, reproductive success, or harvest index in crops as biological success indexes (Dudley 1996; Donovan et al. 2009). However, there is currently still little information available regarding forest tree species (Brendel et al. 2008; Scotti et al. 2010; Ramírez-Valiente et al. 2011; Lamy et al. 2011).

12.3.2 Genetics and Phenotypic Plasticity in Tracking Future Environmental Changes

The key point in the future maintenance of local populations coping with drier environments will be given by both the genetic variability and phenotypic plasticity facilitating adjustment to the new environmental conditions (Abrams 1994; Hamrick 2004). New expectations have emerged during the last years in relation to this issue in the face of climate change. Natural forest tree populations are advocated to the extirpation, migration (by tracking the environmental change), or adaptation in relation to the new environments (see Aitken et al. 2008 and Kremer et al. 2010 for a compressive review). In respect to the adaptive potential, it has been highlighted recently that adaptive differentiation between extant populations of forest tree species can be fast (Kremer et al. 2010). However, the speed of the expected change will likely be faster than generation turnover for most forest tree species. Such a high speed could override the potential to generate new recombinants better adapted to the future conditions. This leaves phenotypic plasticity within populations as one of the main evolutionary mechanisms to cope with the new climate contexts (Nicotra et al. 2010). Thriving under the new climatic contexts at the local scale will therefore rely on the degree of genetic polymorphism in adaptive traits, phenotypic plasticity or both within populations (Nicotra et al. 2010). Whichever the mechanisms acting first on the process of adaptation, both probably will play an important role for forest tree populations to adjust to more water stressed environments in the future. Albeit, it has been outlined that the extent of phenotypic plasticity blurs in some cases, the importance of local adaptation when considering the functional response to drought (Baquedano et al. 2008, Gimeno et al. 2009), highlighting the major effect of the environment in the phenotypic variance and, particularly, the strong genotype x environment interactions. Thus the capability of a genotype to acclimate to different environments, showing different phenotypes (plasticity), seems to be a noteworthy feature of forest tree species, in which epigenetic factors likely play a key role in adaptive terms further than the mere molecular basis of the acclimation response (see below and Bossdorf et al. 2008 or Raj et al. 2011).

12.3.3 Interaction of Drought With Other Environmental Factors

Differentiation among and within forest tree species in the response to drought is given by the expression of some of the aforementioned traits, which are mainly related to delaying or minimizing dehydration of tissues. The univariate point of view in approaching the water-stress response has been embraced in most studies analyzing drought response, in ecological (Ogaya and Peñuelas 2007; Vilagrosa et al. 2010), functional (Aranda et al. 2000; Brodribb and Holbrook 2003; David et al. 2007), or evolutionary contexts (Eveno et al. 2008; Grivet et al. 2011). This view avoids the biological and ecological reality of the multifactor world perceived by most biological organisms, and tree species in particular (Niinemets 2010). However, this response, as others related with the impact of limiting nutrients, light or salt, are commonly viewed in evolutionary and ecological terms from a single-factor perspective, relating the observed response to the capacity to cope with scarcity of the considered factor. Actually, from a functional and ecological point of view, we must keep in mind there is a complex interplay between different abiotic and biotic factors in shaping the response to drought. An example is given, for instance, by the interaction between light and water stress on some specific mechanisms related to drought tolerance as the osmotic adjustment capacity (Fig. 12.3). There are other examples showing a nonadditive action of factors such as light, water, nutrients, temperature or atmospheric CO_2 concentration over the expected expression, and assemblages of physiological, morphological and even molecular traits (Ellsworth and Reich 1992; Valladares and Pearcy 1997; Mittler 2006; Eller et al. 2011). This adds a new point of complexity to the interpretation of functional and morphological traits responding to water stress. The interaction with biotic agents, indeed potentially relevant, is out of the scope of the present review.

12.4 Molecular Foundations of Drought Response in Forest Tree Species

Drought response in forest tree species relies on changes at the molecular level that represent the lowest scale in the biological integration range. However, it is an additional step in the understanding of the overall response, maintaining important connections with the phenotypic and genetic levels previously described. The complexity of the interrelations between different molecular processes is now beginning to be specifically elucidated for some model forest tree species. However, thanks to the development of high-throughput technologies, and a higher degree of interaction among different fields of research, more knowledge is generated about the complex interplay between phenotype and genotype and water stress responses. In the next years, these advances probably will impact our understanding of the responses to drought by forest tree species in ecological and evolutionary contexts.

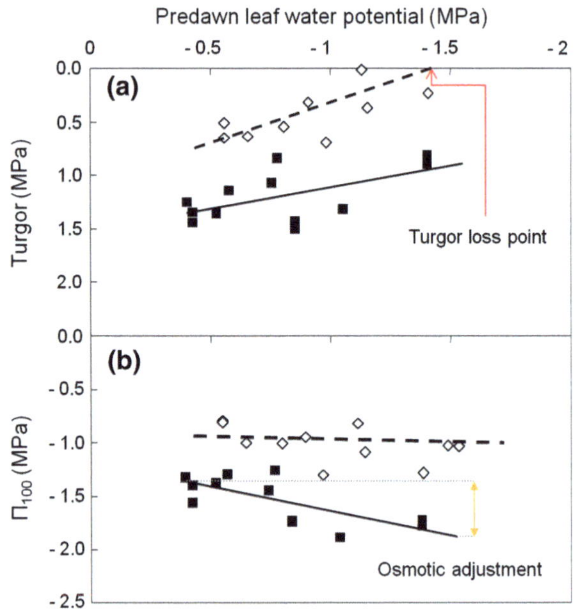

Fig. 12.3 A decrease in osmotic potential at full turgor (π_{100}) has been considered traditionally a mechanism of increasing tolerance of leaf tissues to dehydration. It is present even for temperate forest species such as beech. However, this mechanism of drought tolerance is hindered by light limiting conditions, eventually restraining the capacity to maintain leaf turgor under water stress conditions. In the graph it is shown the negative relationship between osmotic potential at full turgor versus predawn water potential, the last as a proxy of the water stress endured by beech seedlings. Black points represent seedlings growing with high light levels in a gap of a natural stand, and white points represent seedlings growing in the understory (modified from Robson et al. 2009)

12.4.1 Dissecting the Genetic Foundations of Response to Drought

12.4.1.1 Identification of Quantitative Trait Loci and Candidate Genes Involved in Drought Stress Response Using Association Analysis in Forest Tree Species

Drought stress response is a complex biological process. Thus, combination of genomic and physiological studies is required to advance knowledge in this field (Fig. 12.4a, b). Two forward-genetic phenotype-based strategies have been used to start unraveling genetic control of this complex response (Fig. 12.4b). Family based quantitative trait loci (QTL) mapping can be applied to search for associations between markers and phenotypes in pedigrees from crosses among genotypes with a contrasting response to drought. Alternatively, we can use population-based association mapping, with populations of unrelated individuals to examine associations between single nucleotide polymorphisms (SNPs) and phenotypes.

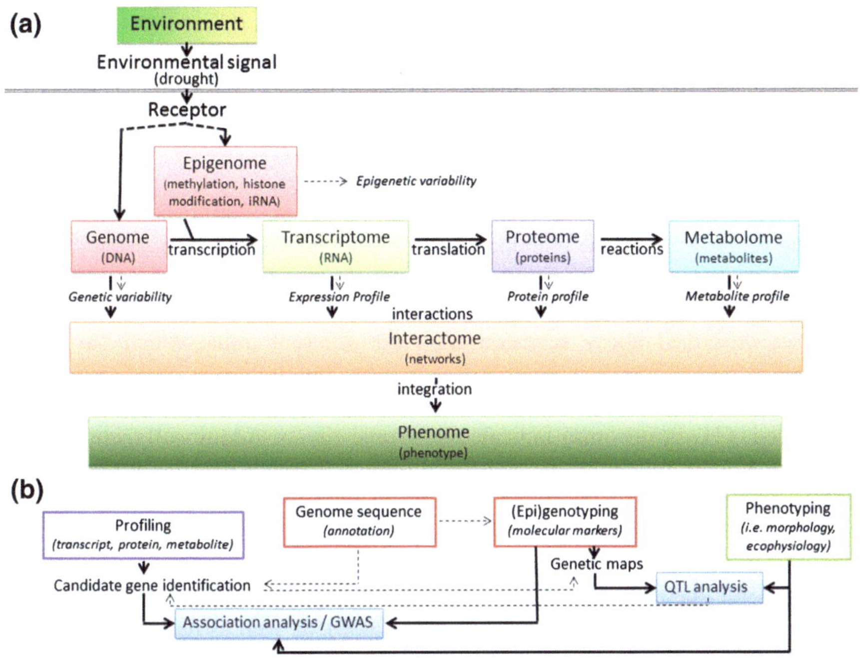

Fig. 12.4 A comprehensive understanding of plant response to environmental modification (i.e., drought stress) will require integrating data from different levels of biological organization, whose study is addressed by different "omics" disciplines. **a** Schematic of the "omic" hierarchy and integration. **b** A flow sheet summarizing the most common strategies employed to identify and validate a candidate gene using forest tree models. Association analysis and genome-wide association studies GWAS; QTL analysis

Detection of QTL depends on the size of the mapping progeny, genetic background, heritability of the trait under study, coverage, and saturation of the genetic map as well as genotype by environment interactions. QTL analysis has been applied to identify genome regions involved in the genetic control of drought response of different model forest tree species, evaluating different phenotypic traits of materials grown in different experimental layouts. QTL have been detected for, among other traits, carbon isotope discrimination in F1 individuals from intra-specific crosses of *Castanea sativa* (Cassasoli et al. 2004) and *Quercus robur* (Brendel et al. 2008), in three consecutive years. Comparative analysis between these two studies, based on a set of orthologous markers, revealed that no QTL for carbon isotope discrimination was conserved between both Fagaceae species (Casasoli et al. 2006). In Salicaceae, response to drought stress was also analyzed in a hybrid F2 progeny of *Salix dasyclados* × *Salix viminalis* (Rönnberg-Wästljung et al. 2005). In poplar, two different studies used an F2 progeny of *Populus trichocarpa* × *P. deltoides*, to analyze QTL for osmotic potential (Tschaplinski et al. 2006), and leaf coloration (chlorophyll and carotenoid content), expansion, and abscission (Street et al. 2006). The latest study also integrated transcriptomics, searching for positional candidate genes (CGs)—genes co-localizing with QTL for the parameter evaluated, in many cases supported by previous functional description in other species. In *Eucalyptus* sp., Teixeira et al. (2011) have detected QTL related to drought response in a hybrid *Eucalyptus grandis* × *E. urophylla* F1 progeny. QTL analysis for drought response in Pinaceae has been approached in *Pinus pinaster*, species characterized by a significant genetic and adaptive diversity (Eveno et al. 2008). Brendel et al. (2002) analyzed an intraspecific F2 progeny whose grandparents showed a differential response to drought in terms of δ^{13}C on growth rings.

A way to validate QTL is through comparative mapping. Most of the QTL analyses described above revealed QTL for different traits, which have been detected across different years. It is important to highlight that overlapping QTL among drought-related traits have been observed. Such co-locations indicate that the shared QTL clusters may indicate pleiotropic effects. Additionally, correlations between drought-related traits have been observed which may be attributable to either pleiotropic effects of single genes or to tight linkage of several genes that individually influence specific traits (Pelgas et al. 2011). However, up to now, comparison of QTL among different genetic backgrounds (families or species) has been difficult due to the limited number of orthologous markers among unrelated mapping families or species (Casasoli et al. 2006), as well as due to the different experimental procedures and phenotypic/physiological parameters used to evaluate drought response. The standardization of experimental procedures together with the use of high-resolution genetic maps will allow generalizing this approach (Neale and Kremer 2011). For instance, the work on *Q. robur* (Brendel et al. 2008) revealed a major QTL for water-use efficiency in the same linkage group where other analysis found a major linkage group for stomatal density (Gailing et al. 2008). Allelic variants, associated with mapped SNPs, may be analyzed in association population to unravel these two effects.

Power for QTL detection depends on the number of generations. Most forest trees are highly heterozygous outbred plant species. Due to their long generation time, QTL analysis of forest trees has been mainly carried out using F1 progenies or, in the better cases, three generation populations, limiting accuracy of these QTL analyses. Different strategies have been developed to increase the power of QTL detection for forest tree species (Plomion et al. 2007). The relatively low number of individuals in several forest tree mapping progenies also negatively influences the accuracy of calculated QTL effects and the power to detect QTL with small effects. However, this may be, to some degree, compensated by the use of ramets of each individual that provide higher precision in the phenotypic screening. Precision of phenotypic measurements is an important factor for QTL mapping, because a high measurement error reduces the estimated heritability and decreases the detection power. Nowadays, the use of high-throughput genotyping techniques allows construction of highly dense genetic maps, most of them integrating high numbers of gene-based markers, including markers transferable across pedigrees. The use of highly dense genetic maps and more precise phenotyping protocols will increase precision of QTL detection by narrowing down the intervals to small regions. Additionally, integration of complementary information such as expression analysis (expressional QTL—eQTL), proteomics, metabolomics, and epigenomics (i.e., Drost et al. 2010; Morreel et al. 2006; Henery et al. 2007; Long et al. 2011), will provide a more comprehensive view of mechanisms underlying drought response in forest tree species, highlight putative CGs and corresponding alleles involved in such a response.

Population-based association mapping searches for associations between genetic markers and adaptive traits in natural populations; thus, analyzing variation accumulated during evolution of the targeted germplasm used as a discovery population. Linkage disequilibrium (LD) usually extends over much shorter distances in association mapping populations than in family mapping population. Therefore, high numbers of genetic markers are needed to ensure adequate coverage to detect linkage between markers and a causal locus. As the cost of genotyping has dropped dramatically, association mapping has rapidly become a very promising approach for the genetic dissection of complex traits in plants (Ingvarsson and Street 2011). Association mapping has been applied to analyze *Pinus taeda* drought response (González-Martínez et al. 2008). It is important to differentiate association mapping based on analysis of CGs from the genome-wide association studies (GWAS). The first one is based on genetic markers genotyped at loci thought to be involved in trait expression, testing for associations between these genetic markers and the phenotype. In plants, this approach has been successful for CGs in relatively simple pathways and for CGs with extensive prior evidence of a role in the phenotype of interest (Thumma et al. 2005; Ingvarsson et al. 2008). However, drought response is a complex trait that involves genetic control of different pathways in different tissues. GWAS implies genotyping enough markers across the genome, so that functional alleles will likely be in LD with at least one of the genotyped markers. The number of markers and their density is defined by genome size and LD decay, and will therefore vary considerably among species, being mainly approached for those species whose genome sequence (or at least exome sequence) is already known.

12.4.1.2 Transcriptomics

A complementary approach to the identification of drought-responsive genes is the analysis of transcriptomic response. As reviewed recently by Vaahtera and Brosché (2011), there are several processes contributing to the response to a specific abiotic stress: post-translational activation and selective nuclear import of transcription factors, regulation of DNA accessibility by chromatin modifying and remodeling enzymes, interplay between response elements, and so on. Most of these factors have been usually analyzed independently, due to the intricacy of the whole process.

Analysis of the expression patterns under stress and recovery allows the identification of expressional CGs, induced by stress. This has been the most common approach followed to explore the molecular basis of the response to biotic and abiotic stresses, and when functional information is lacking, provides the most reliable CGs. Expression profiles reveal not just the genes induced by the stress, but also the genes whose transcription is inhibited. Inhibition of transcription shares importance with transcriptional up-regulation, as suggested by the results of Watkinson et al. (2003) and Lorenz et al. (2006). Nevertheless, first transcriptomic approaches to the analysis of the response to drought stress have focused on the identification of genes overexpressed during the stress.

Prior to the development of current techniques of expression analysis and high-throughput sequencing, researchers could deal only with few genes at a time. Chang et al. (1996) published a pioneer work where four water deficit-induced cDNAs were identified and characterized in loblolly pine (*Pinus taeda*). In an attempt to discover a higher number of genes induced by water deficit, Dubos and Plomion (2003) used cDNA-AFLP and reverse Northern blot in maritime pine (*Pinus pinaster*), and identified 48 putative genes presumably involved in the response to drought stress in roots. Nowadays, accurate techniques such as suppression subtractive hybridization (SSH), together with the capacity to check the expression of high number of genes at a time using microarrays provide noticeable precision for the isolation of induced genes in stressed versus control material. Complementarily, massive sequencing technology is much more affordable, and analysis of the whole transcriptome under control, stress, or recovery conditions is currently feasible. Nevertheless, for many forest tree species (mainly for Gymnosperms) a good annotation is not available yet, and for many genes detected this way, even putative ones, homology-based function has not been proposed.

Among angiosperm forest trees, poplars are the main model for molecular analysis, and even more since the release in 2006 of the complete genome of *Populus trichocarpa* (Tuskan et al. 2006). Poplar species show a comparatively fast growth and a noticeable capability for vegetative propagation, which allows undertaking experimental procedures not applicable to other tree species. Additionally, poplar was one of the first tree species successfully transformed, both using *Agrobacterium* (Fillatti et al. 1987) and gene gun (McCown et al. 1991). In the last years, many works referred to the transcriptomic response to drought stress of *Populus* species. Some of those studies have focused on the identification of genes induced by stress (Caruso et al. 2008; Bae et al. 2010), while others examined the expression pattern and function of specific genes (Bae et al. 2009, 2011; Chen et al. 2011). Exhaustive transcriptome studies using massive sequencing have also been published, describing the intra-specific variation (e.g., Hamanishi and Campbell 2011) or comparing specific genotypes (e.g., Cohen et al. 2010). The effect of drought stress on the transcriptome of the cambial region, where wood cells are developing, has been recently published by Berta et al. (2010). Although not in the same depth as in poplar, transcriptomic response to drought has also been studied in other angiosperm forest tree species. For instance, Gailing et al. (2009) have highlighted the role of *Quercus sp.* as a model for forest tree species, being one of the most important forest genera in the northern hemisphere.

Knowledge is not so advanced for gymnosperms which display peculiar characteristics that make the study of their genetic adaptations difficult. They usually have huge genomes, with a high percentage of repeated sequences and pseudogenes whose functions are not well known. For instance, whereas the genomes of *Arabidopsis thaliana* and *Populus trichocarpa* are approximately 150 and 550 Mbp long, respectively, the *Pinus pinaster* genome is about 30,900 Mbp, 70–75 % of which is made up of highly repeated sequences (Morse et al. 2009), and no complete genome of a reference conifer species is yet available. Furthermore, angiosperms diverged from gymnosperms more than 300 million years ago. Thus, to a large extent, knowledge and molecular tools developed for the former are not readily applicable to the latter. For these reasons, the selection of CGs for the study of adaptation in gymnosperms based solely on their homology with angiosperm genes, without further confirmation of their participation in the stress response, is not fully reliable. Additionally, conifers are highly recalcitrant to transformation and vegetative propagation, limiting the sort of feasible experimental designs.

Picea and *Pinus* are the most common model genera for conifers. In *Picea*, some studies have focused on the expression patterns of specific drought-related genes or on proteomic changes induced by drought (i.e. Blödner et al. 2007). In the case of *Pinus,* several genomic studies of the response to drought have been launched in different species, in an attempt to identify genes induced by water deficit. For instance, Heath et al. (2002) and Watkinson et al. (2003) performed a preliminary analysis of gene expression during drought-induced stress in *P. taeda*. These authors used microarrays that included cDNAs obtained from pre-existing libraries from the xylem, male cones and shoot tips, but did not attempt to exhaustively identify genes induced by water stress. Other noteworthy work is the one by Dubos and Plomion (2003), based on cDNA-AFLP in *P. pinaster*. More recently, several groups have used more precise techniques for the identification of drought-induced genes, such as comparison of EST libraries generated during drought-induced stress and drought recovery as well as from well-watered roots (Lorenz et al. 2006, in *P. taeda*), or SSH libraries, followed by microarray and RT-PCR expression analysis (Perdiguero et al. 2012, in *P. pinaster*).

Genes putatively involved in the response to water deficit, identified by these techniques or by their homology with other known genes, have been used to assess diversity and differentiation at different taxonomical and demographic levels regarding drought response. For instance, DNA sequence variability for drought stress CGs has been analyzed in different populations of *Pinus pinaster* across a latitudinal and precipitation gradient (Eveno et al. 2008). In the same way, different works have focused not on the genomic diversity but on the expressed response. Thus, Yang et al. (2010) performed a proteomic analysis of the response in *Populus kangdingensis* and *Populus cathayana*. Comparison of the transcriptomic response in different populations has also been performed in different species (e.g., Sathyan et al. 2005 in *Pinus halepensis* or Hamanishi and Campbell 2011 in *Populus balsamifera*). In poplar, comparisons at the individual level, among particular genotypes, have been published (Cohen et al. 2010) or even among ramets of the same genotype acclimated to different sites (Raj et al. 2011).

The study of the transcriptomic profiles allows identification of different functional groups of genes related to drought response. A high proportion of genes induced by drought stress are related to metabolism, and mainly to carbohydrate metabolism. The accumulation of sugars has been correlated with the acquisition of desiccation tolerance in plants, probably because sugars protect the structures from mechanical and metabolic stresses during dehydration (Oliver et al. 2010). Within this group genes presumably involved in the detoxification of aldehydes generated by alcohol metabolism are found. This result is consistent with the accumulation of ethanol in conifer seedlings during drought (Manter and Kelsey 2008). This group also includes genes related to aminoacid (e.g., proline), lipid, fatty acid and isoprenoid metabolism, which could be involved in the synthesis and accumulation of compatible solutes and hormones during water stress.

Another important group of genes comprises those related to defence and cell rescue. This category includes genes coding for late embryogenesis abundant (LEA) proteins, heat shock proteins (HSP) or dehydrins, among others. HSP are known to act as molecular chaperones, helping in the correct folding of other proteins and protecting them from unfolding and denaturation, and are involved in different abiotic stresses (Wang et al. 2004). LEA proteins seem to be involved in detoxification and may also act as chaperones, and their participation in desiccation tolerance is well known (Battaglia et al. 2008). Dehydrins are a complex family of proteins directly related to the response to water deficit and included among the LEA proteins. They are thought to act in protecting the cell metabolism during the stress, and have been the subject of different expression analysis in forest tree species (e.g., Bae et al. 2009; Vornam et al. 2011).

Transport processes play an important role in the mobilisation and accumulation of solutes and hormones and in cell detoxification pathways during adaptation to water stress. Thus, sugar transporters, involved in the modification of osmotic pressure under stress, or ABC transporters, which are involved in the response to different biotic and abiotic stresses (Wanke and Kolukisaoglu 2010) have been detected in drought response analysis. Aquaporins deserve special mention. These are channel proteins located in cellular membranes and mediate water flux, maintaining proper water balance. They can be found both in the plasma membrane and in the vacuole membrane, and their expression is induced by water deficit. Several works have focused on this protein family in forest tree species and on its role during hydric stress recovery (e.g. Almeida-Rodriguez et al. 2010; Berta et al. 2010).

Finally, there is a major group of proteins involved in drought stress response that corresponds to regulation processes. Two major pathways have been described in the plant response to abiotic stresses, including water deficit: an ABA-dependent pathway, and an ABA-independent one, with complex interactions (see, for review Shinozaki and Yamaguchi-Shinozaki 2007 or Hirayama and Shinozaki 2010). Several ABA-responsive transcription factors, such as bZIPs, NACs, MYBs and MYCs, have been described (e.g. Olsen et al. 2005). In the ABA-independent pathways other NAC and DREB transcription factors are involved (e.g. Chen et al. 2009, 2011). Other proteins involved in post-transcriptional (alternative splicing) and post-translational (e.g., kinases and phosphorylases) modifications are also included in this group.

Nevertheless, there are still a number of genes responsive to drought stress whose function is unknown and even a putative function cannot be proposed, due to a lack of homology with annotated sequences; this situation is particularly relevant in conifers (see, for instance, Perdiguero et al. 2012).

12.4.2 Epigenetic Foundations of Response to Drought

In addition to the genetic component, epigenetic variation has been suggested to contribute to the phenotypic plasticity and adaptive potential of individuals and populations (Bossdorf et al. 2008; Herrera and Bazaga 2010). Epigenetic mechanisms include heritable, but potentially reversible, changes in gene expression without changing the nucleotide sequence. Three epigenetic processes have been determined: insertion of methyl groups in cytosine bases of DNA, acetylation and methylation of N-terminal histone tails, and generation of regulatory noncoding RNA molecules. DNA methylation is the only epigenetic mark for which the mechanism of inheritance has been well established, so this mark can be considered the driving force for epiallelic variation (Mirouze and Paszkowski 2011).

Regulation of gene expression is important to plant tolerance to stresses. For this reason, epigenetic variation is gaining interest, because it may regulate the expression of genes that have a key role in acclimation responses. Although the number of studies concerning epigenetic variation in plants subjected to stress is increasing, only few studies correspond to forest tree species growing under drought stress.

In the first study of this type with a woody plant (Gourcilleau et al. 2010) six hybrid genotypes of the genus *Populus* were subjected to moderate water deficit. Significant genotype and treatment effects were detected for global DNA methylation and for morphological traits, such as height, stem biomass, leaf number and total leaf area. A positive correlation was demonstrated between DNA methylation percentage and productivity (stem biomass and height) under well-watered conditions but no correlation between DNA methylation and morphological traits was observed in response to water deficit. While there was a general decrease of growth for all genotypes in response to a water deficit, variations in DNA methylation were found, suggesting different responses among hybrids. Raj et al. (2011) studied the influence of recent individual history on the *Populus* transcriptome and global DNA methylation-level response to drought. Plants of the same genotype established in contrasting locations showed different transcript patterns and DNA methylation levels in response to drought, which suggests an epigenomic base for the clone history-dependent transcriptome divergence.

Trees of the white mangrove (*Laguncularia racemosa*) can occur naturally in salt marsh and riverside, exhibiting significant differences in morphological traits, thus presenting an excellent system to study natural variation in genetic and epigenetic marks and their relation to phenotypic variation and plasticity in trees (Lira-Medeiros et al. 2010). Analysis of DNA methylation patterns and nucleotide sequences detected greater epigenetic than genetic variation within and between populations in contrasting environments, which indicates that epigenetic variation in natural populations plays an important role in long-term adaptation to different environments.

12.4.3 Metabolomics

Modifications in the profiling of metabolites represent one of the first outcomes at the molecular level brought about by drought. Some molecular changes such as accumulation of compounds with osmotic effect, triggers as a consequence of a runway sensing system that tend to compensate the negatives effects of water stress on processes such as turgor maintenance or growth of plants. In this respect, the amount and kind of certain metabolites represent a by-product resulting from the direct impact of water stress at a higher level of organization (i.e., organ or individual). Probably, osmotic adjustment is one of the mechanisms affected in higher degree by the accumulation of different metabolites, as it involves the active accumulation of osmotic compounds acting directly on turgor maintenance (Chaves et al. 2003, and Fig. 12.2 on this chapter). Increase in metabolites such as nonstructural carbohydrates (NSC) is a classic example of response to drought related with the adaptation to water stress of forest tree species. Thus, the increase in NSC is a typical response for those species with a higher tolerance to water stress (Piper 2011). Cyclitol is other important group of compounds related with osmotic adjustment, and especially quercitol has found to increase in greater proportion for stressed plants in several species of *Quercus* sp. and *Eucalyptus* sp. (Merchant et al. 2006; Passarinho et al. 2006). However, important changes in other compounds lacking of osmotic role have been observed as well, and playing an important role in cellular processes further than the osmotic adjustment as those involved in photorespiration, stability of membranes, or changes in compounds of the cellular wall (Warren et al. 2011).

Differences in the specific metabolites expressed, even under constitutive conditions without water stress, are related to ecological and evolutionary aspects of forest tree species such as tolerance to drought (Merchant et al. 2007a, b; Warren et al. 2012), or may result from different phylogenetic pathways (Warren et al. 2011). In this respect, even within the same genus, such as for example *Eucalyptus* sp., it is possible to find species with a trend to accumulate certain metabolites compared to others species that do not (Fig. 12.5).

Although metabolic profiling depends on the species considered and the degree of water stress endured, not less important is the change in metabolic profiling in

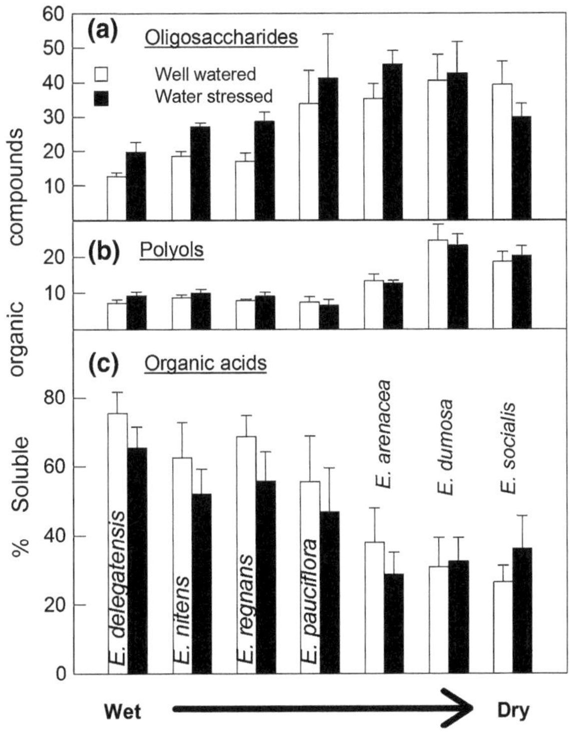

Fig. 12.5 Relative soluble organic compound to total polar organic metabolites in seven Eucalyptus species submitted to 2 months of water stress ($\Psi_{pd} \sim -2$ MPa) ordered by annual potential water balance. Upper panel shows relative proportion of oligosaccharides (fructose, glucose, sucrose); middle panel shows polyols (agrupping myo inositol, scyllo inisitol, galactinol; proto quercitol and others cyclohexanepentols) and lower panel shows organic acids (shikimic acid, quinic acid, malic acid, gallic acid and citric acid). These main metabolites accounted for over 90 % of organic soluble metabolites in Eucalyptus leaves. Open and closed bars represent well watered and water stressed plants, respectively. Means and standard deviation (n = 5). Redrawn from data in Warren et al. (2011, 2012)

the recovering phase after drought release, which shows different patterns according to the ecology of forest tree species. In a recent report, Warren et al. (2012) observed a different pattern in the osmotic potential and osmotically active solutes during the re-watering phase after a drought cycle for two *Eucalyptus* species. The mesic *E. pauciflora* showed an increase in osmotic potential at full turgor and decrease of osmotically active solutes, possibly related to a recovery of growth; the semi-arid *E. dumosa* maintained an increase in osmotic compounds such as some sugars, even after water stress relief, and that would allow plants to endure new water stress cycles.

Until recently, assessment of metabolites involved in osmotic adjustment or other physiological processes related to water stress was limited to the analysis of a few specific sugars or amino acid compounds. With the fast development of more insightful techniques, a more detailed approach to metabolites profile is possible, with hundreds of different compounds analyzed with the same platform. In the last, modifications in biochemical products result from a complex regulation of molecular networks responding to an endogenous and exogenous signaling process in response to the water stress suffered by the plant, which will be responsible of the resulting phenotype (Fig. 12.4).

12.5 Facing the Future of the Study of Drought Response in Trees: "Omics" and Interplay with Higher Scales of Biological Integration

High-throughput technologies have made it possible to see organisms as complex interactive systems. This new approach represents a shift toward a more holistic perspective. Integrative functional genomics combines the molecular components ((epi)genome, transcripts, proteins, metabolites, and their interactions) of an organism and incorporates them into functional networks or models designed to describe the dynamic activities of that organism which are responsible for the observed phenotype (Fig. 12.4a).

The first studies of the response to drought stress at the transcriptomic level addressed simple experimental conditions, with a single stress level. However, an integrated knowledge of the response to water deficit needs experimental designs mimicking field conditions, in order to detect the various pathways initiated by different levels of stress, as suggested by Watkinson et al. (2003). Trees in nature often must face more than one stressful factor at a time. Interaction among stresses, both biotic and abiotic, must be elucidated at different scales of biological integration from ecophysiological (Aranda et al. 2005) to the molecular response (Mittler 2006). For instance, Duan et al. (2008) analyzed the combined effects of UV-B, ABA and water status in allometric patterns of *Populus yunnanensis*. In this sense, natural populations can offer a valuable (although with evident practical hindrances) resource for these experiments, as proven in the work of Brosché et al. (2005). Advances in the study of drought response in forest tree species using other "omics" have been recently reported. In this respect, the study of metabolomic profiling together with water relations, at the leaf or plant level, deserves new efforts in order to better integrate both scales of study in the knowledge of acclimation and adaptation to water stress (Sardans et al. 2011). Different responses can take place in the diverse organs and even tissues of the tree. Although organ-specific analysis has been performed, development of laser microdissection capture techniques will allow in the future the analysis of the response in differentiated tissues and cells.

Knowledge of the role played by the adaptive genetic and epigenetic variability on the observed phenotypic variation in traits related with drought tolerance is still limited for most forest tree species, and not conclusive when framed in an evolutionary context. Hence, disentangling the potential adaptive value, and its molecular control, for those characters involved in adaptation to drought will be of prime importance in the future (Neale and Kremer 2011). Epigenetic and genetic studies will probably open new avenues to the understanding of the evolution of forest tree species, and the always complex interplay between genotype and environment as the main driver in the adaptation process under water stress (Raj et al. 2011). In this regard, and taking into account the rate of climate change, epigenetic modifications operating as molecular basis of phenotypic plasticity, which may be heritable, will likely play a key role that will be unraveled in the forthcoming years (see Sect. 12.3.2).

In this game of scales, the integration of ecophysiological studies with the characterization of the molecular response will break the traditional oversimplification from the functional ecology approach focused on the species-specific response, as well as the reductionist view that emerges from most of molecular analysis. Probably, one clear example of oversimplification in ecophysiological studies is the frequent tendency to forget species comprises groups of local populations. Furthermore, it is well known that in many cases, one unique forest population embraces a high intraspecific genetic variability for quantitative traits related with growth, phenology, or morpho-functional traits. The genetic variability has its counterpart in a high degree of phenotypic alternatives to the same environmental challenge.

In the future, and according to the expectations about the intensification of droughts in wide areas all round the world, the ecophysiological studies will have to center efforts in the analysis of those traits more related to the fitness and survival of forest tree species facing water stress. In the next years, more detailed studies will be necessary to improve our understanding of complex functional traits such as the stomatal control of water losses (Leonardi et al. 2006; Voltas et al. 2008; De Miguel et al. 2012), net photosynthesis (Major and Johnsenn 1996; Benowicz et al. 2000; Scotti et al. 2010), or hydraulic resistance to drought-induced cavitation (Corcuera et al. 2011; Lamy et al. 2011).

Acknowledgments This work was supported by funding from the Spanish projects ECOFISEPI AGL2011-25365, MAPINSEQ AGL2009-10496, SUM2008-62500004-C03-01, as well as the transnational project Plant-KBBE PLE2009-0016. We thank Drs. J. Voltas and F. Ewers for a previous critical reading of this work. Authors are grateful to the anonymous reviewer that improved a first version of this contribution.

References

Aasamaa K, Sober A, Rahi M (2001) Leaf anatomical characteristics associated with shoot hydraulic conductance, stomatal conductance and stomatal sensitivity to changes of leaf water status in temperate deciduous trees. Funct Plant Biol 28:765–774

Abrams MD (1988) Sources of variation in osmotic potentials with special reference to North American tree species. Forest Sci 34:1030–1046

Abrams MD (1994) Genotypic and phenotypic variation as stress adaptations in temperate tree species: a review of several case studies. Tree Physiol 14:833–842

Adams HD, Guardiola-Claramonte M, Barron-Gafford GA et al (2009) Temperature sensitivity of drought-induced tree mortality: implications for regional die-off under global-change-type drought. Proc Natl Acad Sci U S A 106:7066–7070

Aitken SN, Yeaman S, Holliday JA, Wang T, Curtis-McLane S (2008) Adaptation, migration or extirpation: climate change outcomes for tree populations. Evol Appl 1:95–111

Allen CD, Macalady AK, Chenchouni H, Bachelet D, McDowell N, Vennetier M, Kitzberger T, Rigling A, Breshears DD, Hogg EH (Ted), Gonzalez P, Fensham R, Zhangm Z, Castro J, Demidova N, Lim J-H, Allard G, Running SV, Semerci A, Cobb N (2010) A global overview of drought and heat-induced tree mortality reveals emerging climate change risks for forests. For Ecol Manag 259:660–684

Almeida-Rodriguez AM, Cooke JE, Yeh F, Zwiazek JJ (2010) Functional characterization of drought-responsive aquaporins in *Populus balsamifera* and *Populus simonii* × *balsamifera* clones with different drought resistance strategies. Physiol Plantarum 140:321–333

Anekonda TS, Lomas MC, Adams WT, Kavanagh KL, Aitken SN (2002) Genetic variation in drought hardiness of coastal Douglas-fir seedlings from British Columbia. Can J For Res 32:1701–1716

Aranda I, Gil L, Pardos JA (2000) Water relations and gas exchange in (*Fagus sylvatica* L. and *Quercus petraea* (Mattuschka) Liebl. in a mixed stand at the southern limit of distribution in Europe. Trees Struct Funct 14:344–352

Aranda I, Gil L, Pardos JA (2001) Effects of thinning in a *Pinus sylvestris* L. stand on foliar water relations of *Fagus sylvatica* L. seedlings planted within the pinewood. Trees Struct Funct 15:358–364

Aranda I, Gil L, Pardos JA (2002) Physiological responses of Fagus sylvatica L. seedlings under *Pinus sylvestris* L. and *Quercus pyrenaica* Willd. overstories. Forest Ecol Manag 162:153–164

Aranda I, Gil L, Pardos JA (2004) Osmotic adjustment in two temperate oak species [*Quercus pyrenaica* Willd and *Quercus petraea* (Matt.) Liebl] of the Iberian Peninsula in response to drought. Forest Syst (formerly Investigación Agraria: Sistemas y Recursos Forestales) 13:339–345

Aranda I, Castro L, Pardos M, Gil L, Pardos J (2005) Effects of the interaction between drought and shade on water relations, gas exchange and morphological traits in cork oak (*Quercus suber* L.) seedlings. Forest Ecol Manag 210:117–129

Aranda I, Alía R, Ortega U, Dantas A, Majada J (2010) Intra-population variability in biomass partitioning and carbon isotopic discrimination under moderate drought stress in four maritime pine (*Pinus pinaster* L.) populations. Tree Genet Genomes 6:169–170

Arend M, Kuster T, Günthardt-Goerg MS, Dobbertin M (2011) Provenance-specific growth responses to drought and air warming in three European oak species (*Quercus robur*, *Q. petraea* and *Q. pubescens*). Tree Physiol 31:287–297

Ares A, Fownes JH, Sun W (2000) Genetic differentiation of intrinsic water-use efficiency in the hawaiian native Acacia koa. Inte J Plant Sci 161:909–915

Aspelmeier S, Leuschner C (2006) Genotypic variation in drought response of silver birch (*Betula pendula* Roth): leaf and root morphology and carbon partitioning. Trees Struct Funct 20:42–52

Baaziz BK, Lopez D, Rabot A, Combes D, Gousset A, Bouzid S, Cochard H, Sakr S, Venisse J-S (2012) Light-mediated Kleaf induction and contribution of both the PIP1s and PIP2s aquaporins in five tree species: walnut (*Juglans regia*) case study. Tree Physiol 32:423–434

Bae EK, Lee H, Lee JS, Noh EW (2009) Differential expression of a poplar SK2-type dehydrin gene in response to various stresses. BMB Rep 42:439–443

Bae EK, Lee H, Lee JS, Noh EW (2010) Isolation and characterization of osmotic stress-induced genes in poplar cells by suppression subtractive hybridization and cDNA microarray analysis. Plant Physiol Bioch 48:136–141

Bae EK, Lee H, Lee JS, Noh EW (2011) Drought, salt and wounding stress induce the expression of the plasma membrane intrinsic protein 1 gene in poplar (*Populus alba* × *P. tremula* var. glandulosa). Gene 483:43–48

Baquedano FJ, Valladares F, Castillo FJ (2008) Phenotypic plasticity blurs ecotypic divergence in the response of *Quercus coccifera* and *Pinus halepensis* to water stress. Eur J For 127:495–506
Barnard HR, Ryan MG (2003) A test of the hydraulic limitation hypothesis in fast-growing *Eucalyptus saligna*. Plant, Cell Environ 26:1235–1245
Bassman JH, Zwier JC (1991) Gas exchange characteristics of *Populus trichocarpa*, *Populus deltoides* and *Populus trichocarpa* × *P. deltoides* clones. Tree Physiol 8:145–159
Battaglia M, Olvera-Carrillo Y, Garciarrubio A, Campos F, Covarrubias AA (2008) The enigmatic LEA proteins and other hydrophilins. Plant Physiol 148:6–24
Becker P, Meinzer FC, Wullschleger SD (2000) Hydraulic limitation of tree height: a critique. Funct Ecol 14:4–11
Benito M, Alía R, Robson TM, Zavala MA (2011) Intra-specific variability and plasticity influence potential tree species distributions under climate change. Global Ecol Biogeogr 20:766–778
Benowicz A, Guy RD, El-Kassaby YA (2000) Geographic pattern of genetic variation in photosynthetic capacity and growth in two hardwood species from British Columbia. Oecologia 123:168–174
Berta M, Giovannelli A, Sebastiani F, Camussi A, Racchi ML (2010) Transcriptome changes in the cambial region of poplar *Populus alba* L. in response to water deficit. Plant Biol 12:341–354
Bhaskar R, Valiente-Banuet A, Ackerly DD (2007) Evolution of hydraulic traits in closely related species pairs from mediterranean and non-mediterranean environments of North America. New Phytol 176:718–726
Bigras FJ (2005) Photosynthetic response of white spruce families to drought stress. New For 29:135–148
Blödner C, Majcherczyk A, Kües U, Polle A (2007) Early drought-induced changes to the needle proteome of Norway spruce. Tree Physiol 10:1423–1431
Bloom AJ, Chapin FS III, Mooney HA (1985) Resource limitation in plants—an economic analogy. Annu Rev Ecol Syst 16:363–392
Borghetti M, Edwards WRN, Grace J, Jarvis PG, Raschi A (1991) The refilling of embolized xylem in *Pinus sylvestris* L. Plant, Cell Environ 14:357–369
Bossdorf O, Richards CL, Pigliucci M (2008) Epigenetics for ecologists. Ecol Lett 11:106–115
Bradshshaw AD (1965) Evolutionary significance of phenotypic plasticity. Adv Genet 13:115–155
Breda N, Huc R, Granier A, Dreyer E (2006) Temperate forest trees and stands under severe drought:a review of ecophysiological responses, adaptation processes and long-term consequences. Ann For Sci 63:625–644
Brendel O, Pot D, Plomion C, Rozenberg P, Guehl JM (2002) Genetic parameters and QTL analysis of $\delta^{13}C$ and ring width in maritime pine. Plant Cell Environ 25:945–953
Brendel O, Le Thiec D, Scotti-Saintagne C, Bodenes C, Kremer A, Guehl JM (2008) Quantitative trait loci controlling water use efficiency and related traits in *Quercus robur* L. TGG 4:263–278
Breshears DD, Cobb NS, Rich PM, Price KP, Allen CD, Balice RG, Romme WH, Kastens JH, Floyd ML, Belnap J, Anderson JJ, Myers OB, Meyer CW (2005) Regional vegetation die-off in response to global-change-type drought. Proc Natl Acad Sci U S A 102:15144–15148
Breshears DD, Myers OB, Meyer CW, Barnes FJ, Zou CB, Allen CD, McDowell NG, Pockman WT (2009) Tree die-off in response to global-change-type drought: mortality insights from a decade of plant water potential measurements. Front Ecol Environ 7:185–189
Brodribb TJ, Cochard H (2009) Hydraulic failure defines the recovery and point of death in water-stressed conifers. Plant Physiol 149:575–584
Brodribb TJ, Holbrook NM (2003) Stomatal closure during leaf dehydration, correlation with other leaf physiological traits. Plant Physiol 132:2166–2173
Brodribb TF, Holbrook NM (2006) Declining hydraulic efficiency as transpiring leaves desiccate: two types of response. Plant Cell Environ 29:2205–2215
Brodribb TJ, Holbrook NM, Zwieniecki MA, Palma B (2005) Leaf hydraulic capacity in ferns, conifers and angiosperms: impacts on photosynthetic maxima. New Phytol 165:839–846
Brodribb TJ, Bowman DJMS, Nichols S, Delzon S, Burlett R (2010) Xylem function and growth rate interact to determine recovery rates after exposure to extreme water deficit. New Phytol 188:533–542

Brosché M, Vinocur B, Alatalo ER, LamminmäkiA, Teichmann T, Ottow EA, Djialianov D, Afif D, Bogeat-Triboulot MB, Altman A, Polle A, Dreyer e, Rudd S, Paulin L, Auvinen P, Kangasjärvi J (2005) Gene expression and metabolite profiling of *Populus euphratica* growing in the Negev desert. Genome Biol 6:R101

Bucci SJ, Scholz FG, Goldstein G, Meinzer FC, Sternberg LDL (2003) Dynamic changes in hydraulic conductivity in petioles of two savanna tree species: factors and mechanisms contributing to the refilling of embolized vessels. Plant Cell Environ 26:1633–1645

Burczyk J, Giertych M (1991) Response of Norway Spruce (*Picea abies* [L] Karst) annual increments to drought for various provenances and locations. Silvae Genetica 40:146–152

Camarero JJ, Bigler C, Linares JC, Gil-Pelegrín E (2011) Synergistic effects of past historical logging and drought on the decline of Pyrenean silver fir forests. For Ecol Manag 262:759–769

Campanello PI, Gatti MG, Goldstein G (2008) Coordination between water-transport efficiency and photosynthetic capacity in canopy tree species at different growth irradiances. Tree Physiol 28:85–94

Canadell J, Jackson RB, Ehleringer JB, Mooney HA, Sala OE, Schulze ED (1996) Maximum rooting depth of vegetation types at the global scale. Oecologia 108:583–595

Caruso A, Chefdor F, Carpin S, Depierreux C, Delmotte FM, Kahlem G, Morabito D (2008) Physiological characterization and identification of genes differentially expressed in response to drought induced by PEG 6000 in *Populus Canadensis* leaves. J Plant Physiol 165:932–941

Casasoli M, Derory J, Morera-Dutrey C, Brendel O, Porth I, Guehl J-M, Villani F, Kremer A (2006) Comparison of quantitative trait loci for adaptive traits between oak and chestnut based on an expressed sequence tag consensus map. Genetics 172:533–546

Cassasoli M, Pot D, Plomion C, Monteverdi MC, Barreneche T, Laureti M, Villani F (2004) Identification of QTLs affecting adaptive traits in *Castanea sativa* Mill. Plant Cell Environ 27:1088–1101

Ceulemans R, Impens I (1980) Leaf gas exchange processes and related characteristics of seven poplar clones under laboratory conditions. Can J For Res 10:429–435

Chambel MR, Climent J, Alia R (2007) Divergence among species and populations of Mediterranean pines in biomass allocation of seedlings grown under two watering regimes. Ann For Sci 64:87–97

Chang S, Puryear JD, Dias MADL, Funkhouser EA, Newton RJ, Cairney J (1996) Gene expression under water deficit in loblolly pine (*Pinus taeda*): isolation and characterization of cDNA clones. Physiol Plantarum 97:139–148

Chaves MM, Pereira JS, Maroco J (2003) Understanding plant response to drought—from genes to the whole plant. Funct Plant Biol 30:239–264

Chen J, Xia X, Yin W (2009) Expression profiling and functional characterization of a DREB2-type gene from *Populus euphratica*. Biochem Bioph Res Co 378:483–487

Chen J, Xia X, Yin W (2011) A poplar DRE-binding protein gene, PeDREB2L, is involved in regulation of defense response against abiotic stress. Gene 483:36–42

Chmura DJ, Anderson PD, Howe GT, Harrington CA, Halofsky JE, Peterson DL, Shaw DC, St.Clair JB (2011) Forest responses to climate change in the northwestern United States: ecophysiological foundations for adaptive management. For Ecol Manag 261:1121–1142

Ciais P, Reichstein M, Viovy N, Granier A, Ogee J, Allard V, Aubinet M, Buchmann N, Bernhofer C, Carrara A, Chevallier F, De Noblet N, Friend AD, Friedlingstein P, Grunwald T, Heinesch B, Keronen P, Knohl A, Krinner G, Loustau D, Manca G, Matteucci G, Miglietta F, Ourcival JM, Papale D, Pilegaard K, Rambal S, Seufert G, Soussana JF, Sanz MJ, Schulze ED, Vesala T, Valentini R (2005) Europe-wide reduction in primary productivity caused by the heat and drought in 2003. Nature 437:529–533

Clearwater M, Goldstein G (2005) Embolism repair and long distance water transport. In: Holbrook NM, Zwieniecki MA (eds) Vascular transport in plants. Elsevier Academic Press, USA

Cochard H, Tyree MT (1990) Xylem dysfunction in *Quercus*: vessel sizes, tyloses, cavitation and seasonal changes in embolism. Tree Physiol 6:393–407

Cochard H, Breda N, Granier A, Aussenac G (1992) Vulnerability to air embolism of three European oak species (*Quercus petraea* (Matt) Liebl., *Q. pubescens* Willd., *Q. robur* L.). Ann For Sci 49:225–233

Cochard H, Venisse JS, Barigah TS, Brunel N, Herbette S, Guilliot A, Tyree MT, Sakr S (2007) Putative role of aquaporins in variable hydraulic conductance of leaves in response to light. Plant Physiol 143:122–133

Cohen D, Bogeat-Triboulot MB, Tisserant E, Balzergue S, Martin-Magniette ML, Lelandais G, Ningre N, Renou JP, Tamby JP, Le Thiec D, Hummel I (2010) Comparative transcriptomics of drought responses in *Populus*: a meta-analysis of genome-wide expression profiling in mature leaves and root apices across two genotypes. BMC Genomics 11:630

Corcuera L, Camarero JJ, Gil-Pelegrin E (2002) Functional groups in *Quercus* species derived from the analysis of pressure–volume curves. Trees Struct Funct 16:465–472

Corcuera L, Camarero JJ, Gil-Pelegrín E (2004a) Effects of a severe drought on *Quercus ilex* radial growth and xylem anatomy. Trees Struct Funct 18:83–92

Corcuera L, Camarero JJ, Gil-Pelegrín E (2004b) Effects of a severe drought on growth and wood anatomical properties of *Quercus faginea*. IAWA J 25:185–204

Corcuera L, Morales F, Gil-Pelegrín E (2005) Seasonal changes in photosynthesis and photoprotection in a *Quercus ilex* subsp *ballota* woodland located in its upper altitudinal extreme in the Iberian Peninsula. Tree Physiol 25:599–608

Corcuera L, Camarero JJ, Sisó S, Gil-Pelegrin E (2006) Radial-growth and wood-anatomical changes in overaged *Quercus pyrenaica* coppice stands: functional responses in a new Mediterranean landscape. Trees Struct Funct 0:91–98

Corcuera L, Cochard H, Gil-Pelegrin E, Notivol E (2011) Phenotypic plasticity in mesic populations of *Pinus pinaster* improves resistance to xylem embolism (P50) under severe drought. Trees Struct Funct 25:1033–1042

Cregg BM, Zhang JW (2001) Physiology and morphology of *Pinus sylvestris* seedlings from diverse sources under cyclic drought stress. For Ecol Manag 154:131–139

David TS, Henriques MO, Kurz-Besson C, Nunes J, Valente F, Vaz M, Pereira JS, Siegwolf R, Chaves MM, Gazarini LC, David JS (2007) Water use strategies in two co-occurring mediterranean evergreen oaks: surviving the summer drought. Tree Physiol 27:793–803

Day ME, Greenwood MS, Diaz-Sala C (2002) Age- and size-related trends in woody plant shoot development: regulatory pathways and evidence for genetic control. Tree Physiol 22:507–513

De Miguel M, Sánchez-Gómez D, Cervera MT, Aranda I (2012) Functional and genetic characterization of gas-exchange and water use efficiency in a full-sib family of *Pinus pinaster* Aitt. in response to drought. Tree Physiol 32:94–103

Demmig-Adams B, Adams WW III (1996) Xanthophyll cycle and light stress in nature; uniform response to excess direct sunlight among higher plants. Planta 198:460–470

Donovan LA, Ludwig F, Rosenthal DM, Rieseberg LH, Dudley SA (2009) Phenotypic selection on leaf ecophysiological traits in *Helianthus*. New Phytol 183:868–879

Drost DR, Benedict CI, Berg A, Novaes E, Novaes CR, Yu Q, Dervinis C, Maia JM, Yap J, Miles B, Kirst M (2010) Diversification in the genetic architecture of gene expression and transcriptional networks in organ differentiation of *Populus*. Proc Natl Acad Sci U S A 107:8492–8497

Duan B, Xuan Z, Zhang X, Korpelainen H, Li C (2008) Interactions between drought, ABA application and supplemental UV-B in *Populus yunnanensis*. Physiol Plantarum 134:257–269

Dubos C, Plomion C (2003) Identification of water-deficit responsive genes in maritime pine (*Pinus pinaster* Ait.) roots. Plant Mol Biol 51:249–262

Ducrey M, Huc R, Ladjal M, Guehl JM (2008) Variability in growth, carbon isotope composition, leaf gas exchange and hydraulic traits in the eastern Mediterranean cedars *Cedrus libani* and *C. brevifolia*. Tree Physiol 28:689–701

Dudley SA (1996) Differing selection on plant physiological traits in response to environmental water availability: a test of adaptive hypotheses. Evolution 50:92–102

Eggemeyer KD, Awada T, Harvey FE, Wedin DA, Zhou X, Zanner W (2009) Seasonal changes in depth of water uptake for encroaching trees *Juniperus virginiana* and *Pinus ponderosa* and two dominant C4 grasses in a semiarid grassland. Tree Physiol 29:157–169

Eller ASD, McGuire KL, Sparks JP (2011) Responses of sugar maple and hemlock seedlings to elevated carbon dioxide under altered above- and belowground nitrogen sources. Tree Physiol 31:391–401

Ellsworth DS, Reich PB (1992) Water relations and gas exchange of *Acer saccharum* seedlings in contrasting natural light and water regimes. Tree Physiol 10:1–20

Esteso-Martinez J, Camarero JJ, Gil-Pelegrin E (2006) Competitive effects of herbs on *Quercus faginea* seedlings inferred from vulnerability curves and spatial-pattern analyses in a Mediterranean stand (Iberian System, northeast Spain). Ecoscience 13:378–387

Eveno E, Collada C, Guevara MA, Léger V, Soto A, Díaz L, Léger P, González-Martínez SC, Cervera MT, Plomion C, Garnier-Géré PH (2008) Contrasting patterns of selection at *Pinus pinaster* Ait. drought stress candidate genes as revealed by genetic differentiation analyses. Mol Biol Evol 25:417–437

Fillatti JJ, Sellmer J, McCown B, Haissig B, Comai L (1987) Agrobacterium mediated transformation and regeneration of *Populus*. Mol Gen Genet 206:192–199

Flanagan LB, Johnsen KH (1995) Genetic variation in carbon isotope discrimination and its relationship to growth under field conditions in full-sib families of *Picea mariana*. Can J For Res 25:39–47

Foster DR, Oswald WW, Faison EK, Doughty ED, Hansen BCS (2006) A climatic driver for abrupt mid-Holocene vegetation dynamics and the hemlock decline in New England. Ecology 87:2959–2966

Gailing O, Langenfeld-Heyser R, Polle A, Finkeldey R (2008) Quantitative trait loci affecting stomatal density and growth in a *Quercus robur* progeny: implications for the adaptation to changing environments. Glob Change Biol 14:1934–1946

Gailing O, Vornam B, Leinemann L, Finkeldey R (2009) Genetic and genomic approaches to assess adaptive genetic variation in plants: forest trees as a model. Physiol Plantarum 137:509–519

Gebre GM, Kuhns MR, Brandle JR (1994) Organic solute accumulation and dehydration tolerance in tree water-stressed *Populus deltoides* clones. Tree Physiol 14:575–587

Gimeno TE, Pías B, Lemos-Filho JP, Valladares F (2009) Plasticity and stress tolerance override local adaptation in the responses of Mediterranean holm oak seedlings to drought and cold. Tree Physiol 29:87–98

González-Martínez SC, Huber D, Ersoz E, Davis JM, Neale DB (2008) Association genetics in *Pinus taeda* L. II. Carbon isotope discrimination. Heredity 101:19–26

Gourcilleau D, Bogeat-Triboulot MB, Le Thiec D, Lafon-Placette C, Delaunay A, El-Soud WA, Brignolas F, Maury S (2010) DNA methylation and histone acetylation: genotypic variations in hybrid poplars, impact of water deficit and relationships with productivity. Ann For Sci 67:208

Grether GF (2005) Environmental change, phenotypic plasticity, and genetic compensation. Am Nat 166:E115–E123

Grivet D, Sebastiani F, Alía R, Bataillon T, Torre S, Zabal-Aguirre M, Vendramin GG, González-Martínez SC (2011) Molecular footprints of local adaptation in two mediterranean conifers. Mol Biol Evol 28:101–116

Grossnickle SC, Fan SH, Russell JH (2005) Variation in gas exchange and water use efficiency patterns among populations of western red cedar. Trees Struct Funct 19:32–42

Guy RD, Holowachuk DL (2001) Population differences in stable carbon isotope ratio of *Pinus contorta* Dougl. ex Loud.: relationship to environment, climate of origin, and growth potential. Can J Bot 79:274–283

Hacke UG, Sauter JJ (1996) Xylem dysfunction during winter and recovery of hydraulic conductivity in diffuse-porous and ring-porous trees. Oecologia 105:435–439

Hacke UG, Sperry JS (2003) Limits to xylem refilling under negative pressure in *Laurus nobilis* and *Acer negundo*. Plant Cell Environ 26:303–311

Hacke UG, Sperry JS, Pockman WT, Davis SD, McCulloh KA (2001) Trends in wood density and structure are linked to prevention of xylem embolism by negative pressure. Oecologia 54:275–280

Hacke UG, Sperry JS, Pittermann JA (2005) Efficiency vs. safety trade-offs for water conduction in angiosperm vessels vs. gymnosperm tracheids. In: Holbrook NM, Zwieniecki M (eds) Vascular transport in plants. Elsevier Inc., Amsterdam, pp 333–353

Hamanishi ET, Campbell MM (2011) Genome-wide responses to drought in forest trees. Forestry 84:273–283

Hamrick JL (2004) Response of forest trees to global environmental changes. For Ecol Manag 197:323–335

Heath LS, Ramakrishnan N, Sederoff RR, Whetten RW, Chevone BI, Struble CA, Jouenne VY, Chen DW, van Zyl L, Grene R (2002) Studying the functional genomics of stress responses in loblolly pine with the Expresso microarray experiment management system. Comp Funct Genom 3:226–243

Henery ML, Moran GF, Foley WJ (2007) Identification of quantitative trait loci influencing foliar concentrations of terpenes and formylated phloroglucinol compounds in *Eucalyptus nitens*. New Phytol 176:82–95

Herrera CM, Bazaga P (2010) Epigenetic differentiation and relationship to adaptive genetic divergence in discrete populations of the violet *Viola cazorlensis*. New Phytol 187:867–876

Hirayama T, Shinozaki K (2010) Research on plant abiotic stress responses in the post-genome era: past, present and future. Plant J 61:1041–1052

Ingvarsson PK, Street NR (2011) Association genetics of complex traits in plants. New Phytol 189:909–922

Ingvarsson PK, Garcia MV, Luquez V, Hall D, Jansson S (2008) Nucleotide polymorphism and phenotypic associations within and around the phytochrome B2 locus in European Aspen (*Populus tremula, Salicaceae*). Genetics 178:2217–2226

IPCC 2007. Climate change 2007:impacts, adaptation and vulnerability. In: Parry ML, Canziani OF, Palutikof JP, van der Linden PJ, Hanson CE (eds) Contribution of working group II to the fourth assessment report of the intergovernmental panel on climate change. Cambridge University Press, Cambridge, p 976

Jansen S, Gortan E, Lens F, Lo Gullo MA, Salleo S, Schloz A, Stein A, Trifilò P, Nardini A (2011) Do quantitative vessel and pit characters account for ion-mediated changes in the hydraulic conductance of angiosperm xylem? New Phytol 189:218–228

Jensen JS (1993) Variation of growth in Danish provenance trials with oak (*Quercus robur* L and *Quercus petraea* Mattuschka Liebl). Ann Sci For 50:203–207

Johnson DM, Woodruff DR, McCulloh KA, Meinzer F (2009) Leaf hydraulic conductance, measured in situ, declines and recovers daily:leaf hydraulics, water potential and stomatal conductance in four temperate and three tropical tree species. Tree Physiol 29:879–887

Jones HG, Sutherland RA (1991) Stomatal control of xylem embolism. Plant Cell Environ 49:607–612

Jump AS, Hunt JM, Peñuelas J (2006) Rapid climate change-related growth decline at the southern range edge of *Fagus sylvatica*. Global Change Biol 12:2163–2174

King DA (1990) The adaptive significance of tree height. Am Nat 135:809–828

König A (2005) Provenance research: evaluation the spatial pattern of genetic variation. In: Geburek T, Turok J (eds) Conservation and management of forest genetic resources in Europe. Arbora Publishers, Zvolen and IPGRI, Rome, pp 275–334

Kremer K (1995) Phenotypic plasticity of the phenology of seven European tree species in relation to climatic warming. Plant Cell Environ 18:93–104

Kremer A (2010) Evolutionary responses of European oaks to climate change. Irish For 67:53–66

Kremer A, Le Corre V, Petit RJ, Ducousso A (2010) Historical and contemporary dynamics of adaptive differentiation in European oaks. In: DeWoody A, Bickham J, Michler C, Nichols K, Rhodes G, Woeste K (eds) Molecular Approaches in natural resource conservation and management. Cambridge University Press, Cambridge, pp 101–122

Lamy J-B, Bouffier L, Burlett R, Plomion C, Cochard H, Delzon S (2011) Uniform selection as a primary force reducing population genetic differentiation of cavitation resistance across a species range. PLoS ONE 6(8):e23476

Langlet O (1971) Two hundred years of genecology. Taxon 20:653–722

Lauteri M, Scartazza A, Guido MC, Brugnoli E (1997) Genetic variation in photosynthetic capacity, carbon isotope discrimination and mesophyll conductance in provenances of *Castanea sativa* adapted to different environments. Funct Ecol 11:675–683
Lenz TI, Wright IJ, Westoby M (2006) Interrelations among pressure–volume curve traits across species and water availability gradients. Physiol Plantarum 127:423–433
Leonardi S, Piovani P, Magnani F, Menozzi P (2006) A simple general method to evaluate intra-specific transpiration parameters within and among seedling families. Oecologia 149:185–193
Li XP, Björkman O, Shih C, Grossman AR, Rosenquist M, Jansson S, Niyogi KK (2000a) A pigment-binding protein essential for regulation of photosynthetic light harvesting. Nature 403:391–395
Li C, Berninger F, Koskela J, Sonninen E (2000b) Drought responses of *Eucalyptus microtheca* provenances depend on seasonality of rainfall in their place of origin. Aust J Plant Physiol 27:231–238
Linares JC, Camarero JJ, Carreira JA (2009) Interacting effects of climate and forest-cover changes on mortality and growth of the southernmost European fir forests. Global Ecol Biogeogr 18:485–497
Linares JC, Camarero JJ, Carreira JA (2010) Competition modulates the adaptation capacity of forests to climatic stress: insights from recent growth decline and death in relict stands of the Mediterranean fir *Abies pinsapo*. J Ecol 98:592–603
Lindner M, Maroschek M, Netherer S, Kremer A, Barbati A, Garcia-Gonzalo J, Seidl R, Delzon S, Corona P, Kolstrom M, Lexer MJ, Marchetti M (2010) Climate change impacts, adaptive capacity, and vulnerability of European forest ecosystems. Forest Ecol Manag 259:698–709
Linhart YB, Grant MC (1996) Evolutionary significance of local genetic differentiation in plants. Annu Rev Ecol Syst 267:237–277
Lira-Medeiros CF, Parisod C, Fernandes RA, Mata CS, Cardoso MA, Ferreira PCG (2010) Epigenetic variation in mangrove plants occurring in contrasting natural environment. PLoS ONE 5:e10326. doi:10.1371/journal.pone.0010326
Lo Gullo MA, Salleo S (1993) Different vulnerabilities of *Quercus ilex* L. to freeze- and summer drought-induced xylem embolism: an ecological interpretation. Plant Cell Environ 16:511–519
Lo Gullo MA, Nardini A, Salleo S, Tyree MT (1998) Changes in root hydraulic conductance (K_R) of *Olea oleaster* seedlings following drought stress and irrigation. New Phytol 140:25–31
Long Y, Xia W, Li R, Wang J, Shao M, Feng J, King GJ, Meng J (2011) Epigenetic QTL mapping in *Brassica napus*. Genetics 189:1093–1102
Lorenz WW, Sun F, Liang C, Kolychev D, Wang HM, Zhao X, Cordonnier-Pratt MM, Pratt LH, Dean JFD (2006) Water stress-responsive genes in loblolly pine (*Pinus taeda*) roots identified by analyses of expressed sequence tag libraries. Tree Physiol 26:1–16
Maherali H, Pockman WT, Jackson RB (2004) Adaptive variation in the vulnerability of woody plants to xylem cavitation. Ecology 85:2184–2199
Major JE, Johnsenn KH (1996) Family variation in photosynthesis of 22-year-old black spruce: a test of two models of physiological response to water stress. Can J For Res 26:1922–1933
Major JE, Johnsenn KH (1999) Shoot water relations of mature black spruce families displaying a genotype x environment interaction in growth rate. II. Temporal trends and response to varying soil water conditions. Tree Physiol 19:375–382
Manter DK, Kelsey RG (2008) Ethanol accumulation in drought-stressed conifer seedlings. Int J Plant Sci 169:361–369
Martínez-Vilalta J, Piñol J (2002) Drought-induced mortality and hydraulic architecture in pine populations of the NE Iberian Peninsula. For Ecol Manag 161:247–256
Martínez-Vilalta J, Prat E, Oliveras I, Piñol J (2002) Xylem hydraulic properties of roots and stems of nine Mediterranean woody species. Oecologia 133:19–29
Mayr S, Rothart B, Daèmon B (2003) Hydraulic efficiency and safety of leader shoots and twigs in Norway spruce growing at the alpine timberline. J Exp Bot 54:2563–2568
McCown BH, McCabe DE, Russell DR, Robinson DJ, Barton KA, Raffa KF (1991) Stable transformation of *Populus* and incorporation of pest resistance by electric discharge particle acceleration. Plant Cell Rep 9:590–594

McDowell NG (2011) Mechanisms linking drought, hydraulics, carbon metabolism, and vegetation mortality. Plant Physiol 155:1051–1059

McDowell N, Pockman WT, Allen CD, Breshears DD, Cobb N, Kolb T, Sperry J, West A, Williams D, Yepez EA (2008) Mechanisms of plant survival and mortality during drought: why do some plants survive while others succumb to drought? Transley review. New Phytol 178:719–739

Meier IC, Leuschner C (2008a) Belowground drought response of European beech: fine root biomass and carbon partitioning in 14 mature stands across a precipitation gradient. Global Change Biol 14:2081–2095

Meier IC, Leuschner Ch (2008b) Genotypic variation and phenotypic plasticity in the drought response of fine roots of European beech. Tree Physiol 28:297–309

Melcher PJ, Goldstein G, Meinzer FC, Yount DE, Jones TJ, Holbrook NM, Huang CX (2001) Water relations of coastal and estuarine *Rhizophora mangle*: xylem pressure potential and dynamics of embolism formation and repair. Oecologia 126:182–192

Merchant A, Tausz M, Arndt SK, Adams MA (2006) Cyclitols and carbohydrates in leaves and roots of 13 *Eucalyptus* species suggest contrasting physiological responses to water deficit. Plant Cell Environ 29:2017–2029

Merchant A, Callister A, Arndt S, Tausz M, Adams M (2007a) Contrasting physiological responses of six *Eucalyptus* species to water deficit. Ann Bot 100:1507–1515

Merchant A, Ladiges PY, Adams MA (2007b) Quercitol links the physiology, taxonomy and evolution of 279 eucalypt species. Global Ecol Biogeogr 16:810–819

Mirouze M, Paszkowski J (2011) Epigenetic contribution to stress adaptation in plants. Curr Opin Plant Biol 14:267–274

Mittler R (2006) Abiotic stress, the field environment and stress combination. Trends Plant Sci 11:15–19

Monclus R, Dreyer E, Villar M, Delmotte FM, Delay D, Petit J-M, Barbaroux C, Le Thiec D, Bréchet C, Brignolas F (2006) Impact of drought on productivity and water-use efficiency in 29 genotypes of *Populus deltoides* × *Populus nigra*. New Phytol 169:765–777

Monk C (1966) Ecological importance of root/shoot ratios. B Torrey Bot Club 93:402–406

Morales F, Abadía A, Abadía J (2006) Photoinhibition and photoprotection under nutrient deficiencies, drought and salinity. In: Demmig-Adams B, Adams III WW, Mattoo AK (eds) Photoprotection, photoinhibition, gene regulation, and environment. Springer, The Netherlands, pp 65–85

Morreel K, Goeminne G, Storme V, Sterck L, Ralph J, Coppieters W, Breyne P, Steenackers M, Georges M, Messens E, Boerjan W (2006) Genetical metabolomics of flavonoid biosynthesis in *Populus*: a case study. Plant J 47:224–237

Morse AM, Peterson DG, Islam-Faridi MN, Smith KE, Magbanua Z, Garcia SA, Kubisiak TL, Amerson HV, Carlson JE, Nelson CD, Davis JM (2009) Evolution of genome size and complexity in *Pinus*. PLoS ONE 4:e4332

Nardini A, Tyree MT, Salleo S (2001) Xylem cavitation in the leaf of *Prunus laurocerasus* and its impact on leaf hydraulics. Plant Physiol 125:1700–1709

Nardini A, Lo Gullo MA, Salleo S (2011) Refilling embolized xylem conduits: is it a matter of phloem unloading? Plant Sci 180:604–611

Neale DB, Kremer A (2011) Forest tree genomics: growing resources and applications. Nat Rev Genet 12:111–122

Newbery DM, Lingenfelder M (2009) Plurality of tree species responses to drought perturbation in Bornean tropical rain forest. Plant Ecol 201:147–167

Nicotra AB, Atkin OK, Bonser SP, Davidson A, Finnegan EJ, Mathesius U, Poot P, Purugganan MD, Richards CL, Valladares F, van Kleunen M (2010) Plant phenotypic plasticity in a changing climate. Trends Plant Sci 15:684–692

Niinemets Ü (2010) Responses of forest trees to single and multiple environmental stresses from seedlings to mature plants: past stress history, stress interactions, tolerance and acclimation. For Ecol Manag 260:1623–1639

Nobel PS, Lee CH (1991) Variations in root water potentials: influence of environmental factors for two succulent species. Ann Bot 67:549–554

Ogaya R, Peñuelas J (2007) Tree growth, mortality, and above-ground biomass accumulation in a holm oak forest under a five-year experimental field drought. Plant Ecol 189:291–299

Oliver MJ, Cushman JC, Koster KL (2010) Dehydration tolerance in plants. In: Sunkar R (ed) Plant stress tolerance, vol 639. Humana Press, New York, pp 3–24

Olsen AN, Ernst HA, Leggio LL, Skriver K (2005) NAC transcription factors: structurally distinct, functionally diverse. Trends Plant Sci 10:79–87

Passarinho JAP, Lamosa P, Baeta JP, Santos H, Ricardo CPP (2006) Annual changes in the concentration of minerals and organic compounds of *Quercus suber* leaves. Physiol Plant 127:100–110

Peguero-Pina JJ, Camarero JJ, Abadía A, Martín E, González-Cascón R, Morales F, Gil-Pelegrín E (2007) Physiological performance of silver-fir (*Abies alba* Mill.) populations under contrasting climates near the south-western distribution limit of the species. Flora 202:226–236

Peguero-Pina JJ, Sancho-Knapik D, Morales F, Flexas J, Gil-Pelegrín E (2009) Differential photosynthetic performance and photoprotection mechanisms of three Mediterranean evergreen oaks under severe drought stress. Funct Plant Biol 36:453–462

Peguero-Pina JJ, Sancho-Knapik D, Cochard H, Barredo G, Villarroya D, Gil-Pelegrín (2011) Hydraulic traits are associated with the distribution range of two closely related Mediterranean firs, Abies alba Mill. and Abies pinsapo Boiss. Tree Physiol 31:1067–1075

Pelgas B, Bousquet J, Meirmans PG, Ritland K, Isabel N (2011) QTL mapping in white spruce: gene maps and genomic regions underlying adaptive traits across pedigrees, years and environments. BMC Genomics 12:145

Pemán J, Voltas J, Gil-Pelegrín E (2006) Morphological and functional variability in the root system of *Quercus ilex* L. subject to confinement: consequences for afforestation. Ann For Sci 63:425–430

Peñuelas J, Lloret F, Montoya R (2001) Severe drought effects on mediterranean woody flora in Spain. Forest Sci 47:214–218

Peñuelas J, Ogaya R, Hunt JM, Jump AS (2008) 20th century changes of tree-ring d13C at the southern range-edge of *Fagus sylvatica*. Increasing water-use efficiency does not avoid the growth decline induced by warming at low altitudes. Glob Change Biol 14:1076–1088

Perdiguero P, Collada C, Barbero MC, García-Casado G, Cervera MT, Soto A (2012) Identification of water stress genes in *Pinus pinaster* Ait. by controlled progressive stress and suppression-subtractive hybridization. Plant Physiol Biochem 50:44–53

Peuke AD, Hartung W, Schraml C, Rennenberg H (2002) Identification of drought sensitive beech ecotypes by physiological parameters. New Phytol 154:373–388

Piper FI (2011) Drought induces opposite changes in the concentration of non-structural carbohydrates of two evergreen *Nothofagus* species of differential drought resistance. Ann For Sci 68:415–424

Pittermann J, Sperry JS, Hacke U, Wheeler JK, Sikkema EH (2006) Inter-tracheid pitting and the hydraulic efficiency of conifer wood: the role of tracheid allometry and cavitation protection. Am J Bot 93:1265–1273

Plomion C, Chagné D, Pot D, Kumar S, Wilcox PL, Burdon RD, Prat D, Peterson DG, Paiva J, Chaumeil P, Vendramin GG, Sebastiani F, Nelson CD, Echt CS, Savolainen O, Kubisiak TL, Cervera MT, de María N, Islam-Faridi MN (2007) Pines. In: Kole C (ed) Forest trees. Genome mapping and molecular breeding in plants, vol 7. Springer, Berlin. Paper 44

Poorter H, Niklas KJ, Reich PB, Oleksyn J, Poot P, Mommer L (2012) Biomass allocation to leaves, stems and roots: meta-analyses of interspecific variation and environmental control. New Phytol 193:30–50

Possen BJHM, Oksanen E, Rousi M, Ruhanen H, Ahonen V, Tervahauta A, Heinonen J, Heiskanen J, Kärenlampi S, Vapaavuori E (2011) Adaptability of birch (*Betula pendula* Roth) and aspen (*Populus tremula* L.) genotypes to different soil moisture conditions. For Ecol Manag 262:1387–1399

Prasolova NV, Xu ZH, Farquhar GD, Saffigna PG, Dieters MJ (2001) Canopy carbon and oxygen isotope composition of 9-year-old hoop pine families in relation to seedling carbon isotope composition, growth, field growth performance, and canopy nitrogen concentration. Can J For Res 31:673–681

Raj S, Bräutigama K, Hamanishi ET, Wilkins O, Thomas BR, Schroederf W, Mansfield SD, Planth AL, Campbell MM (2011) Clone history shapes *Populus* drought responses. Proc Natl Acad Sci U S A 108:12521–12526

Ramírez-Valiente JA, Lorenzo Z, Soto A, Valladares F, Gil L, Aranda I (2009) Elucidating the role of genetic drift and natural selection in cork oak differentiation regarding drought tolerance. Mol Ecol 18:3803–3815

Ramírez-Valiente JA, Valladares F, Delgado A, Granados A, Aranda I (2011) Factors affecting cork oak growth under dry conditions: local adaptation and contrasting additive genetic variance within populations. Tree Gen Genom 7:285–295

Resco V, Ewers BE, Sun W, Huxman TE, Weltzin JF, Williams DG (2009) Drought-induced hydraulic limitations constrain leaf gas exchange recovery after precipitation pulses in the C3 woody legume, *Prosopis velutina*. New Phytol 181:672–682

Robson TM, Rodríguez-Calcerrada J, Sánchez-Gómez D, Aranda I (2009) Summer drought impedes beech seedlings performance more in a sub-Mediterranean forest understory than in small gaps. Tree Physiol 29:249–259

Robson TM, Sánchez-Gomez D, Cano FJ, Aranda I (2012) Differences in functional leaf traits among beech provenances during a Spanish summer reflect the differences in their origin. Tree Gen Genom. doi: 10.1007/s11295-012-0496-5

Rodríguez-Calcerrada J, Cano FJ, Valbuena-Carabaña M, Gil L, Aranda I (2010) Functional performance of oak seedlings naturally regenerated across different understory microhabitats in a sub-Mediterranean forest: influence of light and water availability. New Forest 39:245–259

Rönnberg-Wästljung AC, Glynn C, Weih M (2005) QTL analyses of drought tolerance and growth for a *Salix dasyclados* × *Salix viminalis* hybrid in contrasting water regimes. Theor Appl Genet 110:537–549

Ryan MG, Yoder BJ (1997) Hydraulic limits to tree height and tree growth. Bioscience 47:235–242

Ryan MG, Phillips N, Bond BJ (2006) The hydraulic limitation hypothesis revisited. Plant Cell Environ 29:367–381

Sack L, Holbrook NM (2006) Leaf hydraulics. Annu Rev Plant Biol 57:361–381

Sack L, Cowan PD, Jaikumar NJ, Holbrook NM (2003) The 'hydrology' of leaves: coordination of structure and function in temperate woody species. Plant Cell Environ 26:1343–1356

Sala A, Piper F, Hoch G (2010) Physiological mechanisms of drought-induced tree mortality are far from being resolved. New Phytol 186:274–281

Salleo S, Raimondo F, Trifilo P, Nardini A (2003) Axial-to-radial water permeability of leaf major veins: a possible determinant of the impact of vein embolism on leaf hydraulics? Plant Cell Environ 26:1749–1758

Sánchez-Gómez D, Velasco-Conde T, Cano FJ, Guevara MA, Cervera MT, Aranda I (2011) Stone pine, a genetically homogeneous species that displays inter-clonal variation in functional traits related to differential growth under drought. Environ Exp Bot 70:104–109

Sangsing K, Kasemsap P, Thanisawanyangkura S, Sangkhasila K, Gohet E, Thaler P, Cochard H (2004) Xylem embolism and stomatal regulation in two rubber clones (*Hevea brasiliensis* Muell. Arg.). Trees 18:109–114

Sardans J, Peñuelas J, Rivas-Ubach A (2011) Ecological metabolomics: overview of current developments and future challenges. Chemoecology 21:191–225

Sathyan P, Newton RJ, Loopstra CA (2005) Genes induced by WDS are differentially expressed in two populations of aleppo pine (*Pinus halepensis*). Tree Genet Genom 1:166–173

Savolainen O, Pyhajarvi T, Knurr T (2007) Gene flow and local adaptation in trees. Annu Rev Ecol Syst 38:595–619

Scotti I, Calvo-Vialettes L, Scotti-Saintagne C, Citterio M, Degen B, Bonal D (2010) Genetic variation for growth, morphological, and physiological traits in a wild population of the Neotropical shade tolerant rainforest tree *Sextonia rubra* (Mez) van der Werff (Lauraceae). Tree Genet Genomes 6:319–329

Seager R, Ting M, Held I, Kushnir Y, Lu J, Vecchi G, Huang H-P, Harnik N, Leetmaa A, Lau N-C, Li C, Velez J, Naik N (2007) Model projections of an imminent transition to a more arid climate in southwestern North America. Science 316:1181–1184

Secchi F, Lovisolo C, Uehlein N, Kaldenho R, Schubert A (2007) Isolation and functional characterization of three aquaporins from olive (*Olea europaea* L.). Planta 225:381–392
Shatil-Cohen A, Attia Z, Moshelion M (2011) Bundle-sheath cell regulation of xylem-mesophyll water transport via aquaporins under drought stress: a target of xylem-borne ABA? Plant J 67:72–80
Shinozaki K, Yamaguchi-Shinozaki K (2007) Gene networks involved in drought stress response and tolerance. J Exp Bot 58:221–227
Sperry J, Ikeda T (1997) Xylem cavitation in roots and stems of Douglas fir and white fir. Tree Physiol 17:275–280
Sperry JS, Saliendra NZ (1994) Intra- and inter-plant variation in xylem cavitation in *Betula occidentalis*. Plant Cell Environ 17:1233–1241
Sperry JS, NM Holbrook NM, Zimmermann MH, Tyree MT (1987) Spring filling of xylem vessels in wild grapevine. Plant Physiol 83:414–417
Sperry JS, Stiller V, Hacke UG (2003) Xylem hydraulics and the soil-plant-atmosphere continuum: opportunities and unresolved issues. Agron J 95:1362–1370
Sperry JS, Meinzer FC, McCulloh KA (2008) Safety and efficiency conflicts in hydraulic architecture: scaling from tissues to trees. Plant Cell Environ 31:632–645
Street NR, Skogstrom O, Sjodin A, Tucker J, Rodriguez-Acosta M, Nilsson P, Jansson S, Taylor G (2006) The genetics and genomics of the drought response in *Populus*. Plant J 48:321–341
Teixeira J, Missiaggia A, Dias D, Scarpinati E, Viana J, Paula N, Paula R, Bonine C (2011) QTL analyses of drought tolerance in *Eucalyptus* under two contrasting water regimes. BMC Proc 5:40
Thumma BR, Nolan MR, Evans R, Moran GF (2005) Polymorphisms in Cinnamoyl CoA Reductase (CCR) are associated with variation in microfibril angle in *Eucalyptus* spp. Genetics 171:1257–1265
Topa MA (2004) Tree physiology: root system physiology. In: Burley J, Evans J, Youngquist J (eds) The encyclopedia of forest sciences. Elsevier Science Ltd., London, pp 1606–1616
Tschaplinski TJ, Tuskan GA, Sewell MM, Gebre GM, Todd DE, Pendley CD (2006) Phenotypic variation and quantitative trait locus identification for osmotic potential in an interspecific hybrid inbred F2 poplar pedigree grown in contrasting environments. Tree Physiol 26:595–604
Tsuda M, Tyree MT (1997) Whole-plant hydraulic resistance and vulnerability segmentation in *Acer saccharinum*. Tree Physiol 17:351–357
Tuskan GA, DiFazio S, Jansson S, Bohlmann J et al (2006) The genome of black cottonwood, *Populus trichocarpa* (Torr. & Gray). Science 313:1596–1604
Tyree MT, Ewers FW (1991) The hydraulic architecture of trees and other woody plants. New Phytol 119:345–360
Tyree MT, Sperry JS (1989) Vulnerability of xylem to cavitation and embolism. Ann Rev Plant Physiol Plant Mol Biol 40:19–38
Tyree MT, Zimmermann MH (2002) Xylem structure and the ascent of sap, 2nd edn. Springer, Berlin
Tyree MT, Snyderman DA, Wilmot TR, Machado JL (1991) Water relations and hydraulic architecture of a tropical tree (*Schefflera morototoni*): data, models, and a comparison with two temperate species (*Acer saccharum* and *Thuja occidentalis*). Plant Physiol 96:1105–1113
Tyree MT, Cochard H, Cruiziat P, Sinclair B, Ameglio T (1993) Drought-induced leaf shedding in walnut: evidence for vulnerability segmentation. Plant Cell Environ 16:882–979
Vaahtera L, Brosché M (2011) More than the sum of its parts. How to achieve a specific transcriptional response to abiotic stress. Plant Sci 180:421–430
Valladares F, Pearcy RW (1997) Interactions between water stress, sun-shade acclimation, heat tolerance and photoinhibition in the sclerophyll *Heteromeles arbutifolia*. Plant Cell Environ 20:25–36
Valladares F, Pearcy RW (2002) Drought can be more critical in the shade than in the sun: a field study of carbon gain and photo-inhibition in a Californian shrub during a dry El Niño year. Plant Cell Environ 25:749–759
van Mantgem PJ, Stephenson NL, Byrne JC, Daniels LD, Franklin JF, Fulé PZ, Harmon ME, Larson AJ, Smith JM, Taylor AH, Veblen TT (2009) Widespread increase of tree mortality rates in the western United States. Science 323:521–524

Vander Willigen C, Pammenter NW (1998) Relationship between growth and xylem hydraulic characteristics of clones of *Eucalyptus spp.* at contrasting sites. Tree Physiol 18:595–600

Vilagrosa A, Morales F, Abadía A, Bellot J, Cochard H, Gil-Pelegrin E (2010) Are symplast tolerance to intense drought conditions and xylem vulnerability to cavitation coordinated? An integrated analysis of photosynthetic, hydraulic and leaf level processes in two Mediterranean drought-resistant species. Env Exp Bot 69:233–242

Vitasse Y, Bresson CC, Kremer A, Michalet R, Delzon S (2010) Quantifying phenological plasticity to temperature in two temperate tree species. Funct Ecol 24:1211–1218

Voltas J, Chambel MR, Prada MA, Ferrio JP (2008) Climate-related variability in carbon and oxygen stable isotopes among populations of Aleppo pine grown in common-garden tests. Trees Struct Funct 22:759–769

Von Humboldt A, Bonpland A (2009) Essay on the geography of plants. In: Jackson ST (ed) Translated by Romanowski S. The University of Chicago Press, Chicago, p 274

Vornam B, Gaitling O, Derory J, Plomion C, Kremer A, Finkeldey R (2011) Characterisation and natural variation of a dehydrin gene in *Quercus petraea* (Matt.) Liebl. Plant Biol 13:881–887

Wang WX, Vinocur B, Shoseyov O, Altman A (2004) Role of plant heat-shock proteins and molecular chaperones in the abiotic stress response. Trends Plant Sci 9:244–252

Wanke D, Kolukisaoglu HU (2010) An update on the ABCC transporter family in plants: many genes, many proteins, but how many functions? Plant Biol 12:15–25

Warren CR, Aranda I, Cano FJ (2011) Responses to water stress of gas exchange and metabolites in *Eucalyptus* and *Acacia* spp. Plant Cell Environ 34:1609–1629

Warren CR, Aranda I, Cano FJ (2012) Metabolomics demonstrates divergent responses of two *Eucalyptus* species to water stress. Metabolomics 8:186–200

Watkinson JI, Sioson AA, Vasquez-Robinet C, Shukla M, Kumar D, Ellis M, Heath LS, Ramakrishnan N, Chevone B, Watson LT, van Zyl L, Egertsdotter U, Sederoff RR, Grene R (2003) Photosynthetic acclimation is reflected in specific patterns of gene expression in drought-stressed Loblolly pine. Plant Physiol 133:1702–1716

Yang F, Wang Y, Miao LF (2010) Comparative physiological and proteomic responses to drought stress in two poplar species originating from different altitudes. Physiol Plantarum 139:388–400

Zach A, Schuldt B, Brix S, Hornaa V, Culmsee H, Leuschner C (2010) Vessel diameter and xylem hydraulic conductivity increase with tree height in tropical rainforest trees in Sulawesi, Indonesia. Flora 205:506–512

Zhang JW, Marshall JD (1995) Variation in carbon isotope discrimination and photosynthetic gas exchange among populations of *Pseudotsuga menziessi* and *Pinus ponderosa* in different environments. Funct Ecol 9:402–412

Zhang JW, Marshall JD, Jaquish BC (1993) Genetic differentiation in carbon isotope discrimination and gas exchange in *Pseudotsuga menziesii*. Oecologia 93:80–87

Zhang JW, Fins L, Marshall JD (1994) Stable carbon isotope discrimination, photosynthetic gas exchange, and growth differences among wetern larch families. Tree Physiol 14:531–539

Zufferey V, Cochard H, Ameglio T, Spring JL, Viret O (2011) Diurnal cycles of embolism formation and repair in petioles of grapevine (Vitis vinifera cv. Chasselas). J Exp Bot 62:3885–3894

Zwieniecki MA, Holbrook NM (2009) Confronting Maxwell's demon: biophysics of xylem embolism repair. Trends Plant Sci 14:530–534

Zwieniecki MA, Melcher PJ, Holbrook NM (2001) Hydrogel control of xylem hydraulic resistance. Science 291:1059–1062

Chapter 13
Contribution of Arbuscular Mycorrhizal Symbiosis to Plant Drought Tolerance: State of the Art

JM Ruiz-Lozano, R. Porcel, G. Bárzana, R. Azcón and R. Aroca

Abstract Water deficit is considered one of the most important abiotic constraints limiting plant growth and yield in many areas on the Earth. Indeed, it accounts for over 70 % of yield losses during crop production. Several eco-physiological studies have demonstrated that the arbuscular mycorrhizal (AM) symbiosis improves plant tissue hydration and physiology under drought stress conditions. It is currently accepted that the contribution of AM symbiosis to plant drought tolerance is the result of accumulative physical, nutritional, physiological, and cellular effects. This chapter summarizes, from physiological and molecular points of view, the current knowledge about the mechanisms by which the AM symbiosis is thought to protect host plants against the detrimental effects of water deficit. Thus, the role of fungal hyphae in water uptake and transfer to plant tissues and in modification of soil water retention properties is updated. At the same time, the implication of plant aquaporins in regulation of root hydraulic properties is discussed. The influence of the AM symbiosis on host plant processes such as osmotic adjustment, gas exchange, hormonal balance, and antioxidant capacity is also reviewed.

13.1 Introduction

In nature, plants are constantly exposed to adverse environmental conditions that have a negative effect on plant survival, development, and productivity. Drought is considered the most important abiotic factor limiting plant growth and yield in

J. Ruiz-Lozano (✉) · R. Porcel · G. Bárzana · R. Azcón · R. Aroca
Departamento de Microbiología del Suelo y Sistemas Simbióticos, Estación Experimental del Zaidín (CSIC), Profesor Albareda n° 1, 18008 Granada, Spain
e-mail: juanmanuel.ruiz@eez.csic.es

R. Aroca (ed.), *Plant Responses to Drought Stress*,
DOI: 10.1007/978-3-642-32653-0_13, © Springer-Verlag Berlin Heidelberg 2012

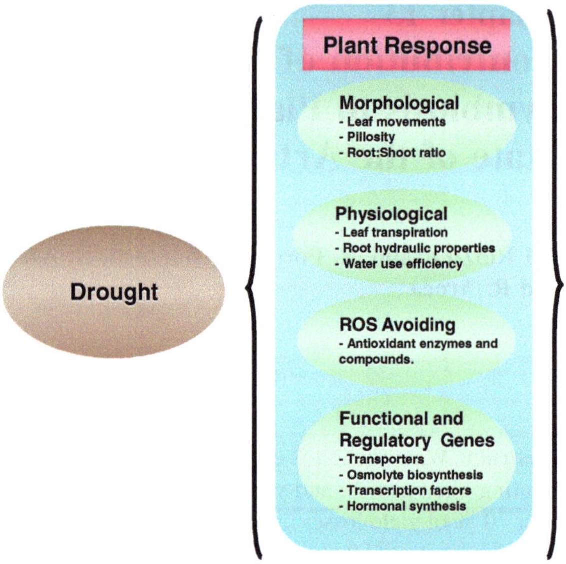

Fig. 13.1 Common plant responses to water deprivation

many areas (Bray 2004). The severity of drought is unpredictable as it depends on many factors such as occurrence and distribution of rainfall, evaporative demands, and moisture storing capacity of soils (Farooq et al. 2009). In addition, the global climate change is contributing to spread the problems of water deficit to regions where drought was negligible in the past. These difficulties will be particularly pronounced in currently semi-arid agricultural zones and/or under conditions of irrigation that often exacerbate soil salinization (Denby and Gehring 2005).

Although different plant species can vary in their sensitivity and response to soil water deficit, it may be assumed that all plants have encoded capability for stress perception, signaling, and response (Bohnert et al. 1995). Plants have developed several mechanisms in order to cope with drought stress (Fig. 13.1), including morphological adaptations, osmotic adjustment, optimization of water resources, antioxidant systems able to diminish the harmful effects of reactive oxygen species (ROS) linked to drought, and induction of a variety of stress-responsive genes and proteins (Farooq et al. 2009). Indeed, under drought conditions, plants trend to increase their root: shoot ratio in order to provide sufficient water to their aerial parts (Turner et al. 2001). At the same time, when plants perceive a decrease of the soil water availability, they close their stomata in order to prevent leaf water lost (Loreto and Centritto 2008). Plants can also accumulate compatible solutes in their tissues, which are involved in osmotic adjustment in order to maintain water content in shoot cells, as well as, a water potential gradient favorable to the entrance of water into the roots (Armengaud et al. 2004). The molecular responses may be initiated through hormonal signals which induce the accumulation of several stress-responsive proteins (Bray 2004).

Apart from to the intrinsic protective systems of plants against stress, plants grow associated with a number of soil microorganisms that can alleviate the stress symptoms. Arbuscular mycorrhizal (AM) fungi are widespread microorganisms able to establish a symbiotic association with the roots of most terrestrial plants. AM plants have improved ability for nutrient uptake and tolerance to biotic and abiotic stresses while the fungus acquires a protected ecological niche and plant photosynthates (Varma 2008). The AM symbiosis is present in all natural ecosystems, even in those affected by adverse environmental conditions (Barea and Jeffries 1995), and it can be defined as a specialized system for nutrient uptake and transfer, more efficient than roots alone (Varma 2008). Nevertheless, the physiological role of the AM symbiosis is not limited to uptake and transfer of nutrients to the host plant. Indeed, it is thought that the transition of plants from aquatic environment to land was possible through the association of plants with mycorrhizal fungi (Wang et al. 2010). In most cases studied, the association between an AM fungi (AMF) and a plant makes the host plant more tolerant to drought (Aroca et al. 2011). The mechanisms involved in such protective effect are reviewed in the following sections.

13.2 Uptake of Water Through the Fungal Hyphae and Transfer to the Host Plant: Implications of Aquaporins

One of the most common explanations for the improved water status and physiology in mycorrhizal plants is the strong increased absorbing surface caused by soil-growing hyphae combined with the fungal capability to take up water from soils with low water potential (Augé 2001; Ruiz-Lozano 2003; Lehto and Zwiazek 2011). Hyphal length density associated with AM root has been estimated to vary from 1 to more than 100 m per gram of soil (Smith et al. 2010). Average diameters of hyphae of AM fungi are in the range of 2–20 μm, which are one or two orders of magnitude narrower than roots. This size difference has important implications for access to water-filled pores, because hyphae may be able to penetrate a much higher proportion of pores than roots. In addition, AM hyphae have been suggested to maintain liquid continuity in the soil and limit the loss of soil hydraulic conductivity caused by air gaps (Allen 2007; Smith et al. 2010).

Studies by Allen (1982) and Hardie (1985) indicated a possible role of AM fungal hyphae in water uptake and transfer to the host plant. AM hyphae can penetrate soil pores inaccessible to root hairs and so absorb water that is not available to non-mycorrhizal plants. Allen (1991) estimated that the rate of water transport by extraradical hyphae to the root was 0.1 μl h^{-1} per hyphal entry point, a level sufficient to modify plant water relations. Faber et al. (1991) measured rates of water transport by hyphae to range from 0.37 to 0.76 μl h^{-1} per hypha. In contrast, negligible rates of water uptake by hyphae were also predicted (George et al. 1992; Koide 1993). Ruiz-Lozano and Azcón (1995) designed an experiment with lettuce plants grown in containers that had three compartments arranged

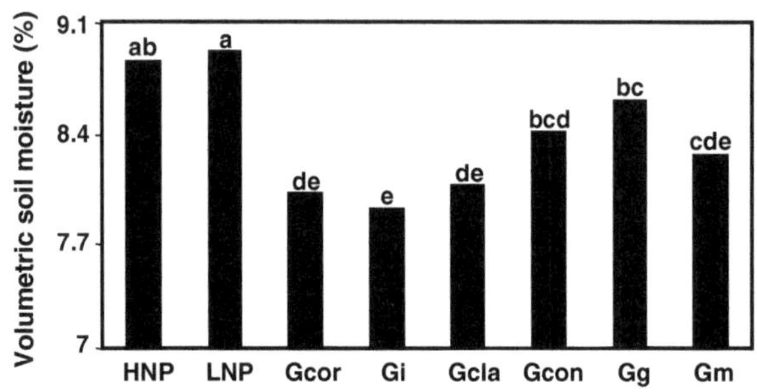

Fig. 13.2 Volumetric soil moisture (%) in lettuce plants fertilized with two rates of N and P (4 mM N 1 mM P, HNP and 2 mM N 0.5 mM P, LNP) or mycorrhizal lettuce plants colonized by *G. coronatum* (*Gcor*), *G. intraradices* (*Gi*), *G. claroideum* (*Gcla*), *G. constrictum* (*Gcon*), *G. geosporum* (*Gg*) or *G. mosseae* (*Gm*). Reproduced from Marulanda et al. (2003) with kind permission from John Wiley and Sons

vertically. The upper compartment (root compartment) was separated from the next compartment by a 50-μm nylon screen that allowed penetration by AM hyphae but not by roots. Addition of water to the lowermost compartment (located at 10 cm from the root compartment) increased plant fresh weight by 150 % in plants colonized by *Glomus fasciculatum*, and by 215 % in those colonized by *G. deserticola*, as compared to a non-inoculated P-fertilized treatment. Similarly, leaf water content and gas exchange increased in mycorrhizal plants with water application to the hyphal compartment. The study demonstrated a fungal activity in water uptake and translocation to the host plant (Ruiz-Lozano and Azcón 1995).

In a subsequent study, Marulanda et al. (2003) evaluated and compared the ability of six AM fungi to alter rates of root water uptake under drought stress conditions. Lettuce plants colonized by *Glomus coronatum*, *G. intraradices*, *G. claroideum*, and *G. mosseae* depleted soil water to a higher extent than comparably sized uninoculated control plants or plants colonized by *G. constrictum* or *G. geosporum*. The differences ranged from 0.6 % volumetric soil moisture for *G. mosseae*-colonized plants to 0.95 % volumetric soil moisture for *G. intraradices*-colonized plants (Fig. 13.2).

These differences in soil moisture were equivalent to 3– 4.75 ml plant^{-1} day^{-1}, respectively, and could not be ascribed to differences in plant size, but to the activity of AM fungi. The AM fungi tested in this study differed in their effectiveness to enhance plant water uptake from soil and this ability seemed to be related to the amount of external mycelium produced by each AM fungus (Fig. 13.3) (Marulanda et al. 2003).

More recently, Khalvati et al. (2005) designed a new split-root hyphae system to quantify the contribution of the hyphae to plant water uptake in mycorrhizal barley plants grown under well-watered and drought conditions. Compared to the non-mycorrhizal treatment, 4 % of water in the hyphal compartment was transferred to the root compartment through the arbuscular mycorrhizal hyphae under drought conditions. However, the much higher hyphal density found in the root

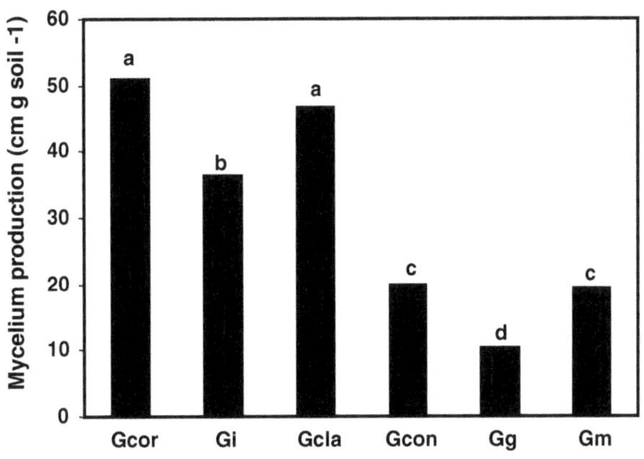

Fig. 13.3 Extraradical mycelium production (cm g soil^{-1}) by mycorrhizal lettuce plants colonized by six different AM fungi. See legend for Fig. 13.2. Reproduced from Marulanda et al. (2003) with kind permission from John Wiley and Sons

compartment than in the hyphal compartment suggested that a larger amount of water uptake by the arbuscular mycorrhizal hyphae could occur in the root compartment. Indeed, the same authors have quantified very recently the hyphal contribution to the total plant water uptake by means of high-resolution online water content sensors (Ruth et al. 2011). The direct and indirect hyphal contribution to the total plant water uptake was estimated to be at least 20 %, which is considerably higher than the 4 % estimated previously (Ruth et al. 2011).

The studies mentioned above demonstrated that increased water uptake by hyphae may be less important when the soil is near saturation and large pores are filled with water as the root surfaces are also in contact with water. However, as the soil dries and water is retained only in smaller pores where fungal hyphae can grow, but roots cannot, the water uptake function of hyphae becomes more significant for survival (Allen 2007; Lehto and Zwiazek 2011).

13.3 Implications of aquaporins in water relations of AM plants

Since AM fungi can transfer water to the root of the host plants, it is expected that the plant must increase its permeability for water and that aquaporin genes should be upregulated in order to allow a higher rate of transcellular water flow (Javot and Maurel 2002). In fact, in the last years it has been demonstrated that AM symbiosis regulates root hydraulic properties, including root hydraulic conductivity (*Lo*), and these effects have been linked to regulation of plant aquaporins (Ruiz-Lozano and Aroca 2010a). Aquaporins are water channel proteins that facilitate and regulate the passive movement of water molecules down a water potential gradient (Kruse et al. 2006). These proteins belong to the large major intrinsic protein (MIP) family of transmembrane proteins and are represented in all kingdoms (Maurel 2007). In plants, aquaporins are subdivided into five evolutionarily distinct subfamilies, the plasma membrane intrinsic proteins (PIPs), the tonoplast intrinsic proteins (TIPs), the small basic intrinsic proteins (SIPs), the nodulin26-like intrinsic proteins (NIPs) (Chaumont et al. 2001; Johanson et al. 2001), and the uncharacterized X intrinsic proteins (XIPs) (Gupta and Sankararamakrishnan 2009), which have been recently shown to transport a variety of uncharged substrates (Bienert et al. 2011).

The discovery of aquaporins in plants has caused a significant change in the understanding of plant water relations and much effort has been concentrated on investigating the function and regulation of aquaporins. High levels of aquaporin expression were shown not only in tissues with high water fluxes across membranes, e.g., in fast growing regions, in shoots, and leaves, but also in roots where water uptake occurs (Otto and Kaldenhoff 2000). Hence, aquaporins seem to play a specifically important role in controlling transcellular water transport in plant tissues (Javot and Maurel 2002; Zhao et al. 2008).

The literature on aquaporins modulation by AM symbiosis has risen significantly in the last decade as reviewed by Uehlein et al. (2007) and Ruiz-Lozano and Aroca (2010a). The first report on the modulation of aquaporin genes by AM symbiosis was provided by Roussel et al. (1997) followed by Krajinski et al. (2000), who found mycorrhiza-induced expression of TIP aquaporins in parsley and *Medicado truncatula*, respectively. Several aquaporin-encoding genes have been shown to be upregulated in ectomycorrhizal poplar plants, and this was correlated with an increased water transport capacity of mycorrhizal poplar roots (Marjanovic et al. 2005). PIP and NIP aquaporin genes from *Medicago truncatula* were also shown to be induced by mycorrhization, while other four aquaporin genes analyzed did not change their expression pattern as consequence of mycorrhization (Uehlein et al. 2007). Authors of this work related the mycorrhiza-induced change in expression of the two genes with physiological changes in the plant roots, i.e., the symbiotic exchange processes located at the pariarbuscular membrane (Uehlein et al. 2007).

However, the studies mentioned above were carried out under well-watered conditions and they did not test the expression of the aquaporin gene in AM plants under water deficit conditions. Thus, genes encoding plasma membrane aquaporins (PIPs) from soybean and lettuce were cloned and their expression pattern studied, in AM and non-AM plants cultivated under well-watered or drought stress conditions (Porcel et al. 2006). As mentioned previously, the starting hypothesis was that if AM fungi can transfer water to the root of the host plants, it is expected that the plant must increase its permeability for water and that aquaporin genes should be upregulated in order to allow a higher rate of transcellular water flow. In contrast, to this hypothesis, results obtained showed that the PIP genes studied were downregulated both in soybean (Fig. 13.4a) and in lettuce (Fig. 13.5) under drought stress and that such downregulation was even more severe in plants colonized by *G. mosseae* than in non-AM plants (Porcel et al. 2006).

When the expression of *GmPIP2* gene from soybean was analyzed in a time course (Fig. 13.4b), it was clearly visible that AM plants already downregulated that gene significantly at 5 days after inoculation (dai) and 12 dai, while both non-AM control plants still maintained *GmPIP2* gene expression almost unaltered. At

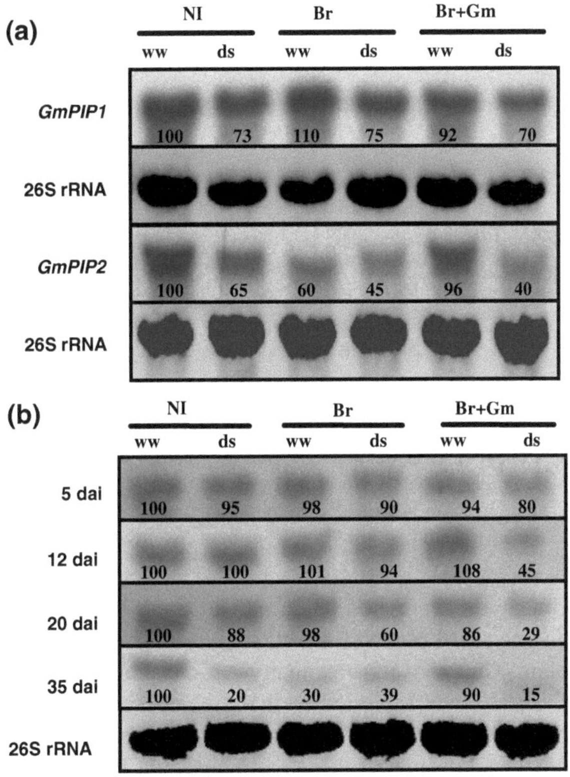

Fig. 13.4 **a** Northern blot of total RNA from soybean roots using *GmPIP1* and *GmPIP2* gene probes. **b** Northern blot of total RNA from soybean roots harvested 5, 12, 20, or 35 days after inoculation (dai) using *GmPIP2* gene probe. Treatments are designed as NI, non-inoculated controls; Br, *Bradyrhizobium japonicum*; Br + Gm, *B. japonicum* plus *G. mosseae*. Plants were either well watered (*ww*) or drought stressed (*ds*). The percentage of gene expression is indicated by numbers close to each northern. Reproduced from Porcel et al. (2006), with kind permission of Springer Science and Business Media

20 dai, the more intense downregulation of that gene in AM plants than in both non-AM plants was still clearly visible. Finally, at 35 dai all treatments had the same level of *GmPIP2* gene expression.

The effect of the AM symbiosis anticipating the downregulation of *GmPIP2* gene may have a physiological importance to help AM plants to cope with drought stress. In fact, according to Aharon et al. (2003) and Jang et al. (2007), the overexpression of PIP aquaporins in transgenic tobacco and Arabidopsis improves plant vigor under favorable growth conditions, but the overexpression of such PIP genes was negative during drought stress, causing fast wilting. Hence, the decreased expression of plasma membrane aquaporin genes during drought stress in AM plants can be a regulatory mechanism to limit the water lost from the cells (Barrieu et al. 1999; Porcel et al. 2006).

To further illustrate the complexity of the response of aquaporin genes to AM fungi, the expression of four PIP aquaporin genes from *Phaseolus vulgaris* (Aroca et al. 2007) was analyzed in mycorrhizal and non-mycorhizal plants subjected to three different osmotic stresses: drought, cold, or salinity. Three of these PIP genes showed differential regulation by AM symbiosis under the specific conditions of each stress applied (Fig. 13.6). In fact, *PvPIP1;1* was slightly inhibited by *G. intraradices* under drought stress conditions, while non-mycorrhizal plants did not

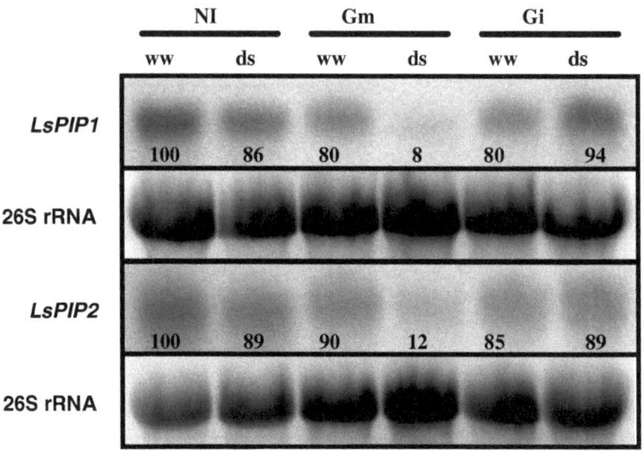

Fig. 13.5 Northern blot of total RNA from lettuce roots, using *lsPIP1* and *lsPIP2* gene probes. Treatments are designed as NI, non-inoculated controls; *Gm*, *Glomus mosseae* and *Gi*, *Glomus intraradices*. Plants were either wellwatered (*ww*) or drought stressed (*ds*). The percentage of gene expression is indicated by numbers close to each northern. Reproduced from Porcel et al. (2006), with kind permission of Springer Science and Business Media

change its expression pattern. Cold stress inhibited its expression similarly in AM and non-AM plants. Finally, salinity raised the gene expression in both groups of plants, but the enhancement was considerably higher in AM plants. The gene *PvPIP1;2* was inhibited by the three stresses in the same way in AM and non-AM plants. In contrast, *PvPIP1;3* showed important differences in AM and non-AM plants according to the stress imposed. This gene was clearly induced in non-AM plants under drought stress but inhibited in AM plants.

Under cold stress the behavior was the opposite since it was inhibited in non-AM plants and induced in AM ones. Finally, under salinity it was also induced in both groups of plants, especially in AM ones. The gene *PvPIP2;1* was induced in non-AM plants under drought stress but resulted downregulated in AM plants. The response of this gene to cold stress was not significant for any of the two plant groups and, again, the gene was considerably upregulated under salinity, especially in AM plants.

When *Lo* was measured, authors found that AM symbiosis regulated root hydraulic properties and enhanced *Lo* tolerance to drought, cold, and salinity stresses. The regulation of root hydraulic properties by AM symbiosis was strongly correlated with the regulation of PIP2 protein amount and phosphorylation state.

Thus, the results obtained so far on regulation of PIP aquaporins by the AM symbiosis show that the effects of the symbiosis on PIP genes depend on the own intrinsic properties of the osmotic stress, on the plant species studied, and on the specific aquaporin gene analyzed. In any case, the induction or inhibition of particular aquaporins by AM symbiosis should result in a better regulation of plant water status and contribute to the global plant resistance to the stressful conditions (Barrieu et al. 1999; Jang et al. 2004) as evidenced by their better growth and water status under conditions of water deficit. In addition, the results obtained

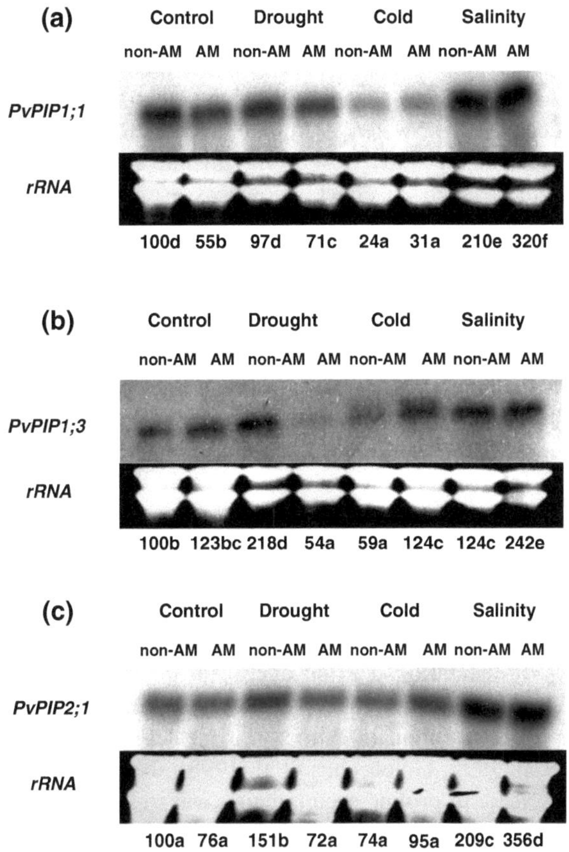

Fig. 13.6 Northern blots analysis using 3′ UTR as probes of *PvPIP1;1* (**a**), *PvPIP1;3* (**b**), and *PvPIP2;1* (**c**) in total RNA of *Phaseolus vulgaris* roots not inoculated (non-AM) or inoculated (AM) with *Glomus intararadices*. Plants were either kept as control or subjected to drought, cold, or salinity stresses. Control value of NI roots was referred as 100. Reproduced from Aroca et al. (2007), with permission from John Wiley and Sons

by Uehlein et al. (2007) suggest that the role of aquaporins in the AM symbiosis could be more complex than simply regulating plant water status. In fact, they described the induction by the AM symbiosis of specific PIP and NIP aquaporin isoforms exhibiting permeability to water and to ammonia, respectively. Authors suggested that these aquaporins could be involved in the symbiotic exchange processes between the fungus and the plant, which opens new perspectives in the study of aquaporins in the AM symbiosis.

To finish this topic, it is noticeable that Aroca et al. (2009) cloned the first aquaporin from an AM fungus (*GintAQP1*). Authors found some evidences supporting the idea that fungal aquaporins could compensate the downregulation of host plant aquaporins caused by drought. Also, they found that *GintAQP1* expression was upregulated in the mycelium part not osmotically stressed when the other mycelium part was stressed by NaCl. So, a possible communication between unstressed and stressed mycelium could be taken place. Obviously, the study of more fungal aquaporins is needed to completely understand this possible communication and the role of AM fungal aquaporins under conditions of water deficit.

13.4 Osmotic Adjustment

As soil dries out and soil water potential becomes more negative, plants must decrease their water potential to avoid cell dehydration and to maintain a favorable gradient for water flow from soil into roots. The most important mechanism to achieve such an effect, known as osmotic adjustment or osmoregulation, is to decrease the plant osmotic potential by active accumulation of organic ions or solutes (Morgan 1984; Hoekstra et al. 2001). Osmotic adjustment allows cells to maintain turgor and the processes that depend on it, such as cellular expansion and growth, stomatal opening, and photosynthesis, as well as keeping a gradient of water potential favorable to water entrance into the plant. The solutes which participate in osmotic adjustment are inorganic ions (mainly K^+ and Cl^-) or uncharged organic compounds such as proline or glycine betaine, as well as carbohydrates such as sucrose, pinitol or mannitol.

Proline is a non-protein amino acid that accumulates in most tissues subjected to water stress and, together with sugars, it is readily metabolized upon recovery from drought (Singh et al. 2000). In addition to acting as an osmoprotectant, proline also serves as a sink for energy to regulate redox potentials, as a hydroxyl radical scavenger, as a solute that protects macromolecules against denaturation, and as a means of reducing acidity in the cell (Kishor et al. 1995).

Investigations carried out so far on osmoregulation in the AM symbiosis are scarce and somewhat contradictory. While some studies have shown an increase in proline accumulation in mycorrhizal plants subjected to drought (Ruiz-Lozano et al. 1995; Goicoechea et al. 1998), the same studies also demonstrated that the increase in proline accumulation was quite variable depending on the AM fungus involved. For instance, while plants colonized by *G. deserticola* accumulated 120 nmol of proline per g fresh weight, plants colonized by *G. intraradices* only accumulated 41 nmol proline per g fresh weight (Ruiz-Lozano et al. 1995). It has also been shown that under low Ca in the medium AM plants accumulated more proline than non-AM plants when subjected to PEG-induced drought stress, while under high Ca in the medium this was not so (Ruiz-Lozano and Azcón 1997). Recently, Bheemareddy and Lakshman (2011) have examined the accumulation of metabolites in AM and non-AM *Triticum aestivum* plants and found that proline, carbohydrates, and proteins accumulated more in droughted AM plants than in the corresponding non-AM ones. In contrast, other studies regarding drought (Porcel et al. 2004; He et al. 2011) or salt stress (Ruiz-Lozano et al. 1996a; Jahromi et al. 2008; Sheng et al. 2011) have shown a lower proline accumulation in AM plants than in non-AM ones.

To further illustrate the complex response of AM plants in terms of solute accumulation, it is noticeable that in a recent study with lettuce plants (Ruiz-Lozano et al. (2011) found that under drought stress non-AM plants accumulated more proline in shoots than AM plants. In contrast, in roots, AM plants subjected to drought accumulated more proline than non-AM plants. This suggest that in root tissues AM plants accumulate more proline in order to cope with the low water potential of drying soil and to keep a water potential gradient favorable to water entrance into the roots, as was observed in soybean plants (Porcel and Ruiz-Lozano 2004). In this way, AM plants would have a better water status than non-AM plants and their shoots would be less strained by drought stress. By that reason shoots of AM plants would need to accumulate less proline (Ruiz-Lozano et al. 2011). Moreover, Sheng et al. (2011) found enhanced reducing sugar accumulation in AM maize plants subjected to salt stress, while the content of proline was lower than in non-AM plants. Authors proposed that the high levels of sugars in AM plants may be the result of an increase in photosynthetic capacity of plants and that these sugars contributed to the osmotic adjustment of the plants (Sheng et al. 2011).

It has also been shown that mycorrhizal colonization and drought interact in modifying free amino acid, starch, and sugar pools in roots (Augé et al. 1992; Schellembaum et al. 1998). Similarly, a greater osmotic adjustment has been reported in leaves of mycorrhizal basil plants than in nonmycorrhizal ones during a lethal drought period (Kubikova et al. 2001). More recently, Wu et al. (2007) also described in citrus plants that AM seedlings accumulated more soluble sugars, soluble starch, and total non-structural carbohydrates in leaves and roots than corresponding non-AM seedlings in response to drought stress, thus providing better osmotic adjustment in AM seedlings, which correlated with glucose, fructose, and sucrose accumulation.

The regulation of proline biosynthesis by the AM symbiosis has also been studied at the molecular level, by studying the expression pattern of Δ^1-pyrroline-5-carboxylate synthetase (P5CS) genes in lettuce and soybean plants subjected to drought stress (Porcel et al. 2004). Accumulation of proline is mainly due to de novo synthesis, although a reduced rate of catabolism has also been observed (Kishor et al. 1995). The first two steps of proline biosynthesis are catalyzed by P5CS by means of its γ-glutamil kinase and glutamic-γ-semialdehyde dehydrogenase activities. Subsequently, the Δ^1-pyrroline-5-carboxylate (P5C) formed is reduced by P5C reductase (P5CR) to proline (Hu et al. 1992). The rate-limiting step in this pathway is represented by the γ-glutamil kinase activity of P5CS, which is sensitive to feedback inhibition by relatively low levels of proline. In addition, in *Arabidopsis*, the P5CS-encoding gene is induced by drought stress, salinity, and ABA, but P5CR is not (Yoshiba et al. 1995). Hence, the P5CS-encoding gene is of key importance for the biosynthesis of proline in plants (Ábrahám et al. 2003).

A P5CS-encoding gene from *Glycine max* (*gmp5cs*) and another from *Lactuca sativa* (*lsp5cs*) were cloned and their contribution to the response against drought in mycorrhizal soybean and lettuce plants was analyzed. Results showed that *gmp5cs* and *lsp5cs* genes responded to drought and were upregulated in drought-stressed treatments, suggesting that they are important for the plant response against water deficit. Nevertheless, the expression of *gmp5cs* and *lslp5cs* genes decreased in drought-stressed AM plants as compared to non-inoculated plants. A similar result was obtained more recently with *lsp5cs* gene (Ruiz-Lozano et al. 2011). This was probably due to a decrease in ABA level in AM plants and to the fact that AM plants were less strained by drought stress than non-AM plants by primary drought-avoidance mechanisms. The results suggest that the induction of *p5cs* gene does not seem to be a mechanism by which the AM symbiosis protects their host plant (Porcel et al. 2004).

13.5 Enhancement of Plant Gas Exchange

An important number of studies have demonstrated that, during soil drying, mycorrhizal plants often maintain higher gas exchange rates than nonmycorrhizal plants of similar size and nutrient status (reviewed by Augé 2001; Ruiz-Lozano and Aroca 2010b). Currently, it is not clear how AM effects on stomatal behavior vary among AM fungi or host plants. Indeed, the literature shows that *gs* of some Citrus species are little affected by AM colonization, while soybean, sunflower, lettuce, or rose plants show frequent alteration of stomatal behavior by AM symbiosis (Augé 2001). Stomatal conductance rates are usually higher in AM than in non-AM plants, which implies that AM plants have a lower resistance to vapor transfer from inside the leaves to the atmosphere when exposed to the same water conditions. These higher *gs* rates in AM plants have been associated with lower xylem-sap abscisic acid (ABA) and lower ABA fluxes to leaves in AM plants (Ebel et al. 1997). AM and non-AM root systems exposed to the same stress and constrained to similar soil volumes may not necessarily suffer similar strain (see topic 2 above) that would explain the higher *gs* rates in AM plants (Duan et al. 1996).

Recently, Augé et al. (2008) have studied if the changes in *gs* induced by the AM symbiosis are accompanied by changes in leaf hydraulic conductance and in gradients of leaf water potential. They observed that under amply watered conditions the fungus *Glomus intraradices* enhanced by 27 % *gs* of squash plants, but leaf hydraulic conductance did not increase in these plants and, consequently, the water potential gradient across the leaf was higher in AM leaves. Authors concluded that the higher gradients of water potential in leaves of AM plants were consistent with the higher rates of gas exchange found in AM plants and that, presumably, are necessary to supply the carbon needs of the fungal symbiont.

It is also noticeable that AM and non-AM plants have shown different critical points or thresholds of stomatal behavior during drought episodes (Augé et al. 1986). Stomatal conductance in AM plants has also remained unaffected by declines in available soil moisture longer than in non-AM plants (Osundina 1995; Duan et al. 1996). Furthermore, AM plants showed higher *gs* than non-AM plants when only a portion of a root system was dried and before drying affected leaf water status and these effects occurred independently of plant size or phosphorus nutrition through non-hydraulic signals (Augé et al. 1992, 1994, 1995). Thus, it has been proposed that root dehydration alters root metabolism leading to the production of a non-hydraulic, chemical signal that moves to leaves where it induces stomatal closure (Augé 2000). Root-to-shoot regulation of stomatal behavior during soil drying may occur via a multiple chemical signal, including cytokinins and ABA (Druge and Schonbeck 1992; Goicoechea et al. 1997). In fact,

Goicoechea et al. (1997) noted that higher *gs* and transpiration of mycorrhizal alfalfa plants was associated with altered ABA/cytokinins ratios in their leaves.

It has been also shown that the influence of AM fungi on *gs* can be mediated by the host plant life style and the above-mentioned results can vary when the host plant shows a water conservative strategy as is the case of *Rosmarinus officinalis*. *R. officinalis* is a perennial Mediterranean plant well adapted to dry conditions that avoids drought stress by reducing its transpiration rate, as well as its *gs* (Munné-Bosch et al. 1998). The behavior of these plants against drought stress was studied by Sánchez-Blanco et al. (2004). Authors showed that, both under well-watered and under drought stress conditions, AM symbiosis enhanced root hydraulic conductivity and plant water status, but these effects did not correlate with *gs* since no significant differences in *gs* between AM and non-AM plants were found. On the contrary, significant increases of *gs* were found in two arid land woody plants such as *Olea europaea* and *Rhamnus lycioides* inoculated with the AM fungus *Glomus intraradices* (Caravaca et al. 2003; Querejeta et al. 2003). In spite of the common enhancement of *gs* in both plants, WUE was significantly enhanced only in *Olea europaea*, but not in *Rhamnus lycioides* (Caravaca et al. 2003; Querejeta et al. 2003). Authors related this differential behavior with the own plant lifestyle. *Olea* is a long-lived, slow-growing evergreen tree with a conservative water use strategy, while *Rhamnus* is a drought-deciduous shrub with a shorter lifespan.

Reduced *gs* under drought stress conditions in AM *Anthyllis cytisoides* plants has been observed by Goicoechea et al. (2004). *A. cytisoides* is a deciduous shrub plant that exhibits a slow growth rate and behaves as a drought avoider. It exhibits a resting period and one of its strategies to withstand prolonged water deficit is defoliation. When these plants were recovered from drought, young leaves from previously stressed mycorrhizal plants continued showing lower *gs* values than those from non-AM plants, concomitantly with a higher rate of leaf abscission. Leaf shedding can be a beneficial adaptation that reduces water loss and redistributes resources in order to favor plant survival under drought stress (Munné-Bosch and Alegre 2004). Thus, authors considered that this behavior could minimize both water loss by transpiration and whole plant respiration, preventing mycorrhizal plants from suffering desiccation. They concluded that, in this way, AM symbiosis conferred a greater responsiveness of *A. cytisoides* to drought.

The question whether native, drought-adapted AMF improve host plant performance to a greater extent than non-native AMF was addressed by Querejeta et al. (2006). They conducted a field experiment in which seedlings of two native wild shrub species were inoculated with either native or non-native *Glomus* species before outplanting in an agricultural land. The results obtained showed that native AMF enhanced *gs* values to a greater extent than non-native AMF in both shrub species.

AM symbiosis not only modifies *gs* values. There are also several reports in the literature showing an increase of plant photosynthetic activity and of water use efficiency (WUE) by the AM symbiosis both under well-watered and osmotic stress conditions (Ruiz-Lozano et al. 1995a; Wu and Xia 2006; Khan et al. 2008).

13.6 Regulation of Plant Hormonal Balance

The alteration in the homeostasis of plant hormones has been implicated in the enhanced drought tolerance exhibited by AM plants (Augé 2001; Ruiz-Lozano 2003). There are a large number of publications showing that the levels of plant hormones such as cytokinins, jasmonates, auxins, and abscisic acid (ABA) actually change upon the establishment of AM symbiosis (Hause et al. 2007). The knowledge of how plant hormones control and regulate the establishment and functioning of AM symbiosis has been recently summarized by López-Raez et al. (2011) as follows: (1) they redirect assimilates toward the root tissue colonized by the fungus and regulate mycorrhizal plant growth; (2) they are involved in the formation of fungal structures and the establishment of functional symbiosis; and (3) they regulate defence mechanisms and stress alleviation in mycorrhizal plants. However, there are still huge gaps in our knowledge on how both partners in the symbiosis can contribute to the regulation of hormone production (Ludwig-Müller 2010). Moreover, results by Grunwald et al. (2009) indicate that hormonal changes could be highly dependent on the symbiotic fungus used for inoculation.

ABA has been the most studied phytohormone in the field of the AM symbiosis. In fact, ABA is considered the most important signal transduction pathway among all the plant responses to stresses (Zhang et al. 2006; Hirayama and Shinozaki 2007). Some studies have suggested that an enhancement of ABA production in mycorrhizal roots could contribute to the suppression of the salicylic acid-dependent defence mechanisms activated during AMF colonization of roots (Herrera-Medina et al. 2007). These studies have also demonstrated that ABA participates in the susceptibility of tomato to infection by AM fungi, and that it seems to play an important role in the development of the complete arbuscule and its functionality (Herrera-Medina et al. 2007). More recently, an antagonistic interaction between ABA and ethylene, and different roles of each of the two hormones during AM formation has been found (Martín-Rodriguez et al. 2011).

Apart from its role during the formation of the AM symbiosis, ABA must be also considered in relation to plant responses to drought stress. Indeed, ABA modulates plant water status through regulation of important plant processes such as root hydraulic conductivity (*L*) (Schraut et al. 2005; Aroca 2006) and transpiration rate (Wilkinson and Davies 2002; Zhang et al. 2006), as well as, by induction of genes that encode enzymes and other proteins involved in cellular dehydration tolerance (Zhang et al. 2006; Hirayama and Shinozaki 2007). Several reports have shown that mycorrhization can alter ABA levels in the host plant, although contradictory results were also obtained. Studies of plants growing under normal conditions showed increased ABA levels in AM as compared to non-AM maize (Danneberg et al. 1992), soybeans (Murakami-Mizukami et al. 1991; Meixner et al. 2005), as well as *Medicago truncatula* and *Tropaeolum majus* (Ludwig-Müller 2010). On the contrary, Allen et al. (1982) reported decreased ABA levels in leaves from AM *Bouteloua gracilis* plants and an unchanged ABA content in their roots. ABA has also been detected in fungal hyphae at higher levels than in roots (Esch et al. 1994) and it was suggested that ABA in AM fungi may control the flow of water and mineral salts from the soil to the hyphae or from the fungus to the root cells. On the other hand, Duan et al. (1996) tested whether AM symbiosis changes the movement of ABA from roots to shoots at a particular degree of drought or, if contrarily, the stomata of leaves of AM and non-AM plants differ in their sensitivity to ABA. They found that AM symbiosis did not alter stomatal sensitivity to ABA, but that the amount of ABA moving in xylem was lower in AM than in similar-sized non-AM plants in dry soil (Duan et al. 1996). Other studies have also shown that when plants are subjected to drought stress, the levels of ABA are lower in AM than in non-AM plants (Goicoechea et al. 1997; Estrada-Luna and Davies 2003). Similarly, ectomycorrhizal larch plants subjected to osmotic stress due to the addition of polyethylene glycol had lower ABA levels than their non-mycorrhizal counterparts (Rincon et al. 2005). Thus, the combined influence of AM symbiosis and exogenous ABA application on the responses of lettuce plants during drought was studied (Aroca et al. 2008b). The results showed that the application of exogenous ABA had contrasting effects on the physiological responses of AM and non-AM plants and on the expression of several stress-related genes (*lea*, *p5cs* or *nced*). The results obtained in that study suggested that AM plants regulate better and faster their ABA levels than non-AM plants, allowing a more adequate balance between leaf transpiration and root water movement during drought and recovery (Aroca et al. 2008b). The changes in stress and ABA related genes have also been attributed to the different ABA levels in AM and non-AM plants (Ruiz-Lozano et al. 2006).The fact that a key gene for ABA biosynthesis is also differentially expressed in AM roots (Jahromi et al. 2008; Aroca et al. 2008b) corroborates the importance of ABA during AM symbiosis.

Other investigations aimed at elucidating whether the effects of the AM symbiosis on plant tolerance in relation to water deficit are mediated by the endogenous ABA content of the host plant and whether the exogenous application of ABA modifies such effects (Aroca et al. 2008a). The mycorrhization of two tomato plant lines (wild-type and *sitiens*) with *G. intraradices* differently regulated the expression of several drought-induced genes in wild-type and in *sitiens* plants, which suggested that the effects of the AM symbiosis on plant responses to water deficit were also mediated by the plant ABA phenotype (endogenous plant ABA content).

More recently, another study has evaluated the combined influence of AM symbiosis and exogenous ABA application on plant root hydraulic properties and on PIP aquaporin gene expression and protein levels after both a drought and a recovery period (Ruiz-Lozano et al. 2009). Results showed first that the application of exogenous ABA enhanced *Lo* in all plant, regardless of water conditions, and second that AM plants showed lower *Lo* values than non-AM plants, especially when plants were supplied with exogenous ABA, which accentuated the differences in *Lo* between AM and non-AM plants. This effect was clearly correlated with the accumulation pattern of the different PIPs analyzed, since most of them reduced their expression and protein levels in AM plants fed with ABA as compared to their non-AM counterparts. As a whole, results suggested that the combination of exogenous ABA and AM symbiosis inhibits the expression of PIP aquaporins as a strategy of water conservation in the host plant, which allowed these plants to maintain higher shoot RWC (Ruiz-Lozano et al. 2009). Thus, it can be concluded that ABA is necessary for sustained root colonization by AM fungi (Fester and Hause 2007) and to improve the symbiotic efficiency under drought stress conditions (Aroca et al. 2008a, b; Ruiz-Lozano et al. 2009).

13.7 Changes in Soil Water Retention Properties and Soil Hyphal Density

In the absence of a clear plant-based explanation for mycorrhizal influence on stomatal conductance and other leaf water relations, Augé et al. (2001) proposed that AM and non-AM plants may behave differently during drought because the symbiosis affects soil water retention properties. Authors found that 7 months of mycorrhization by *G. intraradices* in a Sequatchie loam altered the soil moisture characteristic curve relative to "nonmycorrhizal" soils with similar rooting densities. The study revealed that the "mycorrhizal" soil had significantly more water stable aggregates and substantially higher extraradical hyphal densities than the "nonmycorrhizal" soil which correlated well with the improved moisture retention properties of the "mycorrhizal" soil. The effect of AM hyphae in stabilizing soil structure by enmeshing soil particles and stabilizing aggregates has also been shown more recently (Hallet et al. 2009). In addition, the secretion by AM fungi of hydrophobic proteinaceous substances, referred to as glomalin (Rillig et al. 2002), also contributes to soil stability and water retention (Bedini et al. 2009).

Augé and co-workers tested also whether the effects of AM fungi occurred mostly via mycorrhization of roots, mycorrhization of soil or both. They used wild-type (myc+) and noncolonizing (myc−) bean mutants planted into soils previously produced using AM or non-AM sorghum plants in order to partition mycorrhizal influence on *gs* and drought resistance into soil and root components (Augé 2004; Augé et al. 2004). The studies conducted showed that about half of the considerable promotion of *gs* by two AM fungi was attributable to soil colonization and about half to root colonization. Moreover, merely growing in a mycorrhizal soil resulted in promotion of *gs* of non-AM plants under both amply watered and droughted conditions. That means that AM soil can affect the physiology of non-AM plants when grown in that soil. To explain this surprising result, authors suggested a nonhydraulic basis for the AM influence on the non-AM plant and proposed that the biochemical effects of the different soil organisms that interact with AM fungi in the rhizosphere (including P solubilizers, N_2 fixer, producers of hormones or siderophores, etc.) may contribute to the effects of AM soil on *gs* of host and nonhost plants (Augé et al. 2004).

In order to better understand why merely growing in a mycorrhizal soil resulted in promotion of *gs* of non-AM plants, Augé et al. (2007) further studied a possible relationship between the degree of soil mycorrhization (estimated as soil hyphal density) and variations in plant *gs* before and during drought. The objectives of the study were to learn if the amount of soil hyphae would also contribute to variation in stomatal behavior and soil drying and to test if soil colonization would make stronger contributions than root colonization to these variables. Results obtained showed that both hypotheses were true. Firstly, less hyphae in soil were related to higher *gs* values. Secondly, soil colonization had more influence on *gs* values than did root colonization. Thus, authors concluded a role for mycorrhization of soil itself in contributing to the regulation of stomatal behavior of host plants (Augé et al. 2007).

13.8 Protection Against the Oxidative Damage Generated by Drought

During drought stress, different metabolic pathways are uncoupled and electrons, that have a high-energy state, are transferred to molecular oxygen to form ROS. At saturating CO_2 concentration, the leakage rate to oxygen is very low [3.5 mmol/mg chlorophyll/h, (Steiger and Beck 1981)], but under shortage of carbon dioxide up to 50 % of the entire photosynthetic electron flow may end up as O_2^- (Biehler and Fock 1996). ROS, such as $^1O_2$, H_2O_2, O_2^- and $HO^{\cdot}$, are toxic molecules capable of causing oxidative damage to proteins, DNA, and lipids (Miller et al. 2010). On the other hand, ROS can act as signaling molecules for stress responses and generation of ROS is an early event in plant response to stress (Singh et al. 2011). Plants have evolved several antioxidants mechanisms to avoid oxidative damage linked to stressful conditions. These mechanisms can be divided as enzymatic or nonenzymatic. Enzymatic antioxidants include superoxide dismutase (SOD), glutathione reductase (GR), catalase (CAT), ascorbate- or thiol-dependent peroxidases, and the enzymes of the ascorbate–glutathione pathway. Non-enzymatic mechanisms include compounds able to scavenge directly several ROS as are ascorbic acid (AsA), glutathione (GSH), or α-tocopherol (for a recent review of antioxidant systems in plants see Scheibe and Beck 2011). Briefly, the efficient destruction of O_2^- and H_2O_2 requires the action of several antioxidant enzymes acting in synchrony. Superoxide is rapidly converted to H_2O_2 by the action of SOD. However, since H_2O_2 is a strong oxidant that rapidly oxidizes thiol groups, it cannot be allowed to accumulate to excess. CATs convert H_2O_2 to water and molecular oxygen in peroxisomes. An alternative mode of H_2O_2 destruction is via peroxidases, which are found throughout the cell and which have a much higher affinity for H_2O_2 than CAT. Plants also contain high activities for the enzymes of the ascorbate–glutathione cycle in which H_2O_2 is scavenged. In the first step of this pathway ascorbate peroxidase (APX), which is the most important peroxidase in H_2O_2 detoxification, catalyzes the reduction of H_2O_2 to water by ascorbate, and the resulting monodehydroascorbate and dehydroascorbate are reduced back to ascorbate by monodehydroascorbate reductase (MR) and by dehydroascorbate reductase (DR) plus GR, respectively (Noctor and Foyer 1998; Miller et al. 2010; Scheibe and Beck 2011).

Information about the activity of antioxidant enzymes in the AM symbiosis has focused mainly on the role of superoxide dismutases (SODs). Pioneer studies on this subject have shown that the AM fungus *G. mosseae* possesses CuZn-SOD activity and that mycorrhizal clover roots exhibit two additional SOD isoforms as compared to non-mycorrhizal roots: a mycCuZn-SOD and a Mn-SOD (Palma et al. 1993). Mycorrhizal lettuce plants subjected to drought have increased SOD activity compared to non-mycorrhizal controls (Ruiz-Lozano et al. 1996b) and molecular analyses have confirmed this response at the transcriptional level (Ruiz-Lozano et al. 2001a). Three cDNAs putatively encoding two Mn-SODs and one Fe-SOD were used to follow gene expression in lettuce roots. The most interesting results were obtained in relation to the *Mn-sod II* gene. The expression of this gene in P-fertilized plants was unaffected by drought stress. In contrast, changes in transcript accumulation occurred in mycorrhizal plants both as a consequence of fungal presence and of drought stress. Under well-watered conditions, fungal presence decreased *Mn-sod II* gene expression (52 % in the case of *G. mosseae* and 29 % in the case of *G. intraradices*) relative to non-mycorrhizal plants. In contrast, when the plants were subjected to drought stress, both mycorrhizal treatments induced a significant increase in *Mn-sod II* transcript accumulation. This increase was 50 % in *G. mosseae*-colonized plants and 138 % in *G. intraradices*-colonized roots relative to non-mycorrhizal plants. Both the increase in SOD activity (Ruiz-Lozano et al. 1996b) and the increase in *Mn-sod II* gene expression (Ruiz-Lozano et al. 2001a) were related to enhanced tolerance to drought, in terms of plant growth maintenance, by both mycorrhizal treatments.

In addition, other studies have shown that the AM symbiosis can alleviate drought-induced nodule senescence in legume plants (Ruiz-Lozano et al. 2001b). The most remarkable observation was the substantial reduction in oxidative damage to lipids and proteins in nodules of mycorrhizal plants subjected to drought as compared to the nodules of non-mycorrhizal plants. Such reduction in oxidative damage to biomolecules was considered the main mechanism by which the AM symbiosis protected root nodules in legume plants against premature nodule senescence induced by drought stress. Moreover, AM symbiosis can considerably increase the GR activity both in roots and nodules of soybean plants subjected to drought stress (Porcel et al. 2003). The GR is an important component of the ascorbate–glutathione cycle since it is the enzyme that regenerates oxidized glutathione into its reduced form (Noctor and Foyer 1998).

More recently there are few reports showing regulation of host antioxidant compounds by the AM symbiosis under drought stress. Indeed, Wu et al. (2006a, b); Wu and Zou (2009) and Ruiz-Sánchez et al. (2011) found an increase of AsA content in AM plants. Ascorbic acid is an important non-enzymatic antioxidant compound since it is involved in the removal of H_2O_2 by ascorbate peroxidases, which use AsA as electron donor, and is closely related to GSH in the ascorbate–glutathione cycle (Noctor and Foyer 1998). An increase in GSH content in AM

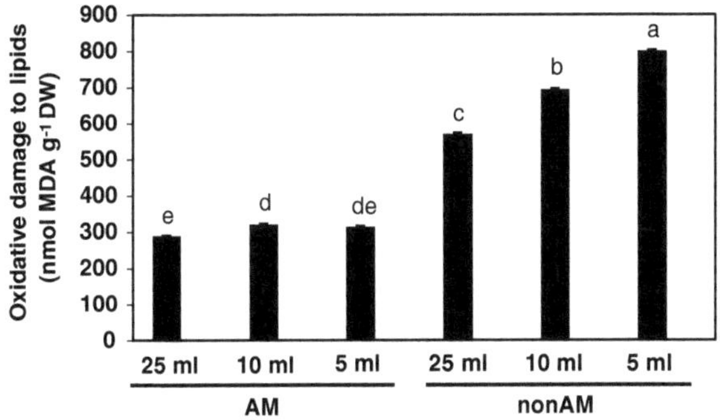

Fig. 13.7 Oxidative damage to lipids of rice plants inoculated or not with the AM fungus *G. intraradices*. Plants were subjected to drought stress by irrigation with 10 or 5 mL of aqueous nutrient solution or remained as well-watered control receiving 25 mL of aqueous nutrient solution. Adapted from Ruiz-Sánchez et al. (2010) with kind permission from Elsevier

plants has also been found (Wu et al. 2006a; Wu and Zou 2009; Ruiz-Sánchez et al. 2010), concomitantly with a reduced oxidative damage to lipids (Fig. 13.7). GSH has not only the functions of scavenging peroxides or regenerating AsA pool, but it also regulates the expression of photosynthetic genes and may keep the cell pools of reducing power (NADPH) under necessary conditions for plant living cells. No data about other antioxidant compounds regulated by AM symbiosis like α-tocopherol are available in the literature. Interesting results were found by Subramanian et al. (2006), who showed an increase in the ascorbic acid pool content in tomato fruits of AM-colonized plants under several water regimes (from well watered plants to severe droughted ones), without almost affecting plant yield. These results are really interesting in order to improve tomato food quality due to the high importance of ascorbic acid in human nutrition (Ginter 2007). Interestingly, and pointing to the same direction, Baslam and Goicoechea (2012) have recently described positive combined effects of water deficit and AM symbiosis on the accumulation of several antioxidant compounds in lettuce plants. Such compounds include carotenoids, anthocyanins, and phenolics which can improve the nutritional properties of lettuce as they are also beneficial for human health.

All these results suggest that mycorrhizal protection against oxidative stress caused by drought may be one of the most important mechanisms by which the AM symbiosis increases the tolerance of host plants to drought. These observations agree with the proposal by Bartels (2001) that both the prevention of oxidative stress and the elimination of reactive oxygen species are the most effective approaches used by plants to gain tolerance against several abiotic stresses, including drought. In any case, Porcel and Ruiz-Lozano (2004) also proposed that since AM symbiosis prevented leaf dehydration through drought avoidance mechanisms; it also diminished the production of ROS, and therefore the host plant may not need to enhance the activity of the antioxidant enzymes. The same hypothesis was also claimed by other authors (Caravaca et al. 2005; Alguacil et al. 2006). Moreover, Marulanda et al.

(2007) compared the effectiveness of four AM fungal isolates (two autochthonous drought-tolerant *Glomus* spp and two allochthonous drought-sensitive spp) on a drought-adapted plant (*Lavandula spica*) growing under drought conditions. Non-significant differences between antioxidant enzyme activities such as GR, CAT, and SOD in colonized plants were found. The two autochthonous strains of *Glomus intraradices* and *Glomus mosseae* were more effective in terms of shoot and root growth, water content, and nutrients uptake by lavender plants in the drought-stressed soil than similar allochthonous strains. These effects were concomitant with an increase in water content and a decline in H_2O_2 accumulation and antioxidant compounds such as glutathione and ascorbate.

Additional studies have shown that the antioxidant responses of AM plants can be modified by the soil conditions and composition. Thus, Ruiz-Lozano et al. (2011) cultivated AM and non-AM lettuce plants on soil amended or not with a treated SB waste under drought stress conditions. Accumulation of hydrogen peroxide, the oxidative damage to lipids, and the activity of four antioxidant enzymes were measured. Results showed that these parameters varied considerably in AM and non-AM plants depending on the presence or absence of SB waste. Authors related this variation in antioxidant enzyme activities with the availability of micronutrients in the growing substrate, e.g., CAT, APX, and SOD are metalloenzymes and their activity can be determined by the availability of the metals they utilize (Alguacil et al. 2003). Thus, both excess and deficiency of micronutrients can modulate the activity of these metalloenzymes. Taken together, these results pointed out the multiple factors involved in the drought tolerance enhancement by AM symbiosis.

13.9 Conclusions

There are clear evidences that the AM symbiosis enhances host plant tolerance to drought stress. This effect is achieved by alteration of several physiological or ecological processes, which are summarized in Fig. 13.8. These include the improvement of soil water retention properties and the ability of AMF hyphae to take up water from sources inaccessible to roots and the transfer of such water to the host plant. The contribution of hyphae to total plant water uptake has been estimated to be up to 20 %. In this process, the regulation of host plant aquaporins and root hydraulic properties is also involved. The AM symbiosis may also improve the plant osmotic adjustment by accumulation of different compounds such as proline, sugars, free amino acids, etc., although this effect may differ according to the plant tissue considered. Mycorrhizal plants exhibit enhanced gas exchange parameters and, usually, improved WUE, which also contributes to maintain plant growth and productivity under drought stress conditions. This effect has been closely related to the regulation of plant hormonal balance, being ABA the plant hormone most commonly modulated by the AM symbiosis under drought conditions. Finally, it is becoming clear that the AM symbiosis protects the host

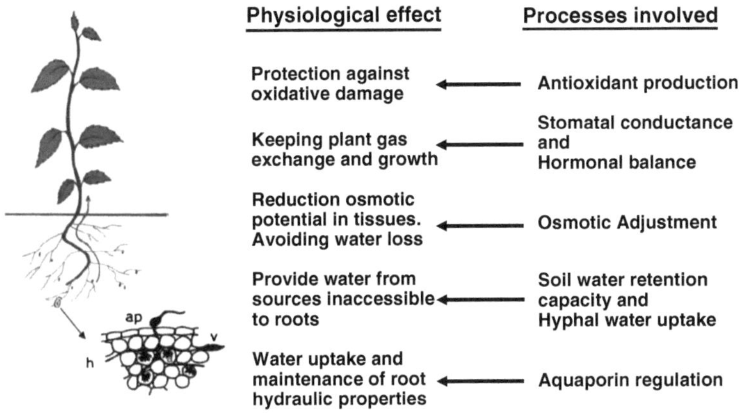

Fig. 13.8 Schematic representation of the integral physiological processes by which AM fungi can improve the tolerance of the host plant to drought. Adapted from Evelin et al. (2009), with kind permission from Oxford University Press

plant against the detrimental effects of ROS generated by drought. Thus, improved antioxidant enzymatic activities and/or accumulation of non-enzymatic antioxidant compounds result in reduced oxidative damage to AM plants.

13.10 Perspectives

Although in recent years there has been an increase in the understanding of the water relations of AM plants and the processes involved in enhanced tolerance of mycorrhizal plants to water limitation, there are still many unknown aspects which must be elucidated. Indeed, the role of plant hormones such as jasmonates or ethylene in such a protective effect, as well as, the signaling pathways involved should be elucidated. For that, the use of plant mutants altered in the production of these hormones could be an important tool. The relationship between aquaporin regulation and root hydraulic properties also needs further attention. Moreover, the possible role of the own fungal aquaporins should be the subject of future studies. Photosynthetic efficiency and WUE are dependent on the activity of carbon assimilatory enzymes such as Rubisco. Thus, studies dealing with the effects of AM symbiosis on carbon assimilatory enzymes during drought stress are also needed. Finally, the participation of other antioxidant compounds such as α-tocopherol, carotenoids, or anthocyanins in the reduction of oxidative damage by AM symbiosis should be investigated, with special emphasis on the role of AMF on the accumulation of these compounds and the improvement of nutritional quality of host plant species of interest for human nutrition.

Acknowledgments This work is part of a MICINN-FEDER project (Project AGL2008-00898).

References

Ábrahám E, Rigó G, Székely G, Nagy R, Koncz C, Szabados L (2003) Light-dependent induction of proline biosynthesis by abscisic acid and salt stress is inhibited by brassinosteroid in Arabidopsis. Plant Mol Biol 51:363–372

Aharon R, Shahak Y, Wininger S, Bendov R, Kapulnik Y, Galili G (2003) Overexpression of a plasma membrane aquaporins in transgenic tobacco improves plant vigour under favourable growth conditions but not under drought or salt stress. Plant Cell 15:439–447

Alguacil MM, Hernández JA, Caravaca F, Portillo B, Roldán A (2003) Antioxidant enzyme activities in shoot from three mycorrhizal shrub species afforested in a degraded semi-arid soil. Physiol Plant 118:562–570

Alguacil MM, Caravaca F, Díaz-Vivancos P, Hernández JA, Roldán A (2006) Effect of arbuscular mycorrhizae and induced drought stress on antioxidant enzyme and nitrate reductase activities in *Juniperus oxycedrus* L. grown in a composted sewage sludge-amended semi-arid soil. Plant Soil 279:209–218

Allen MF (1982) Influence of vesicular-arbuscular mycorrhizae on water movement through *Bouteloua gracilis* Lag ex Steud. New Phytol 91:191–196

Allen MF (1991) The ecology of mycorrhizae. Cambridge University Press, Cambridge

Allen MF (2007) Mycorrhizal fungi: highways for water and nutrients in arid soils. Vadose Zone J 6:291–297

Allen MF, Moore TS, Christensen M (1982) Phytohormone changes *Bouteloua gracilis* infected by vesicular-arbuscular mycorrhizal fungi. II. Altered levels of gibberellin-like substances and abscisic acid in the host plant. Can J Bot 60:468–471

Armengaud P, Thiery L, Buhot N, Grenier-De March G, Savoure A (2004) Transcriptional regulation of proline biosynthesis in *Medicago truncatula* reveals developmental and environmental specific features. Physiol Plant 120:442–450

Aroca R (2006) Exogenous catalase and ascorbate modify the effects of abscisic acid (ABA) on root hydraulic properties in *Phaseolus vulgaris* L. plants. J Plant Growth Reg 25:10–17

Aroca R, Porcel R, Ruiz-Lozano JM (2007) How does arbuscular mycorrhizal symbiosis regulate root hydraulic properties and plasma membrane aquaporin in *Phaseolus vulgaris* under drought, cold or salinity stresses? New Phytol 173:808–816

Aroca R, Alguacil M, Vernieri P, Ruiz-Lozano JM (2008a) Plant responses to drought stress and exogenous ABA application are differently modulated by mycorrhization in tomato and an ABA-deficient mutant (sitiens). Microb Ecol 56:704–719

Aroca R, Vernieri P, Ruiz-Lozano JM (2008b) Mycorrhyzal and non-mycorrhizal Lactuca sativa plants exhibit contrasting responses to exogenous ABA during drought stress and recovery. J Exp Bot 59:2029–2041

Aroca R, Bago A, Sutka M, Paz JA, Cano C, Amodeo G, Ruiz-Lozano JM (2009) Expression analysis of the first arbuscular mycorrhizal fungi aquaporin described reveals concerted gene expression between salt-stressed and nonstressed mycelium. Mol Plant-Microbe Interact 22:1169–1178

Aroca R, Porcel R, Ruiz-Lozano JM (2011) Plant drought tolerance enhancement by arbuscular mycorrhizal symbiosis. In: Fulton SM (ed) Mycorrhizal Fungi. Nova Science Publishers Inc, New York, pp 229–240

Augé RM (2000) Stomatal behaviour of arbuscular mycorrhizal plants. In: Kapulnik Y, Douds D (eds) Arbuscular mycorrhizas: physiology and function. Kluwer Academic Publishers, Dordrecht, pp 201–237

Augé RM (2001) Water relations, drought and vesicular-arbuscular mycorrhizal symbiosis. Mycorrhiza 11:3–42

Augé RM (2004) Arbuscular mycorrhizae and soil/plant water relations. Can J Soil Sci 84:373–381

Augé RM, Scheckel KA, Wample RL (1986) Greater leaf conductance of well-watered VA mycorrhizal rose plants is not related to phosphorus nutrition. New Phytol 103:107–116

Augé RM, Stodola AJW, Brown MS, Bethlenfalvay GJ (1992) Stomatal responses of mycorrhizal cowpea and soybean to short-term osmotic stress. New Phytol 120:117–125

Augé RM, Duan X, Ebel RC, Stodola AJ (1994) Nonhydraulic signaling of soil drying in mycorrhizal maize. Planta 193:74–82
Augé RM, Stodola AJ, Ebel RC, Duan XR (1995) Leaf elongation and water relations of mycorrhizal sorghum in response to partial soil drying: two *Glomus* species at varying phosphorus fertilization. J Exp Bot 46:297–307
Augé RM, Stodola AJW, Tims JE, Saxton AM (2001) Moisture retention properties of a mycorrhizal soil. Plant Soil 230:87–97
Augé RM, Sylvia DM, Park SJ, Buttery BR, Saxton AM, Moore JL, Cho K (2004) Partitioning mycorrhizal influence on water relations of *Phaseolus vulgaris* into soil and plant components. Can J Bot 82:503–514
Augé RM, Toler HD, Moore JL, Cho K, Saxton AM (2007) Comparing contributions of soil versus root colonization to variations in stomatal behavior and soil drying in mycorrhizal *Sorghum bicolor* and *Cucurbita pepo*. J Plant Physiol 164:1289–1299
Augé RM, Toler HD, Sams CE, Nasim G (2008) Hydraulic conductance and water potential gradients in squash leaves showing mycorrhiza-induced increases in stomatal conductance. Mycorrhiza 18:115–121
Barea JM, Jeffries P (1995) Arbuscular mycorrhizas in sustainable soil-plant systems. In: Varma A, Hock B (eds) Mycorrhiza: structure, function, molecular biology and biotechnology. Springer, Berlin, pp 521–561
Barrieu F, Marty-Mazars D, Thomas D, Chaumont F, Charbonnier M, Marty F (1999) Desiccation and osmotic stress increase the abundance of mRNA of the tonoplast aquaporin BobTIP26-1 in cauliflower cells. Planta 209:77–86
Bartels D (2001) Targeting detoxification pathways: an efficient approach to obtain plants with multiple stress tolerance. Trends Plant Sci 6:284–286
Baslam M, Goicoechea N (2012) Water deficit improved the capacity of arbuscular mycorrhizal fungi (AMF) for inducing the accumulation of antioxidant compounds in lettuce leaves. Mycorrhiza 22:347–359
Bedini S, Pellegrino E, Avio L, Pellegrini S, Bazzoffi P, Argese E, Giovannetti M (2009) Changes in soil aggregation and glomalinrelated soil protein content as affected by the arbuscular mycorrhizal fungal species *Glomus mosseae* and *Glomus intraradices*. Soil Biol Biochem 41:1491–1496
Bheemareddy VS, Lakshman HC (2011) Effect of AM fungus *Glomus fasciculatum* on metabolite accumulation in four varieties of *Triticum aestivum* L. under short-term water stress. Vegetos 24:41–49
Biehler K, Fock H (1996) Evidence for the contribution of the Mehler-peroxidase reaction in dissipating excess electrons in drought-stressed wheat. Plant Physiol 112:265–272
Bienert GP, Désirée-Bienert M, Jahn TP, Boutry M, Chaumont F (2011) Solanaceae XIPs are plasma membrane aquaporins that facilitate the transport of many uncharged substrates. Plant J 66:306–317
Bohnert HJ, Nelson DE, Jensen RG (1995) Adaptations to environmental stresses. Plant Cell 7:1099–1111
Bray EA (2004) Genes commonly regulated by water-deficit stress in *Arabidopsis thaliana*. J Exp Bot 55:2331–2341
Caravaca F, Díaz E, Barea JM, Azcón-Aguilar C, Roldán A (2003) Photosynthetic and transpiration rates of *Olea europaea* subsp. sylvestris and *Rhamnus lycioides* as affected by water deficit and mycorrhiza. Biol Plant 46:637–639
Caravaca F, Alguacil MM, Hernández JA, Roldán A (2005) Involvement of antioxidant enzyme and nitrate reductase activities during water stress and recovery of mycorrhizal *Myrtus communis* and *Phillyrea angustifolia* plants. Plant Sci 169:191–197
Chaumont F, Barrieu F, Wojcik E, Chrispeels MJ, Jung R (2001) Aquaporins constitute a large and highly divergent protein family in maize. Plant Physiol 125:1206–1215
Danneberg G, Latus C, Zimmer W, Hundeshagen B, Schneider-Poetsch HJ, Bothe H (1992) Influence of vesicular-arbuscular mycorrhiza on phytohormone balance in maize (*Zea mays* L.). J Plant Physiol 141:33–39

Denby K, Gehring C (2005) Engineering drought and salinity tolerance in plants: lessons from genome-wide expression profiling in Arabidopsis. Trends Biotechnol 23:547–552
Druge U, Schonbeck F (1992) Effect of vesicular arbuscular mycorrhizal infection on transpiration, photosynthesis and growth of flax (*Linum usitatissimum* L.) in relation to cytokinin levels. J Plant Physiol 141:40–48
Duan X, Newman DS, Reiber JM, Green CD, Saxton AM, Augé RM (1996) Mycorrhizal influence on hydraulic and hormonal factors implicated in the control of stomatal conductance during drought. J Exp Bot 47:1541–1550
Ebel RC, Duan X, Still DW, Augé RM (1997) Xylem sap abscisic acid concentration and stomatal conductance of mycorrhizal *Vigna unguiculata* in drying soil. New Phytol 135:755–761
Esch H, Hundeshagen B, Schneider-Poetsch HJ, Bothe H (1994) Demonstration of abscisic acid in spores and hyphae of the arbuscular-mycorrhizal fungus *Glomus* and in the N_2-fixing cyanobacterium *Anabaena variabilis*. Plant Sci 99:9–16
Estrada-Luna A, Davies FT (2003) Arbuscular mycorrhizal fungi influence water relations, gas exchange, abscisic acid and growth of micropropagated chile ancho pepper (*Capsicum annuum*) plantlets during acclimatization and post-acclimatization. J Plant Physiol 160:1073–1083
Evelin H, Kapoor R, Giri B (2009) Arbuscular mycorrhizal fungi in alleviation of salt stress: a review. Ann Bot 104:1263–1280
Faber BA, Zasoski RJ, Munns DN (1991) A method for measuring hyphal nutrient and water uptake in mycorrhizal plants. Can J Bot 69:87–94
Farooq M, Wahid A, Kobayashi N, Fujita D, Basra SMA (2009) Plant drought stress: effects, mechanisms and management. Agron Sustain Dev 29:185–212
Fester T, Hause B (2007) Drought and symbiosis-why is abscisic acid necessary for arbuscular mycorrhiza? New Phytol 175:383–386
George E, Häuser KU, Vetterlein D, Gorgus E, Marschner H (1992) Water and nutrient translocation by hyphae of *Glomus mosseae*. Can J Bot 70:2130–2137
Ginter E (2007) Chronic vitamin C deficiency increases the risk of cardiovascular diseases. Bratislava Med J 108:417–421
Goicoechea N, Antolin MC, Sánchez-Díaz M (1997) Gas exchange is related to the hormone balance in mycorrhizal or nitrogen-fixing alfalfa subjected to drought. Physiol Plant 100:989–997
Goicoechea N, Szalai G, Antolín MC, Sánchez-Díaz M, Paldi E (1998) Influence of arbuscular mycorrhizae and *Rhizobium* on free polyamines and proline levels in water-stressed alfalfa. J Plant Physiol 153:706–711
Goicoechea N, Merino S, Sánchez-Díaz M (2004) Contribution of arbuscular mycorrhizal fungi (AMF) to the adaptations exhibited by the deciduous shrub *Anthyllis cytisoides* under water deficit. Physiol Plant 122:453–464
Grunwald U, Guo W, Fischer K, Isayenkov S, Ludwig-Müller J, Hause B, Yan X, Küster H, Franken P (2009) Overlapping expression patterns and differential transcript levels of phosphate transporter genes in arbuscular mycorrhizal, Pi-fertilised and phytohormone-treated *Medicago truncatula* roots. Planta 229:1023–1034
Gupta AB, Sankararamakrishnan R (2009) Genome-wide analysis of major intrinsic proteins in the tree plant *Populus trichocarpa*: Characterization of XIP subfamily of aquaporins from evolutionary perspective. BMC Plant Biol 9:134
Hallet PD, Feeney DS, Bengough AG, Rillig MC, Scrimgeour CM, Young IM (2009) Disentangling the impact of AM fungi versus roots on soil structure and water transport. Plant Soil 314:183–196
Hardie K (1985) The effect of removal of extraradical hyphae on water uptake by vesicular-arbuscular mycorrhizal plants. New Phytol 101:677–684
Hause B, Mrosk C, Isayenkov S, Strack D (2007) Jasmonates in arbuscular mycorrhizal interactions. Phytochemistry 68:101–110
He X, Gao L, Zhao L (2011) Effects of AM fungi on the growth and drought resistance of *Seriphidium minchiinense* under water stress. Acta Ecol Sinica 31:1029–1037
Herrera-Medina MJ, Steinkellner S, Vierheilig H, Ocampo JA, García-Garrido JM (2007) Abscisic acid determines arbuscule development and functionality in tomato arbuscular mycorrhiza. New Phytol 175:554–564

Hirayama T, Shinozaki K (2007) Perception and transduction of abscisic acid signals: keys to the function of the versatile plant hormone ABA. Trends Plant Sci 12:343–351

Hoekstra FA, Golovina EA, Buitink J (2001) Mechanisms of plant desiccation tolerance. Trends Plant Sci 6:431–438

Hu CA, Delaunew AJ, Verma DPS (1992) A bifunctional enzyme (Δ^1-pyrroline-5-carboxylate synthetase) catalyzes the first two steps in proline biosynthesis in plants. Proc Natl Acad Sci USA 89:9354–9358

Jahromi F, Aroca R, Porcel R, Ruiz-Lozano JM (2008) Influence of salinity on the in vitro development of *Glomus intraradices* and on the in vivo physiological and molecular responses of mycorrhizal lettuce plants. Microbial Ecol 55:45–53

Jang JY, Kim DG, Kim YO, Kim JS, Kang H (2004) An expression analysis of a gene family encoding plasma membrane aquaporins in response to abiotic stresses in *Arabidopsis thaliana*. Plant Mol Biol 54:713–725

Jang JY, Lee SH, Rhee JY, Chung GC, Ahn SJ, Kang H (2007) Transgenic Arabidopsis and tobacco plants overexpressing an aquaporin respond differently to various abiotic stresses. Plant Mol Biol 64:621–632

Javot H, Maurel C (2002) The role of aquaporins in root water uptake. Ann Bot 90:301–313

Johanson U, Karlsson M, Johansson I, Gustavsson S, Sjovall S, Fraysse L, Weig AR, Kjellbom P (2001) The complete set of genes encoding major intrinsic proteins in Arabidopsis provides a framework for a new nomenclature for major intrinsic proteins in plants. Plant Physiol 126:1358–1369

Khalvati MA, Hu Y, Mozafar A, Schmidhalter U (2005) Quantification of water uptake by arbuscular mycorrhizal hyphae and its significance for leaf growth, water relations, and gas exchange of barley subjected to drought stress. Plant Biol 7:706–712

Khan IA, Ayub N, Mirza SN, Nizami SN, Azam M (2008) Yield and water use efficiency (WUE) of *Cenchrus ciliaris* as influenced by vesicular arbuscular mycorrhizae (VAM). Pak J Bot 40:931–937

Kishor PB, Hong Z, Miao GH, Hu CA, Verma DPS (1995) Overexpression of Δ^1-pyrroline-5-carboxilate synthetase increases proline production and confers osmotolerance in transgenic plants. Plant Physiol 108:1387–1394

Koide R (1993) Physiology of the mycorrhizal plant. Adv Plant Pathol 9:33–54

Krajinski F, Biela A, Schubert D, Gianinazzi-Pearson V, Kaldenhoff R, Franken P (2000) Arbuscular mycorrhiza development regulates the mRNA abundance of *Mtaqp1* encoding a mercury-insensitive aquaporin of *Medicago truncatula*. Planta 211:85–90

Kruse E, Uehlein N, Kaldenhoff R (2006) The aquaporins. Genome Biol 7:206

Kubikova E, Moore JL, Ownlew BH, Mullen MD, Augé RM (2001) Mycorrhizal impact on osmotic adjustment in *Ocimum basilicum* during a lethal drying episode. J Plant Physiol 158:1227–1230

Lehto T, Zwiazek JJ (2011) Ectomycorrhizas and water relations of trees: a review. Mycorrhiza 21:71–90

López-Raez JA, Pozo MJ, García-Garrido JM (2011) Strigolactones: a cry for help in the rhizosphere. Botany 89:513–522

Loreto F, Centritto M (2008) Leaf carbon assimilation in a water-limited world. Plant Biosyst 142:154–161

Ludwig-Müller J (2010) Hormonal responses in host plants triggered by arbuscular mycorrhizal fungi. In: Kapulnik Y, Koltai H (eds) Arbuscular mycorrhizas: physiology and function. Springer, Berlin, pp 169–190

Marjanovic Z, Uehlein N, Kaldenhoff R, Zwiazek JJ, Wei*β* M, Hampp R, Nehls U (2005) Aquaporins in poplar: what a difference a simbiont makes! Planta 222:258–268

Martín-Rodríguez JA, León-Morcillo R, Vierheilig H, Ocampo JA, Ludwig-Müller J, García-Garrido JM (2011) Ethylene-dependent/ethylene-independent ABA regulation of tomato plants colonized by arbuscular mycorrhiza fungi. New Phytol 190:193–205

Marulanda A, Azcón R, Ruiz-Lozano JM (2003) Contribution of six arbuscular mycorrhizal fungal isolates to water uptake by *Lactuca sativa* L. plants under drought stress. Physiol Plant 119:526–533

Marulanda A, Porcel R, Barea JM, Azcón R (2007) Drought tolerance and antioxidant activities in lavender plants colonized by native drought-tolerant or drought-sensitive *Glomus* species. Microb Ecol 54:543–552

Maurel C (2007) Plant aquaporins: novel functions and regulation properties. FEBS Lett 581:2227–2236

Meixner C, Ludwig-Müller J, Miersch O, Gresshoff P, Staehelin C, Vierheilig H (2005) Lack of mycorrhizal autoregulation and phytohormonal changes in the supernodulating soybean mutant nts1007. Planta 222:709–715

Miller G, Suzuki N, Ciftci-Yilmaz S, Mittler R (2010) Reactive oxygen species homeostasis and signalling during drought and salinity stresses. Plant Cell Environ 33:453–467

Morgan JM (1984) Osmoregulation and water stress in higher plants. An Rev Plant Physiol 33:299–319

Munné-Bosch S, Alegre L (2004) Die and let live: leaf senescence contributes to plant survival under drought stress. Funct Plant Biol 31:203–216

Munné-Bosch S, Nogués S, Alegre L (1998) Daily patterns of photosynthesis of two Mediterranean Shrubs response to water deficit. In: Garab G (ed) Photosynthesis: mechanisms and effects. Kluwer Academic Publishers, Dordrecht, pp 4015–4018

Murakami-Mizukami Y, Yamamoto Y, Yamaki S (1991) Analyses of indole acetic acid and abscisic acid contents in nodules of soybean plants bearing VA mycorrhizas. Soil Sci Plant Nut 37:291–298

Noctor G, Foyer CH (1998) Ascorbate and glutathione: keeping active oxygen under control. Rev Plant Physiol Plant Mol Biol 49:249–279

Osundina M (1995) Responses of seedlings of *Parkia biglobes* (African locust bean) to drought and inoculation with vesicular-arbuscular mycorrhiza. Nigerian J Bot 8:1–10

Otto B, Kaldenhoff R (2000) Cell-specific expression of the mercury-insensitive plasma-membrane aquaporin NtAQP1 from *Nicotiana tabacum*. Planta 211:167–172

Palma JM, Longa MA, del Rio LA, Arines J (1993) Superoxide dismutase in vesicular-arbuscular red clover plants. Physiol Plant 87:77–83

Porcel R, Ruiz-Lozano JM (2004) Arbuscular mycorrhizal influence on leaf water potential, solute accumulation, and oxidative stress in soybean plants subjected to drought stress. J Exp Bot 55:1743–1750

Porcel R, Barea JM, Ruiz-Lozano JM (2003) Antioxidant activities in mycorrhizal soybean plants under drought stress and their possible relationship to the process of nodule senescence. New Phytol 157:135–143

Porcel R, Azcón R, Ruiz-Lozano JM (2004) Evaluation of the role of genes encoding for Δ1-pyrroline-5-carboxylate synthetase (P5CS) during drought stress in arbuscular mycorrhizal *Glycine max* and *Lactuca sativa* plants. Physiol Mol Plant Pathol 65:211–221

Porcel R, Aroca R, Azcón R, Ruiz-Lozano JM (2006) PIP aquaporin gene expression in arbuscular mycorrhizal *Glycine max* and *Lactuca sativa* plants in relation to drought stress tolerance. Plant Mol Biol 60:389–404

Querejeta JI, Barea JM, Allen MF, Caravaca F, Roldan A (2003) Differential response of delta ^{13}C and water use efficiency to arbuscular mycorrhizal infection in two aridland woody plant species. Oecologia 135:510–515

Querejeta JI, Allen MF, Caravaca F, Roldán A (2006) Differential modulation of host plant δ^{13}C and δ^{18}O by native and nonnative arbuscular mycorrhiazl fungi in semiarid environment. New Phytol 169:379–387

Rillig MC, Wright SF, Nichols KA, Schmid WF, Torn MS (2002) The role of arbuscular mycorrhizal fungi and glomalin in soil aggregation: comparing effects of five plant species. Plant Soil 238:325–333

Rincon A, Priha O, Lelu-Walter MA, Bonnet M, Sotta B, Le Tacon F (2005) Shoot water status and ABA responses of transgenic hybrid larch *Larix kaempferi* X *L. decidua* to ectomycorrhizal fungi and osmotic stress. Tree Physiol 25:1101–1108

Roussel H, Bruns S, Gianinazzi-Pearson V, Hahlbrock K, Franken P (1997) Induction of a membrane intrinsic protein-encoding mRNA in arbuscular mycorrhiza and elicitor-stimulated cell suspension cultures of parsley. Plant Sci 126:203–210

Ruiz-Lozano JM (2003) Arbuscular mycorrhizal symbiosis and alleviation of osmotic stress. New Perspectives Molecular Stud Mycorrhiza 13:309–317

Ruiz-Lozano JM, Aroca R (2010a) Modulation of aquaporin genes by the arbuscular mycorrhizal symbiosis in relation to osmotic stress tolerance. In: Seckbach J, Grube M (eds) Symbioses and stress: joint ventures in biology, cellular origin, life in extreme habitats and astrobiology. Springer, Berlin, pp 359–374

Ruiz-Lozano JM, Aroca R (2010b) Host response to osmotic stresses: stomatal behaviour and water use efficiency of arbuscular mycorrhizal plants. In: Koltai H, Kapulnik Y (eds) Arbuscular mycorrhizas: physiology and function, 2nd edn. Springer, Berlin, pp 239–256

Ruiz-Lozano JM, Azcón R (1995) Hyphal contribution to water uptake in mycorrhizal plants as affected by the fungal species and water status. Physiol Plant 95:472–478

Ruiz-Lozano JM, Azcón R (1997) Effect of calcium application on the tolerance of mycorrhizal lettuce plants to polyethylene glycol-induced water stress. Symbiosis 23:9–22

Ruiz-Lozano JM, Azcón R, Gómez M (1995) Effects of arbuscular mycorrhizal *Glomus* species on drought tolerance: physiological and nutritional plant responses. Appl Environ Microbiol 61:456–460

Ruiz-Lozano JM, Azcón R, Gómez M (1996a) Alleviation of salt stress by arbuscular-mycorrhizal *Glomus* species in *Lactuca sativa* plants. Physiol Plant 98:767–772

Ruiz-Lozano J, Azcón R, Palma JM (1996b) Superoxide dismutase activity in arbuscular-mycorrhizal *Lactuca sativa* L. plants subjected to drought stress. New Phytol 134:327–333

Ruiz-Lozano JM, Collados C, Barea JM, Azcón R (2001a) Clonig of cDNAs enconding SODs from lettuce plants which show differential regulation by arbuscular mycorhizal symbiosis and by drought stress. J Exp Bot 52:2241–2242

Ruiz-Lozano JM, Collados C, Barea JM, Azcón R (2001b) Arbuscular mycorrhizal symbiosis can alleviate drought-induced nodule senescence in soybean plants. New Phytol 151:493–502

Ruiz-Lozano JM, Porcel R, Aroca R (2006) Does the enhanced tolerance of arbuscular mycorrhizal plants to water deficit involve modulation of drought-induced plant genes? New Phytol 171:693–698

Ruiz-Lozano JM, Alguacil MM, Bárzana G, Vernieri P, Aroca R (2009) Exogenous ABA accentuates the differences in root hydraulic properties between mycorrhizal and non mycorrhizal maize plants through regulation of PIP aquaporins. Plant Mol Biol 70:565–579

Ruiz-Lozano JM, Perálvarez MC, Aroca R, Azcón R (2011) The application of a treated sugar beet waste residue to soil modifies the responses of mycorrhizal and non mycorrhizal lettuce plants to drought stress. Plant Soil 346:153–166

Ruíz-Sánchez M, Aroca R, Munoz Y, Polon R, Ruíz-Lozano JM (2010) The arbuscular mycorrhizal symbiosis enhances the photosynthetic efficiency and the antioxidative response of rice plants subjected to drought stress. J Plant Physiol 167:862–869

Ruíz-Sánchez M, Armada E, Munoz Y, García de Salamone I, Aroca R, Ruíz-Lozano JM, Azcón R (2011) *Azospirillum* and arbuscular mycorrhizal colonization enhanced rice growth and physiological traits under well-watered and drought conditions. J Plant Physiol 168:1031–1037

Ruth B, Khalvati M, Schmidhalter U (2011) Quantification of mycorrhizal water uptake via high-resolution on-line water content sensors. Plant Soil 342:459–468

Sánchez-Blanco MJ, Ferrández T, Morales MA, Morte A, Alarcón JJ (2004) Variations in water status, gas exchange, and growth in *Rosmarinus officinalis* plants infected with *Glomus deserticola* under drought conditions. J Plant Physiol 161:675–682

Scheibe R, Beck E (2011) Drought, desiccation, and oxidative stress. In: Lüttge U et al (eds) Plant desiccation tolerance, ecological studies 215. Springer, Berlin, pp 209–231

Schellembaum J, Muller J, Boller T, Wiemken A, Schuepp H (1998) Effects of drought on non-mycorrhizal and mycorrhizal maize: changes in pools of non-structural carbohydrates, in the activities of invertase and trehelase, and in the pools of amino acids and imino acids. New Phytol 138:59–66

Schraut D, Heilmeier H, Hartung W (2005) Radial transport of water and abscisic acid (ABA) in roots of *Zea mays* under conditions of nutrient deficiency. J Exp Bot 56:879–886

Sheng M, Tang M, Zhang F, Huang Y (2011) Influence of arbuscular mycorrhiza on organic solutes in maize leaves under salt stress. Mycorrhiza 21:423–430

Singh DK, Sale PWG, Pallaghy CK, Singh V (2000) Role of proline and leaf expansion rate in the recovery of stressed white clover leaves with increased phosphorus concentration. New Phytol 146:261–269

Singh LP, Gill SG, Tuteja N (2011) Unravelling the role of fungal symbionts in plant abiotic stress tolerance. Plant Signal Behav 6:175–191

Smith SE, Facelli E, Pope S, Smith FA (2010) Plant performance in stressful environments: interpreting new and established knowledge of the roles of arbuscular mycorrhizas. Plant Soil 326:3–20

Steiger HM, Beck E (1981) Formation of hydrogen peroxide and oxygen dependence of photosynthetic CO_2 assimilation by intact chloroplasts. Plant Cell Physiol 22:561–576

Subramanian KS, Santhanakrishnan P, Balasubramanian P (2006) Responses of field grown tomato plants to arbuscular mycorrhizal fungal colonization under varying intensities of drought stress. Scientia Hortic 107:245–253

Turner NC, Wright GC, Siddique KHM (2001) Adaptation of grain legumes (pulses) to water-limited environments. Adv Agron 71:123–231

Uehlein N, Fileschi K, Eckert M, Bienert GP, Bertl A, Kaldenhoff R (2007) Arbuscular mycorrhizal symbiosis and plant aquaporin expression. Phytochemistry 68:122–129

Varma A (2008) Mycorrhiza. State of the art, genetics and molecular biology, eco-funcion, biotecnology, eco-physiology, structure and systematics, 3rd edn. Springer, Berlin

Wang B, Yeun LH, Liu Y, Xue JY, Ane JM, Qiu YL (2010) Presence of three mycorrhizal genes in the common ancestor of land plants suggests a key role of mycorrhizas in the colonization of land by plants. New Phytol 186:514–525

Wilkinson S, Davies WJ (2002) ABA-based chemical signalling: the co-ordination of responses to stress in plants. Plant Cell Environ 25:195–210

Wu QS, Xia RX (2006) Arbuscular mycorrhizal fungi influence growth, osmotic adjustment and photosynthesis of citrus under well-watered and water stress conditions. J Plant Physiol 163:417–425

Wu QS, Zou YN (2009) Mycorrhiza has a direct effect on reactive oxygen metabolism of drought-stressed citrus. Plant Soil Environ 55:436–442

Wu QS, Xia RX, Zou YN (2006a) Reactive oxygen metabolism in mycorrhizal and non-mycorrhizal citrus (*Poncirus trifoliate*) seedlings subjected to water stress. J Plant Physiol 163:1101–1110

Wu QS, Zou YN, Xia RX (2006b) Effects of water stress and arbuscular mycorrhizal fungi on reactive oxygen metabolism and antioxidant production by citrus (*Citrus tangerine*) roots. Eur J Soil Biol 42:166–172

Wu QS, Xia RX, Zou YN, Wang GY (2007) Osmotic solute responses of mycorrhizal citrus (*Poncirus trifoliata*) seedlings to drought stress. Acta Physiol Plant 29:543–549

Yoshiba Y, Kiyosue T, Katagiri T, Ueda H, Mizoguchi T, Yamaguchi-Shinozaki K, Wada K, Harada Y, Shinozaki K (1995) Correlation between the induction of a gene for Δ^1-pyrroline-5-carboxylate synthetase and the accumulation of proline in *Arabidopsis thaliana* under osmotic stress. Plant J 7:751–760

Zhang J, Jia W, Yang J, Ismail AM (2006) Role of ABA in integrating plant responses to drought and salt stresses. Field Crop Res 97:111–119

Zhao C-X, Shao H-B, Chu L-Y (2008) Aquaporin structure-function relationships: water flow through plant living cells. Col Surf B 62:163–172

Chapter 14
Future Environmental Conditions will Limit Yield in N_2 Fixing Alfalfa

Gorka Erice, Alvaro Sanz-Sáez, Iker Aranjuelo, Juan José Irigoyen and Manuel Sánchez-Díaz

Abstract Drought is recognised as the major environmental factor that constrains productivity and stability of plants. Crop yield under future climatic conditions has increased the interest in "water stress physiology". Plant development under limited water availability together with increasing atmospheric CO_2 concentration is of primary interest to ensure crop production under the projected climate scenarios. The expected reduction in precipitation and rising evapotranspiration rates will limit plant growth either by restricting stomatal conductance and photosynthesis or by restricting leaf expansion. Furthermore, alfalfa is a legume that establishes a symbiotic relationship with N_2-fixing bacteria and hence drought may indirectly compromise plant production via alterations in nodule performance. The effects of water stress on nodules include not only reduction in nodule mass but decreases in nodule functioning. Furthermore, previous studies have confirmed that the performance of nodules is conditioned by their active interaction with other organs like leaves and roots. After long-term exposure to elevated CO_2, photosynthetic downregulation may limit leaf N demand and hence, nodule activity. Moreover, as observed for leaves, nodule responses to water deficit may be altered by the way that drought limitation is imposed. When water shortage is imposed by controlling irrigation levels, plants acclimatise their water status and growth and therefore nodule activity is usually unaffected. In contrast, after progressive drought treatment by withholding water, nodules show significant decreases in nitrogenase activity.

G. Erice (✉) · A. Sanz-Sáez · J. J. Irigoyen · M. Sánchez-Díaz
Departamento de Biología Vegetal, Sección Biología Vegetal (Unidad Asociada al CSIC, EEAD, Zaragoza e ICVV, Logroño), Facultades de Ciencias y Farmacia, Universidad de Navarra, c/Irunlarrea 1, Pamplona, Navarra, Spain
e-mail: gerisor@unav.es

I. Aranjuelo
Instituto de Agrobiotecnología, Universidad Pública de Navarra-CSIC-Gobierno de Navarra, Campus de Arrosadía, E-31192, Mutilva Baja, Spain

R. Aroca (ed.), *Plant Responses to Drought Stress*,
DOI: 10.1007/978-3-642-32653-0_14, © Springer-Verlag Berlin Heidelberg 2012

14.1 Introduction

During the millennium prior to the Industrial Revolution, the atmospheric CO_2 concentration was stable at about 270 μmol mol^{-1}. However, the concentration has been increased by economic development and progressive industrialisation to its current 389 μmol mol^{-1}. According to the predictions of the Intergovernmental Panel on Climate Change, the levels may reach between 660 and 790 μmol mol^{-1} by 2090 (IPCC 2007). With the increase in greenhouse gases, the models project a global surface temperature rise from 1.4 to 5.8 °C over the period 1990–2100. Increasing temperature will enhance evapotranspiration rates and exacerbate low water availability problems commonly observed in Mediterranean environments, in which the annual potential evapotranspiration is often nearly twice the amount of rainfall (Sábate et al. 2002). In addition, drought is considered the major stress factor limiting plant species distribution and growth in Mediterranean regions of the world (Mooney 1983) and constraining crop productivity and stability (Araus et al. 2002). Thus, the need to alleviate practical problems such as crop yield under drought conditions has increased interest in "water stress physiology" (Lawlor and Tezara 2009).

Several studies have observed that instantaneous increases in atmospheric CO_2 concentration stimulated photosynthetic rates through higher carboxylation efficiency and substrate concentration at the primary carboxylating enzyme, rubisco (Drake et al. 1997). However, there is abundant evidence that in the long-term, photosynthesis acclimatises to elevated CO_2 (Jifon and Wolfe 2002; Long et al. 2004; Aranjuelo et al. 2005, 2006). After long-term exposure to elevated CO_2 conditions, the photosynthetic efficiency of nodulated alfalfa plants tends to decrease as a consequence of rubisco activity and protein content depletion (Aranjuelo et al. 2005). This reduction was observed to parallel carbohydrate build-up and N availability limitations. The decreased N availability problems observed in elevated CO_2 treatments might affect plant growth and their capacity to develop new sinks. Drake et al. (1997) observed that the capability of wheat and pea to form additional sinks during early vegetative growth could be limited if N supply was limiting. Imbalanced source/sink relations enhanced concentrations of leaf structural carbohydrate and through increased hexose cycling within the leaf, potentially suppressed the expression of genes encoding for the photosynthetic apparatus, resulting in decreased photosynthetic capacity (Drake et al. 1997; Wolfe et al. 1998; Moore et al. 1999). Under elevated CO_2 conditions, the efficiency of carboxylation by rubisco is increased and the subsequent carbon requirement for the protein synthesis is reduced (Webber et al. 1994). As calculations have suggested that 35 % of the rubisco could be lost from the leaf before rubisco co-limits photosynthesis (Long and Drake 1992), Sage (1994) proposed that a significant fraction of leaf N could be reallocated into other more N-limited photosynthetic and non-photosynthetic processes. This implies a physiological adjustment conducive to higher photosynthetic N use efficiency, i.e. an increase in the rate of carbon assimilation per unit N in the foliage. As suggested by Drake et al. (1997), the key point of studies analysing plant responses to elevated CO_2 is not the removal of a limitation (rubisco) but an increase in nutrient use efficiency.

The inhibitory effects of elevated temperatures target photosynthesis, water status and N_2 fixation in the case of N_2-fixing plants (Bushby 1982). Many authors have reported a decrease in photosynthetic rates in Mediterranean species during summer drought due to stomatal closure (Peñuelas et al. 1998; Llusiá and Peñuelas 2000) and metabolic limitation (rubisco activity, RuBP regeneration, etc.) (Tretiach et al. 1997; Larcher 2000). High temperatures cause reversible alterations in the physicochemical state of biomembrane and the conformation of protein molecules (Larcher 2003). Because thylakoid membranes are especially sensitive to heat, photosynthesis is one of the first indicators of heat stress. Changes to thylakoid membrane fluidity and damaged Calvin cycle enzymes may affect CO_2 fixation through altered excitation energy distribution.

In general, drought also decreases photosynthetic rates due to stomatal closure and nonstomatal processes (Lawlor and Cornic 2002; Aranjuelo et al. 2007; Lawlor and Tezara 2009; Chaves et al. 2009) as a primary symptom of water stress (Chaves et al. 2009). It is considered that the reduction in CO_2 assimilation under water deficit is frequently associated with stomatal closure (Arnau et al. 1997; Chaves et al. 2003). The higher stomatal conductance to water usually increases CO_2 diffusion into the leaf and hence incorporation into biomass. Under mild or moderate water stress it is accepted that stomatal closure is the main factor limiting photosynthetic activity but after severe water limiting conditions metabolic impairment takes place (Medrano et al. 2002) including damage to photosynthetic enzymes (Erice et al. 2011).

Alfalfa (*Medicago sativa* L.) is one of the most widely distributed forage crops in the world, adapted to grow from the equator to almost Arctic Circle regions (Michaud et al. 1988). It is cultivated in more than 80 countries in an area exceeding 35 million ha (Radovic et al. 2009) and world production was around 436 million tons in 2006 (FAO 2006). It is capable of establishing symbiotic relationships with *Sinorhizobium* genus bacteria forming root nodules. This competence for fixing atmospheric N_2 has been recognised as conferring growth advantageous over non-N_2-fixing plants (Serraj et al. 1998; Lüscher et al. 2002; Rogers et al. 2006). Some authors have demonstrated that in response to elevated CO_2, N_2-fixing species show a larger stimulation of growth and photosynthetic rate than non-N_2-fixing species (Ainsworth and Rogers 2007). Nodule activity depends on photosynthate supplied by the plant, which are used by the nitrogenase enzyme as a source of energy and reducing power to fix N_2 (Larrainzar et al. 2009; Aranjuelo et al. 2011). This cycle coupling causes nitrogenase activity in plants to be regulated by photosynthesis (carbon supply), N availability (N source strength) and N demand (N sink strength) (Aranjuelo et al. 2011). Similarly, the products of N_2 fixation are exported throughout the plant via the xylem to other organs where N is required; for example, protein and/or osmoregulant synthesis (Ladrera et al. 2007; Aranjuelo et al. 2011). This growth enhancement associated with exposure to elevated CO_2 may lead to a higher N demand, so it is expected that the plant response to CO_2 may be limited by N availability (Hartwig 1998; West et al. 2005).

The sensitivity of legume-*Sinorhizobium* symbiosis to ambient stresses like temperature and low water availability is well known (Streeter 2003; Curtis et al. 2004; Thomas et al. 2004). Moreover, some authors (Castellanos et al. 1996; Thomas et al. 2004) suggest that the effect of water deficit on N accumulation and N_2 fixation is larger than on biomass accumulation. Carbon availability is considered as a limiting factor in N_2 fixation and assimilation that otherwise requires large amounts of carbohydrates (Heytler et al. 1985). It has long been known that elevated CO_2 stimulates legume growth and N_2 fixation (Wilson et al. 1933). As mentioned above, the greater photosynthetic rate in legumes grown under high CO_2 conditions would imply that there is a larger supply of organic carbon to nodules and this is used in turn by the bacteroid nitrogenase enzyme inside the nodules as a source of energy and reducing power to fix N_2 (Arrese-Igor et al. 1999; Cabrerizo et al. 2001). Such N availability enhancement is usually accompanied by an increase in the number and weight of nodules (Finn and Brun 1980; Norby 1987; Arnone and Gordon 1990; Zanetti et al. 1996; González et al. 2001), or by enhanced nodule enzyme activity (Hardy and Havelka 1976; Ryle et al. 1992).

14.2 N_2-Fixing Alfalfa and Increasing Growth Temperature

According to some authors (Drake and Leadley 1991; Long 1991; Bowes 1993; Morison and Lawlor 1999), increased temperature enhances photosynthesis and thus plant growth when it does not exceed the growth optimum value. This phenomenon could be explained by kinetic properties of rubisco (Long 1991) and the better utilisation of the end products of photosynthesis through increased sink metabolism at elevated temperatures (Farrar and Williams 1991). However, when increased temperature exceeds the plant growth optimum, CO_2 fixation is decreased (Aranjuelo et al. 2007) by an altered excitation energy distribution, as mentioned previously. This could be due to the fact that high temperatures may decrease photosynthetic rates, decreasing CO_2/O_2 solubility and coupled with diminished specificity of rubisco for CO_2 (Jordan and Ögren 1984; Brooks and Farquhar 1985).

Legumes and their associated N_2-fixing bacteria display a remarkable resilience to temperature regimes (Dixon and Wheeler 1983). Upper temperature limits for nodulation are higher in tropical than in temperate species, and precise effects vary with species of host plant and N_2 fixing bacteria. Power and Zachariassen (1993) showed that *Trifolium repens,* a temperate legume species, possessed higher N_2 fixation and N uptake when grown between 10 and 20 than at 30 °C. Pankhurst and Sprent (1976) showed that the optimum temperature range for root-nodule symbiosis is between 25 and 33 °C, however, when the temperature is above 28 °C N_2 fixation is reduced (Hungria and Franco 1993). The elevated temperature could have direct and indirect effects. Direct inhibition of N_2-fixation by elevated temperature may be a consequence of decreased

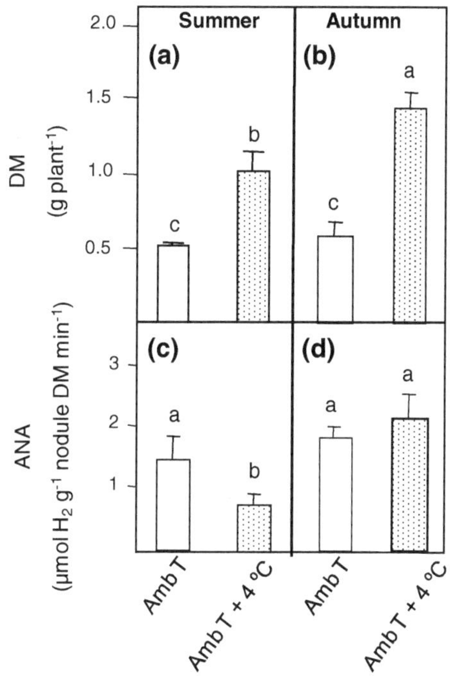

Fig. 14.1 Effect of temperature (ambient and ambient + 4 °C) on plant dry matter (*DM*) and apparent nitrogenase activity (*ANA*) in summer (**a**, **c**) and autumn (**b**, **d**). The different letters indicate significant differences ($p < 0.05$) among treatments as determined by t-student test. Adapted from Sanz-Sáez et al. (2012) (http://www.sciencedirect.com/science/article/pii/S009884721100298X)

nodule development (Dart and Mercer 1965; Piha and Munns 1987), functionality (activity) (Meyer and Anderson 1959; Piha and Munns 1987; Hernández-Armenta et al. 1989) and accelerated senescence. Within functionality factors, the decreased photosynthesis under high temperature conditions could cause the diminishment of N_2-fixation as a consequence of a drop in the photosynthate content (Aranjuelo et al. 2007). Indirect inhibition by elevated temperature is related to a depression in root hair formation, reduction in nodulation sites (Jones and Tisdale 1921; Frings 1976) and modified adherence of bacteria to root hairs (Frings 1976).

The optimum temperature for N_2 fixation in alfalfa grown in growth chambers is between 20 and 25 °C, with 28 °C being the temperature above which fixation is reduced (Aranjuelo et al. 2007). This reduction was caused by carbohydrate shortage in nodules, as a consequence of decreased photosynthesis at 28 °C (Aranjuelo et al. 2007). These authors reported that plants grown at 28 °C showed a diminishment in bacteroid enzymatic activity [malate dehydrogenase (MDH) and isocitrate dehydrogenase (ICDH)], due to a deficient supply of total soluble sugars caused by the fall in photosynthesis (Aranjuelo et al. 2007). Not only is the nodule bacteroid fraction affected, but MDH activity from the nodule plant fraction also drops in plants exposed to high temperature, suggesting that malate availability should also be reduced. Malate depletion implies that less carbon is redirected to mitochondria and bacteroides, and therefore, there is less energy available for N_2 fixation (Aranjuelo et al. 2007).

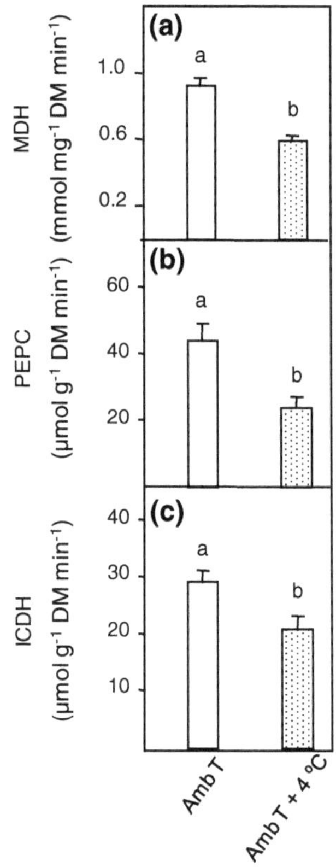

Fig. 14.2 Effect of temperature (ambient and ambient + 4 °C) on malate dehydrogenase (*MDH*) (**a**), phosphoenolpyruvate carboxylase (*PEPC*) (**b**), isocitrate dehydrogenase (*ICDH*) (**c**) in alfalfa nodule plant fraction. Otherwise as for Fig. 14.1. Adapted from Aranjuelo et al. (2008) (www.publish.csiro.au/nid/102/paper/FP07296.htm)

Usually, experiments conducted in controlled environments (growth chambers) overestimate the effect of different treatments such as temperature or CO_2. The use of temperature gradient greenhouses (TGGs) allows the study of increased temperature effects in near field conditions (Aranjuelo et al. 2005; Erice et al. 2006a, 2006b). In a recent study (Sanz-Sáez et al. 2012) that was conducted with alfalfa in TGGs, it was observed that elevated temperature treatment (ambient temperature +4 °C) increased plant dry matter (DM) production in summer and autumn. Elevated temperatures reached during summer (mean temperature 17.3 °C; maximum mean temperature 24.2 °C) resulted in more production in autumn, the cooler season (Fig. 14.1 a, b) (Sanz-Sáez et al. 2012). In summer, elevated temperature decreased N_2 fixation (Fig. 14.1 c, d), measured as apparent nitrogenase activity (ANA), causing N shortage, a usual symptom in high temperature conditions (Aranjuelo et al. 2007; Power and Zachariassen 1993) leading to a reduction in DM production in comparison with autumn.

Therefore, the high temperature does not imply nitrogen fixation reduction via a decline in photosynthesis, but could directly affect nodule development and activity, decreasing the MDH, ICDH and phosphoenolpyruvate carboxylase (PEPC) of the nodule vegetal fraction (Fig. 14.2) (Aranjuelo et al. 2008). PEPC

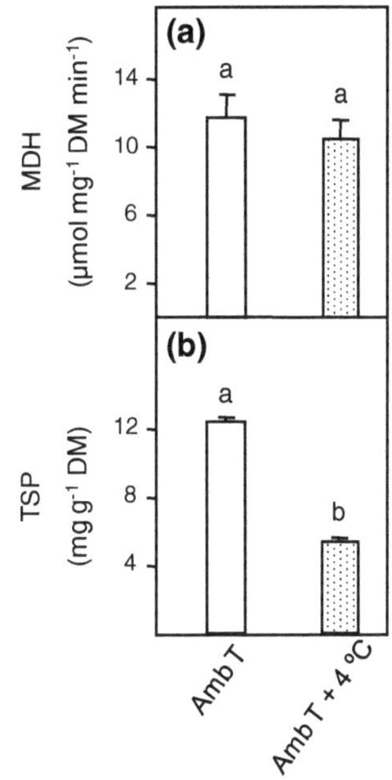

Fig. 14.3 Effect of temperature (ambient and ambient + 4 °C) on malate dehydrogenase (*MDH*) (**a**), total soluble proteins (*TSP*) (**b**) in alfalfa nodule bacteroid fraction. Otherwise as for Fig. 14.1. Adapted from Aranjuelo et al. (2008) (www.publish.csiro.au/nid/102/paper/FP07296.htm)

catalyses the combination of respiratory CO_2 with phosphoenolpyruvate to produce oxaloacetate, which is converted to malate by MDH. Malate is considered to be the main carbon supply for N_2 fixation (Udvardi and Day 1997). As shown in Fig. 14.2 (Aranjuelo et al. 2008), inhibition of PEPC and MDH total activities observed in plants grown at elevated temperature is going to be reflected in diminished malate production. Malate depletion implies that less carbon and energy is available for bacteroid consumption, and also that less carbon is redirected to mitochondria for ammonia assimilation. Carbon skeletons provided by the plant fraction are tightly linked to nitrogenase activity. It is estimated that up to 25 % of the carbon required for nitrogenase activity and N assimilation may be supplied by PEPC (King et al. 1986), which means that decreased MDH and PEPC activities might also have affected nitrogenase activity negatively and therefore reduced N_2 fixation. The ICDH activity is mainly associated with acquiring the reduction power used to produce organic acids (malate) and ammonia assimilation derived from N_2 fixation. Therefore, diminished activities of ICDH, MDH and PEPC observed in plants exposed to elevated temperature, may contribute to decreased nodule metabolism and N_2 fixation. In addition, elevated temperature decreased the total soluble protein content (TSP) in the bacteroidal fraction (Fig. 14.3). It is precisely in bacteroides where nitrogenase catalyses N_2 fixation, so decreased TSP at elevated temperature could imply the fall of nitrogenase content and thus N_2 fixation reduction.

In conclusion, when increased temperature exceeds the optimum growth temperature, N_2 fixation and plant DM production decrease. However, the optimum growth temperature range varies among plant and *Rhizobium* species. In alfalfa grown in controlled conditions (climatic chambers), this range is estimated between 20 and 25 °C, with 28 °C being the maximum temperature. During summer, plants produced less DM in comparison to autumn, due to diminishment of N_2 fixation as a consequence of higher temperatures. In this case, elevated temperature affects N_2 fixation through decreased photosynthesis, nodule activity (MDH, ICDH and PEPC) and bacteroid TSP content.

14.3 Nodulated Plants Affected by Water Availability and Elevated CO_2

Many studies have considered CO_2 and water availability as separate factors. However, it is known that the effect of combined stresses on plant growth causes alterations that cannot be predicted if they are analysed separately, such as those resulting from synergistic and antagonistic phenomena (Valladares and Pearcy 1997).

As described previously (Erice et al. 2006a, 2006b), regardless of water availability, exposure of nodulated alfalfa plants to elevated CO_2, induced photosynthetic downregulation that was caused by a depleted rubisco protein content, and which was the result of carbohydrate build-up and a reallocation of N away from leaves (Aranjuelo et al. 2005). The lack of sufficient sink capacity for incoming photosynthates leads to a negative feedback effect and a downward acclimatisation of photosynthetic capacity (Wolfe et al. 1998). Many aspects of plant photosynthetic acclimatisation to elevated CO_2 may be explained by a source-sink conceptual model in which the capacity for C assimilation is up- or down-regulated depending on the balance between supply and utilisation of photosynthates. Relative growth rate (RGR) determinations in well irrigated and droughted plants of alfalfa revealed that when applied at ambient temperature, elevated CO_2 did not modify its values (Fig. 14.4a). The absence of differences in gas exchange parameters and leaf relative water content (RWC) (Aranjuelo et al. 2006) suggest that, as observed by several authors (Ogaya and Peñuelas 2003; Aranjuelo et al. 2006), droughted alfalfa plants tend towards a quick acclimatisation to a maintained lower soil water availability.

Together with the physiological response to drought there are plant responses affecting leaf morphology. A reduction in the leaf area ratio (LAR), which measures leafiness of the plant, and an increase in the specific leaf area (SLA), an indicator of leaf thickness (Marcelis et al. 1998; Erice et al. 2010), allow the plant to have higher leaf protein density and, consequently, greater photosynthetic capacity. Some studies carried out with nodulated alfalfa have revealed that decreased LAR induced by drought was accompanied by a drop in leaf weight ratio (LWR) but also occurred with SLA increases. In such plants, the combined effect of water limited growth and elevated CO_2 resulted in intermediate plant characteristics in terms of SLA and LWR (Fig. 14.4b, c, d).

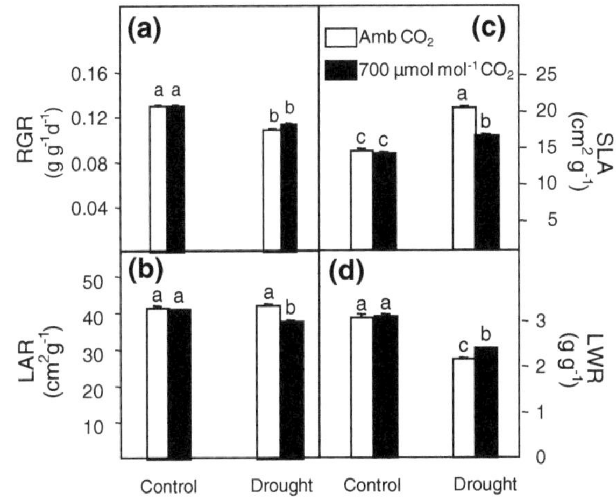

Fig. 14.4 Effect of CO_2 and water availability on relative growth rate (*RGR*) (**a**), leaf area ratio (*LAR*) (**b**), specific leaf area (*SLA*) (**c**) and leaf weight ratio (*LWR*) (**d**), in nodulated alfalfa. The different letters indicate significant differences ($p < 0.05$) among treatments as determined by LSD

Table 14.1 The effect of CO_2 and water availability (H_2O) on total N content (N, mg N plant^{-1}), total N use efficiency (NUE_t mg DM mg^{-1} N_t) and leaf N use efficiency (NUE_l, mg DM mg$^{-1}N_l$) in nodulated alfalfa

Treatments H_2O-CO_2		N	NUE_t	NUE_l
Control-	Ambient CO_2	27.68 ± 0.4 a	105.1 ± 3.0 b	316.2 ± 20 b
	Elevated CO_2	27.06 ± 1.0 a	106.0 ± 2.1 b	488.2 ± 30 ab
Drought-	Ambient CO_2	12.7 ± 0.4 b	100.5 ± 0.4 c	362.3 ± 50 b
	Elevated CO_2	13.5 ± 0.3 b	121.9 ± 1.0 a	496.5 ± 20 a

Each value represents the mean ± SE ($n = 8$). The different letters indicate significant differences ($p < 0.05$) between the treatments, as determined by LSD

As mentioned before, nutrient limitation as well as drought may reduce growth and hence sink strength, driving plants to photosynthetic acclimatisation. This is particularly interesting in the case of plants with the ability for N_2 acquisition, which is usually described as key factor limiting crop yield (Kirkby 1981). Determinations of total N content in alfalfa have shown that elevated CO_2 had no effect on it, whereas, under water limited conditions, the lower growth meant the lower the quantity of N (Table 14.1). Nevertheless, these results revealed that the combination of elevated CO_2 and drought increased plant N use efficiency calculated on a total matter basis (NUE_t) or leaf matter basis (NUE_l). Such plants are consequently able to maintain larger growth due to their capability of using N in a more efficient way, as also observed by other authors (Davey et al. 1999; Osborne et al. 1998; Zerihun and BassiriRad 2000).

As shown above, elevated CO_2 affects nodule metabolism negatively. Decreased leaf N concentration associated with limitations to water availability in plants with equal or similar plant N concentration indicates that the lack of differences in plant N might be a consequence of reduced nodule enzyme activity (Aranjuelo et al. 2006). Once again, as reported before, the analyses of several key enzymes of the plant nodule and bacteroid fraction revealed that elevated CO_2 affected nodule metabolism negatively. Under elevated CO_2 concentration (700 μmol mol^{-1}), PEPC and MDH total activities decreased, which may imply reduced malate production (Fig. 14.5a, c) (Aranjuelo et al. 2008). The inhibitory effect of elevated CO_2 on glutamate-oxaloacetate transaminase (GOT), under both water regimes, and ICDH activity, in the case of fully watered plants, might have contributed to low N availability (Fig. 14.5b, d). The inhibition of ICDH, an enzyme mainly associated with the production of reducing power, might have exacerbated a reduction in organic acid (as malate) production and ammonia assimilation (De Vries et al. 1980). Under drought, improved specific activities of MDH, PEPC and ICDH were shown (Fig. 14.5e, g, h). It is probable that drought growth limitation, which favours higher NUE, leads to maintenance of higher leaf soluble protein concentrations even at elevated CO_2 (Erice et al. 2006b), and which in turn usually induces leaf carbohydrate accumulation (Erice et al. 2006b) followed by the observed SLA decrease. As commented before, the early response of plants to elevated CO_2 is a rise in photosynthetic rates (Erice et al. 2006a). So it is expected that larger amounts of photoassimilates would be delivered to the nodules (Cabrerizo et al. 2001). However, nodule total soluble sugar and starch concentration measurements conducted in the present study showed that instead of increasing its values, elevated CO_2 did not affect soluble sugar carbon availability (Fig. 14.6b, c) (Aranjuelo et al. 2008). This means that the nodules of plants exposed to elevated CO_2 could have had an insufficient supply of carbohydrates and consequently undergone a decline in N_2 fixation (Arrese-Igor et al. 1999).

In parallel to what happened in the plant fraction, analyses of the nodule bacteroid fraction confirmed that elevated CO_2 decreased MDH total activity regardless of the water availability (Fig. 14.6d) (Aranjuelo et al. 2008). Despite a drop in total soluble protein under elevated CO_2, the enhanced MDH specific activity emphasises the importance of malate production. A similar result has been shown in *Anthyllis* nodules where the MDH and PEPC activities were not significantly affected by drought (Goicoechea et al. 2005). In this case the authors related the stable enzyme activity to the maintenance of aerobic metabolism and/or that the reduction in oxygen availability induced the PEP-MDH pathway. These plants were subjected to withholding irrigation until the desired soil water content of one-third of control field capacity was achieved and this resulted in significant decreases in water potential but not in RWC. These results show that plants are able to acclimatise to water shortages. *Anthyllis* is a shrub recognised as a drought avoider (Haase et al. 2000), exhibiting defoliation and distribution of the water absorbed by roots to a smaller number of metabolically active leaves, thus maintaining a higher RWC (Goicoechea et al. 2004). In contrast, alfalfa is drought resistant and limits its growth to the available soil water to keep a stable RWC (Erice et al. 2006a). Elevated CO_2 applied at low water availability treatments maintained MDH activity at control levels due to the ability to enhance specific activity (Fig. 14.6e). As has been observed before at the leaf level (Erice et al. 2006a, 2006b), plant and bacteroid fraction TSP reduction suggests that, elevated CO_2 may also induce downregulation at the nodule level. Because a great

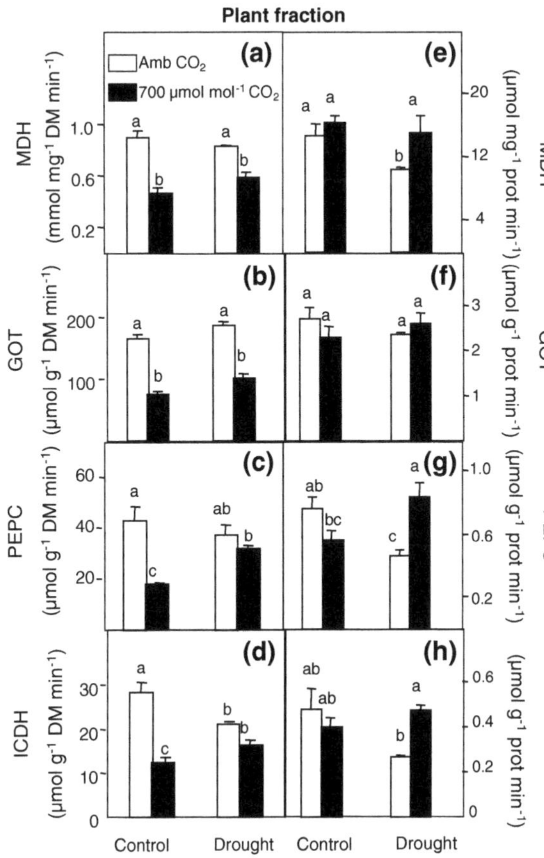

Fig. 14.5 Effect of CO_2 and water availability on total and specific malate dehydrogenase (*MDH*) (**a**, **e**), glutamate-oxaloacetate transaminase (*GOT*) (**b**, **f**), phosphoenolpyruvate carboxylase (*PEPC*) (**c**, **g**) and isocitrate dehydrogenase (*ICDH*) (**d**, **h**) of alfalfa nodule plant fraction. Otherwise as for Fig. 14.4. Adapted from Aranjuelo et al. (2008) (www.publish.csiro.au/nid/102/paper/FP07296.htm)

percentage of N is delivered to TSP formation, its depletion implies that more N is available for other processes that might contribute to increase sink capacity and consequently counteract acclimatisation processes.

In conclusion, enhanced NUE together with stimulation of nodule enzyme specific activities demonstrates that plants grown under elevated CO_2 conditions were capable of using N in a more efficient way, especially under water limited conditions. Analyses of nodule plant and bacteroid metabolism have revealed that elevated CO_2 negatively affected nodule enzymatic activity in all treatments, but this was partially overcome in combination with drought in the case of PEPC and ICDH. The CO_2 inhibitory effect on nodule total activity was a consequence of the depletion of TSP being especially drastic in the bacteroid fraction. Depletions of leaf and nodule TSP suggest that elevated CO_2 induced nodule downregulation in order to release N to more limiting processes that might contribute to the adjustment of source/sink relations. It is also notable that drought limiting plant growth may contribute to increase TSP in the nodule bacteroid fraction under elevated CO_2 conditions.

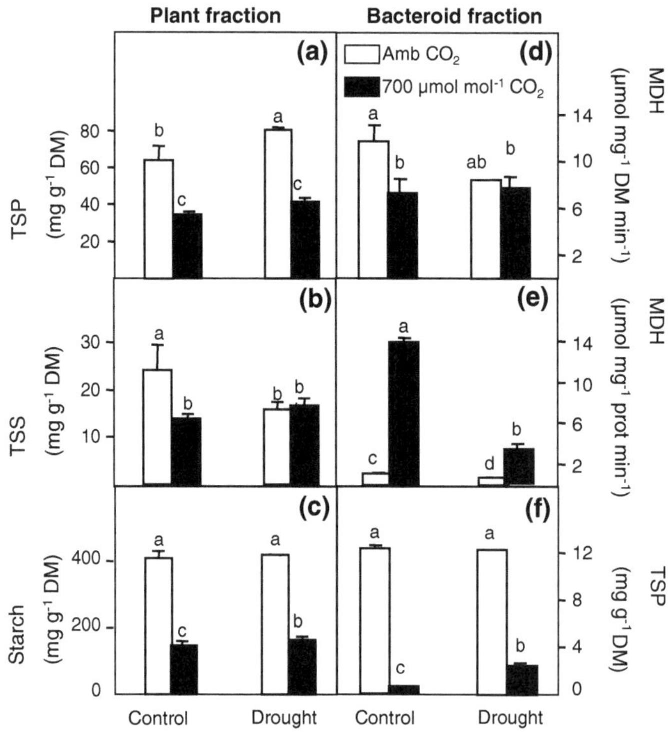

Fig. 14.6 Effect of CO_2 and water availability on total soluble protein (*TSP*) (**a**), total soluble sugars (*TSS*) (**b**) and starch (**c**) concentrations of alfalfa nodule plant fraction; total (**d**) and specific (**e**) malate dehydrogenase (*MDH*) and total soluble protein (*TSP*) (**f**) of alfalfa nodule bacteroid fraction. Otherwise as for Fig. 14.4. Adapted from Aranjuelo et al. (2008) (www.publish.csiro.au/nid/102/paper/FP07296.htm)

14.4 The Way Drought Treatments are Imposed Conditions Nodule Responses in Experiments

Drought is usually considered as a set of climate pressures that are produced by several phenomena (Monneveux and Belhassen 1996) including heat shocks, low air hygrometry, insolation, salinity or, as in the present chapter, water deficit. Many studies have been focussed on the effect of drought without taking into account the wide range of time scales in which plant processes may occur. Short-term responses to the rapid imposition of drought comprise stomatal response, gene expression and heat shock or dehydrin protein production, leaf movement, wilting, osmotic adjustment and responses to ABA. At the other end of the time scale, long-term strategies include canopy development, leaf senescence and root system modification (Passioura 1996). As a consequence, the longer time that is available to carry out the long-term strategies results in plant acclimatisation to drought.

Drought studies may be divided into those in which drought treatment is applied as progressive withholding of water until severe stress is attained, and those in which sustained limited water availability is applied thus making acclimatisation possible. When alfalfa is subjected to water withholding, decreases in leaf RWC are shown (Erice et al. 2010; Aranjuelo et al. 2011) as also reported in other legumes like *Lupinus* (Pinheiro et al. 2004). Nevertheless, there are studies where alfalfa plants were exposed to sustained water stress (50 % of control soil volumetric water content) over one month, allowing plants to acclimatise (Erice et al. 2006a; Aranjuelo et al. 2006, 2009) and reaching RWC values similar to control levels.

For drought tolerance the biological classification permits to underlie some adaptation strategies frequently used. For most annual plants it is favourable for the reproductive period to have ended before the drought season, but for perennials such as alfalfa the development of physiological mechanisms is frequent. The inhibitory effect of drought on photosynthetic activity has been widely described and is mainly associated with stomatal and metabolic limitations (Lawlor and Tezara 2009). As reported by Erice et al. (2011) alfalfa subjected to mild water stress due to water withholding showed decreases in leaf conductance, but it was after severe stress when the damage to the photosynthetic apparatus was revealed. It is also interesting that the same study revealed that in alfalfa it is possible to identify differences in drought sensitivity in relation to different genotypes, probably due the wide distribution of this species. The diminishment of photosynthesis is also accompanied by a reduction in nitrogenase activity after withholding water (Aranjuelo et al. 2011) resulting in a lower plant N availability (Ladrera et al. 2007; Larrainzar et al. 2009). Such a decrease confirmed that legume-*Sinorhizobium* symbiosis is very sensitive to water stress (Aranjuelo et al. 2007; Ladrera et al. 2007; Larrainzar et al. 2009) and the general assumption that symbiotic nitrogen fixation is more affected than CO_2 photosynthetic assimilation rates (Serraj et al. 1999; Aranjuelo et al. 2011).

Nevertheless, in studies carried out with alfalfa maintained under limited water availability and with enough time to acclimatise, plants were observed to develop less leaf area, leading to maintain specific transpiration (Erice et al. 2006a). In a subsequent study (Erice et al. 2006b), neither leaf conductance nor photosynthetic rates were affected in alfalfa growing under a stable soil volumetric content of 50 %, showing similar values to the earlier study at field capacity. As described in the previous section, due to the enhancement of nitrogen use efficiency (Table 14.1), nodule metabolism after acclimatisation to such drought conditions remained relatively unchanged.

Apart from the adjustments affecting shoots, drought-adapted plants are also characterised by deep and vigorous root systems in terms of depth, length and resistance to water flow (Taylor and Klepper 1978). Water-limited alfalfa has shown to increase its root to shoot ratio (Erice et al. 2007). Shoot and root biomass accumulations are affected differently by drought, with the shoots being more prone to reductions in biomass (Antolín et al. 1995; Yin et al. 2004) because the transpiring parts of the plant develop greater and longer water deficits (Kramer and

Table 14.2 Effect of the way drought treatments are imposed (withholding water or sustained limited water) on water status (RWC and leaf water potential), gas exchange (photosynthetic rate and leaf conductance) and nodule activity parameters

Parameter	Withholding water	Sustained limited water
Relative water content (RWC)	Decreased	Unchanged by the contribution of osmotic adjustment.
Leaf water potential	Decreased	Decreased by the osmotic adjustment.
Photosynthetic rate and leaf conductance	Decreased	Unchanged, plants develop less leaf area maintaining specific transpiration. Gas exchange parameters remain stable by changes in leaf morphology (LAR and SLA) and root to shoot ratio.
Nodule activity	Decreased nitrogenase activity	Unchanged nodule activity, drought growth limitation favoured higher NUE to maintain leaf protein concentration.

Boyer 1995). Even if expansion of the belowground system may enhance soil water extraction, larger roots may also increase drought tolerance via reserve enhancement. The alfalfa taproot is the main storage organ of the plant containing the most important carbon and nitrogen pools such as starch, amino acids and proteins (Avice et al. 2003). Compared to plants grown at field capacity, plants acclimatised to drought have been shown (1) to remobilise large amounts of nitrogen from shoots towards roots tissues and (2) to accumulate N as proteins and specially in the form of vegetative storage proteins (VSP) (Erice et al. 2007). The higher VSP concentration has been related to early regrowth after defoliation (Bewley 2002), and greater forage production as well as the N economy may represent an indirect action in stress tolerance (Erice et al. 2007).

Drought also modifies morphological adaptations to modulate stomatal or residual transpiration including the pilosity of leaves, the colour of transpirative organs and leaf rolling or falling (Monneveux and Belhassen 1996). Alfalfa under water stress is capable of modifying leaf morphological features like LAR and SLA (Marcelis et al. 1998) in order to maintain leaf protein density and photosynthesis. This faculty to alter SLA or LAR may be genotypically specific, and in this sense Mediterranean cultivars have been shown to be more plastic than Nordic-type cultivars (Erice et al. 2010).

While plants are facing water stress, the turgor begins to decrease and osmoregulatory responses are initiated. Low molecular weight organic substances accumulate in the cell compartments and the cytosol via synthesis, conversion of starch and translocation (Larcher 2003) with energy costs associated with these processes. This osmotic adjustment is considered a trait associated with yield stability (Morgan 1983; Morgan and Condon 1986). Alfalfa subjected to the withholding of water showed significantly decreased leaf and nodule osmotic potential (Aranjuelo et al. 2011) and this has been attributed to the deterioration of leaf water status. In drought acclimatised alfalfa, leaf water potential values dropped (Erice et al. 2006a) as consequence of osmotic regulation. An opposing pattern of changes was shown in the case of RWC, which did not decrease, and it is precisely the osmotic adjustment that may contribute to RWC maintenance.

In summary (Table 14.2), when alfalfa is subjected to water stress by irrigation suppression, leaf RWC usually diminishes with consequent decreases in photosynthetic and leaf conductance rates. Furthermore, nitrogenase activity is also reduced as it is featured in the decline of osmotic potential in nodules and leaves. Nevertheless, when the combination between the time-scale and stress harshness allows plants to acclimatise to drought, as typically occurs under sustained water limitation, plants develop strategies to grow in such conditions. In acclimatised alfalfa, an enhanced root to shoot ratio with limited leaf area development enables plants to maintain RWC as well as specific transpiration and gas exchange parameters (conductance and photosynthesis). Moreover, the larger root system could increase drought tolerance as it is the main storage organ in alfalfa for carbon and especially nitrogen, and this conditions the regrowth rate after alfalfa defoliation. Osmotic adjustment was also revealed in acclimatised plants with decreased leaf water potential reaching values similar to fully watered plants.

Acknowledgments This work was supported by Ministerio de Economía y Competitividad (MINECO BFU2011-26989), Fundación Universitaria de Navarra (PIUNA-2008) and Fundación Caja Navarra. Gorka Erice was the recipient of a research ANABASi+D contract from Gobierno de Navarra. The assistance of Amadeo Urdiain and Mónica Oyarzun is also appreciated.

References

Ainsworth EA, Rogers A (2007) The response of photosynthesis and stomatal conductance to rising [CO_2]: mechanisms and environmental interactions. Plant Cell Environ 30:258–270

Antolín MC, Yoller J, Sánchez-Díaz M (1995) Effects of temporary drought on nitrate-fed and nitrogen-fixing alfalfa plants. Plant Sci 107:159–165

Aranjuelo I, Pérez P, Hernández L, Irigoyen JJ, Zita G, Martínez-Carrasco R, Sánchez-Díaz M (2005) The response of nodulated alfalfa to water supply, temperature and elevated CO_2: photosynthetic down regulation. Physiol Plantarum 123:348–358

Aranjuelo I, Irigoyen I, Perez P, Martínez-Carrasco R, Sánchez-Díaz M (2006) Response of nodulated alfalfa to water supply, temperature and elevated CO_2: productivity and water relations. Environ Exp Botany 55:130–141

Aranjuelo I, Irigoyen JJ, Sánchez-Díaz M (2007) Effect of elevated temperature and water availability on CO_2 exchange and nitrogen fixation of nodulated alfalfa plants. Environ Exp Botany 59:99–108

Aranjuelo I, Irigoyen JJ, Sánchez-Díaz M, Nogués S (2008) Carbon partitioning in N2 fixing Medicago sativa plants exposed to different CO2 and temperature conditions. Funct Plant Biol 35:306–317

Aranjuelo I, Irigoyen JJ, Sánchez-Díaz M, Nogués S (2009) Elevated CO_2 and water-availability effect on gas exchange and nodule development in N_2-fixing alfalfa plants. Environ Exp Botany 65:18–26

Aranjuelo I, Molero G, Erice G, Avice J-C, Nogués S (2011) Plant physiology and proteomics reveals the leaf response to drought in alfalfa (*Medicago sativa* L.). J Exp Bot 62:111–123

Araus JL, Slafer GA, Reynolds MP, Royo C (2002) Plant breeding and water relation in C3 cereals: what to breed for? Ann Bot 89:925–940

Arnau G, Monneveux P, This D, Alegre L (1997) Photosynthesis of six barley genotypes as affected by water stress. Photosynthetica 34:67–76

Arnone JA II, Gordon JC (1990) Effect of nodulation, nitrogen fixation and CO_2 enrichment on the physiology, growth and dry mass allocation of seedlings of *Alnus rubra* Bong. New Phytol 116:55–66

Arrese-Igor C, González EM, Gordon AJ, Minchin FR, Gálvez L, Royuela M, Cabrerizo PM, Aparicio-Tejo PM (1999) Sucrose synthase and nodule nitrogen fixation under drought and other environmental stresses. Symbiosis 27:1–24

Avice J-C, Le Dily F, Goulas E, Noquet C, Meuriot F, Volenec JJ, Cunningham SM, Sors TG, Dhont C, Castonguay Y, Nadeau P, Bélanger G, Chalifour FP, Ourry A (2003) Vegetative storage proteins in overwintering storage organs of forage legumes: roles and regulation. Can J Bot 81:1198–1212

Bewley JD (2002) Root storage proteins, with particular reference to taproots. Can J Botany 80:321–329

Bowes G (1993) Facing the inevitable: plants and increasing atmospheric CO_2. Annu Rev Plant Physiol Plant Mol Biol 44:309–332

Brooks A, Farquhar GD (1985) Effect of temperature on the CO_2/O_2 specificity of ribulose-1, 5-bisphosphate carboxylase/oxygenase and the rate of respiration in the light. Planta 165:397–406

Bushby HVA (1982) Ecology. In: Broughton WJ (ed) Nitrogen fixation: volume 2 Rhizobium. Claredon Press, Oxford, pp 35–75

Cabrerizo PM, Gonzalez EM, Aparicio-Tejo PM, Arrese-Igor C (2001) Continuous CO_2 enrichment leads to increased nodule biomass, carbon availability to nodules and activity of carbon-metabolising enzymes but does not enhance specific nitrogen fixation in pea. Physiol Plantarum 113:33–40

Castellanos JZ, Pena-Cabriales JJ, Costa-Gallegos JAA (1996) ^{15}N-determined dinitrogen fixation capacity of common bean (*Phaseolus vulgaris* L.) cultivars under water stress. J Agr Sci 126:327–333

Chaves MM, Maroco JP, Pereira JS (2003) Understanding plant responses to drought from genes to the whole plant. Funct Plant Biol 30:239–264

Chaves MM, Flexas J, Pinheiro C (2009) Photosynthesis under drought and salt stress: regulation mechanisms from whole plant to cell. Ann Bot 103:551–560

Curtis J, Shearer G, Khol DH (2004) Bacteroid proline catabolism affects N_2 fixation rate of drought-stressed soybeans. Plant Physiol 136:3313–3318

Dart PJ, Mercer FV (1965) The effect of growth temperature, level of ammonium nitrate and light intensity on the growth and nodulation of cowpea (*Vigna sinensis* Endl. ex. Hassk.). Austr J Agr Res 16:321–345

Davey PA, Parsons AJ, Atkinson L, Wadge K, Long SP (1999) Does photosynthetic acclimation to elevated CO_2 increase photosynthetic nitrogen use efficiency? A study of three native UK grassland species in open-top chambers. Funct Ecol 13:21–28

De Vries GE, In'tVeld P, Kijne JW (1980) Production of organic acids in *Pisum sativum*, root nodules as a result of oxygen stress. Plant Sci 20:115–123

Dixon ROD, Wheeler CT (1983) Biochemical, physiological and environmental aspects of symbiotic nitrogen fixation. In: Gordon JC, Wheeler CT (eds) Biological nitrogen fixation in forest ecosystems: foundations and applications. Nijhoff/junk Publishers, The Hague, pp 107–171

Drake BG, Leadley PW (1991) Canopy photosynthesis of crops and native plants communities exposed to long-term elevated CO_2. Plant Cell Environ 14:853–860

Drake BG, González-Meler MA, Long SP (1997) More efficient plants: a consequence of rising atmospheric CO_2? Annu Rev Plant Physiol Mol Biol 48:609–639

Erice G, Irigoyen JJ, Pérez P, Martínez-Carrasco R, Sánchez-Díaz M (2006a) Effect of elevated CO_2, temperature and drought on dry matter partitioning and photosynthesis before and after cutting of nodulated alfalfa. Plant Sci 170:1059–1067

Erice G, Irigoyen JJ, Pérez P, Martínez-Carrasco R, Sánchez-Díaz M (2006b) Effect of elevated CO_2, temperature and drought on photosynthesis of nodulated alfalfa during a cutting regrowth cycle. Physiol Plantarum 126:458–468

Erice G, Irigoyen JJ, Sánchez-Díaz M, Avice JC, Ourry A (2007) Effect of drought, elevated CO_2 and temperature on accumulation of N and vegetative storage proteins (VSP) in taproot of nodulated alfalfa before and after cutting. Plant Sci 172:903–912

Erice G, Louahlia S, Irigoyen JJ, Sánchez-Díaz M, Avice J-C (2010) Biomass partitioning, morphology and water status of four alfalfa genotypes submitted to progressive drought and subsequent recovery. J Plant Physiol 167:114–120

Erice G, Louahlia S, Irigoyen JJ, Sánchez-Díaz M, Thami Alami I, Avice J-C (2011) Water use efficiency, transpiration and net CO_2 exchange of four alfalfa genotypes submitted to progressive drought and subsequent recovery. Environ Exp Bot 72:123–130

FAO (2006) FAOSTAT. Food and Agriculture Organization of the United Nations

Farrar JF, Williams ML (1991) The effects of increased atmospheric carbon dioxide and temperature on carbon partitioning, source-sink relations and respiration. Plant Cell Environ 14:819–830

Finn GA, Brun WA (1980) Water stress effects on CO_2 assimilation, photosynthate partitioning, stomatal resistance, and nodule activity in soybean. Crop Sci 20:431–434

Frings JFJ (1976) The rhizobium-pea symbiosis as affected by high temperatures. Ph.D. Thesis. Wageningen Agricultural University, Netherlands

Goicoechea N, Merino S, Sánchez-Díaz M (2004) Contribution of arbuscular mycorrhizal fungi (AMF) to the adaptations exhibited by the deciduous shrub *Anthyllis cytisoides* L. under water deficit. Physiol Plantarum 122:453–464

Goicoechea N, Merino S, Sánchez-Díaz M (2005) Arbuscular mycorrhizal fungi can contribute to maintain antioxidant and carbon metabolism in nodules of *Anthyllis cytisoides* L. subjected to drought. J Plant Physiol 162:27–35

González EM, Gálvez L, Royuela M, Aparicio-Tejo PM, Arrese-Igor C (2001) Insights into regulation of nitrogen fixation in pea nodules: lessons from drought, abscisic acid and increased photoassimilate availability. Agronomie 21:607–613

Haase P, Pugnaire FI, Clarck SC, Incoll LD (2000) Photosynthetic rate and canopy development in the drought-deciduous shrub *Anthyllis cytisoides* L. J Arid Environ 46:79–91

Hardy RWF, Havelka UD (1976) Photosynthate as a major factor limiting nitrogen fixation by field-grown legumes with emphasis on soybeans. In: Nutman PS (ed) Symbiotic nitrogen fixation. Cambridge University Press, Cambridge, pp 421–439

Hartwig UA (1998) The regulation of symbiotic N_2 fixation: a conceptual model of N feedback from ecosystem to the gene expression level. Perspect Plant Ecol Evol Syst 1:92–120

Hernandez-Armenta R, Wien HC, Eaglesham ARJ (1989) Maximum temperature for nitrogen fixation in common bean. Crop Sci 29:1260–1265

Heytler PG, Reddy GS, Hardy RWF (1985) In vivo energetics of symbiotic nitrogen fixation in soybeans. In: Ludden PL, Berris JE (eds) Nitrogen fixation and CO_2 metabolism. Elsevier, New York, pp 283–292

Hungria M, Franco AA (1993) Effects of high temperature on nodulation and nitrogen fixation by *Phaseolus vulgaris* L. Plant Soil 149:95–102

IPCC (2007) Climate Change 2007: impacts, adaptation and vulnerability. Working Group II Contribution to the Intergovernmental Panel on Climate Change. Fourth Assessment Report. Summary for Policymakers. Brussels

Jifon JL, Wolfe DW (2002) Photosynthetic acclimation to elevated CO_2 in *Phaseolus vulgaris* L. is altered by growth response to nitrogen supply. Global Change Biol 8:1018–1027

Jones FR, Tisdale WB (1921) Effect of soil temperature upon the development of nodules on the roots of certain legumes. J Agr Res 22:17–37

Jordan DB, Ögren WL (1984) The CO_2/O_2 specificity of ribulose 1, 5 bisphosphate carboxylase/ oxygenase. Planta 161:308–313

King BJ, Layzell DB, Canvin DT (1986) The role of dark carbon dioxide fixation in root nodules of soybean. Plant Physiol 81:200–205

Kirkby EA (1981) Plant growth in relation to nitrogen supply. In: Clark FE, Rosswall T (eds) Terrestrial nitrogen cycles: processes, ecosystem strategies and management impacts, vol 33., ScopeEcological Bulletins, Stockholm, pp 249–267

Kramer PJ, Boyer JS (1995) Water relations of plants and soils. Academic Press, London

Ladrera R, Marino D, Larrainzar E, González EM, Arrese-Igor C (2007) Reduced carbon availability to bacteroides and elevated ureides in nodules, but not in shoots, are involved in the nitrogen fixation response to early drought in soybean. Plant Physiol 145:539–546

Larcher W (2000) Temperature stress and survival ability of Mediterranean schlerophyll plants. Plant Biosyst 134:279–295

Larcher W (2003) Physiological Plant Ecology. Springer, Berlin

Larrainzar E, Wienkoop S, Scherling C, Kempa S, Ladrera R, Arrese-Igor C, Weckwerth W, Gonzalez EM (2009) Carbon metabolism and bacteroid functioning are involved in the regulation of nitrogen fixation in *Medicago truncatula* under drought and recovery. Molecular Plant Microbe In 22:1565–1576

Lawlor DW, Cornic G (2002) Photosynthetic carbon assimilation and associated metabolism in relation to water deficits in higher plants. Plant Cell Environ 25:275–294

Lawlor DW, Tezara W (2009) Causes of decreased photosynthetic rate and metabolic capacity in water-deficient leaf cells: a critical evaluation of mechanisms and integration of processes. Ann Bot 103:543–549

Llusià J, Peñuelas J (2000) Seasonal patterns of terpene content and emission from seven Mediterranean woody species in field conditions. Am J Bot 87:133–140

Long SP (1991) Modification of the response of photosynthetic productivity to rising temperature by atmospheric CO_2 concentration: has its importance been underestimated? Plant Cell Environ 14:729–739

Long SP, Drake BG (1992) Photosynthetic CO_2 assimilation and rising CO_2 atmospheric CO_2 concentrations. In: Baker NR, Thomas N (ed) Photosynthesis: crop photosynthesis: spatial and temporal determinations. Elsevier Science, Amsterdam, pp 69–107

Long SP, Ainsworth EA, Rogers A, Ort DR (2004) Rising atmospheric carbon dioxide: plants FACE the future. Annu Rev Plant Biol 55:591–628

Lüscher A, Hartwig UA, Suter D, Nösberger J (2002) Direct evidence that symbiotic N_2 fixation in fertile grassland is an important trait for a strong response of plants to elevated atmospheric CO_2. Global Change Biol 6:655–662

Marcelis LFM, Heuvelink E, Goudriaan J (1998) Modelling biomass production and yield of horticultural crops: a review. Scientia Hort 74:83–111

Medrano H, Escalona JM, Bota J, Gulias J, Flexas J (2002) Regulation of photosynthesis of C3 plants in response to progressive drought: stomatal conductance as a reference parameter. Ann Bot 89:895–905

Meyer DR, Anderson AJ (1959) Temperature and symbiotic nitrogen fixation. Nature 183:161

Michaud R, Lehmen WF, Rumbaugh MD (1988) World distribution and historical development. In: Hanson AA, Barnes DK, Hill RR (eds) Alfalfa and alfalfa improvement. American Society of Agronomy, Madison, pp 25–91

Monneveux P, Belhassen E (1996) The diversity of drought adaptation in the wide. Plant Growth Regul 20:85–92

Mooney HA (1983) Carbon-gaining capacity and allocation patterns of Mediterranean climate plants. In: Kruger FJ, Mitchel DT, Jarvis JUM (eds) Mediterranean type ecosystems: the role of nutrients. Springer, Berlin, pp 103–119

Moore BE, Cheng SH, Sims D, Seemann JR (1999) The biochemical and molecular basis for acclimation to elevated CO_2. Plant Cell Environment 22:567–582

Morgan JM (1983) Osmoregulation as a metabolism criterion for drought tolerance in wheat. Austr J Agr Res 34:607–614

Morgan JM, Condon AG (1986) Water use, grain yield and osmoregulation in wheat. Austr J Plant Physiol 13:523–532

Morison JIL, Lawlor DW (1999) Interactions between increasing CO_2 concentration and temperature on plant growth. Plant Cell Environ 22:659–682

Norby RJ (1987) Nodulation and nitrogenase activity in nitrogen-fixing woody plants stimulated by CO_2 enrichment of the atmosphere. Physiol Plantarum 71:77–82

Ogaya R, Peñuelas J (2003) Comparative field studies of *Quercus ilex* and *Phillirea latifolia*: photosynthetic response to experimental drought conditions. Environ Exp Botany 50: 137–148

Osborne CP, LaRoche J, Garcia RL, Kimball BA, Wall GW, Pinter PJ Jr, LaMorte RL, Hendrey GR, Long SP (1998) Does leaf position within a canopy affect acclimation of photosynthesis to elevated CO_2? 1. Analysis of a wheat crop under free-air CO_2 enrichment. Plant Physiol 117:1037–1045

Pankhurst CE, Sprent JI (1976) Effects of temperature and oxygen tension on the nitrogenase and respiratory activities of turgid and water-stressed soybeans and French bean root nodules. J Exp Bot 27:1–9

Passioura JB (1996) Drought and drought tolerance. Plant Growth Regul 20:79–83

Peñuelas J, Filella I, Llusià J, Piñol J (1998) Comparative field study of spring and summer leaf gas exchange and photobiology of the Mediterranean trees *Quercus ilex* and *Phillyrea latifolia*. J Exp Bot 49:229–238

Piha MI, Munns DN (1987) Sensitivity of the common bean (*Phaseolus vulgaris*) symbiosis to high soil temperature. Plant Soil 98:183–194

Pinheiro C, Passarinho JA, Ricardo CP (2004) Effect of drought and rewatering on the metabolism of *Lupinus albus* organs. J Plant Physiol 161:1203–1210

Power PF, Zachariassen JF (1993) Relative nitrogen utilization by legume cover crop species at three soil temperatures. Agron J 85:134–140

Radovic J, Sokolovic D, Markovic J (2009) Alfalfa-most important perennial forage legume in animal husbandry. Biotech Anim Husbandry 25:465–475

Rogers A, Gibon Y, Stitt M, Morgan PB, Bernacchi CJ, Ort DR, Long SP (2006) Increased C availability at elevated carbon dioxide concentration improves N assimilation in legume. Plant Cell Environ 29:1651–1658

Ryle GJA, Powell CE, Davidson IA (1992) Growth of white clover, dependent on N_2 fixation, in elevated CO_2 and temperature. Ann Bot 70:221–228

Sábate S, Gracia CA, Sánchez A (2002) Likely effects of climate change on growth of *Quercus ilex*, *Pinus halepensis*, *Pinus pinaster*, *Pinus sylvestris* and *Fagus sylvatica* forests in the Mediterranean region. Forest Ecol Manag 162:23–37

Sage RF (1994) Acclimation of photosynthesis to increasing atmospheric CO_2: the gas exchange perspective. Photosynth Res 39:351–368

Sanz-Sáez A, Erice G, Aguirrolea J, Irigoyen JJ, Sánchez-Díaz M (2012) Alfalfa yield under elevated CO_2 and temperature depends on the *Sinorhizobium* strain and growth season. Environ Exp Bot 77:267–273

Serraj R, Sinclair TR, Allen LH (1998) Soybean nodulation and N_2 fixation response to drought under carbon dioxide enrichment. Plant Cell Environ 21:491–500

Serraj R, Sinclair TR, Purcell LC (1999) Symbiotic N_2 fixation response to drought. J Exp Bot 50:143–155

Streeter JG (2003) Effects of drought on nitrogen fixation in soybean root nodules. Plant Cell Environ 26:1199–1204

Taylor HM, Klepper B (1978) The role of rooting characteristics in the supply of water to plants. Adv Agron 30:99–128

Thomas Robertson MJ, Fukai S, Peoples MB (2004) The effect of timing and severity of water deficit on growth, development, yield accumulation and nitrogen fixation of mungbean. Field Crop Res 86:67–80

Tretiach M, Bolognini G, Rondi A (1997) Photosynthetic activity of *Quercus ilex* at the extremes of a transect between Mediterranean and submediterranean vegetation (Trieste, NE, Italy). Flora 192:369–378

Udvardi MK, Day DA (1997) Metabolic transport across symbiotic membranes of legume nodules. Annu Rev Plant Physiol Plant Mol Biol 48:493–523

Valladares F, Pearcy RW (1997) Interactions between water stresses, sun-shade acclimation, heat tolerance and photo inhibition in the schlerophyll *Heteromeles arbutifoliar*. Plant Cell Environ 20:25–36

Webber AN, Nie G-Y, Long SP (1994) Acclimation of photosynthetic proteins to rising atmospheric CO_2. Photosynth Res 39:401–412

West JB, HilleRisLambers J, Lee TD, Hobbies SE, Reich PE (2005) Legume species identity and soil nitrogen supply determine symbiotic nitrogen fixation responses to elevated atmospheric [CO_2]. New Phytol 167:523–530

Wilson PW, Fred EB, Salmon MR (1933) Relation between carbon dioxide and elemental nitrogen assimilation in leguminous plants. Soil Sci 35:145–165

Wolfe DW, Gifford RM, Hilbert D, Luo Y (1998) Integration of photosynthetic acclimation to CO_2 at the whole-plant level. Global Change Biol 4:879–893

Yin C, Duan B, Wang X, Li C (2004) Morphological and physiological responses of two contrasting poplar species to drought stress and exogenous abscisic acid application. Plant Sci 167:1091–1097

Zanetti S, Hartwig UA, Lüscher A, Hebeisen T, Frehner M, Fischer BU, Hendrey GR, Blum H, Nösberger J (1996) Stimulation of symbiotic N_2 fixation in *Trifolium repens* L. under elevated atmospheric pCO_2 in a grassland ecosystem. Plant Physiol 112:575–583

Zerihun A, BassiriRad H (2000) Photosynthesis of *Helianthus annuus* does not acclimate to elevated CO_2 regardless of N supply. Plant Physiol Biochem 38:897–903

Chapter 15
Enhancement of Plant Drought Tolerance by Microbes

Young-Cheol Kim, Bernard R. Glick, Yoav Bashan and Choong-Min Ryu

Abstract Plants are limited to protect themselves against environmental stresses including drought. Thus, plants develop a wide range of strategies to cope with stress situations. Under conditions of water deficiency, drought escape and drought tolerance are two important strategies to ensure plant growth. There is limited reported information dealing with the role of microbes on the improvement of drought tolerance. Here, we provide an overview of current knowledge on the general features of the induction of drought tolerance mediated by inoculation of plants with fungi, bacteria, and viruses, and several bacterial determinants and plant signaling transduction pathways revealed by classic physiological or morphological observations and recent "Omics" technology. Overall, the application of microbes provides new insights into novel protocols to improve plant defense responses to drought, an important component of agricultural production systems affected by a changing climate.

Y.-C. Kim
Institute of Environmentally-Friendly Agriculture, College of Agriculture and Life Sciences, Chonnam National University, 500-757 Gwangju, South Korea

B. R. Glick
Department of Biology, University of Waterloo, Waterloo, ON, N2L 3G1, Canada

Y. Bashan
Environmental Microbiology Group, Northwestern Center for Biological Research (CIBNOR), 23090 La Paz, BCS, Mexico

C.-M. Ryu (✉)
Molecular Phytobacteriology Laboratory, Systems and Synthetic Biology Research Center, KRIBB, 305-806 Daejeon, South Korea
e-mail: cmryu@kribb.re.kr

R. Aroca (ed.), *Plant Responses to Drought Stress*,
DOI: 10.1007/978-3-642-32653-0_15,

15.1 Introduction

In the natural environment, plants are continuously subjected to a large number of both biotic and abiotic stresses. These stresses include extremes of temperature, high light, flooding, drought, the presence of toxic metals and environmental organic contaminants, radiation, wounding, insect predation, high salt, and various pathogens, including viruses, bacteria, and fungi (Abeles et al. 1992; Glick et al. 2007a). All of these environmental stresses result in the inhibition of plant growth compared to the level that it might attain in their absence. In fact, throughout its life a plant may be subjected to a number of non-lethal stresses. In each instance, the stress generally limits plant growth for some period of time until either the stress is removed or the plant is able to adjust its metabolism to overcome the stress. The net result of this discontinuous exposure to a range of environmental stresses is that plant growth may consist of periods of high growth interspersed with periods of growth inhibition. Since a plant cannot physically remove itself from its environment the way an animal can (e.g. by migrating to a different locale during periods of drought), a plant responds to various environmental stresses by modifying its metabolism and physiology including the synthesis of a wide range of defense-related proteins. In addition, some soil microorganisms can help plants to partially overcome the deleterious effects of a number of different environmental stresses. For example, the so-called plant growth-promoting bacteria (PGPB) may facilitate plant growth either by promoting the acquisition of nutritional resources, such as nitrogen, phosphorus, or iron; preventing or decreasing the damage to plants by pathogenic organisms (mainly fungi and bacteria); or by directly stimulating plant growth by either providing plant hormones, such as auxin, cytokinin, or gibberellin, or lowering plant ethylene levels through the action of the enzyme 1-aminocyclopropane-1-carboxylate (ACC) deaminase (Glick 1995; Glick et al. 1998).

As a consequence of global industrialization and exponential growth of world population, the world's climate is rapidly changing (Wang 2005). Every year, the world loses more and more agricultural land to drought and increased salinity (Wang 2005; Gamalero et al. 2009). Should this trend continue unabated, it will be impossible for the countries of the world to adequately feed all of their people by the end of this century. Obviously, a problem of this magnitude can only be addressed by dealing with a variety of issues, including resource consumption and utilization, human population levels, and agricultural productivity and practices. For a start, it may be possible to significantly decrease the use of chemical fertilizers, pesticides, insecticides, and herbicides in modern agricultural practice and instead, in part through the judicious use of PGPB, develop more sustainable agricultural practices including under drought conditions (Antoun and Prévost 2005). In recent years, several microbial inoculants have been formulated, produced, marketed, and applied successfully by an increasing number of growers worldwide (Reed and Glick 2004).

15.1.1 Plant Growth-Promoting Bacteria

As the name implies, the term PGPB includes a wide variety of different bacterial strains that can facilitate plant growth (Bashan and Holguin 1998). These organisms include plant rhizosphere-binding bacteria, endophytes that colonize the interior tissues of plants, symbiotic bacteria such as *Rhizobia* spp. that only interact with specific plants and stimulate the formation of and occupy nodules on the roots of specific plants, and cyanobacteria (Bashan and Holguin 1998).

PGPB can facilitate plant growth and development in two different ways: indirectly or directly (Glick 1995). Indirect promotion of plant growth occurs when these bacteria decrease some of the deleterious effects of a phytopathogenic organism (often a fungus or a bacterium) by one or more different mechanisms (discussed below). The direct promotion of plant growth often includes providing the plant with a compound that is synthesized by the bacterium or facilitating the uptake of nutrients from the environment.

Those PGPB that are likely to be most effective at helping plants to overcome environmental stresses such as drought probably utilize several different mechanisms to aid plant growth. In facilitating plant nutrient acquisition, PGPB may fix atmospheric nitrogen and supply it to plants; they may synthesize siderophores that can solubilize and sequester iron from the soil and provide it to plant cells; and they may solubilize phosphate from the soil and provide it to the plant (Rodriguez and Fraga 1999).

Not all PGPB are diazotrophic. In addition, many of the PGPB that are diazotrophic fix only limited amounts of nitrogen. Other diazotrophic PGPB can fix high levels of nitrogen; however, most researchers believe that (except for symbiotic PGPB) even these organisms generally provide the plant with only small amounts of fixed nitrogen (Bashan and Levanony 1990). Most likely, for the majority of PGPB, nitrogen fixation is only a minor component of the benefit that the bacterium provides to the plant, although under specific circumstances, some diazotrophic PGPB may provide plants with a significant portion of the fixed nitrogen that they require. The effect on plant growth of a PGPB providing a nutrient such as fixed nitrogen, iron, or phosphorus to the plant, may vary considerably depending upon the soil composition so that PGPB may have little or no measurable effect on plant growth when plants are grown in nutrient-rich soil under optimal conditions.

Despite the fact that iron is the fourth most abundant element on earth (Ma 2005), in soil it is generally unavailable for direct assimilation by plants and microorganisms. This is because ferric ion, Fe^{3+}, the predominant form of iron in nature, is only sparingly soluble ($\sim 10^{-18}$ M at pH 7.4); this level is too low to support either plant or microbial growth. Consequently, bacteria and plants produce low molecular weight compounds that bind to and sequester the available iron from the soil. Bacteria synthesize low molecular-weight siderophores, molecules with a very high affinity for Fe^{+3} (K_d ranging from 10^{-23} to 10^{-52}) and membrane receptors that can bind the Fe-siderophore complex, thereby enabling iron uptake by bacteria (Neilands 1981). While plant siderophores bind iron with a much lower affinity than bacterial siderophores, iron-bacterial siderophore complexes may be taken up and utilized by plants (Wang et al. 1993). Environmental conditions, including iron concentration and pH, the composition of the soil microbial population, and the iron acquisition mechanisms of the host plant, all influence iron uptake from bacterial siderophores by plants. In addition, plant iron nutrition improvement by soil bacteria may be more important to the plant when it is exposed to an environmental stress.

Phosphorus (P) levels in soil are generally between 400 and 1200 mg kg^{-1} of soil, however, the concentration of soluble P in soil is usually $\sim$1 mg kg^{-1} or less (Goldstein 1994) so that it is considered to be one of the elements that may limit plant growth. To satisfy plants' nutritional requirements, P is usually added to soils as fertilizers synthesized through high-energy-intensive processes.

P in soil is typically present in soil in either mineral or organic forms. Bacteria often solubilize mineral (inorganic) P through the synthesis of low molecular weight organic acids such as gluconic and citric acid (Rodriguez et al. 2004) that bind phosphate with their hydroxyl and carboxyl groups thereby chelating cations and inducing soil acidification, both resulting in the release of soluble phosphate (Bnayahu 1991; Kpomblekou and Tabatabai 1994). The mineralization of organic P occurs through the synthesis of phosphatases (by either the plant or soil microbes), including phosphomonoesterase, phosphodiesterase, and phosphotriesterase, catalyzing the hydrolysis of phosphoric esters (Rodriguez and Fraga 1999; Rodriguez et al. 2006).

Arguably, the major mechanism(s) that PGPB use to promote plant growth, especially in the presence of environmental stresses such as drought, is the manipulation of plant hormone levels, in particular auxin and ethylene (Brown 1974; Patten and Glick 1996, 2002; Glick et al. 1998, 2007b). Through early plant development, PGPB may stimulate plant cell division by providing appropriate levels of plant hormones and help plants to tolerate a variety of environmental stresses such as flooding and drought by lowering potentially detrimental levels of ethylene that can form as a response to these stresses as well as providing growth-stimulatory auxin.

Indole-3-acetic acid (IAA) is the most common and the most studied auxin, and often the terms auxin and IAA are used interchangeably. IAA influences a large number of diverse cellular functions and is considered to be a critically important regulator of plant growth and development. IAA has been implicated in the orientation of root and shoot growth in response to light and gravity, in differentiation of plant vascular tissue, in apical dominance, in the initiation of plant lateral and adventitious roots, in stimulation of plant cell division, and in elongation growth in plant stems and roots. IAA also stimulates seed and tuber germination; xylem and root development; vegetative growth; initiation of lateral and adventitious roots; and responses to light and gravity (Glick 1995).

Several reports are consistent with exogenous IAA alleviating certain abiotic stresses. Thus, bacterial IAA was reported to stimulate increased root and shoot length of wheat seedlings exposed to high levels of salt (Egamberdieva 2009).

An increased tolerance of *Medicago truncatula* against salt stress was also observed in plants nodulated by the IAA-overproducing strain *Sinorhizobium meliloti* DR-64 (Bianco and Defez 2009). In several studies where PGPB were used as part of metal phytoremediation protocols, it was concluded that one of the key factors contributing to the ability of the bacterium to facilitate phytoremediation (i.e., to plant growth in the presence of the metal) was the presence of IAA (Glick 2010).

The plant hormone ethylene controls a wide range of behaviors and responses of vascular plants such as fruit ripening, seed germination, tissue differentiation, the formation of root and shoot primordia, lateral bud development, leaf abscission, flower wilting, and the response of plants to both biotic and abiotic stresses, including responses that both turn on a plant's defenses and those that result in increases in plant senescence (Abeles et al. 1992). When PGPB that express the enzyme ACC deaminase are bound to either the seed coat or root of a developing plant, this may act to ensure that the ethylene level within the plant does not become elevated to the point where plant (either root or shoot) growth is impaired. By facilitating the formation of longer roots and protecting stressed plants from some of the deleterious effects of the phytohormone ethylene, these bacteria can enhance the survival and protect the yield of a variety of plants, especially during the first few days after the seeds are planted and the plants are most vulnerable to damage by various environmental stresses (Glick et al. 1998; Glick 2004). Moreover, based on the model that was developed to describe how PGPB that express the enzyme ACC deaminase facilitate the rooting of young seedlings (Glick et al. 1998), it was predicted that the plant-inhibitory effects of many ethylene-causing environmental stresses (both biotic and abiotic) would be significantly decreased if plants were first treated with ACC deaminase-expressing bacteria.

15.1.2 Arbuscular Mychorrizae

Mycorrhizae are fungi that form a relationship with the roots of most vascular plants. The mycorrhizae colonize plant roots extracellularly in the case of ectomycorrhizae, or intracellularly in the case of arbuscular mycorrhizae (Danneberg et al. 1992). Ectomycorrhizal-plant interactions are common with a large number of different tree species and with a smaller number of vascular plants of agricultural importance. On the other hand, arbuscular mycorrhizal associations, while found with some tree species, are much more common in agricultural plants. Arbuscular mycorrhizae penetrate the cortical cells of the roots of vascular plants forming an obligate symbiosis with its plant host with highly branched arbuscules that are the sites for exchange of water, carbon, phosphorus, and other nutrients (Kendrick 1992).

Some of the mechanisms employed by arbuscular mycorrhizae to help their plant hosts overcome the effects of drought (and salt stress) include (Gamalero et al. 2009): (a) Improved water uptake; arbuscular mycorrhizae effectively extend plant roots making the uptake of water much more efficient. (b) Better mineral nutrition, especially phosphorus, as a consequence of effectively extending plant roots. In fact, arbuscular mycorrhizae may provide the major mechanism for phosphorus uptake by many plants, with PGPB possibly acting as a secondary mechanism or adjunct to mycorrhizae in terms of P uptake. (c) Alterations in root architecture. (d) Modification of some physiological and enzymatic activities, especially those involved in plant antioxidative responses. (e) Induction of the plant hormone abscisic acid (ABA) which can play an important role in mediating some of a plant's response to different stresses including drought (Danneberg et al. 1992). While many of the physiological changes in plants that are induced by its association with arbuscular mycorrhizae, including changes that take place under drought conditions, have been well documented (Spychalla and Desbough 1990; Berta et al. 1993; Marulanda et al. 2003; Khalvati et al. 2005; Aroca et al. 2007), a detailed understanding of biochemical and molecular nature of these physiological changes remains to be elucidated. In addition, the tripartite (plant beneficial) interaction between plants, arbuscular mycorrhizae, and PGPB is not at all well understood (e.g. Gamalero et al. 2010) so that additional studies of this complex relationship would most assuredly facilitate the development of microbial strategies to develop more drought resistant plants.

15.2 Microbial Elicitation of Drought Tolerance

15.2.1 Drought Tolerance by Bacteria

PGPB colonize the rhizosphere of many plant species and confer beneficial effects such as increasing plant growth and reducing disease development caused by plant pathogenic fungi, bacteria, viruses, and nematodes (Kloepper et al. 2004). Some PGPB also elicit physical or chemical changes related to plant defense, a process referred to as induced systemic resistance (ISR) (van Loon et al. 2004). ISR elicited by PGPB can suppress plant diseases in the greenhouse and in the field against a broad range of plant pathogens (Kloepper et al. 2004; van Loon et al. 2004). However, fewer reports have been published on the use of PGPB as elicitors of tolerance to abiotic stresses, including drought, salt, and nutrient deficiency or excess. The subject of PGPB-elicited tolerance to heavy metals has been reviewed recently (Glick 2010), so it is excluded in the discussion here. We propose the term induced systemic tolerance (IST) for the elicitation by PGPB of physical and chemical changes in plants that result in enhanced tolerance to abiotic stress. Biotic stress is excluded from IST, because it is conceptually part of biological control and induced resistance.

Drought stress limits the growth and productivity of crops, particularly in arid and semiarid areas (Kramer and Boyer 1995). Early studies on IST by drought

(Timmusk and Wagner 1999) reported that inoculation of the PGPB *Paenibacillus polymyxa* enhanced drought tolerance in *Arabidopsis*. RNA differential display on parallel RNA preparations from *P. polymyxa*-applied and untreated plants revealed that mRNA transcription of a drought response gene, ERD15, was augmented. Another PGPB strain, *Achromobacter piechaudii* ARV8, that produces ACC deaminase, conferred IST to drought-stressed pepper and tomato plants (Mayak et al. 2004a, b). Under stress conditions, including drought, the plant hormone ethylene endogenously regulates plant homeostasis and results in reducing root and shoot growth (Glick et al. 2007a, b). Degradation of ACC, which is a precursor of ethylene by bacterial ACC deaminase, releases plant stress and rescues normal plant growth (Figueiredo et al. 2008).

Efforts to apply these results include using mixtures of rhizospheric strains with symbiotic nitrogen-fixing rhizobia (Figueiredo et al. 2008) or with mycorrhizal fungi (Kohler et al. 2008). The rhizobia are extremely sensitive to drought stress resulting in a significant decrease of N_2 fixation when faced with low soil water content. Under drought stress, co-inoculation of bean with *Rhizobium tropici* and two strains of PGPB, both *P. polymyxa,* resulted in increased plant height, shoot dry weight, and nodule number (Figueiredo et al. 2008). Interestingly, the effect on IST and higher nodule number was greater when a mixture of two strains of *P. polymyxa* treated with *R. tropici* was applied rather than with a single strain (Figueiredo et al. 2008).

The investigation of plant hormone balances by drought stress revealed an increase in ABA content in the leaves, indicating that the reduction of endogenous cytokinin levels magnifies the changes in the ABA content in the shoot (Cowan et al. 1999). The antagonism between cytokinin and ABA may be the result of metabolic interactions: cytokinins share a common biosynthetic origin with ABA (Cowan et al. 1999). Also, *P. polymyxa* was reported to produce the cytokinin zeatin that is involved in plant growth promotion (Timmusk and Wagner 1999). It will be interesting to determine if cytokinin produced by *P. polymyxa* affects ABA signaling of plants or rhizobia-elicited nodulation.

Co-inoculation with the PGPB *Pseudomonas mendocina* and the arbuscular mycorrhizal fungi, *Glomus intraradices* or *G. mosseae*, increased the phosphatase capacity of lettuce roots, resulting in enhanced proline accumulation in shoots (Kohler et al. 2008). Particularly, PGPB augmented the activity of an antioxidant catalase during severe drought, suggesting that PGPB may be used in inoculants to alleviate some of the oxidative damage elicited by drought.

15.2.2 Drought Tolerance by Fungi

Similar to bacterial-plant interactions, numerous studies have shown that symbiotic (mostly endophytic) fungi promote abiotic stress tolerance, including drought stress, to host plants and they act in a critical role in plant survival under stress conditions (Rodriguez et al. 2009; Singh et al. 2011). Two major classes of fungal symbionts can improve plant drought tolerance: (1) endophytic fungi, which reside entirely inside plant tissues and can interact with roots, stems, and leaves, and (2) mycorrhizal fungi, that reside only in roots and the rhizosphere. The induction of plant drought tolerance by symbiotic endophytes has been thoroughly summarized in a recent review (Singh et al. 2011). In this review, many genera, including *Neotyphodium* spp., *Cuvularia* spp., *Colletotrichum* spp., *Fusarium* spp., and *Alternaria* spp. were reported to elicit plant drought tolerance in tomato, pepper, ryegrass, tall fescue, wheat, and barley. Interestingly, in addition to an increase of drought tolerance, the endophyte-infected plants express a range of adaptations to biotic stress and other abiotic stresses including mineral imbalance and salinity (Latch et al. 1985; Lyons et al. 1990; Belesky and Fedders 1995). Second, arbuscular mycorrhiza fungi (AMF) of the phylum Glomeromycota, order Glomerales, develop, in most cases, nonspecific symbioses with the majority of terrestrial plants. In this obligate association, fungal spores perceive the presence of the host plant through biochemical communication between the two partners and, after approaching the plant roots, AM produce fungal hyphae that grow into the apoplastic space of the host root cortical cells. Highly branched tissues (arbuscules), are the exchange interface for nutrients between the AM and the host plant. It is known that AM fungi can enhance plant growth and production under different conditions including various types of stress. Thus, AM can significantly promote plant growth and yield under drought conditions by formation of an extensive hyphal network and by the production of chemical compounds that induce drought tolerance. Numerous previous reports have indicated that mycorrhizal plants facilitate growth under drought conditions compared to non-mycorrhizal plants. Many plant species, including corn, soybean, wheat, onion, lettuce have been shown to induce drought tolerance as a consequence of AM symbiosis (Augé 2001). Under these conditions, AM augmented root surface area and promoted dense growth, resulting in improved drought tolerance. Moreover, the expanded root system enhanced water movement from the limited amount of water around the roots. Finally, AM colonized plants are able to maintain a higher water use efficiency and increase growth at a faster rate when irrigation is restored. Such adjustment of osmotic potential by AM is one of the most important factors for the increased capacity of a plant to survive under drought conditions. In addition, AM may affect plant water potential by modification of soil structure. AM hyphae can increase soil structure by binding soil particles and through the production of glomalin, an insoluble glue-like substance produced by AM (Augé 2001). The important issue, for plants to obtain drought tolerance by AM colonization, is maintaining water potential. More recently, a new class of endophytic fungi, *Piriformospora indica,* which colonizes a wide range of monocot and dicot plants and convey a beneficial effect to host plants under drought stress has been described (Sherameti et al. 2008; Sun et al. 2010). This endophytic fungus shares many characteristics, like the abilities to promote plant growth and increase plant abiotic stress tolerance, with AM fungi. However, unlike AM fungi, *P. indica* has the potential to grow axenically without the requirement of a living host and can colonize plants of the Brassicaceae, including *Arabidopsis thaliana*, and

Chenopodiaceae families, known to be non-hosts for AM fungi. Another recent study has evaluated the behavior of certain strains of endophytic *Trichoderma* spp. that colonize the *Theobroma cacao* tree following drought stress (Bae et al. 2009). In this regard, colonization of cacao seedlings with *Trichoderma hamatum* strain DIS 219b caused a significant delay in the onset of many drought-induced physiological changes.

15.2.3 Drought Tolerance by Viruses

Viruses are obligate intracellular symbionts that cannot multiply without their host tissue. To date, plant viruses have been studied as important pathogens that cause serious problems in crop plants. Thus, it is difficult to imagine that viruses could provide host plants with any beneficial effects. In 2008, unexpectedly, but very intriguingly, plants infected by various viruses were found to acquire significant drought tolerance. Xu et al. (2008) assessed the drought tolerance of ten different monocot and dicot plant species following infection with four different RNA viruses, *Brome mosaic virus* (BMV), *Cucumber mosaic virus* (CMV), *Tobacco mosaic virus* (TMV), and *Tobacco rattle virus* (TRV). In all cases, viral infection resulted in significantly improved tolerance to drought. This result can be understood in terms of the viral infection inducing a reaction that may be part of an elaborate mechanism used by plants to survive under various abiotic stresses. A subsequent metabolomics study revealed an increase in the levels of several osmoprotectants and antioxidants in BMV-infected rice and CMV-infected beet plants before and after drought stress indicating that the virus elicited a molecular and biochemical modulation of the plant host that made the host more tolerant to abiotic stresses including drought.

15.3 Molecular Mechanisms Involved in Drought Tolerance Induced by Microbes

15.3.1 General Mechanisms of Drought Tolerance in Plants

The ability of a plant to respond to and survive drought stress depends on whole-plant mechanisms that integrate cellular responses. Plant drought tolerance results from key changes to its "omics" architectures. Plants perceive water deficiency by a signal transduction pathway. One of the major signals involved in the recognition of drought stress is the plant hormone, ABA. The perception of drought stress triggers the production of ABA, which in turn regulates downstream signal transduction pathways involved in drought tolerance in plants (Shinozaki and Yamaguchi-Schinozaki 1997) The downstream physiological changes involved in

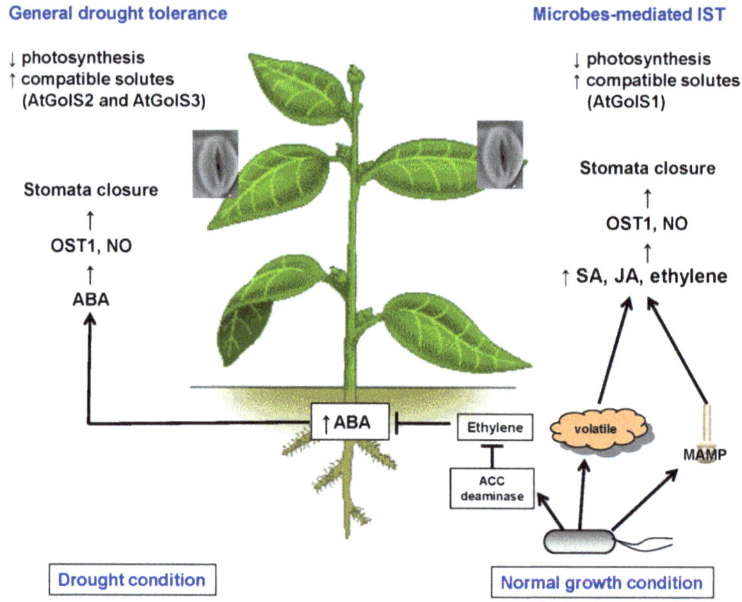

Fig. 15.1 Proposed mechanisms involved in drought tolerance under drought stress (*left*) and involved in rhizosphere-dependent mechanisms underlying induced tolerance to drought stress involving a microbe-mediated priming effect under normal growth condition (*right*). Drought stress triggers the production of ABA, which in turn induces the expression of various genes involved in a signaling cascade, including generation of NO and Ost1 activation, for the regulation of downstream biochemical protective mechanisms, such as stomatal closure and accumulation of compatible solutes. These physiological and biochemical changes reduce foliar transpiration to prevent water loss, and are a primary defense response of the plant against dehydration, but also reduce photosynthesis in leaves. In microbe-mediated IST, the antagonistic effect of ethylene to the ABA effect in drought stress may be diminished by ACC deaminase-producing bacteria. Root colonization by microbes, direct application of volatile compounds, or MAMP in microbes activate plant defense-related plant hormones, such as SA, JA, ethylene, and trigger ABA-independent stomatal closure through an NO and OST1 signal cascade under normal growth conditions. In microbe-mediated IST, compatible solutes, such as galactinol, are accumulated but different signals or genes are involved in general mechanisms. While tolerance responses in drought stress are strong and last long, microbe-mediated IST is a weak and temporally priming state under normal conditions

drought tolerance include stomatal closure, decreased photosynthesis, and accumulations of various compatible solutes. Because many reviews cover the details of the role of ABA and the resulting cellular and molecular events underlying ABA-induced stomatal closure (Shinozaki and Yamaguchi-Schinozaki 1997; Lee and Luan 2012), in this chapter we focus on differences in mechanisms involving general drought tolerance and microbial-dependent induced drought tolerance (Fig. 15.1).

Recent advanced "omic" approaches have identified the metabolic networks at the transcript and protein levels that are linked to the perception of drought stress and the induced responses. In *Arabidopsis*, transcriptome analysis has identified numerous genes responding to drought stress; many have been termed responsive to dehydration (*rd*) and early response to dehydration (*erd*) genes (Shinozaki and Yamaguchi-Shinozaki 1996). At least four regulatory systems for drought tolerance, both ABA-dependent and ABA-independent, are posited (Shinozaki and Yamaguchi-Shinozaki 2000; Nakashima et al. 2009). ABA accumulates in the roots of a drought-stressed plant and moves upward to leaves, where stomatal closure is induced. Mutants such as ABA biosynthesis mutants (*aba* mutants) and some of the ABA response mutants (*abi* mutants) impaired in the regulation of stomatal aperture regulation are very susceptible to water stress (Leung and Giraudat 1998; Schroeder et al. 2001).

ABA signaling is interlinked with other pathways involving nitric oxide (NO), ethylene, salicylic acid, jasmonic acid (JA) to act either antagonistically or synergistically (Manthe et al. 1992; Leung and Giraudat 1998; Evans 2003; Anderson et al. 2004; Suhita et al. 2004; Desikan et al. 2006; Mosher et al. 2010). The triggering of NO generation by ABA is an essential step because scavenging the NO inhibits stomatal closure (Neill et al. 2008). A single exogenous application of ABA, ethylene, methyl jasmonic acid (MeJA), or NO induces stomatal closure. In contrast, stomatal closure is significantly decreased in plants treated with either ABA or NO when mixed with ethylene (Desikan et al. 2006; Munemasa et al. 2007). Generally ABA negatively regulates SA, JA, and ethylene-dependent pathogen resistance (Anderson et al. 2004; Mosher et al. 2010).

15.3.1.1 Stomatal Closure and Compatible Solutes as Key Physiological Modifications in Drought Tolerance

The turgor pressure of guard cells is regulated by water flow based upon the concentration of solutes, including K^+, Cl^-, and malate (Lee and Luan 2012). The ion channels involved in these processes respond differentially to ABA, reactive oxygen species, such as hydrogen peroxide, and NO and calcium ions. This research area has been reviewed recently by Lee and Luan (2012).

During drought stress, plants attempt to maintain their vital turgor through osmotic adjustment by accumulation of compatible solutes (Bartels and Sunkar 2005). The osmolytes include sugars, such as sucrose, trehalose, and raffinose family oligomers (RFO); amino acids such as proline; and amines such as glycine betaine and polyamines (Hoekstra and Buitink 2001). The synthesized compatible solutes substitute for water at the surface of proteins or membranes limiting denaturation (Shinozaki and Yamaguchi-Schinozaki 1999). Overexpression of the compatible solutes in transgenic plants increases drought tolerance (Chen and Murata 2002). Galactinol is a key player in *A. thaliana* where overexpression of drought-inducible galactinol synthase genes (AtGolS2 or AtGolS3) produced plants with enhanced tolerance to drought stress (Taji et al. 2002).

Targets that need protection are proteins involved in the oxygen-evolving photosystem II or those that are embedded in membranes, and the regulators important in changes in gene expression (Koch 1996) under drought stress.

Reduction of photosynthetic activity is one of the biochemical and physiological responses to drought stress due to several factors such as stomatal closure and reduction of photosynthetic enzyme efficacy (Tataeizadeh 1998; Giardi et al. 1996). Photosynthetic system II and its reaction center are very sensitive to the oxidative damage accompanying drought stress (Toivonen and Vidaver 1988).

In summary, the level of ABA has a central role in triggering adaptive responses essential for plant survival under drought stress (Zhu 2001). ABA induces stomatal closure to minimize water loss through transpiration (Leung and Giraudat 1998). Production of activated oxygen species, alkalization, and modified Ca^{2+} and K^{+} fluxes are cited in the guard cells (Mori et al. 2001). The synthesis and accumulation of compatible solutes balance the osmotica and protect sensitive proteins during drought stress (Tataeizadeh 1998).

15.3.2 Specific Mechanisms in Drought Tolerance by Microbes

Stomatal closure is elicited not only by water stress but also in response to plant–microbe interactions (Frommel et al. 1991). When the microbe is not pathogenic, IST to drought has been observed. This process is similar to the ISR against pathogenic challenge. IST for drought stress in plants, is stimulated by several different types of microbes, including the Gram-positive bacterium, *P. polymyxa* (Timmusk and Wagner 1999), and a volatile-producing *Bacillus subtilis* strain (Ryu et al. 2004; Zhang et al. 2010). Gram-negative bacteria that produce ACC deaminase (Mayak et al. 2004a), and an isolate of *Pseudomonas chlororaphis* strain O6 (Cho et al. 2008) known to induce systemic resistance to pathogens, also induce tolerance to drought stress. Additionally, an endophytic fungus, an isolate of *Trichoderma* (Bae et al. 2009), promotes drought stress tolerance. The mechanisms involved in such microbe-mediated IST against drought have not been fully elucidated, although different parts of the overall process have been analyzed.

One mechanisms involved in microbe-mediated IST against drought stress is associated with altered plant ethylene levels. The root colonizing-bacterium, *A. piechaudii* ARV8, lowers the production of ethylene in the root because it deaminates the root-produced precursor of ethylene ACC (Mayak et al. 2004b). The lowering of ethylene levels could be correlated with stomatal closure because antagonism between ethylene and ABA is reduced.

Other triggers inducing stomatal closure are part of the innate immunity response in plants against plant pathogens. The flagellar peptide, flg22, and the lipopolysaccharide (LPS) component of the outer membranes of Gram-negative cells induce stomatal closure when applied to leaves (Melloto et al. 2006). Consequently, they are part of the array of compounds termed bacterial microbe-associated molecular patterns (MAMPs) that are associated with elicitation of fundamental resistance mechanisms in a host (Melloto et al. 2006; Underwood et al. 2007).

Volatiles, such as butanediol, are a third class of compounds inducing stomatal closure by nonpathogenic root colonizing microbes. IST to drought caused by root colonization with *B. subtilis* GB03 (Zhang et al. 2010) and *P. chlororaphis* O6 (Cho et al. 2008) is reproduced by application of this volatile. Both of these bacteria promote plant growth (Ryu et al. 2003) and *P. chlororaphis* O6 has strong biological control activity (Cho et al. 2008). These findings indicate a high degree of interplay between tolerance to drought stress, plant growth, and ISR to pathogens.

Exploration of the signaling pathways required for IST-induction by *P. chlororaphis* O6 indicates that SA-, JA-, and ethylene-pathways, instead of ABA accumulation, is involved (Cho et al. 2008). This observation agrees with the finding that MAMPS activated stomatal closure by local application of MAMPs also requires SA signaling (Melloto et al. 2006). The relationship between signaling by microbial factors and tolerance to drought and salinity (Bezrukova et al. 2001) awaits clarification as does the precise role of ABA. The kinase essential for ABA-induced stomatal closure (OST1 kinase; Mustilli et al. 2002), is required for stomatal closure stimulated by *P. chlororaphis* O6 and MAMPs.

One approach to resolving the signal pathways required for IST is an examination of genes that are activated within the plant. Crossover with the genes activated during water withholding is evinced by expression from ERD15 (Early response to dehydration) in plants colonized by *P. polymyxa* and RD21A (response to dehydration) by *P. chlororaphis* O6. The *P. chlororpahis* O6-protected plants also show expression of PDF1.2 and VSP1 that are marker genes for the JA-regulated pathway (Cho et al. 2011).

Genes involved in osmolyte synthesis are also activated. Genes required for choline biosynthesis, a precursor of the osmolyte glycine betaine, are activated by the IST-inducing *B. subtilis* GB03, and polyamine accumulation occurs in IST accompanying endophytic colonization by *Trichoderma* (Bae et al. 2008). Priming of gene expression to enhance expression of genes associated with stress tolerance is documented. Priming is thought to be a key response enabling a faster reaction time of the plant during its response to pathogen challenge. Kim et al. (2008) find AtGolS1 involved in galactinol synthesis to be primed by colonization with *P. chlororaphis* O6. Overexpression of AtGolS1 showed ISR against plant pathogens and IST against drought tolerance (Cho et al. 2010; Kim et al. 2008), again showing the value of a common defense measure for both pathogens and abiotic stress.

Interestingly, some genes involved in chlorophyll biosynthesis and photosynthesis are downregulated upon root colonization with *P. chlororaphis* O6 (Cho et al. 2011). Are these photosynthetically related changes part of the same regulatory scheme that affects photosynthesis during drought stress? How do these changes correlate with the observed growth seen with *P. chlororaphis* O6 and colonization with other rhizobacteria?

The induced closure of stomatas is not only involved in limiting water loss from the plant but also in restricting invasion, especially pathogenic bacteria. However, pathogenic bacteria thwart this barrier to penetration by secretion of additional effectors to relax the guard cell aperture (Melloto et al. 2006). As with the induction of plant defenses and with plant tolerance to water stress, interplay between plant growth regulators condition the plant responses. One such effector from the pathogen *Pseudomonas syringae* DC3000, the toxin corantine (COR), mimics the isoleucine-JA complex eliminating stomatal closure by mechanisms involving SA and ABA (Melloto et al. 2006). Other pseudomonad pathogenic strains produce different effectors that influence stomatal opening; a type III secretion system is responsible for delivery of such regulators (Goel et al. 2008). One effector termed HopAM1 enhances the rate of ABA-dependent stomatal closure in water-stressed plants and impairs defense responses (Goel et al. 2008). Another effector, a nonribosomally synthesized peptide, syringolinA, eliminates stomatal closure; here there is involvement with the proteasome as a regulatory system (Additionally, a plant resistance factor, RIN4, regulates stomatal closure) and effectors from pathogens disrupt this mechanism. Additionally, a plant resistance factor RIN4 regulates stomatal closure and effectors from pathogens disrupt this mechanism (Liu et al. 2009).

15.4 Mitigation of Drought by Plant Growth-Promoting Bacteria: Greenhouse and Field Applications

In recent years, scientists have gained considerable knowledge regarding the use of PGPB to alleviate a variety of abiotic and biotic stresses of ecologically important plants. One of the newest approaches is the employment of PGPB alone, as part of a microbial consortium or in combination with AM fungi to combat some of the negative effects of drought (Linderman 1988; Barea et al. 2002; Saleem et al. 2007; Yang et al. 2009). In this regard, the implementation of laboratory-based knowledge at the greenhouse level, and later in agricultural practice in the field, is still in an embryonic stage (Bashan 1998; Bashan and de-Bashan 2010a).

15.4.1 Use of AM Fungi and PGPB in Drought Affected Agriculture

15.4.1.1 Greenhouse Studies

PGPB Alone

A number of studies have demonstrated that the effects of drought stress may be mitigated to a significant extent by application of various PGPB including: *Azospirillum, Bacillus, Pseudomonas, Achromobacter*, and rhizobia. Some examples of this plant-protective behavior are presented below.

In the greenhouse, inoculation of maize (*Zea mays*) seedlings with *Azospirillum brasilense* was found to mitigate many of the negative effects of drought stress (Casanovas et al. 2002). In these experiments, inoculation with the PGPB, under a 75 % water supply reduction, improved the relative water content, and prevented a significant drop of water potential of leaves. Inoculation also increased root growth, total foliage biomass, and foliar area, and promoted proline accumulation in both leaves and roots (Casanovas et al. 2002). However, different strains of *Azospirillum* may behave differently and thereby affect the growth of drought-stressed plant differently. For example, in one report, three separate strains of *A. lipoferum* were used to inoculate wheat seedlings (*Triticum aestivum*) exposed to three different drought levels for five different time periods over a 25-day-period. While the stress intensity significantly affected the bacterial population and wheat growth, stress longevity only affected wheat water potential and water content. Interestingly, compared to uninoculated treatments, one strain (that fixed N, produced the highest level of auxin, and had phosphate-solubilizing and ACC-deaminase activities) significantly increased wheat yields where another strain (that produced siderophores) was the most resistant to the drought stress (Arzanesh et al. 2011).

Another recent study evaluated the effect of a strain of *Pseudomonas fluorescens* on the growth of the medicinal plant *Catharanthus roseus* and its production of the alkaloid ajmalicine under drought stress. In this study, it was observed that drought decreased plant growth and also increased the plant's content of the alkaloid. Inoculation of the plant with *P. fluorescens* under drought stress enhanced growth, partially ameliorating the drought induced growth inhibition, and significantly increasing the plant fresh and dry weights. Surprisingly, the alkaloid content of drought-stressed plants was significantly increased when the plants were inoculated with the bacterium (Jaleel et al. 2007). Similarly, one pot culture study revealed that green gram plants (also called mung bean; *Vigna radiata*) had greater resistance to water stress, compared to untreated plants, when they were treated with *P. fluorescens* (Saravanakumar et al. 2011). In another study, three indigenous bacterial strains (*P. putida*, *P.* sp., and *Bacillus megaterium*), isolated from soil that had been affected by drought were able to stimulate plant growth under dry conditions. These bacteria seem to have developed mechanisms to sustain drought stress because all strains exhibited osmotic tolerance. Their inoculation under drought conditions increased shoot and root biomass and water content. Bacterial IAA production under stressed conditions may explain their effectiveness in promoting plant growth and shoot water content, both of which increase plant drought tolerance. When these strains were tested in the field, *B. megaterium* was the most effective species under drought conditions in successive harvests (Marulanda et al. 2009).

A PGPB, *A. piechaudii,* isolated from an arid desert environment and having ACC deaminase activity as a major PGP trait, significantly increased the fresh and dry weights of tomato (*Lycopersicom esculentum*) and pepper (*Capsicum annuum*) seedlings exposed to transient water stress. Following water stress, this bacterium reduced the production of ethylene by tomato seedlings and significantly improved the recovery of plants when watering was resumed. Consequently, inoculation of tomato plants with this strain of *A. piechaudii* yielded a continuous pattern of plant growth during transient periods of drought and irrigation (Mayak et al. 2004b).

Investigations have been carried out to explore the effects of different degrees of water stress in cowpea (*Vigna unguiculata)* in the presence and absence of added *Bradyrhizobium* spp., and to evaluate the physiological responses of cowpea to stress. When the soil matric potential was lowered, this had a negative effect on N_2-fixation. Under conditions of severe drought, inoculated plants formed associations of greater symbiotic efficiency which helped the cowpea plants to withstand the drought stress better than the uninoculated plants (Figueiredo et al. 1999). In addition, the added strains of *Bradyrhizobium* were superior to the indigenous strains, emphasizing the importance of inoculation in stressed plants. Additionally, a greenhouse experiment showed the positive mitigation potential of *R. tropici* when it was jointly inoculated with each of two *P. polymyxa* strains, on growth, nitrogen content, phytohormone levels, and nodulation of common bean (*Phaseolus vulgaris*) under three levels of drought. Joint inoculation yielded increased plant growth, nitrogen content, and nodulation compared to inoculation with *Rhizobium* alone. This was especially evident at the severest levels of drought (Figueiredo et al. 2008).

AM Fungi-Alone

Efficient use of water and enhanced drought resistance are the major goals for increasing plant productivity in drought-prone environments. Achieving these goals may be facilitated through the use of AM fungi. Several factors are involved in the mycorrhizal increase of plant tolerance to drought. These include the volume and the new-formed architecture of the root system, both of which positively affected the efficiency of water and nutrient uptake. Little is known about the physiological basis for differences in the influence of various AM fungi in plant adaptation to drought and only limited information is available regarding greenhouse and field conditions. Moreover, all approaches to study these interactions have been exclusively focused on the genus *Glomus.*

The AM fungus *Glomus fasciculatum* can enhance the growth of nonirrigated drought-stressed soybean (*Glycine max*) plants by improving their drought tolerance. In these experiments, in plants harvested immediately after water stress there was a significant reduction in both mycorrhizal infection and soybean nodulation. The root-to-shoot ratio of stressed plants inoculated with AM fungi decreased by 24 % compared with non-inoculated plants, suggesting enhanced host drought tolerance by *G. fasciculatum.* Moreover, the total seed weight of stressed plants increased 10 % as a result of reduced pod abortion when the plants were infected with AM fungi. Overall, the enhanced yield of AM fungi infected plants may have been the result of more extensive soil water extraction (Busse and Ellis 1985).

PGPB and AM Fungi

A number of greenhouse studies have suggested that selected free-living bacteria, including many PGPB, should be co-inoculated with AM fungi to optimize the formation and functioning of the AM symbiosis in both normal and adverse/stressful environments.

Monitoring the development and physiology of the interaction between lettuce (*Lactuca sativa*) and the AM fungi *G. mosseae* and *G. intraradices,* after induced drought stress, indicated that in AM-associated plants, inoculation with *Bacillus* sp. resulted in a greater enhancement of fungal, as compared to plant, development and metabolism. Under non-stressed conditions, dual inoculation with *G. intraradices* and *Bacillus* sp. increased all plant physiological parameters to a greater extent than inoculation of the plant with only the AM fungus. Under stress conditions, the bacterium had an important stimulatory effect on *G. intraradices* development. Under these conditions, the effects of the dual inoculation on the plant's photosynthetic rate, water use efficiency, and stomatal conductance differed depending upon the particular fungal species employed. Drought stress was especially detrimental to the plant in the presence of *G. intraradices* without the bacterium, reducing intra and extraradical mycelium growth and vitality, and plant-gas exchange (Vivas et al. 2003). On the other hand, *Bacillus* sp. inoculation improved all measured plant and fungal parameters to the same level as in non-stressed plants. Moreover, the highest level of living and active AM mycelium for both fungi was obtained after co-inoculation with *Bacillus* sp. (Vivas et al. 2003). When the effects of different species of the *Glomus* fungus on lettuce growth were examined following progressive drought stress periods, plant growth responses were greatly influenced by the specific *Glomus* species colonizing the roots with a highly variable response among strains. This result suggests that the ability of different fungal endophytes to protect the host plant against progressive drought stress is not linked to either a specific physiological mechanism or to the colonizing ability of individual AM fungal strains (Ruiz-Lozano et al. 1995).

The effects of interactions between *B. thuringiensis*, a drought-adapted bacterium, and two isolates of *G. intraradices* on *Retama sphaerocarpa*, a drought-adapted legume, were investigated (Marulanda et al. 2006). This was done to show the effectiveness of this rhizosphere bacterium, either by itself or acting in concert with an AM fungus, in increasing plant water uptake. The fungal strains that were tested included an indigenous drought-tolerant and a nonindigenous drought-sensitive strain. In 30 days, *Retama* plants that were colonized by dual inoculation with *G. intraradices* and *B. thuringiensis* attained a shoot length that was similar to the shoot length attained by non-inoculated *Retama* plants after 150 days. While inoculation of these plants with this bacterium increased root growth by 200 %, maximum root development was obtained by joint inoculation with *B. thuringiensis* and the indigenous *G. intraradices.* The relative water uptake was always higher in inoculated than in non-inoculated plants. *G. intraradices*-colonized plants attained a similar shoot length irrespective of the origin of the fungus despite the large differences in relative water uptake by plants colonized by different fungi. Indigenous *G. intraradices*-colonized roots showed the highest intensity and arbuscule richness when associated with *B. thuringiensis*. The interactive microbial effects on *Retama* plants were more pronounced when indigenous microorganisms were involved. Dual inoculation of autochthonous microorganisms resulted a 42 % reduction in the amount of water needed to produce 1 mg of shoot biomass (Marulanda et al. 2006).

15.4.1.2 Field Studies

A very limited amount of scientific information is available concerning the effects of PGPB and AM fungi on agricultural crops in the field. This may be the case as a consequence of the fact that the majority of these experiments have been performed by commercial enterprises, and the results of these experiments have never been disseminated except in commercial brochures intended for farmers (Penna C., Gonzalez, G., personal communication). For example, the limited information provided on labels of some commercial PGPB products states that wheat and maize seedlings inoculated with strains of *Azospirillum* and later exposed to water stress will be better off than non-inoculated plants.

The effect on dryland *sorghum* (*Sorghum bicolor)* growth, water status and yield (without irrigation) in semi-arid fields in Israel with limited soil moisture, inoculated with *A. brasilense* was examined in five field experiments. In this case, inoculation led to an average increase of 20 % in total straw dry-matter yield, as a result of higher rates of dry-matter accumulation through improved water utilization during the early stages of plant growth. *Azospirillum* inoculation caused a 15–18 % increase in grain yield in all experiments (Sarig et al. 1984, 1988). This increase was associated with a greater number of seeds per panicle and enhanced mineral content (N, P, and K). The water status of *sorghum* plants was also improved by inoculation, as seen in their higher leaf water potential, lower canopy temperatures, and greater stomatal conductance and transpiration. Total extraction of soil moisture by inoculated plants was greater (by about 15 %) and occurred from deeper soil layers, compared with non-inoculated controls (Sarig et al. 1984, 1988). In other experiments, in Argentina, *A. brasilense*-inoculated wheat seeds were sown in nonirrigated and control plots. Significantly higher water content, relative water content, water potential, apoplastic water fraction, and lower cell wall modulus of elasticity values were obtained in *Azospirillum*-inoculated plants following drought conditions. Grain yield loss due to drought was 26 %; this was reduced to 14 % in *Azospirillum*-inoculated plants. In addition, the amount of Mg and K in grain was diminished in nonirrigated and non-inoculated plots. However, grains harvested from *Azospirillum*-inoculated plants had significantly higher Mg, K, and Ca contents than non-inoculated plants. This study suggested that a superior water status and an additional "elastic adjustment" in *Azospirillum*-inoculated wheat plants could be crucial in promoting higher grain yield and mineral quality at harvest, particularly when drought strikes during anthesis (Creus et al. 2004).

Some of the measures that have been suggested in India for amelioration of the effects of drought in rice include application of nitrogen fertilizer and plant growth regulators. Since PGPB produce plant growth substances in vitro and sometimes in situ as well, the effect of foliar application of the PGPB *Klebsiella* sp. on rice in the field subjected to different periods of drought was examined. Bacterial foliar inoculation and nitrogen fertilization was found to quadruple grain yield of acutely water stressed plants along with an improvement in several growth and yield variables. However, the application of synthetic hormones, such as IAA, kinetin, and GA_3 (applied as a mixture) were less effective than inoculation with a living PGPB in alleviating drought effects. This suggests either the amounts of these added compounds were suboptimal, or that the PGPB provides other benefits in addition to these three hormones. The adverse effects of water stress on respiration and photosynthesis were also alleviated by these treatments. While uptake of K, Mg, Ca, Fe, and Mo was increased, Na content decreased, accompanied by an increase in plant proline content (Razi and Sen 1996).

15.4.2 Use of PGPB and AM: Fungi in Solving Environmental Problems of Drought

By definition, the desert is a habitat where plants suffer from a chronic shortage of water, and most plants grow under drought conditions most of the time, even in undisturbed deserts. The abundance of water in deserts is a rare event. When deserts are disturbed, a process called desertification, where physical, chemical, and biological soil damage can occur simultaneously, there are major disturbances to the plant communities and an increase of drought effects for the surviving plants. Such soil degradation limits reestablishment of the natural plant cover. Specifically, desertification causes a disturbance of the plant–microbe interactions that might otherwise act as a fundamental ecological factor in restoration of a degraded ecosystem. Natural revegetation in deforested deserts is extremely slow (Bashan and de-Bashan 2010b). When combating desertification, reforestation programs are one of the proposed solutions to reduce encroaching deserts. Some projects, such as the tree-belt of northeastern China, are notable for their very large scale (Moore and Russell 1990). Consequently, all experiments to combat drought in the environment with plant-PGPB combination have been performed under desert conditions where water is in short supply.

Only in recent years have PGPB been used for the purposeful restoration of vegetation in abandoned, dryland habitats (Bashan and de-Bashan 2005; Bashan et al. 2012; de-Bashan et al. 2012; Trejo et al. 2012). Although desert plants can survive as seedlings without inoculation with PGPB, maximizing water uptake by various mechanisms and also by shading one another (Bacilio et al. 2011), inoculation with PGPB provides a significant advantage for further growth. To date, artificial inoculation with PGPB has been employed mainly with cacti, and to a lesser extent with several legume trees and shrubs.

15.4.2.1 Greenhouse Studies

PGPB Alone

Cardon cacti have a widespread, finely branched, subsurface root system that stabilizes desert soil. Initially, these cacti were inoculated with strains of the PGPB *A. brasilense* of agricultural origin. Differential responses were observed, as *A. brasilense* Cd significantly reduced and *A. brasilense* Sp245 significantly increased seed germination at the same inoculum concentration. Both strains improved several plant-growth parameters of cardon related to seedling survivability, such as height, diameter, volume, the volume/surface ratio, and persistence of seedling spines. Under drought conditions, both strains survived in the cactus rhizosphere for about 300 days after seed inoculation (Puente and Bashan 1993). The effect of *A. brasilense* on plant growth increased linearly as the level of soil nutrients declined. In nutrient-rich resource island soil, *A. brasilense* had no effect on cardon growth, but in the poor soil from bare areas, shoot dry mass increased by almost 60 % and root length by more than 100 % as a result of inoculation (Carrillo-Garcia et al. 2000). A mechanism proposed to explain how this plant–bacterium association increased growth of seedlings in poor desert soils was largely based on acidification of the rhizosphere of cactus seedlings after inoculation, with a consequent increase in phosphorus availability (Carrillo et al. 2002). Similarly, when used in conjunction with compost, *A. brasilense* Cd had little effect on cardon growth, despite the fact that the bacterium significantly affected plant growth in the absence of any amendment (Bacilio et al. 2006).

Studies using PGPB obtained from the roots of several species of cacti growing in rocks where soil is absent open another venue for desert restoration under drought conditions. When cardon seeds or seeds of the small cactus *Mammillaria fraileana* were inoculated with several rhizoplane or endophytic PGPB strains capable of weathering rocks, the seedlings grew in pulverized rock in the greenhouse for at least a year without fertilization and without showing distress. Cacti that were not inoculated grew less vigorously and often died. Both types of bacteria–plant associations released significant amounts of nutrients from the pulverized rock substrate. When endophytic bacteria were eliminated from the seeds using antibiotics, the development of seedlings stopped. Upon reinoculation with the endophytes, plant growth was restored (Puente et al. 2004a, b, 2009a, b; Lopez et al. 2011, 2012).

The effect of PGPB inoculation on a number of desert plants was demonstrated under greenhouse conditions. Upon inoculation of the root system of mesquite amargo, *Prosopis articulata*, one of the main nurse trees (a plant that supports the growth of seedlings of other plants under its canopy) of the Sonoran Desert in northwestern Mexico and the southwestern USA with *A. brasilense* Cd, the seedlings grew taller, were greener, had longer leaves, and were heavier than uninoculated plants (Leyva and Bashan 2008). Three slow-growing legume trees used for desert reforestation and urban gardening in the Sonoran Desert were evaluated for their response to treatment with PGPB, AM fungi, and common compost under regular screen-house cultivation. Mesquite amargo and the yellow palo verde *Parkinsonia microphylla* showed positive responses to several of the treatments, while the blue palo verde *P. florida* did not respond (Bashan et al. 2009a). Two tree species, the drought-avoiding pine, *Pinus halepensis*, and the drought-tolerant oak, *Quercus coccifera,* were evaluated for water stress responses under moderate or severe water stress of their seedlings after inoculation with the PGPB *P. fluorescens* and a consortia of soil microorganisms. Both inoculation with PGPB and native soil microorganisms improved tree growth, and their interactions had either additive or synergistic effects. The study demonstrated that microbial associates of roots can significantly influence the response of tree seedlings to drought, but the magnitude and nature of this effect appears to depend on the water-use strategy of each tree species (Rincón et al. 2008).

AM Fungi Alone or Combined with PGPB

In Australia, revegetation on the mineral sand mining sites aimed to re-establish the original plant species on the topsoil. Initially, three pot experiments were conducted to evaluate whether the growth of the tree *Acacia concurrens* in the returned topsoil from two mineral sand mines could be facilitated by either the addition of phosphorus or inoculation with AM fungi. In soils from a former mine (i.e. leftover tailings sand), the dry weight of *A. concurrens* shoots increased as a consequence of either P application or inoculation with AM fungi; however, in less adsorptive topsoil, plant dry weight was only increased at low levels of applied P. On the other hand, when *A. concurrens* was grown in a layer of topsoil placed over tailings sand, shoot dry weight increased, in response to inoculation with AM fungi. In topsoil from a different mine, the dry weight of shoots (at low rates of applied P) was increased up to four times by inoculation with AM fungi. The response to inoculation in both experiments was due to increases in the uptake of P by the plants. Species of AM fungi differed in their ability to increase plant growth. However, in soils from both mines, the same fungal species were effective (Jasper et al. 1989). The effect of AM fungi on growth and drought resistance of *A. nilotica* and *Leucaena leucocephala* seedlings was studied. The growth promotion effect of AM fungi was equivalent to the effect of phosphorus fertilization after 12 weeks. Drought treatment reduced seedling biomass and nodulation. Differences between plant species were found in relation to growth improvements due to AM inoculation and/or phosphorus fertilization under drought stress conditions (Michelsen and Rosendahl 1990).

Plant growth characteristics, nutrient absorption, and water relations were compared in non-inoculated and inoculated mimosa trees (*L. leucocephala*) with the AM fungi *G. fasciculatum*. Mycorrhizal plants had substantially greater shoot and root dry weight, leaf area and root length than non-mycorrhizal plants, and also exhibited greater P, K, and Ca uptake. In addition, mycorrhizal plants had a five times greater leaf area and a nearly double rate of leaf water conductance compared to plants without mycorrhizae. The difference between xylem pressure potential and soil water potential were considerably less in mycorrhizal plants than in non-mycorrhizal plants and stomatal responses to the humidity deficits of the air during the day were nearly twice as much in mycorrhizal plants. However, leaflet folding and their orientation responses to avoid direct sunlight during the day were much greater in non-mycorrhizal plants. The latter parameters lead to less absorption of sunlight, and this response may contribute significantly to the reduction in growth rate observed in non-mycorrhizal plants (Huang et al. 1985). Several AM fungi, rhizobia, and PGPB that were isolated from a desertified semi-arid ecosystem in the south–east of Spain were screened for effectiveness on *Anthyllis cytisoides*, a pioneer legume, dominant in this Mediterranean ecosystem. The microbial strains were tested alone and in various combinations in comparison to similar but exotic strains. Overall, native strains were more effective than the exotic strains. Many microbial combinations were effective in improving plant development, nutrient uptake, N_2-fixation, or growth of the root system. The study concluded that despite the difficulty in selecting a multifunctional microbial inoculum, appropriate microbial combinations can be recommended for improvement of plant performance in arid area (Requena et al. 1997).

15.4.2.2 Field Studies

To date, there have been only a few attempts to move beyond greenhouse studies to test PGPB, AM fungi, or a combination of both types of organisms in the field. In Mexico, survival and development of cactus transplants in urban, disturbed areas in the State of Baja California Sur was monitored. Young plants of three species of pachycereid cacti (*Pachycereus pringlei, Stenocereus thurberi,* and *Lophocereus schottii*) inoculated with the PGPB *A. brasilense* had a high rate of survival and developed more rapidly, compared to uninoculated control plants over a 3.5-year period. Soil erosion in the inoculated experimental area diminished. A small, but significant amount of soil accumulated in association with the growth of cactus roots in the form of wind-deposited dust. One demonstrated mechanism for stabilizing the dust was the upward growth of small roots during the rainy season. *A. brasilense* survived well in the rhizosphere of these cacti for 2 years, but not in root-free soil (Bashan et al. 1999). To evaluate the feasibility of long-term desert reforestation technology using mixed vegetation, cardon cactus seedlings from indoor and outdoor nurseries were planted in the field adjacent to one seedling of various potential legume nurse trees: mesquite amargo, yellow palo verde, and blue palo verde. Additionally, the combinations of legume tree and cactus were inoculated with either a consortium of desert AM fungi, PGPB (the diazotroph *A. brasilense* Cd, and the phosphate solubilizer *P. chitinolyticus*), or a mixture of all of these. Some of the planting holes were also supplemented with common dairy cow compost (Bacilio et al. 2006). The field trials were monitored over 5 years for survival and growth. Cardon cactus that had been reared in an outdoor screen-house survived better in the field than cardon that was reared in a controlled growth chamber and later hardened outdoors. Association with any legume nurse tree increased survival and enhanced growth of untreated cardon. For cardon growing alone, application of compost, AM fungi, or all the treatments combined increased survival. Detailed assessment, after 30 months of cultivation showed that all microbial treatments positively affected cardon growth when growing alone or in combination with mesquite (Bashan et al. 2009b). Reforestation of highly eroded desert land in the southern Sonoran Desert, after it had lost its natural capacity for self-revegetation, was attempted with three native legume trees (mesquite amargo, yellow palo verde, and blue palo verde) in seven field trials. Reforestation was aided by inoculation with two species of PGPB, *A. brasilense* and *Bacillus pumilus*, native mycorrhizal fungi, small amounts of compost, and very limited irrigation that maintained the fields under continuous drought stress. Survival of the trees was marginally affected by the various treatments after 30 months and was in the range of 60–90 %, depending on the plant species. With respect to growth, mesquite amargo and yellow palo verde responded positively to inoculation with PGPB, AM fungi, and supplements of compost, whereas blue palo verde did not respond to most treatments. Some combinations of tree/inoculant/amendment resulted in small negative effects or no response when evaluated over extended periods of time of up to 5 years (Bashan et al. 2012). These studies demonstrate that reforestation of severely eroded desert lands is possible with native legume trees aided by microbial agents and compost to increase soil fertility. Introduction of indigenous plants associated with a managed microbial community should prove a successful biotechnological approach to aid recovery of desertified ecosystems and degraded deserts (Bashan and de-Bashan 2010b).

In two long-term experiments in a desertified Mediterranean ecosystem in Spain, inoculation with indigenous AM fungi and rhizobia not only enhanced establishment of key plant species, but also increased soil fertility and quality. The dual symbioses increased soil nitrogen and organic matter content, stabilized soil aggregates, and enhanced nitrogen transfer from nitrogen-fixing to non-fixing species associated within the natural succession (Herrera et al. 1993, Requena et al. 2001). The re-establishment of native shrub species in an abandoned semiarid agricultural Mediterranean basin restored the characteristic biodiversity and prevented the processes of soil erosion and desertification. A field experiment assessed the effectiveness of mycorrhizal inoculation, with a mixture of native AM fungi and an exotic AM fungus (*Glomus claroideum*) on the establishment of several shrubs. One year after planting, the shoot biomass of several inoculated shrubs was up to 630 % greater than that of non-inoculated plants. Inoculation with a mixture of native AM fungi was the most effective treatment for increasing shoot biomass, and the N, P, and K contents of shoot tissues. This study suggested that the use of native mycorrhizae as a potential source of AM inoculum may be a preferential inoculation strategy to guarantee the successful re-establishment of native shrub species in a semiarid degraded soil (Caravaca et al. 2002).

15.5 Conclusion

In recent years, there have been a large number of studies, both in the laboratory and in the field, that have demonstrated that the use of plant-associated microbes is beneficial to plant growth. Entophytic and epiphytic bacteria, fungi, and even virus enhanced tolerance to a wide range of environmental stresses including drought stress. A lot remains to be learned about how different plants, different conditions, and treatment with different beneficial microbes such as PGPB and mycorrhizal fungi may be combined to optimally confer plant tolerance against drought stress referred to as IST. Nevertheless, the fundamental understanding of some of the key mechanisms of IST that has developed up to the present time, makes us optimistic that we are on the verge of an age where the purposeful use of microorganisms will begin to replace the use of energy-intensive and often polluting chemicals in agricultural practice.

References

Abeles FB, Morgan PW, Saltveit ME Jr (1992) Ethylene in plant biology. Academic, New York

Anderson JP, Badruzsufari E, Schnek PM, Manners JM, Desmond OJ, Ehlert C, Maclean DJ, Ebert PR, Kazan K (2004) Antagonistic interaction between abscisic acid and jasmonate-ethylene signaling pathways modulates defense gene expression and disease resistance in *Arabidopsis*. Plant Cell 16:3460–3479

Antoun H, Prévost D (2005) Ecology of plant growth promoting rhizobacteria. In: Siddiqui ZA (ed) PGPR: biocontrol and biofertilization. Springer, Dordrecht, pp 1–38

Aroca R, Porcel R, Ruiz-Lozano JM (2007) How does arbuscular mycorrhizal symbiosis regulate root hydraulic properties and plasma membrane aquaporins in *Phaseolus vulgaris* under drought, cold or salinity stresses? New Phytol 173:808–816

Augé RM (2001) Water relations, drought and vesicular-arbuscular mycorrhizal symbiosis. Mycorrhiza 11:3–42

Arzanesh MH, Alikhani HA, Khavazi K, Rahimian HA, Miransari M (2011) Wheat (*Triticum aestivum* L.) growth enhancement by *Azospirillum* sp. under drought stress. World J Microbiol Biotechnol 27:197–205

Bacilio M, Hernandez J-P, Bashan Y (2006) Restoration of giant cardon cacti in barren desert soil amended with common compost and inoculated with *Azospirillum brasilense*. Biol Fertil Soils 43:112–119

Bacilio M, Vazquez P, Bashan Y (2011) Water versus spacing: a possible growth preference among young individuals of the giant cardon cactus of the Baja California Peninsula. Environ Exp Bot 70:29–36

Bae H, Kim S-H, Kim MS, Sicher RC, Lary D, Strem MD, Natarajan S, Bailey BA (2008) The drought response of *Theobroma cacao* (cacao) and the regulation of genes involved in polyamine biosynthesis by drought and other stresses. Plant Physiol Biochem 46:174–188

Bae H, Sicher RC, Kim MS, Kim SH, Strem MD, Melnick RL (2009) The beneficial endophyte *Trichoderma hamatum* isolate DIS 219b promotes growth and delays the onset of the drought response in *Theobroma cacao*. J Exp Bot 60:3279–3295

Barea J-M, Azcon R, Azcon-Aguilar C (2002) Mycorrhizosphere interactions to improve plant fitness and soil quality. Anton Leeuw Int J G 81:343–351

Bartels D, Sunkar R (2005) Drought and salt tolerance in plants. Crit Rev Plant Sci 24:1–36

Bashan Y (1998) Inoculants of plant growth-promoting bacteria for use in agriculture. Biotechnol Adv 16:729–770

Bashan Y, de-Bashan LE (2005) Bacteria/plant growth-promotion. In: Hillel D (ed) Encyclopedia of soils in the environment, vol 1. Elsevier, Oxford, pp 103–115

Bashan Y, de-Bashan LE (2010a) How the plant growth-promoting bacterium *Azospirillum* promotes plant growth-a critical assessment. Adv Agron 108:77–136

Bashan Y, Levanony H (1990) Current status of *Azospirillum* inoculation technology: *Azospirillum* as a challenge for agriculture. Can J Microbiol 36:591–608

Bashan Y, Rojas A, Puente ME (1999) Improved establishment and development of three cacti species inoculated with *Azospirillum brasilense* transplanted into disturbed urban desert soil. Can J Microbiol 45:441–451

Bashan Y, Salazar B, Puente ME (2009a) Responses of native legume desert trees used for reforestation in the Sonoran desert to plant growth-promoting microorganisms in screen house. Biol Fertil Soils 45:655–662

Bashan Y, Salazar B, Puente ME, Bacilio M, Linderman RG (2009b) Enhanced establishment and growth of giant cardon cactus in an eroded field in the Sonoran desert using native legume trees as nurse plants aided by plant growth-promoting microorganisms and compost. Biol Fertil Soils 45:585–594

Bashan Y, Salazar BG, Moreno M, Lopez BR, Linderman RG (2012) Restoration of eroded soil in the Sonoran desert with native leguminous trees using plant growth-promoting microorganisms and limited amounts of compost and water. J Environ Manag 102: 26–36. doi:10.1016/j.jenvman.2011.12.032

Bashan Y, de-Bashan LE (2010b) Chapter 6: Microbial populations of arid lands and their potential for restoration of deserts. In: Dion P (ed) Soil biology and agriculture in the tropics. Soil biology series 21. Springer, Berlin, pp 109–137

Bashan Y, Holguin G (1998) Proposal for the division of plant growth-promoting rhizobacteria into two classifications: biocontrol-PGPB (plant growth-promoting bacteria) and PGPB. Soil Biol Biochem 30:1225–1228

Berta G, Fusconi A, Trotta A (1993) VA mycorrhizal infection and the morphology and function of root systems. Environ Exp Bot 33:159–173

Belesky DP, Fedders JM (1995) Tall fescue development in response to *Acremonium coenophialum* and soil acidity. Crop Sci 35:529–553

Bezrukova MV, Kakhabutdinova R, Fakhutinova RA, Kyldiarova I, Shakirova F (2001) The role of hormonal changes in protective action of salicylic acid on growth of wheat seedlings under water deficit. Agrochemiya 2:51–54

Bianco C, Defez R (2009) *Medicago truncatula* improves salt tolerance when nodulated by an indole-3-acetic acid-overproducing *Sinorhizobium meliloti* strain. J Exp Bot 60:3097–3107

Bnayahu BY (1991) Root excretions and their environmental effects: influence on availability of phosphorus. In: Waisel Y, Eshel A, Kafkafi U (eds) Plant roots: the hidden half. Marcel Decker, New York, pp 529–557

Brown ME (1974) Seed and root bacterization. Ann Rev Phytopathol 12:181–197

Busse MD, Ellis JR (1985) Vesicular-arbuscular mycorrhizal (*Glomus fasciculatum*) influence on soybean drought tolerance in high phosphorus soil. Can J Bot 63:2290–2294

Caravaca F, Barea JM, Palenzuela J, Figueroa D, Alguacil MM, Roldan A (2002) Establishment of shrub species in a degraded semiarid site after inoculation with native or allochthonous arbuscular mycorrhizal fungi. Appl Soil Ecol 22:103–111

Carrillo-Garcia A, Bashan Y, Diaz-Rivera E, Bethlenfalvay GJ (2000) Effects of resource—island soils, competition, and inoculation with *Azospirillum* on survival and growth of *Pachycereus pringlei,* the giant cactus of the Sonoran desert. Restor Ecol 8:65–73

Carrillo AE, Li CY, Bashan Y (2002) Increased acidification in the rhizosphere of cactus seedlings induced by *Azospirillum brasilense*. Naturwissenschaften 89:428–432

Casanovas EM, Barassi CA, Sueldo RJ (2002) *Azospirillum* inoculation mitigates water stress effects in maize seedlings. Cereal Res Commun 30:343–350

Chen TH, Murata N (2002) Enhancement of tolerance of abiotic stress by metabolic engineering of betaines and other compatible solutes. Curr Opin Biotechnol 5:250–257

Cho SM, Kang BR, Han SH, Anderson AJ, Park J-Y, Lee Y-H, Cho BH, Yang K-Y, Ryu C-M, Kim YC (2008) 2R,3R-butaneidol, a bacterial volatile produced by *Pseudomonas chlororaphis* O6, is involved in induction of systemic tolerance to drought in *Arabidopsis thaliana*. Mol Plant-Microbe Interact 21:1067–1075

Cho SM, Kim SH, Kim YC, Yang KY, Kim KS, Choi YS, Cho BH (2010) Galactinol is involved in induced systemic resistance against bacterial infection and environmental stresses. Korean J Plant Res 23:248–255

Cho SM, Park JY, Han SH, Anderson AJ, Yang KY, Gardener BM, Kim YC (2011) Identification and transcriptional analysis of priming genes in *Arabidopsis thaliana* induced by root colonization with *Pseudomonas chlororaphis* O6. Plant Pathol J 27:272–279

Cowan AK, Cairns ALP, Bartels-Rahm B (1999) Regulation of abscisic acid metabolism: towards a metabolic basis for abscisic acid-cytokinin antagonism. J Exp Bot 50:595–603

Creus CM, Sueldo RJ, Barassi CA (2004) Water relations and yield in *Azospirillum* inoculated wheat exposed to drought in the field. Can J Bot 82:273–281

Danneberg G, Latus C, Zimmer W, Hundeshagen B, Schneider-Poetsch HJ, Bothe H (1992) Influence of vesicular-arbuscular mycorrhiza on phytohormone balances in maize (*Zea mays* L.). J Plant Physiol 141:33–39

de-Bashan LE, Hernandez J-P, Bashan Y (2012) The potential contribution of plant growth-promoting bacteria to reduce environmental degradation-a comprehensive evaluation. Appl Soil Ecol. doi:10.1016/j.apsoil.2011.09.003

Desikan R, Last K, Harret-Williams R, Tagliavia C, Harter K, Hooley R, Hancock JT, Neil SJ (2006) Ethylene-induced stomatal closure in *Arabidopsis* occurs via atrbohF-mediated hydrogen peroxide synthesis. Plant J 47:907–917

Egamberdieva D (2009) Alleviation of salt stress by plant growth regulators and IAA producing bacteria in wheat. Acta Physiol Plant 31:861–864

Evans NH (2003) Modulation of guard cell plasma membrane potassium currents by methyl jasmonate. Plant Physiol 131:8–11

Figueiredo MVB, Burity HA, Martínez CR, Chanway CP (2008) Alleviation of drought stress in the common bean (*Phaseolus vulgaris* L.) by co-inoculation with *Paenibacillus polymyxa* and *Rhizobium tropici*. Appl Soil Ecol 40:182–188

Figueiredo MVB, Vilar JJ, Burity HA, de Franca FP (1999) Alleviation of water stress effects in cowpea by *Bradyrhizobium* spp. inoculation. Plant Soil 207:67–75

Frommel M, Nowak J, Lazarovits G (1991) Growth enhancement and development modifications of in vitro grown potato (*Solanum tuberosum* ssp. *tuberosum*) as affected by a nonfluorescent *Pseudomonas* sp. Plant Physiol 96:928–936

Gamalero E, Berta G, Glick BR (2009) The use of microorganisms to facilitate the growth of plants in saline soils. In: Khan MS, Zaidi A, Musarrat J (eds) Microbial strategies for crop improvement. Springer, Berlin, pp 1–22

Gamalero E, Berta G, Massa N, Glick BR, Lingua G (2010) Interactions between *Pseudomonas putida* UW4 and *Gigaspora rosea* BEG9 and their consequences on the growth of cucumber under salt stress conditions. J Appl Microbiol 108:236–245

Giardi MT, Cona A, Geiken D, Kucera T, Masojidek J, Mattao AK (1996) Long-term drought stress induced structural and functional reorganization of photosystem II. Planta 199:118–125

Glick BR (1995) The enhancement of plant growth by free-living bacteria. Can J Microbiol 41:109–117

Glick BR (2004) Bacterial ACC deaminase and the alleviation of plant stress. Adv Appl Microbiol 56:291–312

Glick BR (2010) Using soil bacteria to facilitate phytoremediation. Biotechnol Adv 28:367–374

Glick BR, Biljana T, Czarny J, Cheng Z, Duan J, McConkey B (2007a) Promotion of plant growth by bacterial ACC deaminase. Crit Rev Plant Sci 26:227–242

Glick BR, Cheng Z, Czarny J, Duan J (2007b) Promotion of plant growth by ACC deaminase-containing soil bacteria. Eur J Plant Pathol 119:329–339

Glick BR, Penrose DM, Li J (1998) A model for the lowering of plant ethylene concentrations by plant growth promoting bacteria. J Theor Biol 190:63–68

Goel AK, Lundberg D, Torres MA, Mattews R, Akimoto-Tomiyama C, Farmer L, Dangl JL, Grant SR (2008) The *Pseudomonas syringae* type III effector HopAM1 enhances virulence on water-stressed plants. Mol Plant Microbe Interact 21:361–370

Goldstein AH (1994) Involvement of the quinoprotein glucose dehydrogenase in the solubilization of exogenous phosphates by gram-negative bacteria. In: Torriani-Gorini A, Yagil E, Silver S (eds) Phosphate in microorganisms: cellular and molecular biology. ASM Press, Washington, pp 197–203

Herrera MA, Salamanca CP, Barea JM (1993) Inoculation of woody legumes with selected arbuscular mycorrhizal fungi and rhizobia to recover desertified Mediterranean ecosystems. Appl Environ Microbiol 59:129–133

Hoekstra FA, Buitink J (2001) Mechanisms of plant desiccation tolerance. Trends Plant Sci 8:431–438

Huang R-S, Smith WK, Yost RS (1985) Influence of vesicular-arbuscular mycorrhiza on growth, water relations, and leaf orientation in *Leucaena leucocephala* (Lam.) de Wit. New Phytol 99:229–243

Jaleel CA, Manivannan P, Sankar B, Kishorekumar A, Gopi R, Somasundaram R, Panneerselvam R (2007) *Pseudomonas fluorescens* enhances biomass yield and ajmalicine production in *Catharanthus roseus* under water deficit stress. Colloid Surf B 60:7–11

Jasper DA, Abbott LK, Robson AD (1989) Acacias respond to additions of phosphorus and to inoculation with VA mycorrhizal fungi in soils stockpiled during mineral sand mining. Plant Soil 115:99–108

Kendrick WB (1992) The fifth kingdom, 2nd edn. Focus Texts, Newburyport, pp 262–286

Khalvati MA, Hu Y, Mozafar A, Schmidhalter U (2005) Quantification of water uptake by arbuscular mycorrhizal hyphae and its significance for leaf growth, water relations, and gas exchange of barley subjected to drought stress. Plant Biol 7:706–712

Kim MS, Cho SM, Kang EY, Im YJ, Hwangbo H, Kim YC, Ryu C-M, Yang KY, Chung GC, Cho BH (2008) Galactinol is a signaling component of the induced systemic resistance caused by *Pseudomonas chlororaphis* O6 root colonization. Mol Plant-Microbe Interact 21:1643–1653

Kloepper JW, Ryu CM, Zhang S (2004) Induced systemic resistance and promotion of plant growth by *Bacillus* species. Phytopathol 94:1259–1266

Koch KE (1996) Carbohydrate-modulated gene expression in plants. Annu Rev Plant Physiol Plant Mol Biol 47:509–540

Kohler J, Hernández JA, Caravaca F, Roldán A (2008) Plant-growth-promoting rhizobacteria and arbuscular mycorrhizal fungi modify alleviation biochemical mechanisms in water-stressed plants. Funct Plant Biol 35:141–151

Kpomblekou K, Tabatabai MA (1994) Effect of organic acids on release of phosphorus from phosphate rocks. Soil Sci 158:442–453

Kramer PJ, Boyer JS (1995) Water relations of plants and soils. American Press, San Diego

Latch GCM, Hunt WF, Musgrave DR (1985) Endophytic fungi affect growth of perennial ryegrass. N Zeal J Agric Res 28:165–168

Lee SC, Luan S (2012) ABA signal transduction pathway at the crossroad of biotic and abiotic stress response. Plant Cell Environ 35:53–60

Leung J, Giraudat J (1998) Abscisic acid signal transduction. Annu Rev Plant Physiol Plant Mol Biol 49:199–222

Leyva LA, Bashan Y (2008) Activity of two catabolic enzymes of the phosphogluconate pathway in mesquite roots inoculated with *Azospirillum brasilense* Cd. Plant Physiol Biochem 46:898–904

Linderman RG (1988) Mycorrhizal interactions with the rhizosphere microflora: the mycorrhizosphere effect. Phytopathology 78:366–371

Liu J, Elmore JM, Funglsang AT, Palmgren MG, Staskawicz BJ, Coaker G (2009) RIN4 functions with plasma membrane H+-ATPase to regulate stomatal apertures during pathogen attack. PLoS Biol 7:e1000139

Lopez BR, Bashan Y, Bacilio M (2011) Endophytic bacteria of *Mammillaria fraileana*, an endemic rock-colonizing cactus of the southern Sonoran desert. Arch Microbiol 193:527–541
Lopez BR, Tinoco C, Bacilio M, Mendoza A, Bashan Y (2012) Endophytic bacteria of the rock-dwelling cactus *Mammillaria fraileana* affect plant growth and mobilization of elements from rocks. Environ Exp Bot 81:26–36. doi:10.1016/j.envexpbot.2012.02.014
Lyons PC, Evans JJ, Bacon CW (1990) Effects of the fungal endophyte *Acremonium coenophialum* on nitrogen accumulation and metabolism in tall fescue. Plant Physiol 92:726–732
Ma JF (2005) Plant root responses to three abundant soil minerals: silicon, aluminum and iron. Crit Rev Plant Sci 24:267–281
Manthe B, Schulz M, Schnabl H (1992) Effects of salicylic acid on growth and stomatal movement of *Vicia faha* L; evidence for salicylic acid metabolism. J Chem Ecol 18:1525–1539
Marulanda A, Barea J-M, Azcon R (2009) Stimulation of plant growth and drought tolerance by native microorganisms (AM fungi and bacteria) from dry environments: mechanisms related to bacterial effectiveness. J Plant Growth Regul 28:115–124
Marulanda A, Barea JM, Azcon R (2006) An indigenous drought-tolerant strain of *Glomus intraradices* associated with a native bacterium improves water transport and root development in *Retama sphaerocarpa*. Microb Ecol 52:670–678
Marulanda A, Azcon R, Ruiz-Lozano JM (2003) Contibution of six arbuscular mycorrhizal fungal isolates to water uptake by *Lactuca sativa* plants under drought stress. Physiol Plant 119:526–533
Mayak S, Tirosh T, Glick BR (2004a) Plant growth-promoting bacteria that confer resistance to water stress in tomatoes and peppers. Plant Sci 166:525–530
Mayak S, Tirosh T, Glick BR (2004b) Plant growth-promoting bacteria confer resistance in tomato plants to salt stress. Plant Physiol Biochem 42:565–572
Melloto M, Underwood W, Koczan J, Nomura K, He SH (2006) Plant stomata function in innate immunity against bacterial invasion. Cell 126:969–980
Michelsen A, Rosendahl S (1990) The effect of VA mycorrhizal fungi, phosphorus and drought stress on the growth of *Acacia nilotica* and *Leucaena leucocephala* seedlings. Plant Soil 124:7–13
Mori IC, Pinontoan R, Kawano T, Muto S (2001) Involvement of superoxide generation in salicylic acid-induced stomatal closure in *Vicia faba*. Plant Cell Physiol 42:1383–1388
Moore R, Russell R (1990) The 'Three Norths' forest protection system—China. Agrofor Syst 10:71–88
Mosher S, Moeder W, Nishimura N, Jikumaru Y, Joo SH, Urquhart W, Klessig DF, Kim SK, Nambara E, Yoshioka K (2010) The lesion-mimic mutant cpr22 shows alterations in abscisic acid signaling and abscisic acid insensitivity in a salicylic acid-dependent manner. Plant Physiol 152:1901–1913
Munemasa S, Oda K, Watanebe-Sugimoto M, Nakamura Y, Shimoishi Y, Murata Y (2007) The coronatine-insensitive 1 mutation reveals the hormone signaling interaction between abscisic acid and methyl jasmonate in *Arabidopsis* guard cells. Specific impairment of ion channel activation and second messenger production. Plant Physiol 143:1398–1407
Mustilli AC, Merlot S, Vavasseur A, Fenzi F, Giraudat J (2002) *Arabidopsis* OST1 protein kinase mediates the regulation of stomatal aperture by abscisic acid and acts upstream of reactive oxygen species production. Plant Cell 14:3089–3099
Nakashima K, Ito Y, Yamaguchi-Shinozaki K (2009) Transcriptional regulatory networks in response to abiotic stresses in *Arabidopsis* and grasses. Plant Physiol 149:88–95
Neilands JB (1981) Iron adsorption and transport in microorganisms. Annu Rev Nut 1:27–46
Neill S, Barros R, Bright J, Desikan R, Hancock J, Harrision J, Morris P, Ribeiro D, Wilson I (2008) Nitric oxide, stomatal closure, and abiotic stress. J Exp Bot 59:165–176
Patten C, Glick BR (1996) Bacterial biosynthesis of indole-3-acetic acid. Can J Microbiol 42:207–220

Patten CL, Glick BR (2002) The role of bacterial indoleacetic acid in the development of the host plant root system. Appl Environ Microbiol 68:3795–3801

Puente M-E, Bashan Y (1993) Effect of inoculation with *Azospirillum brasilense* strains on the germination and seedlings growth of the giant columnar Cardon cactus (*Pachycereus pringlei*). Symbiosis 15:49–60

Puente ME, Bashan Y, Li CY, Lebsky VK (2004a) Microbial populations and activities in the rhizoplane of rock-weathering desert plants. I. Root colonization and weathering of igneous rocks. Plant Biol 6:629–642

Puente ME, Li CY, Bashan Y (2004b) Microbial populations and activities in the rhizoplane of rock-weathering desert plants. II. Growth promotion of cactus seedlings. Plant Biol 6:643–650

Puente ME, Li CY, Bashan Y (2009a) Rock-degrading endophytic bacteria in cacti. Environ Exp Bot 66:389–401

Puente ME, Li CY, Bashan Y (2009b) Endophytic bacteria in cacti seeds can improve the development of cactus seedlings. Environ Exp Bot 66:402–408

Razi SS, Sen SP (1996) Amelioration of water stress effects on wetland rice by urea-N, plant growth regulators, and foliar spray of a diazotrophic bacterium *Klebsiella* sp. Biol Fertil Soils 23:454–458

Reed MLE, Glick BR (2004) Applications of free living plant growth-promoting rhizobacteria. Anton Van Leeuwen 86:1–25

Requena N, Jimenez I, Toro M, Barea JM (1997) Interactions between plant-growth-promoting rhizobacteria (PGPR), arbuscular mycorrhizal fungi and *Rhizobium* spp. in the rhizosphere of *Anthyllis cytisoides*, a model legume for revegetation in Mediterranean semi-arid ecosystems. New Phytol 136:667–677

Requena N, Perez-Solis E, Azcon-Aguilar C, Jeffries P, Barea J-M (2001) Management of indigenous plant–microbe symbioses aids restoration of desertified ecosystems. Appl Environ Microbiol 67:495–498

Rincón A, Valladares F, Gimeno TE, Pueyo JJ (2008) Water stress responses of two Mediterranean tree species influenced by native soil microorganisms and inoculation with a plant growth promoting rhizobacterium. Tree Physiol 28:1693–1701

Rodriguez H, Fraga R (1999) Phosphate solubilizing bacteria and their role in plant growth promotion. Biotechnol Adv 17:319–339

Rodriguez H, Fraga R, Gonzalez T, Bashan Y (2006) Genetics of phosphate solubilization and its potential applications for improving plant growth-promoting bacteria. Plant Soil 287:15–21

Rodriguez H, Gonzalez T, Goire I, Bashan Y (2004) Gluconic acid production and phosphate solubilization by the plant growth-promoting bacterium *Azospirillum* spp. Naturwissenschaften 91:552–555

Rodriguez RJ, White JF Jr, Arnold AE, Redman RS (2009) Fungal endophytes: diversity and functional roles. New Phytol 182:314–330

Ruiz-Lozano JM, Gomez M, Azcon R (1995) Influence of different *Glomus* species on the time-course of physiological plant responses of lettuce to progressive drought stress periods. Plant Sci 110:37–44

Ryu C-M, Farag MA, Hu C-H, Reddy MS, Kloepper JW, Pare PW (2004) Bacterial volatiles induce systemic resistance in *Arabidopsis*. Plant Physiol 134:1017–1026

Ryu C-M, Farag MA, Hu C-H, Reddy MS, Wei H-X, Pare PW, Kloepper JW (2003) Bacterial volatiles promote growth in *Arabidopsis*. Proc Natl Acad Soc USA 100:4927–4932

Saleem M, Arshad M, Hussain S, Bhatti AS (2007) Perspective of plant growth promoting rhizobacteria (PGPR) containing ACC deaminase in stress agriculture. J Ind Microbiol Biotechnol 34:635–648

Saravanakumar D, Kavino M, Raguchander T, Subbian P, Samiyappan R (2011) Plant growth promoting bacteria enhance water stress resistance in green gram plants. Acta Physiol Plant 33:203–209

Sarig S, Blum A, Okon Y (1988) Improvement of the water status and yield of field-grown grain *sorghum* (*Sorghum bicolor*) by inoculation with *Azospirillum brasilense*. J Agric Sci 110:271–277

Sarig S, Kapulnik Y, Nur I, Okon Y (1984) Response of non-irrigated *Sorghum bicolor* to *Azospirillum* inoculation. Exp Agr 20:59–66

Schroeder JI, Kwak JM, Allen GJ (2001) Guard cell abscisic acid signaling and engineering drought hardiness in plants. Nature 410:327–330

Sherameti I, Tripathi S, Varma A, Oelmüller R (2008) The root-colonizing endophyte *Piriformospora indica* confers drought tolerance in *Arabidopsis* by stimulating the expression of drought stress-related genes in leaves. Mol Plant-Microbe Interact 21:799–807

Shinozaki K, Yamaguchi-Schinozaki K (1997) Gene expression and signal transduction in water-stress response. Plant Physiol 115:327–334

Shinozaki K, Yamaguchi-Schinozaki K (1999) Molecular responses to cold, drought, heat and salt stress in higher plants. R.G. Landes Company, Austin, p 169

Shinozaki K, Yamaguchi-Shinozaki K (1996) Molecular responses to drought and cold stress. Curr Opin Biotechnol 7:161–167

Shinozaki K, Yamaguchi-Shinozaki K (2000) Molecular responses to dehydration and low temperature: differences and cross-talk between two stress signaling pathways. Curr Opin Plant Biol 3:217–223

Singh LP, Gill SS, Tuteja N (2011) Unraveling the role of fungal symbionts in plant abiotic stress tolerance. Plant Signal Behav 6:175–191

Spychalla JP, Desbough SL (1990) Superoxide dismutase, catalase, and alpha-tocopherol content of stored potato tubers. Plant Physiol 94:1214–1218

Suhita D, Raghavendra AS, Kwak JM, Vavasseur A (2004) Cytoplasmic alkalization procedes reactive oxygen species production during methyl jasmonate- and abscisic acid-induced stomatal closure. Plant Physiol 134:1536–1545

Sun C, Johnson JM, Cai D, Sherameti I, Oelmüller R, Lou B (2010) *Piriformospora indica* confers drought tolerance in Chinese cabbage leaves by stimulating antioxidant enzymes, the expression of drought-related genes and the plastid-localized CAS protein. J Plant Physiol 167:1009–1017

Taji T, Ohsumi C, Luchi S, Seke M, Kasuga M, Kobayashi M, Yamaguchi-Shinozaki K, Shinozaki K (2002) Important roles of drought- and cold-inducible genes for galactinol synthase in stress tolerance in *Arabidopsis thaliana*. Plant J 29:417–426

Tataeizadeh Z (1998) Drought-induced responses in plant cells. Int Rev Cytol 182:193–247

Timmusk S, Wagner EGH (1999) The plant-growth-promoting rhizobacterium *Paenibacillus polymyxa* induces changes in *Arabidopsis thaliana* gene expression: a possible connection between biotic and abiotic stress responses. Mol Plant-Microbe Interact 12:951–959

Toivonen P, Vidaver W (1988) Variable chlorophyll a fluorescence and CO_2 uptake in water stressed white spruce seedlings. Plant Physiol 86:744–748

Trejo A, de-Bashan LE, Hartmann A, Hernandez J-P, Rothballer M, Schmid M, Bashan Y (2012) Recycling waste debris of immobilized microalgae and plant growth-promoting bacteria from wastewater treatment as a resource to improve fertility of eroded desert soil. Environ Exp Bot 75:65–73

Underwood W, Melloto M, He SH (2007) Role of plant stomata in bacterial invasion. Cell Microbiol 9:1621–1629

van Loon LC, Bakker PA, Pieterse CM (2004) Systemic resistance induced by rhizosphere bacteria. Annu Rev Phytopathol 36:453–483

Vivas A, Marulanda A, Ruiz-Lozano JM, Barea JM, Azcon R (2003) Influence of a *Bacillus* sp. on physiological activities of two arbuscular mycorrhizal fungi and on plant responses to PEG-induced drought stress. Mycorrhiza 13:249–256

Wang G (2005) Agricultural drought in a future climate: results from 15 global climate models participating in the IPCC 4th assessment. Clim Dyn 25:739–753

Wang Y, Brown HN, Crowley DE, Szaniszlo PJ (1993) Evidence for direct utilization of a siderophore, ferrioxamine B, in axenically grown cucumber. Plant, Cell Environ 16:579–585

Xu P, Chen F, Mannas JP, Feldman T, Sumner LW, Roossinck MJ (2008) Virus infection improves drought tolerance. New Phytol 180:911–921

Yang J, Kloepper JW, Ryu C-M (2009) Rhizosphere bacteria help plants tolerate abiotic stress. Trends Plant Sci 14:1–4

Zhang H, Murzello C, Sun Y, Kim M-S, Xie X, Jeter RM, Zak JC, Dowd SE, Pare PW (2010) Choline and osmotic–stress tolerance induced in *Arabiodosis* by the soil microbe *Bacillus subtilis* (GB03). Mol Plant-Microbe Interact 23:1097–1104

Zhu JK (2001) Cell signaling under salt, water and cold stresses. Curr Opin Plant Biol 4:401–406

Part V
Field Responses

Chapter 16
Physiology of the Yield Under Drought: Lessons from Studies with Lupin

Jairo A. Palta, Jens D. Berger and Helen Bramley

Abstract The 'Old World' lupin species are unique among grain legume crops in their strong specific adaption to acid sandy soils, and in their extremely recent domestication history. Our understanding of lupin responses to drought is limited; based on studies with elite cultivars, representing only a small fraction of the genetic diversity resident in the species, and subjected to strong selection for early phenology throughout their domesticated history, facilitating drought escape. Lupins appear to have a mix of competitive and conservative water use strategies, with profligate water use and high rates of photosynthesis when water is freely available; coupled with high sensitivity to water deficits, whereby stomatal conductance is reduced, phytohormone concentrations and metabolism modified, well before changes in leaf water potential occur. With high root-shoot ratios and high hydraulic conductance, lupin roots are highly efficient at taking up and transporting water. However, the predominantly apoplastic flow in lupins cannot be regulated as flexibly as the symplastic water transport that is so important in cereals. Low rates of remobilization of pre-anthesis stored C in lupin forces the crop to rely almost completely on current photosynthesis, which exacerbates the effects of terminal drought, given that the species respond by reducing leaf area through abscission. Because of the past narrow focus on domesticated material it is not currently possible to put these observations in an ecophysiological context, to answer which

J. A. Palta (✉) · J. D. Berger
CSIRO, Plant Industry, Private Bag No. 5, Wembley, WA 6913, Australia
e-mail: jairo.palta@csiro.au

J. D. Berger
Faculty of Natural and Agricultural Sciences, School of Plant Biology, The University of Western Australia, 35 Stirling Highway, Crawley, WA 6009, Australia

H. Bramley
The UWA Institute of Agriculture (M082), The University of Western Australia, 35 Stirling Highway, Crawley, WA 6009, Australia

R. Aroca (ed.), *Plant Responses to Drought Stress*,
DOI: 10.1007/978-3-642-32653-0_16, © Springer-Verlag Berlin Heidelberg 2012

of these attributes are characteristic of lupins as a species, and which can be expected to vary in response to environmental selection pressure. To advance our understanding of the species we advocate the study of wild germplasm specifically-adapted to habitats that impose contrasting drought stress, to address both the narrowness and short evolutionary history of the domesticated material, by highlighting responses to millennia of natural selection. By identifying the pros and cons of adaptive traits in an ecophysiological context our capacity to improve elite material will be considerably advanced.

16.1 Introduction

This review focuses on the agriculturally most important Old World lupin species, *L. angustifolius* (narrow-leafed lupin), *L. albus* (white lupin), and *L. luteus*, (yellow lupin), which are of wider scientific interest, and arguably have considerable untapped potential, due to their strong specific adaptation to acid, sandy soils; and their very short history as domesticated crop. Accordingly, the lupins are still minor grain legume crops, with a combined average global production $\sim$1 million t year^{-1}, 3 times smaller than lentil or pigeon pea, and 205 times smaller than soybean (FAO 2010). Despite their relatively minor global role, lupins are the principal grain legume option for acid sandy soils, and therefore Australia dominates world production (FAO 2010), largely due to the Western Australian (WA) contribution (ABARE 2010), reflecting the prevalence of suitable soils in that region.

Australian genotype by environment interaction (G $\times$ E) studies confirm that terminal drought is a consistent yield constraint of lupin in both the Mediterranean climates of Western Australia and summer-rainfall dominant eastern Australia (Berger et al. 2012a), leading to 8–16 fold reductions in grain yield. These data are confirmed by studies with controlled water regime treatments contrasting productivity under rainfed, well-watered, and rainout shelter conditions (Palta and Plaut 1999; Palta et al. 2004, 2007), and demonstrate that the lupin species are more sensitive to drought than cereals such as wheat (Henson et al. 1989a; Palta et al. 2008) and grain legumes such as chickpea and lentil Leport et al. (1998). Why is this so? This chapter explores the reasons underlying lupin sensitivity to drought by reviewing the origins, natural, and domestication history of the crop and its principal adaptive strategies. We highlight the relatively sparse literature that focuses on lupin responses to terminal drought using field and controlled environment experiments, and suggest new avenues to better understand the species and their potential in dryland agriculture.

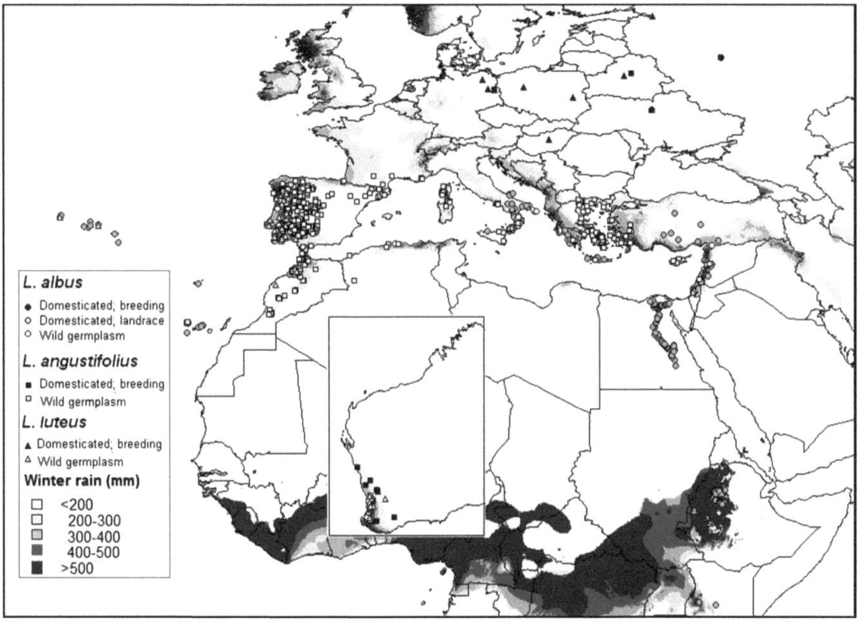

Fig. 16.1 Origins of the agriculturally significant 'Old World' lupin species (*L. albus*, *L. angustifolius*, *L. luteus)* as defined by collection sites sampled by the Australian Lupin Collection (DAFWA), categorized by domestication status. The inset shows Western Australia, world's dominant lupin production region (ABARE 2010), where wild germplasm represents naturalized environmental weed populations, and domesticated material are cultivars developed at various locations

16.2 Lupin Background: Origins, Natural, and Domestication History

The Old World lupin species originate from the Mediterranean basin, where they tend to act as early succession colonizers in neutral-acid, sandy soils of coastal regions (Fig. 16.1) (Cowling et al. 1998a). *L. albus* is somewhat of an exception, also found in inland areas of Turkey and East Africa (Fig. 16.1), including neutral to alkaline clay loams in the Nile Valley and elsewhere (Cowling et al. 1998a). While water-holding capacities of sandy soils are low compared to finer-textured loams and clays, plant available water is relatively high because of low crop water use. As a result, rainfall is readily available, even in dry sandy soils, but can also be quickly lost due to drainage. There are strong west coastal winter rainfall gradients along the Mediterranean basin lupin distribution, particularly in Iberia, the Balkans and Eastern Mediterranean (Fig. 16.1); overlaid with decreasing rainfall, and increasing temperature latitudinal gradients over much of the area. Although there is little published data within and between species, the Old World lupins appear to be typical Mediterranean winter annuals, using appropriate phenology to avoid the key stresses of early season low temperatures and subsequent

Table 16.1 Influence of geography on phenology in old world lupin species

Species	Origin	Reference
Early flowering		
L. albus	Northern Africa: Nile valley (Egypt, Sudan), Ethiopia, Morocco	Christiansen et al. (2000), and Noffsinger and van Santen (2005)
L. albus	Eastern Mediterranean: Anatolian plateau & coastal zones, Israel	Huyghe (1997), and Noffsinger and van Santen (2005)
L. albus	Western Mediterranean: SE Iberia	Simpson (1986), and Huyghe (1997)
L. angustifolius	Eastern Mediterranean: Palestine, S Greece	Troll (1940), and Clements and Cowling (1994)
L. luteus	Western Mediterranean: NW Iberia	Berger et al. (2008)
Late flowering		
L. albus	Western Mediterranean: NW Iberia	(Simpson, 1986; Huyghe, 1997)
L. angustifolius	Eastern Mediterranean: N Greece	Clements and Cowling (1994)
L. luteus	Western Mediterranean: S Iberia	Berger et al. (2008)
L. luteus	Eastern Mediterranean: Israel	Berger et al. (2008)
L. luteus	Northern Africa: Morocco	Berger et al. (2008)

terminal drought. Thus germplasm from terminally drought-prone southern Mediterranean habitats tends to flower and mature early, compared to material collected from cooler, wetter northern regions (Table 16.1), an adaptive strategy that trades off drought escape against biomass production and reproductive potential (Grime 1979).

Lupins are indeterminate, with maximum growth rates occurring after flowering, such that >80 % of carbon accumulation may take place post-anthesis (Pate et al. 1980). Despite their indeterminate nature, lupin architecture comprises an acropetal series of determinate shoots, starting with the main stem, followed by the primary lateral branches which form synchronously with main stem anthesis, and so on up the branch orders, depending on season length (Farrington and Pate 1981; Munier-Jolain et al. 1996). In the first 5 weeks after main stem flowering the stem and roots represent the principal C sink, where after pods and seeds become increasingly dominant (Pate et al. 1980). However, because there is considerable overlap in the development of reproductive and vegetative sinks within and between sequential branch orders, there is competition for C, and the probability of fruit-set decreases acropetally; increasing when earlier set basal pods on the same branch order, or subtending lateral branches are removed (Pate et al. 1980; Farrington and Pate 1981; Pate and Farrington 1981). The growing season is terminated by water stress, associated with leaf drop and the attendant decline in photosynthesis, and increase in respiration (Pate et al. 1980). This is important because lupin is heavily dependent on current photosynthesis; only <3 % of pre-anthesis C is transferred to seeds, where it contributes to the protein fraction (Pate et al. 1980). As a result lupin is very sensitive to environmental factors that reduce photosynthesis, especially if this is associated with terminal leaf drop, and cannot be recovered. Accordingly, in Mediterranean climates early planting allows more time for the development of higher order branches and pod fill within branch orders prior to the onset of terminal drought (Perry 1975).

All the Old World lupin crops species are recent to very recent domesticates; largely sharing a tortuous evolutionary path as crops, characterized by populations of small effective size, isolated in both space and time, subject to strong bottlenecks and selection pressures, as outlined below. *L. albus* is by far the oldest, with Bronze Age remains identified in a number of Aegean locations, and more recent Roman sites throughout Egypt (Zohary and Hopf 2000), and is the only member of the group that could be considered to be a traditional Mediterranean pulse crop. *L. albus* became a temperate crop in the eighteenth century, after Frederick the Great of Prussia took a personal interest in introducing it to the acid sandy soils of his homeland, but these efforts failed because of the inability of the crop to mature in a timely manner (Hondelmann 1984). Subsequently *L. luteus* was successfully introduced as a green manure crop, and became widespread in the north German lowlands from the mid nineteenth century, whereupon *L. angustifolius* was also taken up (Hondelmann 1984). Systematic plant breeding began in Germany and Poland in the early 1900s; rapid early growth and early maturity remaining key breeding priorities, as attested by cultivar names such as Pflugs Allerfrüheste (plough's earliest), a narrow-leafed lupin variety released in 1920. In the 1920–1930s, advances in analytical chemistry facilitated the search for alkaloid-free 'sweet lupins' independently in Germany and Russia, and the crops passed through a series of bottlenecks as many 100,000 of single plants were screened (Hondelmann 1984). This process was repeated during the search for permeable 'soft' seeds and non-shattering pods (Sengbusch and Zimmermann 1937). After the World War II the acid sandy regions of Central Europe became part of the Eastern Bloc, and therefore western interest in lupin breeding declined. In the 1950–1970s Gladstones in Western Australia fully domesticated the narrow-leafed lupin using European material, stabilizing pod shattering, introducing white flower and seeds as unlinked markers of domestication, and made the crop far earlier in warm environments by modifying the vernalization response (Gladstones 1970, 1994; Cowling et al. 1998b). Subsequently, the narrow-leafed lupin has become an important component of Australian acid sandy soil cropping systems, which are particularly prevalent in WA; production peaking at almost 2 million t in 1999, representing 93.5 % of the world total (FAO 2010).

16.3 Implications for the Crop

The domestication histories outlined above have had important ramifications for the Old World lupin crops; severely limiting their genetic diversity and selecting very strongly on early phenology. The best evidence for a narrow genetic base exists in narrow-leafed lupin, where Australian and European cultivars and breeding material form a single tight cluster in an ocean of wild germplasm of far greater diversity (Berger et al. 2012a). White lupin genotyping has been far more superficial (Raman et al. 2008), (n = 94 genotypes, compared to 1,343), but shows similar trends with breeding lines and landraces clustering separately. Given that all published research into lupin responses to terminal drought is based on domesticated material, this implies that our current knowledge is based on a very limited sample, which may not be representative of the species as a whole.

Strong selection for early phenology throughout its crop development history has led to highly temperature responsive, early flowering in narrow-leafed lupin (Berger et al. 2012b), equivalent to that of chickpea specifically-adapted to South India, an environment more than twice as warm during the vegetative phase as the northern WA grainbelt (Berger et al. 2011). This temperature responsive phenology facilitates drought escape in the northern WA grainbelt, the world's principal lupin production area, as demonstrated by the contrast of modern vernalization-unresponsive (VU) cultivars with their older, highly vernalization responsive (VR) counterparts (Berger et al. 2012a). Under these conditions, VU types flowered at 76–79 days, compared to 107–114 days among VR types; leading to 2.5-fold yield advantage in the favor of the former. Both phenology and yield differences between VR and VU types disappeared in terminally drought-stressing environments with sufficiently cold vegetative phases to induce a vernalization response (typical of eastern Australia); yields were consistently low in both groups. These results suggest that the principal breeding response to terminal drought has been to modify phenology, first by removing the vernalization response (Gladstones 1994), and then by selecting for higher temperature responsiveness (Berger et al. 2012a, b). This is confirmed by controlled experiments exposing Australian lupin cultivars to water deficit; where yield has been linked with other escape characteristics, such as rapid seed growth, pod fill, and the associated capacity to set and retain pods under stress (Palta et al. 2003, 2004, 2007).

16.4 An Unusual Mix of Water Use Strategies

In an ecophysiological context, narrow-leafed lupin appears to have a curious mix of conservative and competitive water use strategies. It is considered to be a profligate water user when water is freely available, a competitive strategy that facilitates water capture before it is lost to drainage, or is taken up by neighboring plants (Turner and Henson 1989). However, higher water use in comparison to other crops such as wheat or other cool-season grain legumes depends on timing within the season, particularly after flowering (Farrington et al. 1992; Siddique et al. 2001), with total water use generally being similar for the entire season (Hamblin and Tennant 1987; Gallardo et al. 1996; Eastham and Gregory 2000; Siddique et al. 2001). Conversely, narrow-leafed lupin is very sensitive to drying soils, reducing stomatal conductance well before changes in leaf water potential (Turner and Henson 1989); a strategy that is likely to conserve water when soil water content is relatively low, and drainage unlikely to be an issue. It may be that this mix of water use strategies is the ideal adaptation of lupin to sandy soils. Deep sandy soils have with limited water-holding capacity allowing lupin crops to profligate when there is a high risk of losing water to drainage, and to be conservative when there is not. This is speculative because it is not possible to put these trends into an ecophysiological context, because we simply do not know about the range of water use strategies in the species as a whole, because material specifically-adapted to contrasting environments has not been studied. However, these strategies contrast with cereal crops such as wheat (Henson et al. 1989a) that are more tolerant to terminal drought under the same field conditions, so we will examine these differences more closely in the following section.

16.5 Maintenance of Leaf Water Status

Perhaps the most studied aspect of lupin water relations is the regulation of their stomatal conductance. Stomates of narrow-leafed, yellow, blue, and white lupin behave similarly (Henson and Turner 1991; Correia and Pereira 1994) and are particularly sensitive to changes in the soil water content, rapidly closing when the soil starts to dry (Henson et al. 1989a; Jensen et al. 1989; Correia and Pereira 1994). In pot studies, stomatal closure occurred with partial soil drying (Jensen et al. 1989), associated with small increases in matric potential in the top half of the pots (Henson et al. 1989a). Stomatal conductance, root water uptake, and the fraction of roots in wet soil has been shown to decline linearly with soil water content (Jensen et al. 1989). Stomatal closure with partial soil drying was confirmed for narrow-leafed lupin in the field (Jensen et al. 1998), which appeared to be more related to soil water potential than relative soil water content or the fraction of roots in wet soil (Jensen et al. 1998).

Stomatal closure prevents leaf dehydration under moderate soil water deficit, allowing lupin leaves to maintain high relative water contents and constant water potential (Henson et al. 1989a; Jensen et al. 1989; Correia and Pereira 1995; Rodrigues et al. 1995). However, ongoing severe water deficit causes leaf water status to decline (Henson et al. 1989a). This is in contrast to wheat leaves that tend to have higher bulk turgor (Henson et al. 1989a), lower levels of stomatal conductance (Henson et al. 1990), and more negative leaf water potentials under adequate water conditions. Leaf water potential of wheat is also more closely coupled to soil water content decreasing as the soil dries, reflecting its much lower threshold leaf water potential for stomatal closure (Henson et al. 1989a). The behavior of wheat leaves may however, be dependent on growing conditions and genotypes, as inhibition of stomatal conductance before any decrease in leaf water potential has also been reported (Ali et al. 1999).

The closure of lupin stomata under soil water deficit is associated with increased leaf xylem concentrations of the phytohormone abscisic acid (ABA) (Henson et al. 1989b; Correia and Pereira 1994). ABA was closely negatively related to bulk turgor in wheat leaves, but not in blue lupin leaves, although ABA progressively increased with soil drying in both species (Henson et al. 1989b). There is clear evidence that ABA mediates stomatal conductance in lupin. Feeding detached leaves from well-watered blue, white, or yellow lupin plants with ABA concentrations similar to those found under drought resulted in similar reductions in stomatal conductance as those of attached leaves under drought (Henson et al. 1989b; Henson and Turner 1991; Correia and Pereira 1995). As with wheat leaves, lupin abaxial stomata are more sensitive to drought and ABA (Henson et al. 1989a; Correia and Pereira 1995) and the sensitivity of stomatal conductance to ABA depends on the severity of drought (Correia and Pereira 1995). The origin of the ABA is unclear. A study with saline-stressed white lupin suggests that ABA may be synthesized in roots and transported to shoots, or originates in the shoot and is translocated to roots via phloem where it is re-circulated back to the shoot (Wolf et al. 1990). Other phytohormones and metabolic adjustments have also been implicated in lupins response to soil water deficit before changes in leaf water status (Pinheiro et al. 2011). For example, a 12 % decrease in the soil water content did not affect the water status of white lupin, but caused changes in the phytohormones IAA and cytokinins. IAA increased in leaves, but decreased in roots, while the cytokinin activity was reduced in both roots and shoots (Pinheiro et al. 2011). Metabolic changes also occurred in response to soil water deficit before changes in plant water status, as indicated by changes in carbohydrate (starch and sugars) concentrations in different plant organs. Pinheiro et al. (2011) suggested that these early metabolic changes may be part of a mechanism that triggers adjustments to longer term soil water deficit.

Many species adapted to, or more tolerant to drought, maintain turgor by accumulating solutes in their tissues and increasing tissue elasticity. These features, particularly osmotic adjustment, have been selected for in breeding programs of many important crop species (Blum 2005). The capacity for osmotic adjustment appears to be limited in lupin species (Table 16.2), but could be dependent on growing conditions, stage of plant development when the plant experiences drought and severity of the drought (Table 16.2). Both blue and narrow-leafed lupin show some capacity to reduce their bulk leaf elastic modulus under soil water deficit, which combined with small osmotic adjustment maintained leaf turgor under lower relative water contents (Jensen and Henson 1990). But this is from a single pot study and no measurements appear to have been made under field conditions.

16.6 Root Hydraulic Properties that Favor Rapid Water Uptake Under Water Sufficient Conditions but with Little Capacity to Adjust Rapidly and Reversibly to Prevailing Conditions

Narrow-leafed and yellow lupin roots are highly efficient at taking up and transporting water, which is facilitated in-part by absorption of water along the root length and the abundance of large xylem vessels, respectively (Hamza and Aylmore 1992a; Bramley et al. 2009). They behave like large leaky conduits. In

Table 16.2 Osmotic adjustment in lupin leaves and roots in response to drought

Species	Environment	Timing of drought (DAS)	Degree of osmotic adjustment (MPa)		Reference
			Leaf	Root	
L. albus	Glasshouse, commercial potting mix filled pots	60	0.35		Correia and Pereira, (1995)
L. albus (3 genotypes)	Outdoors, large containers filled with sandy soil—outdoors	Flowering	0		Rodrigues et al. (1995)
L. consentinii	Glasshouse, 1 m soil (coarse field sand) columns	56	Osmotic potential decreased 1 MPa		Gallardo et al. (1994)
L. consentinii	Glasshouse, coarse sand-filled containers—severe drought	27–65	0.3		Jensen and Henson (1990)
L. angustifolius			0.4		
L. consentinii	Growth cabinet, small sand-filled pots	14	0.2		Kuang et al. (1990)
L. angustifolius (5 genotypes)	Field—coarse sandy loam overlying clay loam	107	0		Palta et al. (2007)
L. luteus					
L. angustifolius (2 genotypes)	Glasshouse, sandy/loam filled pots	45	0		Turner et al. (1987)
L. consentinii			0.2		
L. pilosus	Glasshouse, sandy/loam filled pots	40	0.4–0.5		Turner et al. (1987)
L. atlanticus			0.4–0.5		
L. luteus (2 genotypes)			0.2		
L. angustifolius			0.1		
L. angustifolius	Field—deep loamy sand (rainfed v irrigated)	76	0.2–0.3	0.4	Turner et al. (1987)

comparison, cereal roots confine their water absorption to a small region behind root tips (Bramley et al. 2009), but may have the ability to vary which region is involved in water uptake in response to abiotic stress (Bramley et al. 2010). In the field, wheat has several-fold greater root length densities than narrow-leafed lupin (Hamblin and Tennant 1987; Gallardo et al. 1996; Gregory and Eastham 1996), but water uptake per unit root length 2–5-fold lower (Hamblin and Tennant 1987; Gallardo et al. 1996). Despite these differences in root length, as result of larger diameters (Hamblin and Tennant 1987; Bramley et al. 2009) and higher root dry weights (Gregory and Eastham 1996), narrow-leafed lupins invest a greater proportion of their total biomass in the root system (Gregory and Eastham 1996).

So why is specific root water uptake greater in narrow-leafed lupin than wheat when water is not limiting? Lower leaf water potentials in wheat compared with lupin under the same transpirational demands, and soil water contents (Henson et al. 1989a), imply a greater resistance to water transport from roots to shoots in wheat. Measurements on young seedlings (Bramley et al. 2009) and estimations from flowering plants in the field (Gallardo et al. 1996) confirm this supposition. Although the hydraulic resistance of lupin roots is linearly related to xylem development and hence, decreasing axial (longitudinal) resistance with increasing distance from the root tip, the radial resistance is also lower than wheat despite their 2–3-fold larger diameter (Bramley et al. 2009). The transport of water from the root surface to the vasculature is usually the most limiting aspect of root hydraulic resistance (Steudle and Peterson 1998). Water traveling along the radial flow path has to navigate a series of concentric cell layers that are obstacles to water flow in the form of cell membranes and cell walls of varying composition impregnated with hydrophobic barriers such as suberin. In comparison to wheat, water flow through the apoplast is not as obstructed in lupin roots because they do not develop an exodermis (Hartung et al., 2002; Bramley et al. 2009), and suberization and secondary thickening of their endodermis occurs much further from the root tip (Bramley et al. 2009). The absence of these barriers may be the reason that water flow through narrow-leafed and yellow lupin roots preferentially occurs via the apoplast, whereas water crosses membranes in wheat roots with regulation most likely being controlled by aquaporins in the endodermis (Bramley et al. 2009, 2010). Aquaporins are membrane-intrinsic, water-conducting proteins, whose production and pore-gating (opening and closing) can rapidly and reversibly control the rate of water flow across cells, tissues, and organs (Tyerman et al. 2002; Bramley et al. 2007a). Without this facility, changes in root hydraulic resistance are constrained to slow and irreversible growth-related processes. In summary, while high hydraulic conductance facilitates rapid and efficient water flow in and out of lupin roots, the control of these fluxes much weaker and less flexible than in wheat.

The implications of these contrasting root hydraulic properties and water transport strategies may be linked to the strategies of controlling water loss at the shoot level. With aquaporin-regulated water transport, wheat roots can diurnally alter their root hydraulic conductance (reciprocal of resistance) to match transpiration demands (Carvajal et al. 1996) and minimize drop in leaf water potential, whereas this feature has not been observed for lupin species (Passioura and Munns 1984; Hamza and Aylmore 1992b). Lupins apparently rely on controlling leaf hydration by controlling the driving force for water flow through stomatal regulation.

We can only speculate what the implications of the different water transport strategies are in response to soil water deficit because there have been no comparative studies, and indeed no measurements of root hydraulic resistance in lupins in response to drought have been undertaken. We do not know whether lupin roots become more suberized to minimize water loss from roots to dry soils, but there are indications that they do not. Hartung et al. (2002) could not induce an exodermis in narrow-leafed lupin roots when the plants were grown in a mist environment (aeroponics), which have been shown to increase suberization in other plant species (Hose et al. 2001). Without suberization reducing leakiness of the shallower roots, extraction of water at depth as the surface soils dry would become increasingly difficult, as deeper roots become hydraulically isolated, or there is release of water from the wetter to drier soil regions (hydraulic lift) through the leaky surface roots. Lupin roots also dramatically shrink when the soil dries (white lupin, Carminati et al. 2009) or under osmotic stress (narrow-leafed lupin, Hamza et al. 2007). The formation of air gaps around roots associated with root shrinkage (Carminati et al. 2009) is likely to increase the interfacial resistance to water uptake, and would explain the increasing root resistance at low values of water uptake (Jensen et al. 1989). However, lupin roots may exude mucilage to alter the properties of the rhizosphere, and maintain hydraulic connection with the soil (Carminati et al. 2010).

Despite the lack of studies on lupin roots in response to drought, circumstantial evidence indicates that root hydraulic conductance is reduced under soil water deficit. Leaf water potential of blue lupin only partially recovered after transient drought, despite soil water contents returning to control levels and transpiration remaining low (Correia and Pereira 1995). This implies that root hydraulic conductance had decreased without full recovery such that roots could not supply the shoots with sufficient water. There may be a number of reasons for this. Apart from root shrinkage and the obvious loss of hydraulic conductance with root death, embolized xylem vessels would also increase root hydraulic resistance. Lupins may rapidly close stomata in response to soil drying to minimize increasing tensions developing in the xylem and hence, avoid cavitation, as large vessels are more susceptible to cavitation (Tyree and Sperry 1989). This is particularly important in sandy soils that lose water quickly. Narrow-leafed lupins may also have lower capacity to refill embolized vessels than wheat because they have very low root pressures (Bramley et al. 2007b, 2010) and lower solute reflection coefficients (Bramley and Tyerman, "unpublished").

All of the above speculation has been derived from observations on one or two lupin species. The morphology of lupin root systems differs between species (Clements et al. 1993) and possibly growing conditions (Doussan et al. 2006; Garrigues et al. 2006). Different lupin root system architectures (taproot v fibrous) induce different water uptake patterns (Doussan et al. 2006; Garrigues et al. 2006; Oswald et al. 2008) and are likely to have different root hydraulic properties. Moreover, as emphasized in the introduction, the old world lupin distribution comprises a wide range of rainfall and reproductive phase temperatures. More research on these aspects may identify root traits that could be selected for improving lupin tolerance to terminal drought, particularly given that grain yield is correlated with post-flowering water use (Siddique et al. 2001).

16.7 Sensitivity to Drought Spells During Pre-Anthesis

The theme of liberal water use when soil water is plentiful, combined with high sensitivity to water shortages, is reflected in lupin metabolism. Rates of leaf net photosynthesis, leaf expansion, and crop growth are high when soil water is

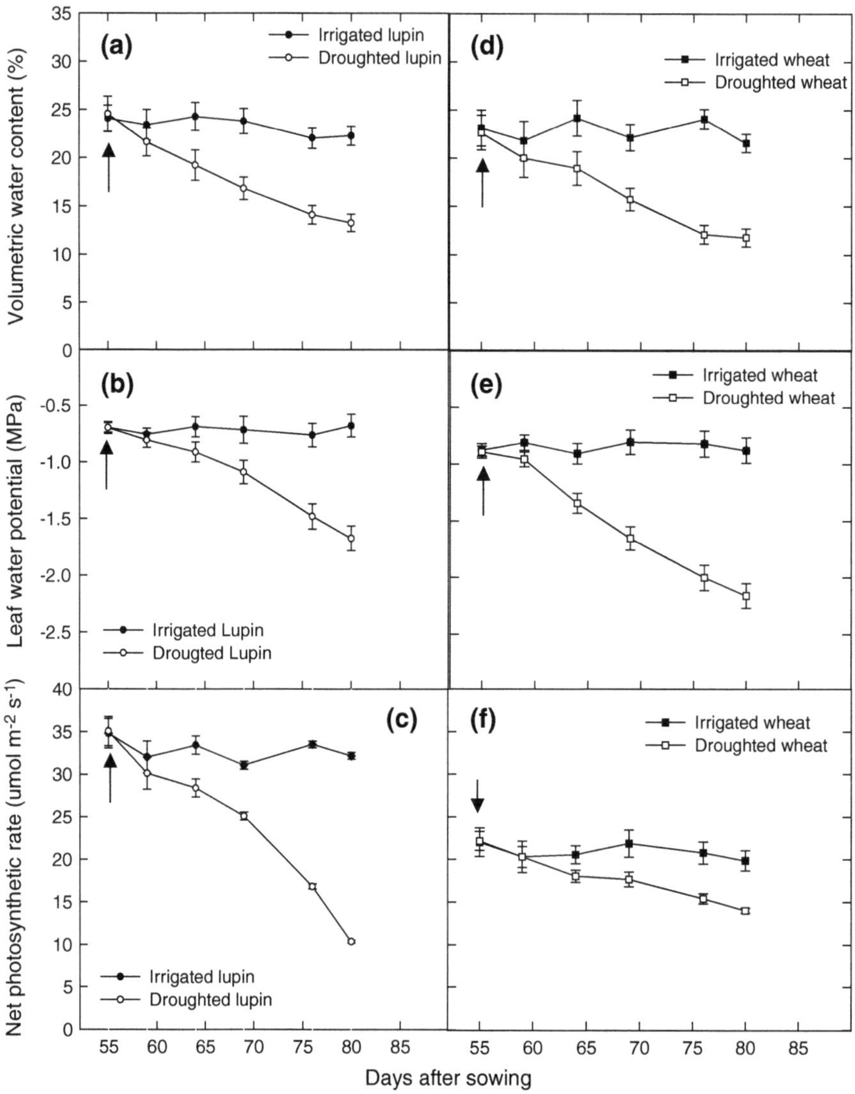

Fig. 16.2 Change with time in the volumetric soil water content, midday leaf water potential, and leaf net photosynthetic rate for narrow-leafed lupin (**a**, **b**, **c**) and spring wheat (**d**, **e**, **f**) under supplemental irrigation (*close symbols*) and under a 25 days spell of drought during pre-anthesis (*open symbols*). Drought was induced by excluding rainfall by positioning a rainout shelter over the crops for 25 days. Bars indicate ± s.e.m. for 6 replicates when larger than the symbol. The time when the drought spell was induced is indicated by the vertical arrows (Palta, "unpublished")

adequate. Compared with spring wheat grown side-by-side on a deep sandy soil under rainfed conditions, the rates of leaf net photosynthesis and crop growth before flowering were 47–59 % and 23–33 % higher in narrow-leafed lupin, respectively (Fig. 16.2c, f; Table 16.3). When rainfall events were excluded by a movable rainout shelter the volumetric soil water content in the top 0.7 m of the soil profile decreased from 24 to 13.2 % (Fig. 16.2b) and leaf net photosynthesis and crop growth rates (CGR) in lupin decreased by 70 and 82 %, respectively (Fig. 16.2c; Table 16.3). Similar changes in the volumetric soil water content reduced wheat leaf net photosynthesis and CGRs by only 36 and 47 %, respectively (Fig. 16.2f). These differences in the reduction in leaf net photosynthesis and CGRs occurred when leaf water potential decreased from −0.7 to −1.7 MPa in narrow-leafed lupin (Fig. 16.2b) and from −0.8 to −2.1 MPa in wheat (Fig. 16.2e) and indicate the high sensitivity of narrow-leafed lupin to soil water shortages occurring before flowering.

Crop growth in lupin under pre-anthesis water shortage is reduced mainly through an adjustment in leaf area by abscission of fully expanded leaves from the mid to lower canopy, and a reduction in leaf expansion of newly formed leaves (Palta, "unpublished"). Reductions in crop growth caused by pre-anthesis drought spells may affect grain yield by reducing nitrogen fixation, limiting biomass, and the capacity to fill pods on the mainstem and develop apical branches (Pate et al. 1980). In most indeterminate grain legumes, like lupin, nitrogen accumulation at anthesis is strongly correlated with grain number per unit area (Pandey et al. 1984; Weisz et al. 1985; Muchow and Sinclair 1986) and grain number per unit area is the main determinant of grain yield (Muchow and Sinclair 1986; Dracup et al. 1998; Purcell and King 1996). Crop nitrogen accumulation in lupin depends on nitrogen fixation as up to 90 % of the accumulated nitrogen at anthesis is derived from atmospheric nitrogen fixation (Ndfa; Unkovich et al. 1994). Ndfa is more sensitive to soil water shortages than either leaf growth or leaf gas-exchange (Sinclair 1986) because Ndfa is affected by shortage of assimilates of carbon, nodule carbon metabolism and oxygen limitation (Serraj et al. 1999). A study in which rainfall was excluded for 25 days from a narrow-leafed lupin crop during its vegetative growth, showed that Ndfa was reduced from 84.2 to 28.5 % (Table 16.3) and this contributed to a 71 % reduction in total crop nitrogen (Palta, "unpublished") (Fig. 16.3).

16.8 Sensitivity to Terminal Drought

Terminal drought or "end-of-season drought" is a common feature in regions with Mediterranean-type climates, where lupin is grown as a crop. Terminal drought develops when rainfall decreases and evaporation and temperature increase in the spring, when lupin enters its reproductive stage (Reader et al. 1995; Palta et al. 2004). The sensitivity of lupin to terminal drought is the major cause of reductions in grain yield and the main deterrent for its adoption to new cropping areas (Palta et al. 2008). Compared with cereals, pulses, and some oilseed crops the grain yield of lupin is more severely affected by terminal drought. The comparison of the yield performance under terminal drought and well-watered treatments showed that while the grain yield of cereals, pulses, and some oilseed crops was reduced by

Table 16.3 Crop growth rate (CGR), nitrogen derived from the atmosphere (Ndfa) and nitrogen accumulation in narrow-leafed lupin (cv. Merrit) and spring wheat (cv. Westonia) crops under 25 d water shortage during their vegetative growth

	CGR g m^{-2} d^{-1}	CGR reduction (%)	%Ndfa	Ndfa (g m^{-2})	Total N (g m^{-2})
		Rainfed			
Narrow-leafed lupin	12.7	–	84.2	13.0	15.5
Spring wheat	8.5	–	–		7.1
		Rainfall excluded for 25 d			
Narrow-leafed lupin	2.3	82	28.5	1.3	4.4
Spring wheat	4.5	47	–	–	4.2
l.s.d (P = 0.05)	1.8	22	31.7	6.8	8.5

Crops were grown side-by-side on a deep sandy soil at Wongan Hills, Western Australia in 1995. Water shortage was induced by positioning over the crops movable rainout shelter to exclude rainfall. The proportion of nitrogen derived from the atmosphere (%Ndfa) was determined using the ^{15}N natural abundance method of Shearer and Kohl (1986). L.s.d (P = 0.05) is for comparison of differences between lupin and wheat and treatments. (J Palta, "unpublished")

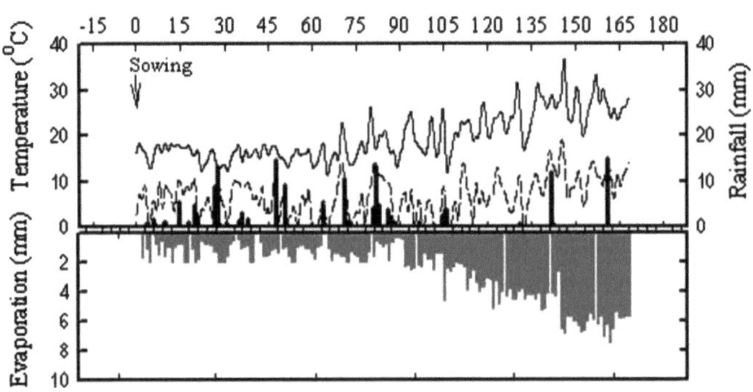

Fig. 16.3 Maximum (—) and minimum (–) air temperatures, total rainfall (*solid histogram*), and evaporation (*shaded histogram*) at Merredin, Western Australia, a region where the incidence of terminal drought is high. Adapted from Palta et al. (2004)

39.7–53.8 % the grain yield of lupin was reduced by 60–93.8 % (Table 16.4). The severe reduction in lupin grain yield under terminal drought raises important questions. What is the underlying physiology that makes it so susceptible? The sensitivity of current assimilation to water deficit and the distribution of recently photosynthesized carbon offer a framework to answer this question.

Lupin grain yield is largely reliant on current assimilation (Pate et al. 1980), and the indeterminate growth habit of the crop introduces competition between vegetative and reproductive sinks for C (Palta and Ludwig 2000). Therefore, any environmental stress that reduces the availability of carbon assimilates after anthesis will largely reduces grain yield (Palta and Ludwig 1996, 2000). A field study in which the availability of carbon assimilates in wheat, narrow-leafed lupin, and pea was reduced by intercepting 90 % of incident solar radiation after anthesis by shading, showed that while pea and wheat reduced their grain yield by 26 and 63 % respectively, lupin grain yield was reduced by 98 % (Saldaña et al. 2009). This indicates that the contribution of pre-anthesis stored assimilates was much lower in lupin than in the other crops. Pate et al. (1980) suggested that less than 3 % of pre-anthesis stored carbon is transferred to the grain, and studies in which canopies of narrow-leafed lupin were fed with $^{13}CO_2$ several times during pre-anthesis confirm that even under terminal drought, remobilization remains low (Table 16.5: 2.8–5.0 %). Compared with spring wheat, which increases pre-anthesis carbon remobilization from 7.7 to 21 % under terminal drought (Palta et al. 1994), these values are very modest, and unlikely to contribute to grain filling in lupin, even though under terminal drought the grain, as opposed to the vegetative tissue, becomes the dominant sink (Table 16.5; French and Turner 1991; Palta and Plaut 1999). ^{13}C feeding studies in narrow-leafed lupin underline the importance of current photosynthesis. C fixed in leaves during the day is either stored as starch and structural carbon compounds, or remains as water-soluble carbohydrates which are completely respired within 48 h (Palta and Ludwig, "unpublished"). This may explain why in some environments abortion of flowers and young pods ($\leq$20 mm) occurs after 2–3 successive overcast days ($\leq$430 μm m^{-2} s^{-1}; PAR) (Berger et al. 2008).

The second question concerns what can be done to make lupin grain yield less severely affected by terminal drought. The identification of adaptive traits in lupin germplasm for terminal drought offers a framework to answer this second question. Lupin yield under terminal drought is reduced through pod and seed abortion caused by a shortage of carbon assimilates (Palta and Ludwig 1996, 2000). As pointed out above, varietal selection has ensured early flowering in narrow-leafed lupin (Berger et al. 2012b), providing more time for podding and pod filling before the severe effects of terminal drought occurs on carbon assimilation (Palta et al. 2004). Modern narrow-leafed lupin cultivars like Belara and Quilinock escape terminal drought by flowering and podding earlier and having faster rates of seed filling than old cultivars like Merrit and Myallie (Table 16.6). However, the escape from terminal drought may not necessarily improve pod retention in lupin. The measured pod retention in the field (Palta et al. 2007) suggests it may be a potential trait conferring high yield in some cultivars like Tanjil under moderate (late) terminal drought, but not under extreme (early) terminal drought conditions, because the carbon resources generated by reduced current assimilation were only enough to fill half of the pods that were retained.

Early flowering may limit further yield improvement by constraining source size. Finishing the growing season early limits the time available for biomass accumulation, generating a tension between drought escape, and maximizing source potential. Studies by Palta and Ludwig (2000) and Saldaña et al. (2009) suggested that grain yield in narrow-leafed lupin was source—rather than sink-limited. When sink size was increased by application of cytokinin to the flowers to increase pod retention (Atkins and Pigeaire 1993; Palta and Ludwig 1996), grain yield did not increase because many pods failed to fill their seeds (Palta and Ludwig 1996). The failure to fill extra pods resulted from the inability of source to

Table 16.4 Terminal drought sensitivity of lupin crops, assessed by a comparison of droughted and well-watered treatments

Crop species	Experiment type	Region	mm per season		Grain yield (t/ha)		Yield reduction (%)	Reference
			Well-watered/ rainfall	Drought/ rainfed	Well-watered/ rainfall	Drought/ rainfed		
Narrow-leafed lupin	MET	Australia, eastern grainbelt	477.8	153.7	3.2	0.2	93.8	Berger et al. (2012b)
Narrow-leafed lupin	MET	Australia, western grainbelt	389.2	173.4	1.7	0.2	88.2	Berger et al. (2012a)
Narrow-leafed lupin	TDM	Australia, western grainbelt	494.7	306.2	3.0	1.2	60.0	Dracup et al. (1998)
Narrow-leafed lupin	YUTD	Australia, western grainbelt	217.0	160.0	1.7	0.4	76.5	Palta et al. (2004)
Spring wheat	MST	Yugoslavia, Novi Sad	352.0	138.6	7.8	4.7	39.7	Dencic et al. (2000)
Durum wheat	MET	Spain, eastern Andalusia	370.0	190.5	4.3	2.5	41.8	Garcia del Moral et al. (2003)
Chickpea	YUTD	Australia, western grainbelt	462.0	275.0	3.7	2.0	46.0	Leport et al. (1999)
Soybean (ind.)	MST	Australia, northern grainbelt	264.6	149.5	2.6	1.2	53.8	Rose et al. (1992)
Common bean	YUTD	Mexico, Durango	357.4	136.7	1.1	0.6	42.0	Acosta-Gallegos and Adams (1991)
Canola	MEY	Australia, western grainbelt	446.0	274.0	1.9	1.0	47.4	Farre et al. (2001)

MET Multi-environment G × E trial
TDM Terminal drought manipulation
YUTD Yield under terminal drought GxE trial
MST Multi-seasonal G × S trial
MEY Multi-environment yield analysis

Table 16.5 ^{13}C accumulated at the commencement of flowering, post-anthesis losses and remobilization to the seeds, pod shells, and developing apical branches

	^{13}C (mg plant^{-1})					
Treatments	Flowering	Post-flowering Losses	Remobilized			Retained
			Seeds	Pod shells	Apical branches	
Well-watered	55.6	27.6 (50)	1.6 (2.8)	0.8	3.1 (6.0)	22.5 (40)
Early terminal droughted	55.3	28.4 (51)	2.8 (5.0)	1.3	1.6 (3.0)	21.2 (38)
Late terminal droughted	55.5	27.0 (48)	2.1 (3.8)	0.9	2.2 (4.0)	23.3 (42)
l.s.d. ($P < 0.05$)	3.1	1.6	0.7	0.2	0.5	0.9

Narrow-leafed lupin canopies were fed with $^{13}CO_2$ three times during pre-anthesis to label the pre-anthesis pool of carbon and the fate of ^{13}C was measured under well-watered conditions and under an early and late starting terminal drought. The values in brackets are the percentage of the ^{13}C available at the commencement of flowering. L.s.d ($P < 0.05$) is for comparison between treatments. (Palta, "unpublished")

Table 16.6 Time to flowering and podding, maximum seed growth rates (MSGR), pod retention, and seed yield of modern and old narrow-leafed cultivars grown in the field under terminal drought

Genotype	Release date	Flowering (DAS)	Podding (DAS)	MSGR (mg seed^{-1}d^{-1})	Pod retention (%)	Seed yield (t/ha)
Merrit	1991	97	105	5.4	62	1.6
Mayllie	1995	96	105	5.8	71	1.7
Kalya	1996	98	106	6.0	75	1.7
Tanjil	1997	95	105	6.2	82	2.0
Belara	1998	92	100	7.0	67	2.2
Quilinock	1999	91	99	7.6	69	2.1
l.s.d ($P = 0.05$)	–	4	2	0.7	4	0.2

L.s.d ($P = 0.05$) is for comparison among genotypes. Adapted from Palta et al. (2004), and (2007)

match sink demands (Palta and Ludwig 2000). When the availability of carbon assimilates (source) was increased by exposing plants with increased number of pods (large sink) to elevated CO_2 during pod filling, grain yield increased by 42 % (Palta and Ludwig 2000). This implies that lupin must accumulate as much biomass as the growing season will allow to meet its sink demands. Under terminal drought this is exacerbated by the importance of current photosynthesis, given the lack of C remobilization, underlining the need to screen for genetic variation for this trait. In wheat, selection for vigorous seedling growth has successfully increased biomass accumulation and yield, particularly in drier environments and seasons (Rebetzke and Richards 1999; Richards and Lukacs 2002), probably because of the capacity of that crop to remobilize earlier fixed C. Clearly lupin needs both: increased early biomass accumulation, and a greater capacity to remobilize it.

16.9 Challenges in Improving Terminal Drought Tolerance in Lupin

The challenge facing the lupin researcher identifying and evaluating traits for improving tolerance to terminal drought is to find physiological and morphological characteristics that are independent of terminal drought escape. To date this has been the primary breeding strategy for adapting the crop to dry areas (Berger et al. 2012a, b; Gladstones 1994). Further selection for even earlier flowering will be difficult, given that modern cultivars are at the upper extreme of temperature responsiveness, and will limit crop yield potential by constraining the time available for biomass accumulation by finishing the growing season early, increasing the trade-off between drought escape and maximizing source potential. Therefore, the task is to identify and evaluate traits that are associated with postponing dehydration and dehydration tolerance strategies.

Narrow-leafed lupin postpones dehydration by high stomatal sensitivity to soil water deficits (Turner and Henson 1989), as outlined previously. These can be delayed by accessing previously unavailable water through the development of deeper roots, and/or by making more efficient use of existing water resources by optimizing hydraulic conductivity. Alternatively, it may be possible to change the threshold at which stomata close, or manipulate leaf area directly, given the variation available between and within species. Finally, osmotic adjustment, the active accumulation of osmotically active solutes in plant tissues as water deficits develop so that cell turgor is maintained at low ψ_L, may improve lupin dehydration tolerance.

All these avenues for manipulating responses to water deficit are hamstrung by common constraints that are unique to the lupin species. Because of their short domesticated history we simply don't know the species' well enough to optimize adaptation to cropping systems. This problem is exacerbated because lupin cultivars are highly inbred, do not represent the genetic diversity available in the species', and have been strongly selected for drought escape. Moreover, our understanding of water use strategies and responses to water deficit in the much older, much more diverse wild germplasm pool is currently extremely rudimentary.

Despite these challenges, there is enough evidence that the search improved adaptation to soil water deficit is worthwhile. Our unpublished studies with wild germplasm exposed to terminal drought demonstrate that both yellow and narrow-leafed lupins collected from dry areas are much more conservative in water use than those from high rainfall areas. Turner and Hartung (2012) showed that isolated roots of *Lupinus* species vary significantly in their synthesis of free and conjugated ABA in response to 50 % dehydration of tips (but found no relationship with root dehydration). There is significant genetic variation in root systems of young lupin plants under controlled environment conditions (Clements et al. 1993; Chen et al. 2011), and leaflet size throughout the growing season in field plots (Buirchell, "personal communication"). Turner and Henson (1989) found significant osmotic adjustment of up to 0.2 MPa in narrow-leafed lupin grown in controlled environments, [but which was not supported by field evaluation of narrow-leafed lupin and white lupin cultivars (Rodriguez et al. 1995; Leport et al. 1998; Palta et al. 2007)].

Clearly, this short track record is a mixed one, which makes it difficult to generalize on lupin adaptive responses to water deficit. The way forward is to undertake ecophysiological studies of wild germplasm specifically-adapted to contrasting drought habitats because this addresses both the narrowness and short evolutionary history of domesticated material, and will highlight the results of millennia of natural selection. With this approach—if nature has selected for it, we will find it! Subsequently it will be important to implement a pre-breeding strategy of crossing with well-adapted elite cultivars to deliver proof of concept under real-world field conditions. This is essential, given the confounding effects of other wild traits, such as pod dehiscence, presence of alkaloids, hard-seeded ness, etc. Given the difficulty of measuring physiological traits in breeding populations this will almost certainly require some form of molecular marker tagging to ensure that the trait of interest is not lost in ongoing hybridization.

As outlined above, improving lupin adaption to water deficits represents a considerable challenge, but one which is very worthwhile, because this journey will not only increase productivity in a grain legume uniquely adapted to acid, sandy soils; it will simultaneously increase our understanding of drought adaptation in annual plants.

References

ABARE (2010) Australian crop report. Australian Bureau of Agricultural and Resource Economics, Canberra

Acosta-Gallegos JA, Adams MW (1991) Plant traits and yield stability of dry bean *(Phaseolus vulgaris)* cultivars under drought stress. J Agric Sci Cambridge 117:213–219

Ali M, Jensen CR, Mogensen VO, Bahrun A (1999) Drought adaptation of field grown wheat in relation to soil physical conditions. Plant Soil 208:149–159

Atkins CA, Pigeaire A (1993) Application of cytokinins to flowers to increase pod set in *Lupinus angustifolius L.* Aust J Agric Res 44:1799–1819

Berger JD, Adhikari KN, Wilkinson D, Buirchell BJ, Sweetingham MW (2008) Ecogeography of the old world lupins. 1 Ecotypic variation in yellow lupin (*Lupinus luteus* L.). Aust J Agric Res 59:691–701

Berger JD, Buirchell B, Luckett DJ, Nelson MN (2012a) Domestication bottlenecks limit genetic diversity and constrain adaptation in narrow-leafed lupin (*Lupinus angustifolius* L.). Theor Appl Genet 124:637–652

Berger JD, Buirchell B, Luckett DJ, Palta JA, Ludwig C, Liu D (2012b) How has narrow-leafed lupin changed in its 1st 40 years as an industrial, broad-acre crop? A G × E-based characterization of yield-related traits in Australian cultivars. Field Crops Res 126:152–164

Berger JD, Milroy SP, Turner NC, Siddique KHM, Imtiaz M, Malhotra R (2011) Chickpea evolution has selected for contrasting phenological mechanisms among different habitats. Euphytica 180:1–15
Blum A (2005) Drought resistance, water-use efficiency, and yield potential—are they compatible, dissonant, or mutually exclusive? Aust J Agric Res 56:1159–1168
Bramley H, Turner DW, Tyerman SD, Turner NC (2007a) Water flow in the roots of crop species: the influence of root structure, aquaporin activity, and waterlogging. In: Sparks DL (ed) Advances in agronomy, vol 96, pp 133–196
Bramley H, Turner NC, Turner DW, Tyerman SD (2007b) Comparison between gradient-dependent hydraulic conductivities of roots using the root pressure probe: the role of pressure propagations and implications for the relative roles of parallel radial pathways. Plant Cell Environ 30:861–874
Bramley H, Turner NC, Turner DW, Tyerman SD (2009) Roles of morphology, anatomy, and aquaporins in determining contrasting hydraulic behavior of roots. Plant Physiol 150:348–364
Bramley H, Turner NC, Turner DW, Tyerman SD (2010) The contrasting influence of short-term hypoxia on the hydraulic properties of cells and roots of wheat and lupin. Funct Plant Biol 37:183–193
Carminati A, Moradi AB, Vetterlein D, Vontobel P, Lehmann E, Weller U, Vogel HJ, Oswald SE (2010) Dynamics of soil water content in the rhizosphere. Plant Soil 332:163–176
Carminati A, Vetterlein D, Weller U, Vogel HJ, Oswald SE (2009) When roots lose contact. Vadose Zone J 8:805–809
Carvajal M, Cooke D, Clarkson D (1996) Responses of wheat plants to nutrient deprivation may involve the regulation of water-channel function. Planta 199:372–381
Chen YL, Dunbabin VM, Postma JA, Diggle AJ, Palta JA, Lynch JP, Siddique KH, Rengel Z (2011) Phenotypic variability and modelling of root structure of wild *Lupinus angustifolius* genotypes. Plant Soil 348:345–364
Christiansen JL, Raza S, Jørnsgård B, Mahmoud SA, Ortiz R (2000) Potential of landrace germplasm for genetic enhancement of white lupin in Egypt. Genet Resour Crop Evol 47:425–430
Clements JC, White PF, Buirchell BJ (1993) The root morphology of *Lupinus angustifolius* in relation to other *Lupinus* species. Aust J Agric Res 44:1367–1375
Clements JC, Cowling WA (1994) Patterns of morphological diversity in relation to geographical origins of wild *Lupinus angustifolius* from the Aegean region. Genet Resour Crop Evol 41:109–122
Correia MJ, Pereira JS (1994) Abscisic-acid in apoplastic sap can account for the restriction in leaf conductance of white lupins during moderate soil drying and after re-watering. Plant Cell Environ 17:845–852
Correia MJ, Pereira JS (1995) The control of leaf conductance of white lupin by xylem ABA concentration decreases with the severity of water deficits. J Exp Bot 46:101–110
Cowling WA, Buirchel B, Tapia ME (1998a) *Lupin. Lupinus* L. promoting the conservaiton and use of underutilized and neglected crops. International Plant Genetic Resources Institute (IPGRI), Rome, Italy
Cowling WA, Huyghe C, Swiecicki W (1998b) Lupin breeding. In: Gladstones JS, Atkins CA, Hamblin J (eds) Lupins as crop plants: biology, production and utilization. CAB International, pp 93–120
Doussan C, Pierret A, Garrigues E, Pages L (2006) Water uptake by plant roots: II Modelling of water transfer in the soil root-system with explicit account of flow within the root system—comparison with experiments. Plant Soil 283:99–117
Dracup M, Reader M, Palta JA (1998) Timing of terminal drought is an important cause of yield variability in lupin grown on duplex soils in southern Australia. Aust J Agric Res 49:799–810
Dencic S, Kastori R, Kobiljski B, Duggan B (2000) Evaluation of grain yield and its components in wheat cultivars and landraces under near optimal and drought condition. Euphytica 113:43–52

Eastham J, Gregory PJ (2000) The influence of crop management on the water balance of lupin and wheat crops on a layered soil in a Mediterranean climate. Plant Soil 221:239–251
Farre I, Robertson MJ, Walton GH, Asseng S (2001) Yield analysis of canola in a variable environment. In: 12th Biennial Australian Research Assembly on Brassicas Victoria, pp 130–134
FAO (2010) FAOSTAT. Food and Agriculture Organization of the United Nations
Farrington P, Pate JS (1981) Fruit Set in *Lupinus angustifolius* cv. unicrop. I Phenology and growth during flowering and early fruiting. Aust J Plant Physiol 8:293–305
Farrington P, Salama RB, Watson GD, Bartle GA (1992) Water-use of agricultural and native plants in a Western Australia wheat-belt catchment. Agric Water Manag 22:357–367
French RJ, Turner NC (1991) Water deficits change dry matter partitioning and seed yield in narrow-leafed lupins (*Lupinus angustifolius* L.). Aust J Agric Res 42:471–484
Gallardo M, Eastham J, Gregory PJ, Turner NC (1996) A comparison of plant hydraulic conductances in wheat and lupins. J Exp Bot 47:233–239
Gallardo M, Turner NC, Ludwig C (1994) Water relations, gas-exchange and abscisic-acid content of *Lupinus cosentinii* leaves in response to drying different proportions of the root-system. J Exp Bot 45:909–918
Garcıa del Moral LF, Rharrabti Y, Villegas D, Royo C (2003) Evaluation of grain yield and its components in durum wheat under Mediterranean conditions: an ontogenic approach. Agron J 95:266–274
Garrigues E, Doussan C, Pierret A (2006) Water uptake by plant roots: I formation and propagation of a water extraction front in mature root systems as evidenced by 2D light transmission imaging. Plant Soil 283:83–98
Gladstones JS (1970) Lupins as crop plants. Field Crop Abstr 23:123–148
Gladstones JS (1994) An historical review of lupins in Australia. In: Dracup M, Palta JA (eds) 1st Lupin technical symposium. Department of Agriculture, Perth, pp 1–38
Gregory PJ, Eastham J (1996) Growth of shoots and roots, and interception of radiation by wheat and lupin crops on a shallow, duplex soil in response to time of sowing. Aust J Agric Res 47:427–447
Grime JP (1979) Plant strategies and vegetation processes. Wiley, Chichester
Hamblin A, Tennant D (1987) Root length density and water uptake in cereals and grain legumes: how well are they correlated? Aust J Agric Res 38:513–527
Hamza MA, Anderson SH, Aylmore LAG (2007) Computed tomographic evaluation of osmotica on shrinkage and recovery of lupin (*Lupinus angustifolius* L.) and radish (*Raphanus sativus* L.) roots. Environ Exp Bot 59:334–339
Hamza MA, Aylmore LAG (1992a) Soil solute concentration and water-uptake by single lupin and radish plant-roots. 1 Water extraction and solute accumulation. Plant Soil 145:187–196
Hamza MA, Aylmore LAG (1992b) Soil solute concentration and water-uptake by single lupin and radish plant-roots. 2 Driving forces and resistances. Plant Soil 145:197–205
Hartung W, Leport L, Ratcliffe RG, Sauter A, Duda R, Turner NC (2002) Abscisic acid concentration, root pH and anatomy do not explain growth differences of chickpea (*Cicer arietinum* L.) and lupin (*Lupinus angustifolius* L.) on acid and alkaline soils. Plant Soil 240:191–199
Henson IE, Jensen CR, Turner NC (1989a) Leaf gas exchange and water relations of lupins and wheat. I Shoot responses to soil water deficits. Aust J Plant Physiol 16:401–413
Henson IE, Jensen CR, Turner NC (1989b) Leaf gas exchange and water relations of lupins and wheat. III Abscisic acid and drought-induced stomatal closure. Funct Plant Biol 16:429–442
Henson IE, Jensen CR, Turner NC (1990) Influence of leaf age and light environment on the gas exchange of lupins and wheat. Physiol Plant 79:15–22
Henson IE, Turner NC (1991) Stomatal responses to abscisic-acid in 3 lupin species. New Phytol 117:529–534
Hondelmann W (1984) The Lupin-ancient and modern crop plant. Theor Appl Genet 68:1–9
Hose E, Clarkson D, Steudle E, Schreiber L, Hartung W (2001) The exodermis: a variable apoplastic barrier. J Exp Bot 52:2245–2264

Huyghe C (1997) White lupin (*Lupinus albus* L.). Field Crops Res 53:147–160

Jensen CR, Henson IE (1990) Leaf water relations characteristics of *Lupinus angustifolius* and *L. cosentinii*. Oecologia 82:114–121

Jensen CR, Henson IE, Turner NC (1989) Leaf gas exchange and water relations of lupins and wheat. II Root and shoot water relations of lupin during drought-induced stomatal closure. Aust J Plant Physiol 16:415–428

Jensen CR, Mogensen VO, Poulsen HH, Henson IE, Aagot S, Hansen E, Ali M, Wollenweber B (1998) Soil water matric potential rather than water content determines drought responses in field-grown lupin (*Lupinus angustifolius*). Aust J Plant Physiol 25:353–363

Kuang J-B, Turner NC, Henson IE (1990) Influence of xylem water potential on leaf elongation and osmotic adjustment of wheat and lupin. J Exp Bot 41:217–221

Leport L, Turner NC, French RJ, Tennant D, Thomson BD, Siddique KHM (1998) Water relations, gas exchange and growth of cool-season grain legumes in a Mediterranean-type environment. Eur J Agron 9:295–303

Leport L, Turner NC, French RJ, Barr MD, Duda R, Davies SL, Tennant D, Siddique KHM (1999) Physiological responses of chickpea genotypes to terminal drought in a Mediterranean-type environment. Eur J Agron 11:279–291

Muchow RC, Sinclair TR (1986) Water and nitrogen. Field Crops Res 15:143–156

Munier-Jolain NM, Ney B, Duthion C (1996) Analysis of branching in spring-sown white lupins (*Lupinus albus* L.): the significance of the number of axillary buds. Ann Bot 77:123–131

Noffsinger SL, van Santen E (2005) Evaluation of *Lupinus albus* L. germplasm for the southeastern USA. Crop Sci 45:1941–1950

Oswald SE, Menon M, Carminati A, Vontobel P, Lehmann E, Schulin R (2008) Quantitative imaging of infiltration, root growth, and root water uptake via neutron radiography. Vadose Zone J 7:1035–1047

Palta JA, Kobata T, Fillery IR, Turner NC (1994) Remobilisation of carbon and nitrogen in wheat as influenced by postanthesis water deficits. Crop Sci 34:118–124

Palta JA, Ludwig C (1996) Pod set and seed yield as affected by cytokinin application and terminal drought in narrow-leafed lupin. Aust J Agric Res 48:81–90

Palta JA, Plaut Z (1999) Yield and components of seed yield of indeterminate narrow-leafed lupin (*Lupinus angustifolius* L.) subjected to transient water deficit. Aust J Agric Res 50:1225–1232

Palta JA, Ludwig C (2000) Elevated CO_2 during pod-filling increased seed yield but not harvest index in indeterminate narrow-leaf lupin. Aust J Agric Res 51:279–286

Palta JA, Turner NC, French RJ (2004) The yield performance of lupin genotypes under terminal drought in a Mediterranean-type environment. Aust J Agric Res 55:449–459

Palta JA, Turner NC, French RJ, Buirchell BJ (2003) Towards improvement of drought resistance in lupin—a crop for acid sandy soils. J Exp Bot 54:19–19

Palta JA, Turner NC, French RJ, Buirchell BJ (2003) Towards improvement of drought resistance in lupin—a crop for acid sandy soils. J Exp Bot 54:19–19

Palta JA, Turner NC, French RJ, Buirchell BJ (2007) Physiological responses of lupin genotypes to terminal drought in a Mediterranean-type environment. Ann Appl Biol 150:269–279

Palta JA, Berger DJ, Ludwig C (2008) The growth and yield of narrow-leafed lupin: myths and realities. In: Palta JA, Berger JD (eds) Lupins for health and wealth. ILA Press, Canterbury, New Zealand, pp 20–25

Pandey RK, Herrera WAT, Pendleton JW (1984) Drought response of grain legumes under irrigation gradient: I Yield and yield components. Agronomuy J 76:549–553

Passioura J, Munns R (1984) Hydraulic resistance of plants. II Effects of rooting medium, and time of day, in barley and lupin. Aust J Plant Physiol 11:341–350

Pate JS, Atkins CA, Perry MW (1980) Significance of photosynthate produced at different stages of growth as carbon source for fruit filling and seed reserve accumulation in *Lupinus angustifolius* L. Aust J Plant Physiol 7:283–297

Pate JS, Farrington P (1981) Fruit-set in *Lupinus angustifolius* cv unicrop. 2 Assimilate flow during flowering and early fruiting. Aust J Plant Physiol 8:307–318

Pinheiro C, Antonio C, Ortuno MF, Dobrev PI, Hartung W, Thomas-Oates J, Ricardo CP, Vankova R, Chaves MM, Wilson JC (2011) Initial water deficit effects on *Lupinus albus* photosynthetic performance, carbon metabolism, and hormonal balance: metabolic reorganization prior to early stress responses. J Exp Bot 62:4965–4974

Perry MW (1975) Field environment studies on lupins. 2 The effects of time of planting on dry matter partition and yield components of *Lupinus angustifolius* L. Aust J Agric Res 26:809–818

Purcell LC, King CA (1996) Drought and nitrogen source effects on nitrogen nutrition, seed growth, and yield in soybean. J Plant Nutr 19:969–993

Raman R, Luckett DJ, Raman H (2008) Estimation of genetic diversity in Albus lupin (*Lupinus albus* L.) using DArT and genic markers. In: Palta JA, Berger JD (eds) Lupins for health and wealth. ILA Press, Canterbury, New Zealand, pp 263–241

Reader MA, Dracup M, Kirby EJM (1995) Time to flowering in narrow-leafed lupin. Aust J Agric Res 46:1063–1077

Rebetzke G, Richards RA (1999) Genetic improvement of early vigour in wheat. Aust J Agric Res 50:291–301

Richards RA, Lukacs Z (2002) Seedling vigour in wheat–sources of variation for genetic and agronomic improvement. Aust J Agric Res 53:41–50

Rodrigues ML, Pacheco CMA, Chaves MM (1995) Soil-plant water relations, root distribution and biomass partitioning in *Lupinus albus* L under drought conditions. J Exp Bot 46:947–956

Rose IA, McWhirter KS, Spurway RA (1992) Identification of drought tolerance in early-maturing indeterminate soybeam [*Glycine Max* (L.) Merr.]. Aust J Agric Res 43:645–657

Saldaña PA, Harcha CI, Calderini DF (2009) Sensitivity of yield and grain nitrogen concentration of wheat, lupin and pea to source reduction during grain filling. A comparative survey under high yielding conditions. Field Crops Res 114:233–243

Sengbusch R, Zimmermann K (1937) Die Auffindung der ersten gelben und blauen Lupine (*Lupinus luteus* und *Lupinus angustifolius*) mit nicht platzenden Hülsen und die damit zusammenhängenden Probleme der Süßlupinenzüchtung. Zuchter 9:57–65

Serraj R, Sinclair T, Purcell L (1999) Symbiotic N_2 fixation response to drought. J Exp Bot 50:143–155

Shearer G, Kohl DH (1986) N2-fixation in field settings: estimations based on natural 15 N abundance. Aust J Plant Physiol 13:699–757

Sinclair TR (1986) Water and nitrogen limitations in soybean grain production. I Model development. Field Crops Res 15:125–141

Siddique KHM, Regan KL, Tennant D, Thomson BD (2001) Water use and water use efficiency of cool season grain legumes in low rainfall Mediterranean-type environments. Eur J Agron 15:267–280

Simpson MJA (1986) Geographical variation in *Lupinus albus* L. I Iberia. Plant Breeding 96:232–240

Steudle E, Peterson C (1998) How does water get through roots? J Exp Bot 49:775–788

Troll HJ (1940) Saatzeitversuche mit Zucht- und Landsorten sowie Wildformen von *L. luteus* und *L. angustifolius*. Pflanzenbau 16:403–430

Turner NC, Stern WR, Evans P (1987) Water relations and osmotic adjustment of leaves and roots of lupins in response to water deficits. Crop Sci 27:977–983

Turner NC, Henson IE (1989) Comparative water relations and gas-exchange of wheat and lupins in the field. In: Kreeb KH, Richter H, Hinckley TM (eds) Structural and functional responses to environmental stresses: water shortage. SPB Academic Publishing, The Hague, pp 293–304

Turner NC, Hartung W (2012) Dehydration of isolated roots of seven Lupinus species induces synthesis of different amounts of free, but not conjugated, abscisic acid. Plant Growth Regul 66:265–269

Tyerman SD, Niemietz CM, Bramley H (2002) Plant aquaporins: multifunctional water and solute channels with expanding roles. Plant Cell Environ 25:173–194

Tyree M, Sperry J (1989) Vulnerability of xylem to cavitation and embolism. Annu Rev Plant Physiol Plant Mol Biol 40:19–36

Unkovich MJ, Pate JS, Sanford P, Armstrong R (1994) Potential precision of the ∂15 N natural abundance method in field estimates of nitrogen fixation by crop and pasture legumes in south–west Australia. Aust J Agric Res 45:119–132
Weisz PR, Denison RF, Sinclair TR (1985) Response to drought stress of nitrogen fixation (acetylene reduction) rates by field-grown soybeans. Plant Physiol 78:525–530
Wolf O, Jeschke WD, Hartung W (1990) Long distance transport of abscisic acid in NaCI-treated intact plants of *Lupinus albus*. J Exp Bot 41:593–600
Zohary D, Hopf M (2000) Lupins: *Lupinus*. Domestication of plants in the old world. Clarendon Press, Oxford, pp 122–124

Chapter 17
Drought in Deciduous Fruit Trees: Implications for Yield and Fruit Quality

Gerardo Lopez, M. Hossein Behboudian, Joan Girona and Jordi Marsal

Abstract In many production areas of the world deciduous fruit trees require irrigation to maximize yield and optimize fruit quality. Drought could therefore have a negative impact on fruit production. This chapter addresses yield and fruit quality responses of deciduous fruits (DF) to drought. The chapter has emphasis on the most important deciduous fruit trees and includes apple, apricot, cherry, peach, pear, and plum. Confronted with a period of drought, fruit growers' success will depend on their ability to modify cultural practices. We evaluated various strategies at different levels of water shortage. We expected that this chapter will introduce enough information to understand the effects of drought in deciduous fruit tree orchards and facilitate fruit marketability under water shortage.

Abbreviations and Acronyms

CI	Conventional irrigation, commercial irrigation
DF	Deciduous fruits
DI	Deficit irrigation
ETc	Crop evapotranspiration
FF	Fruit firmness
IEC	Internal ethylene concentration
NI	No irrigation
RDI	Regulated deficit irrigation
RDM	Relative dry matter
SPI	Starch pattern index

G. Lopez (✉) · M. Hossein Behboudian · J. Girona · J. Marsal
Irrigation Technology, Institut de Recerca i Tecnologia Agroalimentàries (IRTA), Centre UdL-IRTA, Avda. Rovira Roure 191, 25198 Lleida, Spain
e-mail: gerardo.lopez@irta.es

R. Aroca (ed.), *Plant Responses to Drought Stress*,
DOI: 10.1007/978-3-642-32653-0_17,

SSC	Soluble solids concentration
SSP	Severe summer pruning
SWP	Stem water potential
TA	Titratable acidity

17.1 Introduction

Deciduous fruits (DF) include pome (apple, pear, and quince) and stone (peach, nectarine, cherry, plum, and apricot) fruits. Apple is the most widely cultivated DF in the world followed by pear, peach and nectarine (Faostast 2009). Cherry, plum, and apricot are widely cultivated while quince is not very popular (Faostast 2009). DF are consumed fresh or processed. Plums when consumed dried are called prunes. Grower's income depends on yield, which usually is determined by crop load and fruit size at harvest. But fruit quality is also an important issue in fruit production. In many production areas of the world DF trees require irrigation to maximize yield and optimize fruit quality. However, in years with drought, water resources may be insufficient to optimize irrigation and to obtain the maximum yield for the highest return. Water restrictions have been experienced in major fruit growing areas such as Central Washington State, USA (Proebsting et al. 1981), California, USA, (Lampinen et al. 1995), and Catalonia, Spain, (Girona et al. 2010). These problems could worsen in the future due to three interacting factors: (1) water supplies are becoming more limited worldwide and they will not be sufficient to meet the growing demands by 2025 (Postel 1998), (2) irrigation consumes, on a global scale, at least 85 % of all the water used (van Schilfgaarde 1994; Jury and Vaux 2007), and (3) increased competition with urban and industrial users and economic and social pressures to reduce irrigation water (Fereres and Evans 2006). Irrigated agriculture will therefore face water scarcity in the future (Fereres and Soriano 2007). It is therefore important to understand the effects of water stress in DF orchards and to develop tools for drought mitigation.

This chapter addresses yield and quality responses of DF to water stress. Drought in this chapter refers to a lack of rain for an extended period of time resulting in irrigation water being lower than the prevailing crop evapotranspiration (ETc). In drought years deficit irrigation (DI) may have to be imposed in phenological stages that are very sensitive to water stress. This situation is completely different from water stress arising from regulated deficit irrigation (RDI). RDI is a system of managing soil water supply to impose periods of pre-determined tree water deficit that can result in some economic benefit (Behboudian and Mills 1997; Naor 2006a). RDI necessitates the possibility of managing irrigation water during the whole season. This may not be possible in years with drought. Consequently RDI will not be mentioned in this chapter.

Yield and fruit quality responses of DF to water stress vary among crops and within each crop depending on the growth stage water stress occurs. We will start by highlighting some notions about the reproductive characteristics of DF trees. This will allow understanding the susceptibility of crops and cultivars to drought. Then, we will cover the yield aspect followed by fruit quality. In a large number of articles cited in this chapter, a variety of DI has been compared by researchers to a well-watered treatment either called conventional irrigation or commercial irrigation (CI). We conclude the chapter evaluating various irrigation management options and cultural practices at different levels of water shortage. We expect that this chapter will introduce enough information to understand the effects of drought in DF orchards facilitating fruit marketability under water shortage.

17.2 Reproductive Characteristics of Deciduous Fruit Trees

The reproductive characteristics of DF trees are crop specific. There are also differences among cultivars of a given crop. We cannot cover all the specific characteristics of cultivars. We will therefore cover those aspects that are in common among DF crops.

In mature DF trees the growing season starts with the bloom. But the flowers are formed in the previous season. In general floral initiation occurs in early summer (Westwood 1993). Most of the flower organs are present by early autumn and their development continues until low temperatures prevent further growth. This process is defined as floral differentiation (Greene 1996). Then flower buds become dormant and winter chilling is necessary to permit renewed growth the following year (Dennis 2003). When temperature rises again in spring flowers will open (anthesis). Flowers therefore open nearly a year after floral induction occurs (Greene 1996).

Flowers set fruit shortly after full bloom. Unfertilized flowers fall and many fruitlets (small fruits) abscise early in the season (Dennis 2003). The fruitlets that remain on the tree will start to grow. Fruit development is different in stone and pome fruits. Stone fruits have a double sigmoid pattern of growth where three stages can be determined (Chalmers and van den Ende 1975): Stage I, reproductive cell division and initial fruit growth; Stage II, intermediate stage of slow growth and pit hardening; and Stage III, final period of rapid fruit growth that concludes with fruit maturity. The length of the different stages will depend on the cultivar. Figure 17.1a presents the seasonal patterns of fruit growth of four different peach cultivars grown in Lleida, Spain. The three distinct stages of fruit development are more evident in late-maturing cultivars than in early-maturing cultivars. In apricots and cherries harvest occurs early in the season and Stage II is too short to be clearly distinguished. In these crops the post-harvest period is longer than pre-harvest period. Examples of two cherry cultivars grown in the Ebro basin (Spain) are shown in Fig. 17.1b. In pome fruits, due to the absence of pit, the pattern of growth is somewhat different from that of stone fruits. Pome fruit have an expo-

Fig. 17.1 Seasonal patterns of fruit fresh weight for different crops and cultivars in the Ebro basin, Spain

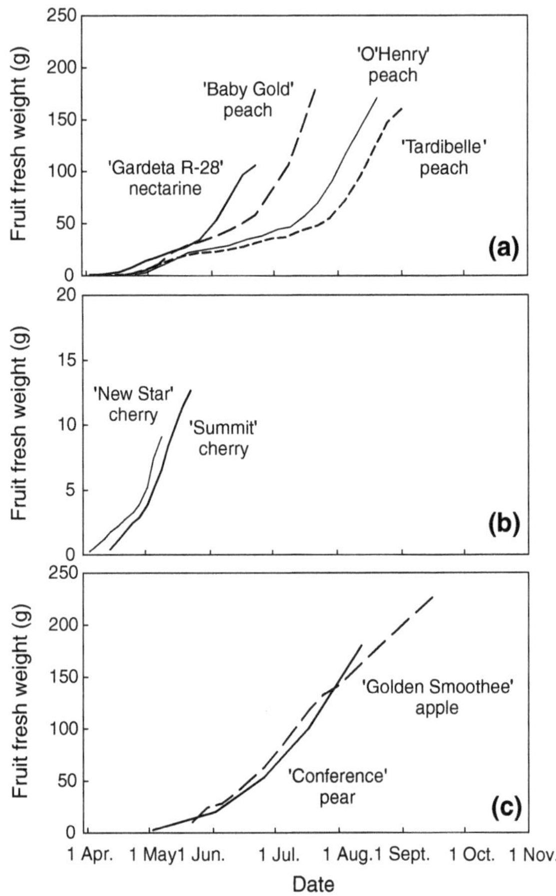

linear growth pattern (Lakso et al. 1995). Two examples for 'Conference' pear and 'Golden Smoothee' apple grown in Lleida (Spain) are presented in Fig. 17.1c.

Since water shortage in water limited regions usually occur in mid-summer, early-maturing cultivars could avoid water stress before harvest. But these cultivars could be exposed to post-harvest water stress (Fig. 17.1). In mid-late maturing cultivars, water stress is more likely to occur during the final stages of fruit development (Fig. 17.1).

17.3 Yield Responses to Drought

In DF total yield is positively related with the number of fruit. Financial return to growers is generally based on size and quantity of fruit produced, with an optimal balance between the two. When fruits are destined for the fresh market, high crop loads will result with many fruits smaller than commercial size and fruit thinning is required to remove the excessive number of fruits. If fruits are destined for processing, fruit size is less important and the financial return to growers could only be dependent on fruit number. A drought event can impair the yield components (fruit number and size) in the current season or during the following years.

17.3.1 Yield Responses in the Current Season

Yield reductions to drought events have been reported for peach (Lopez et al. 2011a), apricot (Torrecillas et al. 2000), cherry and plum (Proesting et al. 1981), pear (Marsal et al. 2010b), and apple (Ebel et al. 2001; Girona et al. 2010). Decreased yield by water stress is mainly due to a reduction in fruit size at harvest. Fruit growth could be limited by two mechanisms: (1) limitation by carbohydrate availability owing to a decrease in photosynthetic rate (Bradford and Hsiao 1982), and (2) direct limitation of fruit growth as a result of a reduction in cell turgor in response to water stress (Kramer and Boyer 1995). Another factor that can reduce yield is pre-harvest fruit drop (Powell 1974).

17.3.2 Yield Responses in the Following Season

Water stress can also reduce yield in the following season due to a reduction in fruit number. Fruit number depends on the number of initiated flowers and on the final fruit set. Since floral initiation occurs in early summer, water stress at this time may decrease return bloom as demonstrated for 'Braeburn' apple (Kilili et al. 1996a; Behboudian et al. 1998). In these experiments fruit thinning was still needed and yield was not affected. Fruit set was reduced in the year following DI in 'O'Henry' peach (Lopez et al. 2007b) and 'Bulida' apricot (Ruiz-Sanchez et al. 1999). In apricot, water stress decreased fruit set because of a reduction in pollen vitality (Ruiz-Sanchez et al. 1999). In peach, water stress decreased fruit set because of a reduction in roots' concentration of carbohydrates (Lopez et al. 2007b). However, yield was not affected because: (1) reduced fruit set notwithstanding, there was still enough fruit set for an adequate commercial crop, and (2) early fruit growth is not sensitive to roots' low concentration of carbohydrates. It seems that reductions in return bloom and fruit set may not have a negative effect on yield when low crop loads are required for promoting commercial fruit size. However, for cultivars whose profitability depends more on fruit number than on fruit size, reductions in return bloom and fruit set could have a negative impact on yield (Marsal et al. 2008a).

If water stress occurs after harvest, yield could decrease in the following season. This has been also related to early reproductive events (flowering and fruit set). Post-harvest water stress decreased crop yield in the subsequent season when the levels of water stress were severe in apricot (Uriu 1963), nectarine (Naor et al.

2005), and peach (Larson et al. 1988; Girona et al. 2003). Johnson et al. (1994) studied how much water stress plum trees can tolerate during post-harvest without reducing productivity. Irrigating 'Red Beaut' plum at 50 % of CI during post-harvest over a 3 year period did not affect flowering, crop load, and yield compared to CI. However, when trees were irrigated at 15 % of CI only for a 3 week period after harvest, subsequent flowering was reduced by 28 % and crop load by 64 %. This caused a significant reduction in yield (Johnson et al. 1994). There was a significant linear relationship between stem water potential (SWP) during post-harvest and subsequent yield (Johnson et al. 1994). Yield (kg $tree^{-1}$) was reduced from 50 at a SWP of −1.0 MPa to 25 at a SWP of −2.5 MPa. In 'Spadona' pear, yield was not affected if midday SWP values were higher than −2.2 MPa (Naor et al. 2006). The corresponding value for 'Conference' was −1.2 MPa (Marsal et al. 2011). If the level of water stress after harvest is not severe no effect on the following season has been reported in peach (Johnson et al. 1992; Gelly et al. 2003, 2004); pear (Naor et al. 2006) and cherry (Marsal et al. 2009, 2010a).

Yield in the year following water stress may be reduced due to other factors different from flower bud development and available reserve carbohydrates. Water stress can also reduce the fruit bearing capacity due to a significant mortality of branches. This has been reported in plum and cherry when the levels of water stress were very severe (Proesbsting et al. 1981; Johnson et al. 1994). Two years may be necessary to recover from this negative effect (Proesbsting et al. 1981).

17.4 Quality Responses to Drought

The definition of quality in DF is complex because multiple customers are involved in the marketing chain. Crisosto and Costa (2008) described the economic relevance of the various quality traits in the peach chain production. As pointed out by Crisosto and Costa (2008), fruit growers are interested in obtaining fruit with large size and reduced number of pickings (homogeneous maturity). Packers, shippers, distributors, and wholesalers are interested in obtaining fruit with the adequate firmness for optimum fruit storage and market life behavior. For retailers it is important to have attractive size, colour, and firmness for appealing to the consumer. For the consumer fruit quality is related with sensory properties (appearance, texture, taste, and aroma). No fruit defects are desired by all the customers in this chain (Crisosto and Costa 2008). Water stress can modify the most important quality traits: size, firmness, colour, concentration of sugars, titratable acidity (TA), sugar/acid ratio, concentrations of dry matter, and aroma volatiles. Water stress can also modify fruit maturity and cold storage behavior. Water stress has been also associated with some fruit disorders. In the following the most important fruit quality attributes are listed and mention is made as how they are affected by water stress.

17.4.1 Fruit Size

Size is one of the most important quality criteria for fresh-market fruit. Less irrigation usually results in smaller fruit in apple (Naor et al. 2008). Many studies in stone fruit (e.g., Lopez et al. 2006) have shown that fruit weight at harvest is reduced under water stress conditions. Fruit growth rate in peach have been linearly related to tree water status expressed as midday SWP (Girona et al. 2006). Low midday SWP arising from reduced irrigation has been also related with smaller pear (Shackel 2007) and apple (Ebel et al. 2001) fruits. Marsal et al. (2010b) determined that because of small size 85 % of 'Conference' pear fruit irrigated at 20 % of CI during Stage II were disqualified for the fresh market.

17.4.2 Fruit Firmness

Fruit firmness (FF) is an important quality index and is strongly influenced by maturity with FF decreasing as the fruit ripens (Kingston 1991; Crisosto 1994). FF is also influenced by fruit size (Ebel et al. 1993) with smaller fruit being generally firmer than larger fruit due to a higher cellular density in the former. Treatments that alter fruit size may therefore alter FF. In an experiment on 'Braeburn' apple, it was shown that FF was higher in fruit for which irrigation was withheld either late in the season or during the entire season (Kilili et al. 1996b). Mpelasoka et al. (2000) applied late-season DI to 'Braeburn' apple and divided the fruit into three size categories of small, medium, and large. DI had a significantly higher FF than CI treatment for all size categories at harvest. For 'Pink Lady' apple, was not affected by DI (O′Connell and Goodwin 2007). Possible cultivar, degree of water stress and crop load differences notwithstanding, it seems that apple FF increases in response to water stress. Higher FF has been also reported in 'Conference' pear that received no irrigation (NI) during the first 3 weeks of Stage II followed by 20 % of ETc. until harvest (Lopez et al. 2011b). In stone fruits, DI fruit had higher firmness than CI fruit in 'Ziraat' sweet cherry (Cigdem et al. 2008), 'O'Henry' peach (Lopez et al. 2010), 'Ryan's Sun' peach, (Lopez et al. 2011a) and 'Fortune' plum (Battilani 2004).

17.4.3 Fruit Color

Color is important because it determines fruit appearance. Color characteristic of the cultivar develops as the fruit ripens (Crisosto et al. 1994). Water stress can modify fruit color. For peach, severe water stress decreased red-skin pigmentation in 'Elberta' (Proebsting and Middleton 1980), 'O'Henry' (Lopez et al. 2010), and 'Ryan's Sun' (Lopez et al. 2011a). However, moderate water stress favored fruit color in 'Alexandra' peach (Mercier et al. 2009) and 'Bulida' apricot (Torrecillas et al. 2000). Mills et al. (1994) and Kilili et al. (1996a, b) found enhancements in red color in 'Braeburn' apple with a lowered water status. This may have been due to the advanced accumulation of sugars measured in these fruit, as sucrose plays an important role in anthocyanin development. Improvement in fruit color in peach has been related to higher exposure of DI fruit to sunlight due to a lower vegetative growth in DI trees (Gelly et al. 2003, 2004).

17.4.4 Sugars, Acids, and Sugar/Acid Ratio

Sugars, acids, and their ratio are very important in DF because they are primarily related with fruit taste. The sugar/acid ratio is related with consumer acceptance in peach (Crisosto and Crisosto 2005). Consumer acceptance seems to increase with increases in this ratio. For pear, Jackson (2003) reported that fruit taste is mainly related to the sugar concentration in the juice if acidity is low. In plum, according to Crisosto et al. (2004), an increase in soluble solids concentration (SSC) with no decrease in acidity would be perceived positively by consumers. Therefore, those treatments that increase SSC and maintain or reduce fruit acidity may be perceived positively by consumers. This seems to be the case of DI. The concentration of the four major soluble sugars in apple (sucrose, glucose, fructose, and sorbitol) normally rises with DI as exemplified for 'Braeburn' (Mills et al. 1994; Kilili et al. 1996a; Mpelasoka et al. 2000). Sugar concentration is often estimated from SSC using refractometers. In general fruit subjected to water stress had a higher SSC than those from CI trees in pear (Lopez et al. 2011b), peach (Lopez et al. 2010), and plum (Naor et al. 2004; Intrigliolo and Castel 2010). Significant linear relationships have been found between tree water status and fruit SSC in 'Barlett' pear (Ramos et al. 1994) and in 'O'Henry' peach (Lopez et al. 2010), with higher levels of water stress leading to higher SSC. This could be in part a result of increasing dry matter concentration (DMC) of the DI fruit (Naor 2006a). The effects of water stress on fruit acidity are not as clear as those reported for sugars. In 'Conference' pear, the responses of TA to water stress were not consistent between years. No effect on TA was observed one season while DI fruit were more acidic than CI fruit in another season (Lopez et al. 2011b). In peach, water stress has been reported to increase (Wu et al. 2002; Lopez et al. 2010, 2011b) or have no effect on fruit acidity (Crisosto et al. 1994; Domingo et al. 2007). No changes in acidity in response to water stress have been reported for plum (Battilani 2004; Naor et al. 2004) and apple (Kilili et al. 1996a; Mpelasoka and Behboudian 2002).

17.4.5 Fruit Dry Matter Concentration

Fruit DMC is very important in deciduous fruit. Mpelasoka et al. (2000) and Kilili et al. (1996b) presented data showing that DMC in DI apple fruit was higher than in CI fruit. In 'O'Henry' peach fruit DMC was strongly correlated with SWP. The DMC of no irrigated fruit was higher than those from CI fruit (Marsal et al. 2006;

Lopez et al. 2007a, 2010). DMC values for DI fruit may indicate partial dehydration of fruit under water stress and increase in sugar concentration. Fruit with higher DMC could be more acceptable by the consumer than fruit with low DMC (Palmer et al. 2010). In prunes an increase in fruit DMC could be very positive. Since the fruit are dried at harvest, a low fresh to dry weight ratio will reduce energy requirements for drying and could improve the economic return to the grower (Shackel et al. 2000). However, a substantial increase in DMC could produce a lack of juiciness in fruit for the fresh-market which may finally impair the quality of the fruit (Lopez et al. 2010, 2011a).

17.4.6 Aroma Volatiles

Aroma volatiles are important quality attributes for DF. Much of the character of fruit depends on their aroma (Jackson 2003). For example, in peach the perception of flavor has been related to the emission of aroma volatiles (Ortiz et al. 2008). However, there is few and contradictory information on how aroma volatiles are affected by water stress. 'Braeburn' apple under late DI had higher concentrations of aroma volatiles than CI fruit at harvest (Behboudian et al. 1998). However, Mpelasoka and Behboudian (2002) applied whole-season DI to 'Braeburn' apple and the total concentration of aroma volatiles and total odour units in the fruit juice was the same at harvest in DI and CI fruit.

17.4.7 Maturity

Maturity is important for DF because it can determine the optimum harvest date. The selection of the appropriate harvest date will have multiple influences on consumer acceptance of fruit (Iglesias and Echeverria 2009). The fruit quality attributes of FF, colour, SSC, and TA could be used as maturity indices. The appropriate index can change with crop. For 'Braeburn' apple, internal ethylene concentration (IEC) and starch pattern index (SPI) have been used to determine the effect of water stress on fruit maturity. Kilili et al. (1996b) applied late DI irrigation as well as NI for the entire season. They reported that late DI and NI fruit had more advanced maturity and earlier ripening than CI fruit in terms of higher IEC. Mpelasoka et al. (2001) applied late DI and had three harvests. In all harvests, the fruit SPI was higher in late DI than in CI. The fruit IEC was the same among the treatments in the first two harvests. But it was higher in late DI in the last harvest. In peach it seems that FF and color are the most relevant maturity indices (Crisosto 1994). Peach grown under water stress during Stage III advanced maturity when the levels of water stress where moderate (midday SWP values about -1.55 MPa, Mercier et al. 2009). However, if more severe water stress occurs during Stage III, advanced maturity is not observed. Peach growth on trees without irrigation during the whole season tended to be arrested, and fruit skin often lacked red pigmentation (Proebsting and Middleton 1980). Midday leaf water potential in NI trees ranged from −4.0 to −3.0 MPa. Delays in fruit maturity have been confirmed in peach grown under severe water stress conditions in ‘O’Henry’ and ‘Ryan’s Sun’ (Lopez et al. 2010, 2011a). It therefore seems that the degree of water stress may play a significant role in fruit maturity. While moderate water stress appears to advance fruit maturity, severe water stress delays this process. It is important to mention that water stress, fruit maturity, and fruit quality attributes are interacting factors. For example, some of the effects of water stress on FF (see Sect. 17.4.2), colour (see Sect. 17.4.3) and SSC, and TA (see Sect. 17.4.4) could be partially explained by the effects of water stress on fruit maturity.

17.4.8 Quality After Cold Storage

In addition to fruit quality at harvest, it is also important to consider the effect of water stress on fruit storage behavior and to explore whether any improvement or loss of fruit quality at harvest is also maintained during the storage. Concentration of aroma volatiles from DI fruit seems to increase during cold storage. Mpelasoka and Behboudian (2002) applied whole-season DI in ‘Braeburn’ apple and evaluated total volatile concentration and odour units on Days 1 and 7 on fruit that had been taken out of cold storage after 17 weeks. The total concentration of aroma volatiles was higher in DI fruit on Days 1 and 7 and significantly so for Day 1. Total odour units were higher for DI fruit on both days. Another positive effect on fruit quality following water stress may be reduction in fruit weight loss during cold storage. This has been reported for peach (Mercier et al. 2009), pear, (Lopez et al. 2011b) and apple (Kilili et al. 1996b; Mpelasoka et al. 2000). The reduction in weight loss in DI fruit could prolong the cold storage life, facilitating marketing for a longer period of time. The difference in weight loss between CI and DI fruit could be due to differences in the structure and/or composition of the skin or the epicuticular waxes covering the skin. Cuticle modification by DI has been reported by Crisosto et al. (1994) who attributed a lower rate of water loss for peach grown under DI to a thicker cuticle and a higher density of trichomes on the skin surface. Another possible reason could be the lower water content of DI fruit. For example, in ‘Conference’ pear, differences in weight loss during storage are function of differences in the original water content (Nguyen et al. 2006).

17.4.9 Sensory Quality

Consumer’s panels and trained panels of tasters to evaluate sensory attributes have not been extensively used in DI studies. A recent study on ‘Ryan’s Sun’ peach evaluated sensory attributes and consumer preferences for CI fruit and NI fruit that had not been watered for 50 days before harvest (Lopez et al. 2011a). A trained taste panel indicated that stopping irrigation increased FF, crispness, and sourness but it reduced sweetness, juiciness, and the intensity of peach flavor. A panel of consumers indicated a reduction of consumer preference for NI fruit. The authors concluded that stopping irrigation before harvest impaired organoleptic peach quality despite the increase by 5 % in SSC. The conditions in the period before harvest are the most important in terms of fruit organoleptic quality (Lopez et al. 2011a).

17.4.10 Fruit Disorders

Fruit disorders, present at harvest or developed in storage, are among the most important aspects of fruit quality. It is difficult to arrive at a general conclusion here. It is prudent to assume that the effect of DI on incidence of disorders would depend on the nature of the disorder, severity, and duration of water stress. For example, in apple increases and decreases of disorders for fruit undergoing DI have been reported. Examples of lower incidence of disorders are for bitter pit (Guelfat'Reich et al. 1974; Lötter et al. 1985; Irving and Drost 1987), scald (Guelfat'Reich et al. 1974; Lötter et al. 1985), and water core (Lötter et al. 1985). Goode et al. (1975) reported increased cracking and russetting in apple fruit grown under DI. Lötter et al. (1985) reported an increase in sunburned apple fruit from DI treatments. O′Connell and Goodwin (2007) reported that in each year of their two-year study with 'Pink Lady' apple, incidence of fruit disorders (sunburn, russet, misshape, markings, frost damage) was not affected by DI. In pear, it seems that DI can reduce incidence of fruit disorders such as alfalfa greening and cork spot and this has been related to lower ratio of N/Ca in DI fruit (Brun et al. 1985). In plum, the effect of water stresson fruit cracking was studied by Uriu et al. (1962) and Milad and Shackel (1992). Those authors noticed a higher susceptibility of plums to cracking when DI was applied and released before harvest. Sekse (1995) indicated that DI before plum harvest is not acceptable because it can also increase cracking if stress is relieved during ripening.

Fruit disorders have been observed following water stress in the previous season. For peach substantial increase of fruit disorders such as double fruits (Johnson et al. 1992; Naor et al. 2005), deep sutures (Handley and Johnson 2000; Johnson and Phene 2008), and external split pits (Johnson and Phene 2008) have been observed. The June-July period (in California) seems to be more sensitive than the August–September period. Naor (2006a) determined a SWP threshold of −2.0 MPa. If tree water status is lower than this threshold an increase in fruit doubles could be observed. The response of plum trees seems to be different to that reported for peach. 'Red Beaut' submitted to moderate and severe levels of water stress presented a low percentage of doubles fruits. Double fruit would occur for less than 2 % if stress happened before fruit thinning and less than 1 % if it did before harvest (Johnson et al. 1994). Other plum cultivars such as 'Ambra' and

'Durado' did not respond to water stress in terms of double fruit formation (Johnson et al. 1994). Contradictory results between studies could partially be explained by interaction between water stress and temperature during the DI period. Naor (2006b) suggested that flower buds exposed to high temperatures during bud development stage acquire the potential of forming double fruit, but the proportion of fruit that realizes this potential was affected by the degree of water stress. However, Beppu and Kataoka (1999) indicated that in 'Satohnishiki' cherry high temperature (>30 °C) is responsible for double fruit formation while water stress did not influence the occurrence of double fruit.

17.5 Management of Deciduous Fruit Orchards Under Drought

The effects of water stress on yield and quality have been extensively investigated in DF trees. It is clear that fruit size is affected negatively in response to early stopping of irrigation (see Sect. 17.4.1). Hence, if water limitations coincide with the final period of fruit growth, fruit growers could have problems in obtaining the fruit sizes desired by the market. Different options for managing fruit trees under water stress conditions have been evaluated. We will cover only management options in deciduous fruit orchard equipped with efficient irrigation systems such as drip irrigation (Bryla et al. 2005). The management options are corrective and they may be only necessary in years with drought. Management strategies under drought should focus on reducing water stress in the plant. Corrective actions include reducing soil evaporation and manipulating the demand of the plant for water by using summer pruning, fruit thinning, and net shading.

17.5.1 Reduction in Soil Evaporation

A reduction in soil evaporation under drought conditions could have a significant positive impact on fruit production. A good example is the study of Girona et al. (2010) in 'Golden Smoothee' apple grown in a deep soil (1 m). These authors compared a drip irrigation system with two drippers per tree (system used by fruit growers) and one dripper per tree. The one-dripper version was more advantageous in terms of irrigation use efficiency than the two-dripper version under this soil conditions. Over the 2003–2005 seasons when DI treatments received 33 % of CI, the fresh-market yield in the system with one dripper was 72 % of CI and that of the system with two dippers was 60 % of CI. Removing one dripper per tree from the irrigation system could have reduced water evaporation from the soil. Applying the same volume of irrigation water, the system with one dripper made more water available to the tree. In years with limited allocation of water, we recommend reducing the number of drippers per tree.

17.5.2 Summer Pruning

Confronted with a period of drought, water stress could be partially alleviated by reducing canopy transpiration by severe summer pruning (SSP). Reducing leaf area by pruning decreases water use immediately (Proebsting et al. 1989; Goodwin et al. 2006). A considerable reduction in water consumption may slow the depletion of water in the soil ultimately improving the tree water status (Lopez et al. 2008a). However, SSP can have a negative impact on fruit growth capacity and total crop yield because it reduces the amount of light intercepted by the tree and the amount of assimilates available to support fruit growth (Lopez et al. 2006). SSP under water stress conditions may cause two opposite effects on fruit growth: (1) beneficial effect due to an improvement in tree water status, and (2) detrimental effect due to a limitation on carbohydrate availability. The combination of both effects may not produce any net profit for fruit growth. A suitable technique for mitigating the adverse effects of severe water stress on fruit growth should improve tree water status and at least maintain the assimilation capacity of the tree. Moderate summer pruning would reduce the negative impact on carbohydrate availability, but would also reduce the improvement in tree water status. It is difficult to reach a compromise between the two objectives by application of summer pruning when the orchard is already being managed for optimal canopy light interception. However, SSP could be an interesting technique when trees have excessive vigour and/or are subjected to extreme levels of water stress. For example, Probesting and Middleton (1980) proposed using heavy pruning (dehorning) to aid plant survival under severe drought conditions. In their study peach trees received NI during the season and only 86 mm through rain. When leaf water potential was below −3.0 MPa in summer, many peach trees died. SSP reduced canopy diameter from 5 m to about 2 m and the height from 4 to 1.5 m. This reduction in the canopy compared to well-watered trees delayed the appearance of drought symptoms until very late in the season and resulted in 100 % survival of peach trees.

17.5.3 Fruit Thinning

Heavy fruit thinning has been tested in DF as a drought management technique. Heavy fruit thinning under water stress conditions improves tree water status due to additional root growth and stomatal closure (Lopez et al. 2008b; Marsal et al. 2008b). The improvement in tree water status along with the reduction in fruit-to-fruit competition in response to heavy fruit thinning increased the growth rate of the remaining fruit. Proebsting and Middelton (1980) indicated that fruit thinning increases the size of fruit if trees were subjected to levels of water stress that threatens their survival. Lopez et al. (2006) working with 'O'Henry' peach explored whether fruit thinning could enhance fruit growth under water stress conditions that limited tree productivity but not tree survival. A reduction in crop load from 300 fruit per tree (commercial load) to 150 (increasing fruit spacing from 20 to 40 cm) under conditions of water stress produced an increase in SWP of 0.30 MPa and increased fruit growth capacity. Although it is well known that heavy fruit thinning can reduce overall yields (Naor et al. 1999), heavily thinned trees presented a higher number of fruits classified into commercial fruit size classes than commercially thinned trees (Lopez et al. 2011c). Consequently, heavily thinned trees had higher marketable yields than commercially thinned trees. This was confirmed In 'Conference' pear grown under water stress conditions. Marketable yield in heavily thinned trees were higher than in trees with commercial load (Marsal el al. 2010b). The intensity of fruit thinning should be adjusted to the severity of irrigation water restrictions (Naor et al. 2008). Models have been developed to determine the thinning intensity for a given water restriction in pear (Marsal et al. 2010b) and apple (Naschitz and Naor 2005). Besides improving fruit size, fruit thinning has the potential of improving fruit quality (Crisosto et al. 1997; Wünsche and Ferguson 2005). However, when heavy thinning is performed late in the season to coincide with the timing of water shortage, no effect on fruit quality could be observed in comparison with commercial thinning (Lopez et al. 2010, 2011b). In order to improve fruit quality, fruit thinning should be applied before irrigation restriction.

17.5.4 Net Shading

Net shading is another horticultural technique that may help reduce water requirements of DF. An example is for apricot growing in the semi-arid conditions of Murcia, Spain. Nicolas et al. (2005) evaluated tree transpiration from sap flow measurements in 2-year-old 'Bulida' apricot over a 3 week period. Sixteen trees were monitored. Eight trees were grown in the open air and the rest were placed under a shading net. All the trees received full irrigation for 9 days and then water was withheld for 5 days in four trees from each treatment. Full irrigation was applied after 5 days of water stress. There were significant differences in tree transpiration between exposed and shaded trees grown under full irrigation. The greatest difference was observed in days with high evaporative demand. DI reduced transpiration in comparison with CI. However, this reduction was delayed by 2 days in shaded trees. After rewatering, tree transpiration recovered faster in shaded trees than in exposed trees.

17.6 Conclusions

Although yield and quality responses to water stress are crop-specific, in this chapter we have covered enough information to reach firm conclusions on the effect of drought on yield and fruit quality in DF orchards. Drought in DF can impair yield in the current season or during the following years. Yield reductions in the current season have been mainly related to a reduction in fruit size at harvest in all DF crops. Pre-harvest fruit drop in response to water stress can be another factor. Drought in DF can reduce subsequent return bloom, fruit set and fruit bearing capacity due to branch mortality. 2 years may be necessary to completely recover from the reduction in the fruit bearing capacity. However, reductions in return bloom and fruit set may not have a negative impact on yield when significant fruit thinning is needed for promoting commercial fruit size. For cultivars whose profitability depends more on fruit number than on fruit size, reductions in return bloom and fruit set could have a negative impact on yield. Regarding fruit quality, less irrigation as consequence of water stress always resulted in smaller fruit size. But less irrigation could improve other fruit quality attributes. Improvements in fruit quality have been reported in terms of FF, colour and concentrations of soluble solids. Reduced irrigation can also enhance fruit maturity, storability, and reduce weight loss in storage. However, if the irrigation season cannot be completed successfully because of drought, DF trees will experience severe levels of water stress and this could have negative effects on fruit maturation and organoleptic fruit quality. Consumers' perception of quality is reduced in response to severe levels of water stress before harvest. Some negative effects will be also observed in the next season for stone fruit. These negative effects are mainly related with fruit disorders (double fruit). A severe drought could therefore create difficulty for fruit marketability. Confronted with a period of drought, fruit grower's success will depend on their ability to manage their orchards accordingly. The management options are corrective and they may only be necessary in years with drought. Reductions in soil evaporation by reducing the number of drippers per tree can partially compensate for the negative effects of water stress on fruit size and yield. Other cultural practices based on the reduction in canopy transpiration have also the potential for mitigating the negative effects of water stress. But the horticultural technique that produced the best results under drought conditions is heavy fruit thinning.

Acknowledgments This work was supported by the Spanish Ministry of Science and Innovation (MICINN) projects CSD2006-00067 of CONSOLIDER-INGENIO 2010 and AGL2009-10237.

References

Battilani A (2004) Regulated deficit irrigation (RDI) effects on growth and yield of plum tree. Acta Hortic 664:55–62

Behboudian MH, Mills TM (1997) Deficit irrigation in deciduous orchards. Hortic Rev 21:105–131

Behboudian MH, Dixon J, Pothamshetty K (1998) Plant and fruit responses of lysimeter-grown 'Braeburn' apple to reduced irrigation. J Hortic Sci Biotech 73:781–785

Beppu K, Kataoka I (1999) High temperature rather than drought stress is responsible for the occurrence of double pistils in 'Satohnishiki' sweet cherry. Sci Hortic 81:125–134

Bradford K, Hsiao TC (1982) Physiological responses to moderate water stress. In: Lange OL, Nobel PS, Osmod CB, Ziegler H (eds) Physiological plant ecology II. Water relations and carbon assimilation. Encyclopedia of Plant Physiology, vol 12B. Springer, Berlin/New York, pp 264–324

Brun CA, Raese JT, Stahly EA (1985) Seasonal response of 'Anjou' pear trees to different irrigation regimes. II. Mineral composition of fruit and leaves, fruit disorders and fruit set. J Am Soc Hortic Sci 110:835–840
Bryla DR, Dickson E, Shenk R, Johnson RS, Crisosto CH, Trout TJ (2005) Influence of irrigation method and scheduling on patterns of soil and tree water status and its relation to yield and fruit quality in peach. HortScience 40:2118–2124
Chalmers DJ, van den Ende B (1975) A reappraisal of the growth and development of peach fruit. Aust J Plant Physiol 2:623–634
Cigdem D, Umran E, Yazgan S, Candogan B, Soylu A (2008) Effects of different irrigation levels on the vegetative growth, flower bud formation and fruit quality of sweet cherry in western part of Turkey. J Food Agr Environ 6:173–177
Crisosto CH (1994) Stone fruit maturity indices: a descriptive review. Postharvest News Info 5:65–68
Crisosto CH, Crisosto GM (2005) Relationship between ripe soluble solids concentration (RSSC) and consumer acceptance of high and low acid melting flesh peach and nectarine (*Prunus persica* (L.) Batsch) cultivars. Postharvest Biol Technol 38:239–246
Crisosto CH, Costa G (2008) Preharvest factors affecting peach quality. In: Layne D, Bassi D (eds) The peach: botany, production and uses. CABI, Wallingford, pp 536–549
Crisosto CH, Johnson RS, Luza JG, Crisosto GM (1994) Irrigation regimes affect fruit soluble solids concentration and rate of water loss of 'O'Henry' peaches. HortScience 29:1169–1171
Crisosto CH, Johnson RS, DeJong TM, Day KR (1997) Orchard factors affecting postharvest stone fruit quality. HortScience 32:820–823
Crisosto CH, Garner D, Crisosto GM, Bowerman E (2004) Increasing 'Blackamber' plum (*Prunus salicina* Lindell) consumer acceptance. Postharvest Biol Technol 34:237–244
Dennis F (2003) Flowering, pollination and fruit set and development. In: Ferree DC, Warrington IJ (eds) Apples: botany, production and uses. CABI Publishing, Wallingford, pp 153–166
Domingo X, Arbones A, Villar J, Villar P, Pascual M, Rufat J (2007) Effects of irrigation strategies and nitrogen fertilisation on mineral nutrition, yield and quality of peach for processing. In: Milenio Editorial (ed) Towards a better efficiency in N use. Lleida, Spain, pp 512–514
Ebel RC, Proebsting EL, Patterson ME (1993) Regulated deficit irrigation may alter apple maturity, quality and storage life. HortScience 28:141–143
Ebel RC, Proebsting EL, Evans RG (2001) Apple tree and fruit responses to early termination of irrigation in a semi-arid environment. HortScience 36:1197–1201
Faostat (2009) Citation database results, FAO. http://faostat.fao.org/faostat
Fereres E, Evans RG (2006) Irrigation of fruit trees and vines: an introduction. Irrig Sci 24:55–57
Fereres E, Soriano MA (2007) Deficit irrigation for reducing agricultural water use. J Exp Bot 58:147–159
Gelly M, Recasens I, Mata M, Arbones A, Rufat J, Girona J, Marsal J (2003) Effects of water deficit during stage II of peach fruit development and postharvest on fruit quality and ethylene production. J Hortic Sci Biotech 78:324–330
Gelly M, Recasens I, Girona J, Mata M, Arbones A, Rufat J, Marsal J (2004) Effects of stage II and postharvest deficit irrigation on peach quality during maturation and after cold storage. J Sci Food Agric 84:561–568
Girona J, Mata M, Arbones A, Alegre S, Rufat J, Marsal J (2003) Peach tree response to single and combined regulated deficit irrigation regimes under shallow soils. J Am Soc Hortic Sci 128:432–440
Girona J, Marsal J, Lopez G (2006) Establishment of stem water potential thresholds for the response of 'O'Henry' peach fruit growth to water stress during stage III of fruit development. Acta Hortic 713:197–201
Girona J, Behboudian MH, Mata M, del Campo J, Marsal J (2010) Exploring six reduced irrigation options under water shortage for 'golden smoothee' apple: responses of yield components over three years. Agric Water Manag 98:370–375
Goode JE, Fuller MM, Hyrycz KJ (1975) Skin cracking of cox's orange pippin apples in relation to water stress. J Hortic Sci 50:265–269

Goodwin I, Whitfield DM, Connor DJ (2006) Effects of tree size on water use of peach (*Prunus persica* L. Batsch). Irrig Sci 24:59–68
Greene DW (1996) Flower development. In: Maib K (ed) Tree fruit physiology: growth and development. Good Fruit Grower, Washington, pp 91–98
Guelfat'Reich S, Assaf R, Bravdo BA, Levin I (1974) The keeping quality of apples in storage as affected by different irrigation regimes. J Hortic Sci 49:217–225
Handley DF, Johnson RS (2000) Late summer irrigation of water-stressed peach trees reduces fruit doubles and deep sutures. HortScience 35:771
Iglesias I, Echeverria G (2009) Differential effect of cultivar and harvest date on nectarine colour, quality and consumer acceptance. Sci Hortic 120:41–50
Intrigliolo DS, Castel JR (2010) Response of plum trees to deficit irrigation under two crop levels: tree growth, yield and fruit quality. Irrig Sci 28:525–534
Irving DD, Drost JH (1987) Effects of water deficit on vegetative growth, fruit growth and fruit quality in cox's orange pippin apple. J Hortic Sci 62:427–432
Jackson JE (2003) In: Jackson JE (ed) Biology of apple and pears. University of Cambridge, UK, p 488
Johnson RS, Phene BC (2008) Fruit quality disorders in an early maturing peach cultivar caused by postharvest water stress. Acta Hortic 792:385–390
Johnson RS, Handley DF, DeJong TM (1992) Long-term response of early maturing peach trees to postharvest water deficits. J Am Soc Hortic Sci 117:881–886
Johnson RS, Handley DF, Day KR (1994) Postharvest water stress of an early maturing plum. J Hortic Sci 69:1035–1041
Jury WA, Vaux HJ (2007) The emerging global water crisis: managing scarcity and conflict between water users. Adv Agron 95:1–76
Kilili AW, Behboudian MH, Mills TM (1996a) Composition and quality of 'Braeburn' apples under reduced irrigation. Sci Hortic 67:1–11
Kilili AW, Behboudian MH, Mills TM (1996b) Postharvest performance of 'Braeburn' apples in relation to withholding of irrigation at different stages of the growing season. J Hortic Sci 71:693–701
Kingston CM (1991) Maturity indices for apple and pear. Hortic Rev 13:407–432
Kramer PJ, Boyer JS (1995) Cell water relations. In: Water relations of plants and soils. Academic, San Diego, pp 42–83
Lakso AN, Grappadelli LC, Barnard J, Goffinet MC (1995) An expolinear model of the growth pattern of the apple fruit. J Hortic Sci 70:389–394
Lampinen BD, Shackel KA, Southwick SM, Olson WH, Yeager JT (1995) Sensitivity of yield and fruit quality of French prune to water deprivation at different fruit growth stages. J Am Soc Hortic Sci 120:139–147
Larson KD, DeJong TM, Johnson RS (1988) Physiological and growth-responses of mature peach trees to postharvest water-stress. J Am Soc Hortic Sci 113:296–300
Lopez G, Mata M, Arbones A, Solans JR, Girona J, Marsal J (2006) Mitigation of effects of extreme drought during stage III of peach fruit development by summer pruning and fruit thinning. Tree Physiol 26:469–477
Lopez G, Girona J, del Campo J, Marsal J (2007a) Effects of relative source-sink position within peach trees on fruit growth under water stress conditions. J Hortic Sci Biotech 82:140–148
Lopez G, Girona J, Marsal J (2007b) Response of winter root starch concentration to severe water stress and fruit load and its subsequent effects on early peach fruit development. Tree Physiol 27:1619–1626
Lopez G, Arbones A, del Campo J, Mata M, Vallverdu X, Girona J, Marsal J (2008a) Responses of peach trees to regulated deficit irrigation during stage II of fruit development and summer pruning. Span J Agric Res 6:479–491
Lopez G, Arbones A, Mata M, Paris C, Girona J, Marsal J (2008b) Root growth following defruiting improves peach tree water status. J Hortic Sci Biotech 83:804–808
Lopez G, Behboudian MH, Vallverdu X, Mata M, Girona J, Marsal J (2010) Mitigation of severe water stress by fruit thinning in 'O'Henry' peach: implications for fruit quality. Sci Hortic 125:294–300

Lopez G, Behboudian MH, Echeverria G, Girona J, Marsal J (2011a) Instrumental and sensory evaluation of fruit quality for 'Ryan's Sun' peach grown under deficit irrigation. HortTechnology 21:712–719
Lopez G, Larrigaudiere C, Girona J, Behboudian MH, Marsal J (2011b) Fruit thinning in 'Conference' pear grown under deficit irrigation: implications for fruit quality at harvest and after cold storage. Sci Hortic 129:64–70
Lopez G, Girona J, Marsal J (2011c) Drought in peach orchards: summer pruning vs. fruit thinning for water stress mitigation and long-term effects of water stress. Acta Hortic 903:1187–1194
Lötter J, Beukes DJ, Weber HW (1985) Growth and quality of apples as affected by different irrigation treatments. J Hortic Sci 60:181–192
Marsal J, Lopez G, Mata M, Girona J (2006) Branch removal and defruiting for the amelioration of water stress effects on fruit growth during Stage III of peach fruit development. Sci Hortic 108:55–60
Marsal J, Lopez G, Girona J (2008a) Recent advances in regulated deficit irrigation (RDI) in woody perennials and future perspectives. Acta Hortic 792:429–439
Marsal J, Mata M, Arbones A, del Campo J, Girona J, Lopez G (2008b) Factors involved in alleviating water stress by partial crop removal in pear trees. Tree Physiol 28:1375–1382
Marsal J, Lopez G, Arbones A, Mata M, Vallverdu X, Girona J (2009) Influence of post-harvest deficit irrigation and pre-harvest fruit thinning on sweet cherry (cv new star) fruit firmness and quality. J Hortic Sci Biotech 84:273–278
Marsal J, Lopez G, del Campo J, Mata M, Arbones A, Girona J (2010a) Postharvest regulated deficit irrigation in 'Summit' sweet cherry: fruit yield and quality in the following season. Irrig Sci 28:181–189
Marsal J, Behboudian MH, Mata M, Basile B, del Campo J, Girona J, Lopez G (2010b) Fruit thinning in 'Conference' pear grown under deficit irrigation to optimise yield and to improve tree water status. J Hortic Sci Biotech 85:125–130
Marsal J, Lopez G, Mata M, Girona J (2011) Postharvest deficit irrigation in 'Conference' pear: effects on subsequent yield and fruit quality. Agric Wat Mang 103:1–7
Mercier V, Bussi C, Lescourret F, Genard M (2009) Effects of different irrigation regimes applied during the final stage of rapid fruit growth on an early maturing peach cultivar. Irrig Sci 27:297–306
Milad RE, Shackel KA (1992) Water relations of fruit end cracking in French prune (*Prunus domestica* L. cv. French). J Am Soc Hortic Sci 117:824–828
Mills TM, Behboudian MH, Tan PY, Clothier BE (1994) Plant water status and fruit quality in 'Braeburn' apples. HortScience 29:1274–1278
Mpelasoka BS, Behboudian MH (2002) Production of aroma volatiles in response to deficit irrigation and to crop load in relation to fruit maturity for 'Braeburn' apple. Postharvest Biol Technol 24:1–11
Mpelasoka BS, Behboudian MH, Dixon J, Neal SM, Caspari HW (2000) Improvement of fruit quality and storage potential of 'Braeburn' apple through deficit irrigation. J Hortic Sci Biotech 75:615–621
Mpelasoka BS, Behboudian MH, Mills TM (2001) Effects of deficit irrigation on fruit maturity and quality of 'Braeburn' apple. Sci Hortic 90:279–290
Naor A (2006a) Irrigation scheduling and evaluation of tree water status in deciduous orchards. Hortic Rev 32:111–165
Naor A (2006b) Irrigation scheduling of peach—deficit irrigation at different phenological stages and water stress assessment. Acta Hortic 713:339–349
Naor A, Klein I, Hupert H, Grinblat Y, Peres M, Kaufman A (1999) Water stress and crop level interactions in relation to nectarine yield, fruit size distribution, and water potentials. J Am Soc Hortic Sci 124:189–193
Naor A, Peres M, Greenblat Y, Gal Y, Ben Arie R (2004) Effects of pre-harvest irrigation regime and crop level on yield, fruit size distribution and fruit quality of field-grown 'Black Amber' Japanese plum. J Hortic Sci Biotech 79:281–288

Naor A, Stern R, Peres M, Greenblat Y, Gal Y, Flaishman MA (2005) Timing and severity of postharvest water stress affect following-year productivity and fruit quality of field-grown 'Snow Queen' nectarine. J Am Soc Hortic Sci 130:806–812
Naor A, Stern R, Flaishman M, Gal Y, Peres M (2006) Effects of post-harvest water stress on autumnal bloom and subsequent-season productivity in mid-season 'Spadona' pear. J Hortic Sci Biotech 81:365–370
Naor A, Naschitz S, Peres M, Gal Y (2008) Responses of apple fruit size to tree water status and crop load. Tree Physiol 28:1255–1261
Naschitz S, Naor A (2005) The effect of crop load on tree water consumption of 'Golden Delicious' apples in relation to fruit size: an operative model. J Am Soc Hortic Sci 130:7–11
Nguyen TA, Dresselaers T, Verboven P, D'hallewin G, Culeddu N, Van Hecke P, Nicolai BM (2006) Finite element modelling and MRI validation of 3D transient water profiles in pears during postharvest storage. J Sci Food Agric 86:745–756
Nicolas E, Torrecillas A, Dell Amico J, Alarcon JJ (2005) Sap flow, gas exchange, and hydraulic conductance of young apricot trees growing under a shading net and different water supplies. J Plant Physiol 162:439–447
O'Connell MG, Goodwin I (2007) Responses of 'Pink Lady' apple to deficit irrigation and partial rootzone drying: physiology, growth, yield, and fruit quality. Aust J Agric Res 58:1068–1076
Ortiz A, Lara I, Graell J, Lopez ML, Echeverria G (2008) Sensory acceptance of CA-stored peach fruit. Relationship to instrumental quality parameters. Acta Hortic 796:225–230
Palmer JW, Harker FR, Tustin DS, Johnston J (2010) Fruit dry matter concentration: a new quality matric for apples. J Sci Food Agric 90:2586–2594
Postel S (1998) Water for food production: will there be enough in 2025? Bioscience 48:629–637
Powell DBB (1974) Some effects of water stress in late spring on apple trees. J Hortic Sci 49:257–272
Proebsting EL, Middleton JE (1980) The behavior of peach and pear trees under extreme drought stress. J Am Soc Hortic Sci 105:380–385
Proebsting EL, Middleton JE, Mahan MO (1981) Performance of bearing cherry and prune trees under very low irrigation rates. J Am Soc Hortic Sci 106:243–246
Proebsting EL, Jerie PH, Irvine J (1989) Water deficits and rooting volume modify peach tree growth and water relations. J Am Soc Hortic Sci 114:368–372
Ramos DE, Weinbaum SA, Shackel KA, Shcwankl LJ, Mitcham EJ, Mitchell FG, Snyder RG, Mayer G, McGourty G (1994) Influence of tree water status and canopy position on fruit size and quality of Bartlett pears. Acta Hortic 367:192–200
Ruiz-Sanchez MC, Egea J, Galego R, Torrecillas A (1999) Floral biology of 'Bulida' apricot trees subjected to postharvest drought stress. Ann Appl Biol 135:523–528
Sekse L (1995) Cuticular fracturing in fruits of sweet cherry (*Prunus avium* L.) resulting from changing soil water contents. J Hortic Sci 63:135–141
Shackel KA (2007) Water relations of woody perennial plant species. J Int Sci Vigne Vin 41:121–129
Shackel KA, Lampinen B, Southwick S, Olson W, Sibbett S, Krueger W, Yeager J (2000) Deficit irrigation in prunes: maintaining productivity with less water. HortScience 35:1063–1066
Torrecillas A, Domingo R, Galego R, Ruiz-Sanchez MC (2000) Apricot tree response to withholding irrigation at different phenological periods. Sci Hortic 85:201–215
Uriu K (1963) Effect of post-harvest soil moisture depletion on subsequent yield of apricots. Proc Am Soc Hortic Sci 84:93–97
Uriu K, Hansen CJ, Smith JJ (1962) The cracking of prunes in relation to irrigation. J Am Soc Hortic Sci 80:211–219
van Schilfgaarde J (1994) Irrigation—a blessing or a curse. Agric Water Manag 25:203–219
Westwood MN (1993) Temperate zone pomology: physiology and culture. In: Timber Press, Portland, OR, p 523
Wu BH, Genard M, Lescourret F, Gomez L, Li SH (2002) Influence of assimilate and water supply on seasonal variation of acids in peach (cv Suncrest). J Sci Food Agric 82:1829–1836
Wünsche JN, Ferguson IB (2005) Crop load interactions in apple. Hortic Rev 31:231–290

Index

R. Aroca (ed.), *Plant Responses to Drought Stress*,
DOI: 10.1007/978-3-642-32653-0, © Springer-Verlag Berlin Heidelberg 2012

MIX
Papier aus verantwortungsvollen Quellen
Paper from responsible sources
FSC® C105338

If you have any concerns about our products,
you can contact us on
ProductSafety@springernature.com

In case Publisher is established outside the EU,
the EU authorized representative is:
Springer Nature Customer Service Center GmbH
Europaplatz 3, 69115 Heidelberg, Germany

Printed by Libri Plureos GmbH
in Hamburg, Germany